Lecture Notes in Networks and Systems

Volume 746

The series "Lecture Notes in Networks and Systems" publishes the latest developments in Networks and Systems—quickly, informally and with high quality. Original research reported in proceedings and post-proceedings represents the core of LNNS.

Volumes published in LNNS embrace all aspects and subfields of, as well as new challenges in, Networks and Systems.

The series contains proceedings and edited volumes in systems and networks, spanning the areas of Cyber-Physical Systems, Autonomous Systems, Sensor Networks, Control Systems, Energy Systems, Automotive Systems, Biological Systems, Vehicular Networking and Connected Vehicles, Aerospace Systems, Automation, Manufacturing, Smart Grids, Nonlinear Systems, Power Systems, Robotics, Social Systems, Economic Systems and other. Of particular value to both the contributors and the readership are the short publication timeframe and the world-wide distribution and exposure which enable both a wide and rapid dissemination of research output.

The series covers the theory, applications, and perspectives on the state of the art and future developments relevant to systems and networks, decision making, control, complex processes and related areas, as embedded in the fields of interdisciplinary and applied sciences, engineering, computer science, physics, economics, social, and life sciences, as well as the paradigms and methodologies behind them.

Indexed by SCOPUS, INSPEC, WTI Frankfurt eG, zbMATH, SCImago.

All books published in the series are submitted for consideration in Web of Science.

For proposals from Asia please contact Aninda Bose (aninda.bose@springer.com).

Paweł Strumiłło · Artur Klepaczko ·
Michał Strzelecki · Dorota Bociąga
Editors

The Latest Developments and Challenges in Biomedical Engineering

Proceedings of the 23rd Polish Conference on Biocybernetics and Biomedical Engineering, Lodz, Poland, September 27–29, 2023

Springer

Editors
Paweł Strumiłło
Institute of Electronics
Lodz University of Technology
Lodz, Poland

Artur Klepaczko
Institute of Electronics
Lodz University of Technology
Lodz, Poland

Michał Strzelecki
Institute of Electronics
Lodz University of Technology
Lodz, Poland

Dorota Bociąga
Institute of Materials Science
and Engineering
Lodz University of Technology
Lodz, Poland

ISSN 2367-3370 ISSN 2367-3389 (electronic)
Lecture Notes in Networks and Systems
ISBN 978-3-031-38429-5 ISBN 978-3-031-38430-1 (eBook)
https://doi.org/10.1007/978-3-031-38430-1

This Springer imprint is published by the registered company Springer Nature Switzerland AG
The registered company address is: Gewerbestrasse 11, 6330 Cham, Switzerland

Preface

We are honored to hand over to the readers the Proceedings of the 23rd Polish Conference on Biocybernetics and Biomedical Engineering, which will be held in Lodz from September 27 to 29, 2023. The conference was organized by the Committee of Biocybernetics and Biomedical Engineering of the Polish Academy of Sciences and hosted by the Lodz University of Technology. Due to the complex and multidisciplinary area of issues covered by biomedical engineering, two TUL units were involved in the organization of this conference, namely the Institute of Electronics and the Institute of Materials Science and Engineering. The conference is a continuation of the cyclical, bi-annual meetings of the biomedical engineers' community, which attract scientists and industry representatives from various fields of engineering, IT, biomaterials, biotechnology, and medicine.

The ongoing and dynamic advancement of AI-based data processing and analysis methods is playing an increasingly vital role in medicine. These methods find application in various areas, such as disease diagnosis, prediction, and monitoring, particularly through the utilization of image data analysis algorithms. Other areas of application include personalized medicine, where multimodal patient data is acquired and analyzed, as well as robot-assisted surgery and clinical decision support.

These Proceedings contain 35 publications on the above issues as well as other relevant hot topics regarding the most important challenges of modern biomedical engineering. The papers are organized in the following five chapters:

- Biomedical Imaging & Analysis
- Modeling and Machine Learning
- Signal Processing
- Telemonitoring & Measurement
- Biomaterials and Implants.

The editors would like to express their gratitude to the authors for their submissions and to all the reviewers for their meticulous evaluation of the papers and valuable comments, which undoubtedly enhanced the scientific merit of the accepted papers. We believe that through this collaborative effort, these Proceedings will serve as a significant scientific resource for the biocybernetics and biomedical engineering community.

Lodz, Poland

Paweł Strumiłło
Artur Klepaczko
Michał Strzelecki
Dorota Bociąga

Contents

Telemonitoring & Measurement

Biomaterials and Implants

Biomedical Imaging & Analysis

Modified CNN-Watershed for Corneal Endothelium Segmentation: Image-to-Image Versus Sliding-Window Comparison

Adrian Kucharski and Anna Fabijańska

Abstract This paper considers the problem of corneal endothelium image segmentation using a method that combines a CNN model with a watershed transform. Specifically, first CNN predicts cell bodies, edges, and centers. Next, cell centers are used as markers that guide the watershed transform performed concerning the cell edge probability maps inferred by the CNN to outline cell edges. Different variants of the method are considered. Specifically, a downscaled U-Net is compared with the Attention U-Net in the image-to-image and sliding window setup. Results show that using a marker-driven watershed transform to post-process cell edge probability maps allows for replacing the sliding window setup with an image-to-image setup, reducing prediction time while maintaining similar or better segmentation accuracy. Also, when used as a backbone, Attention U-Net outperforms classical U-Net in determining cell morphometric parameters with high accuracy.

Keywords Corneal endothelial cell · Image segmentation · U-Net · Attention U-Net · Convolutional neural network

1 Introduction

Over time, segmentation methods for corneal endothelial cells have progressed from unsupervised to supervised machine learning systems. Recently, supervised machine learning methods using deep learning solutions based on convolutional neural net-

A. Kucharski (✉) · A. Fabijańska
Lodz University of Technology Institute of Applied Computer Science, 18 Stefanowskiego Str., 90-537 Lodz, Poland
e-mail: 222486@edu.p.lodz.pl

A. Fabijańska
e-mail: anna.fabijanska@p.lodz.pl

P. Strumiłło et al. (eds.), *The Latest Developments and Challenges in Biomedical Engineering*, Lecture Notes in Networks and Systems 746,
https://doi.org/10.1007/978-3-031-38430-1_1

works (CNNs) have become state-of-the-art. The U-Net model [1] has emerged as a popular choice for this task, with the downscaled version applied to overlapping image tiles in a patch-based setup outperforming the full image-based approach [2, 3]. Using a CNN for overlapping image tiles cut out of an image increases the number of training samples, improving the model's ability to detect weak and blurred cell edges due to a smaller image region being analyzed [4]. However, despite these benefits, these methods still have limitations in determining weak and fuzzy boundaries. They also result in additional computational overhead and increased inference time.

Recent studies have addressed these limitations by integrating convolutional encoder-decoder models with conventional image processing methods, including the watershed transform, to improve the post-processing of cell boundaries [5, 6]. Another alternative explored in [7] is extending the U-Net model by build-in submodules to better adapt to weak and fuzzy cell boundaries. This paper adopts this method and proposes an image segmentation technique for corneal endothelium that utilizes the Attention U-Net in conjunction with CNN-Watershed [5]. This approach addresses the challenges posed by weak and discontinuous cell edges. Specifically, we apply a marker-based watershed transform to the cell edge probability maps inferred by the Attention U-Net to outline endothelial cells precisely. We also test the approach in patch-based and full-image scenarios to determine whether additional computational overhead from the patch-based setup can be avoided using the watershed transform in the post-processing step.

2 Dataset Details

The study utilized the Rotterdam dataset [8], which is publicly available and consists of 52 confocal corneal microscopy images. The resolution of these images varies from 324×385 pixels to 763×525 pixels. To handle varying sizes, we resized each image to 384×384 pixels. Although the dataset does not include ground truth results for cell edges, it does contain manually marked cell centers. We used these centers to generate ground truth segmentation results by applying marker-guided watershed segmentation, followed by manual correction as required. The resulting segmentation masks contain three classes: cell bodies, edges, and centers. Sample images from the Rotterdam dataset used in this study are displayed in Fig. 1.

3 Methods

3.1 General Idea

The CNN-watershed method for corneal endothelial cell segmentation is a hybrid approach that combines a CNN with the watershed transform to improve accuracy.

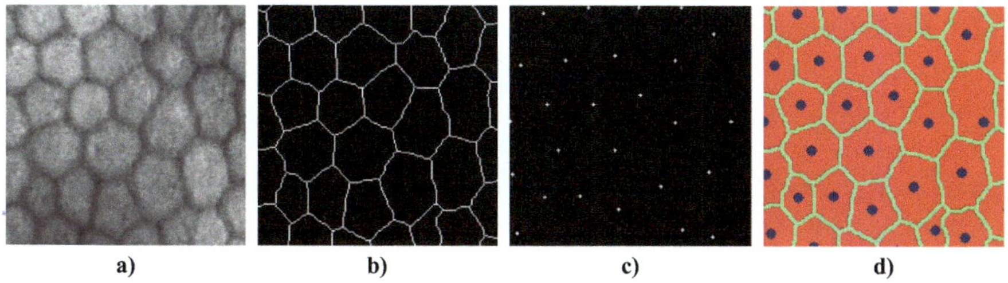

Fig. 1 A sample corneal endothelial image from the Rotterdam dataset. **a** a grayscale microscopy image, **b** cell edges, **c** manually marked cell centers, **d** a mask that depicts cell edges, cell centers, and cell bodies, with each class represented by a different color

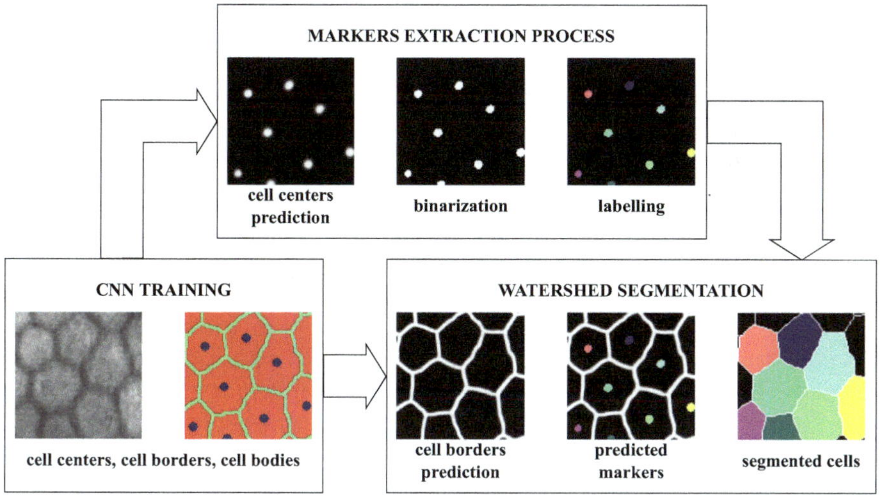

Fig. 2 The general workflow of the CNN-watershed

The process is summarized in Fig. 2. Firstly, the CNN predicts the positions of cell bodies, edges, and centers. Next, the predicted cell centers are used as markers that guide the watershed segmentation. Unlike the traditional setup, the watershed transform is not performed concerning the image gradient but with the cell edge probability maps inferred by the CNN. The resulting watershed dams created around the markers are always continuous and correspond to the resulting cell edges.

The baseline CNN-watershed approach [5] employs the downscaled U-Net model in a sliding window setup. In our study, we modify the baseline approach by using the Attention U-Net in both the image-to-image (I2I) and the sliding window (SW) setup.

3.2 Full Image Versus Sliding Window

In the I2I approach, a CNN was trained to output a segmentation mask directly for a given input image. Input images of size 384 × 384 pixels were considered. This approach considers global information about the image, which can be helpful in cases where the cells are irregularly shaped or arranged in complex patterns.

For the SW approach, input images were divided into overlapping tiles of size 64 × 64 pixels. A CNN model was trained to perform image segmentation in each window, and the predicted segmentation masks were merged to obtain a seamless cell edge probability map. Incorporating multiple image tiles during the prediction stage increases computational complexity. Furthermore, the SW approach may encounter challenges when dealing with large cell clusters or variations in cell size and shape within an image.

3.3 Convolutional Neural Networks

U-Net The configuration of the U-Net model varied depending on the setup. The input to the SW U-Net model (see Fig. 3) was an image of size 64 × 64 × 1 pixels and 384 × 384 × 1 pixels for the I2I model. The contracting path included (three for the SW and four for the I2I architecture) downsampling levels with three 2D convolutional layers (with a filter size 3 × 3) per level, each followed by a ReLU activation function. The number of filters in each downsampling level was (32, 64, 128) for the SW and (64, 128, 256, 512) for the I2I. Max-pooling 2D with a pool size of 2 × 2 was performed after each downsampling block (except the last one). The expansive path had three for the SW and four for the I2I upsampling levels, with three Conv2D layers per level. The number of filters in each upsampling level was [128, 64, 32], and a ReLU activation function followed each convolutional layer. After each upsampling layer, the corresponding feature maps from the contracting and expansive paths were concatenated and passed through three 2D convolutional layers, each followed by a ReLU activation function. The number of filters in each

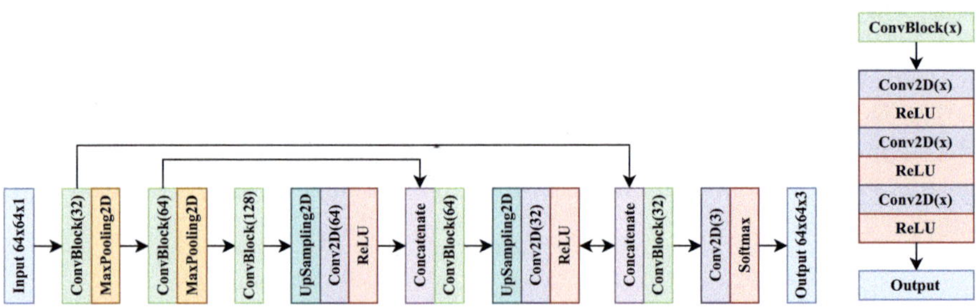

Fig. 3 The architecture of the SW U-Net model

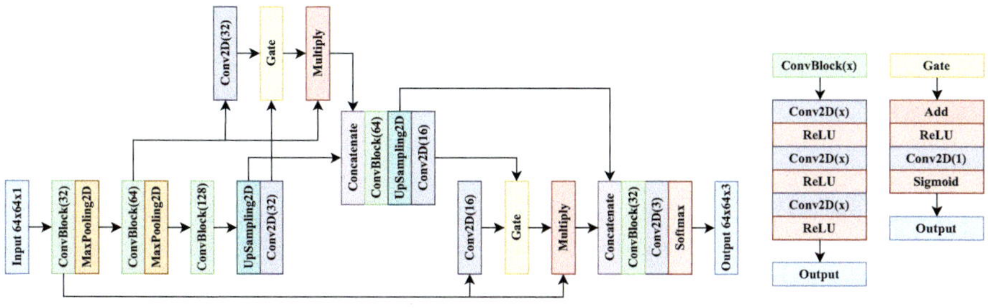

Fig. 4 The architecture of the SW attention U-Net model

upsampling level was (128, 64, 32) for the SW and (512, 256, 128, 64) for the I2I. The final layer of the model was a 2D convolutional layer with one filter of size 1×1 and softmax activation, which outputted the predicted segmentation mask. The number of output labels was three, with the probabilities of the cell edges, cell centers, and cell bodies.

Attention U-Net The Attention U-Net [9] was derived from the baseline U-Nets for both I2I and SW setups. The contracting path remained unaltered, while the expansive path was modified to include an attention mechanism that helped the model focus on critical features in the input. At each upsampling block, the attention mechanism was incorporated by concatenating the relevant feature maps from the contracting and expansive paths and subsequently utilizing a weighting mechanism to assess the significance of the features in the concatenation. The *add* attention mechanism was used (see Fig. 4).

3.4 Data Augmentation

The Rotterdam dataset was augmented to increase its variability using a set of geometric transformations with randomized parameters. The following transformations were applied to each image in the dataset:

- A shear operation introduces distortion along both axes to mimic natural deformation in the corneal endothelium (random value between –0.3 and 0.3).
- A scale transformation to simulate cell size and spacing variations (random scaling factor from 0.5 to 1.5).
- A rotation transformation to simulate random rotational changes in the microscope's imaging plane (random angle from –0.15 to 0.15).
- A vertical flipping operation with a 50% probability and a horizontal flipping operation with a 50% probability to increase dataset variability.

After each training epoch, the transformation parameters were randomized again.

3.5 Training

The I2I and SW models were trained using the Adam optimizer with a learning rate of 1e-4 and utilized categorical cross-entropy loss as the loss function.

For the image-to-image Attention U-Net and U-Net models, the training was conducted for 100 epochs with a batch size of 4 and 34 steps per epoch.

Meanwhile, the sliding-window models were trained for 100 epochs with a batch size of 128 and 102 steps per epoch. The training dataset was augmented using the data augmentation process described in Sect. 3.4, and patches of size 64×64 were extracted from the augmented images. A total of 13,000 patches were extracted per training epoch.

3.6 Prediction and Watershed Postprocessing

Trained models generated three probability maps for cell borders, centers, and bodies (see Fig. 5). Cell bodies were not used for further processing.

While I2I Attention U-Net and U-Net models process a whole image simultaneously, SW Attention U-Net and U-Net models were applied to consecutive image patches of size 64×64 in a sliding window setup. The patches overlapped by a stride of 4 pixels in horizontal and vertical directions to obtain seamless probability maps. The model's predictions for overlapping patch regions were averaged.

Predicted cell centers were then used to generate markers for watershed markers-controlled segmentation. To transform a cell center's probability maps to 1-pixel size seeds, a mean filter of size 3×3 was first applied. Then, the results were binarized via the minimum cross-entropy approach proposed by [10]. Next, morphological erosion with a square structuring element of size 3×3 pixels was applied to the binary image. Finally, the centroid's positions were calculated for each connected region to generate 1-pixel size markers, and markers-controlled watershed segmentation was performed on cell borders probability maps outputted by CNN's models and prepared seeds.

a) b) c) d)

Fig. 5 Probability maps outputted by I2I U-Net models. **a** an original image, **b** cell bodies, **c** cell borders, **d** cell centers

4 Results

4.1 General Assessment Procedure

The quality of endothelial cell segmentation was evaluated visually and quantitatively using image segmentation accuracy measures and measures derived from the cells' morphology. The assessment was performed using a three-fold cross-validation approach. The available corneal endothelium image data for each setup was randomly divided into two approximately equal subsets. Two subsets were used to train the models, while the third subset was used to evaluate their performance. The assessment was repeated thrice with different training and testing fold configurations.

4.2 Visual Results

Visual results of the CNN-Watershed used in I2I and SW setup and running at the top of the U-Net and the Attention U-Net models are presented in Fig. 6. Specifically, the top panel presents the resulting probability maps, with intensities of different colors denoting probabilities of cell edges, cell centers, and cell bodies. The middle panel presents the resulting cell edges overlaid on an original sample image. Finally, the bottom panel compares the ground truth edges in green with the inferred edges in red. Overlapping edges are shown in white.

Additionally, Fig. 7 visualizes the resulting attention maps of the Attention U-Net used in the sliding window and image-to-image setup.

4.3 Image Segmentation Quality Assessment

The DICE coefficient (see Eq. 1) was computed between the resulting $\mathcal{P}$ and the ground truth $\mathcal{T}$ edges. Prior to the DICE calculation, the edges, which were only one pixel wide, were dilated using a square structural element of size 4×4.

$$DICE = \frac{2|\mathcal{T} \bigcap \mathcal{P}|}{|\mathcal{T}| + |\mathcal{P}|} \tag{1}$$

The longest distance between a pixel of ground truth edges and the nearest pixel of predicted edges was quantified with the modified Hausdorff distance [11]. The ideal case MHD value between two images is 0. High MHD values suggest the presence of false or missing edges.

The summary of cell segmentation accuracy measures obtained for each considered version of the CNN-watershed for each testing fold is shown in Table 1. The best scores are shown in bold.

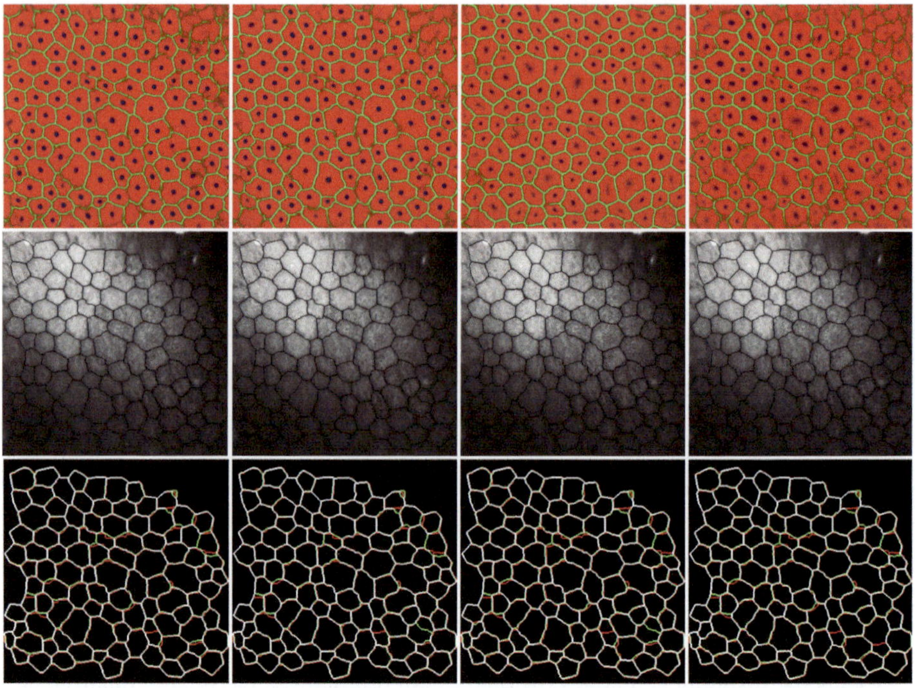

a) b) c) d)

Fig. 6 Visual results of segmentation of corneal endothelial images. Results obtained with **a** I2I Attention U-Net, **b** I2I U-Net, **c** SW attention U-Net, **d** SW U-Net. Top—results obtained by CNNs. Middle-predicted edges overlaid on the original image. Bottom—comparison between the ground truth and edges generated by the CNN-watershed algorithm. The edges identified as ground truth are highlighted in red, the edges generated by the CNN-watershed algorithm are highlighted in green, and overlapping edges are highlighted in white

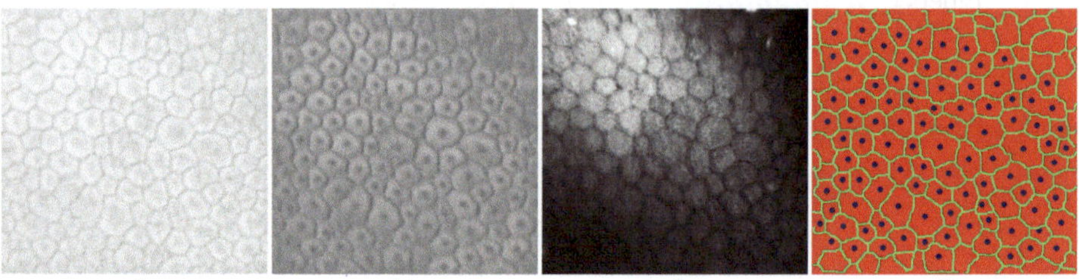

a) b) c) d)

Fig. 7 Visual results of attention maps outputted by SW and I2I attention U-Nets. **a** Attention maps obtained with SW Attention U-Net (pixels value from 0.62 to 0.93), **b** attention maps obtained with I2I Attention U-Net (pixels value from 0.47 to 0.68), **c** an original image, **d** a target ground truth (red: cell bodies, green: cell boundaries, blue: cell centers)

Table 1 Image segmentation accuracy measures for each testing fold (F1, F2, F3). DSC—the DICE coefficient, MHD—the modified Hausdorff distance

Model	DSC			MHD		
	F1	F2	F3	F1	F2	F3
I2I attention U-Net	0.875	0.869	0.848	0.352	0.379	0.480
SW attention U-Net	0.877	0.871	0.848	0.390	0.374	0.585
I2I U-Net	**0.885**	**0.883**	**0.860**	**0.329**	**0.345**	**0.477**
SW U-Net	0.869	0.873	0.831	0.382	0.368	0.636

4.4 Cell Morphometry Assessment

The Pearson correlation coefficient (PCC) [12], the mean absolute error of the number of cell neighbors (MAE_N), and the relative error of cell hexagonality (RE_H) were used to assess the resulting cell morphology.

To compare the number of cell neighbors in the ground truth $\mathcal{T}$ and predicted $\mathcal{P}$ images, the mean absolute error between the number of neighbors (see Eq. 2) was calculated. The reference number of cells and their positions in both images were based on the $\mathcal{T}$ image.

$$MAE_N = \frac{1}{N_{\mathcal{T}}} \sum_{i=1}^{N_{\mathcal{T}}} |\mathcal{T}_i - \mathcal{P}_i| \tag{2}$$

the number of cells in the $\mathcal{T}$ image is denoted by $N_{\mathcal{T}}$, whereas $\mathcal{T}i$ and $\mathcal{P}i$ refer to the numbers of neighbors of the i-th cell in both images being compared.

The relative hexagonality error RE_H (Eq. 3) between $\mathcal{T}$ and $\mathcal{P}$ images was calculated to evaluate cells' hexagonality. The hexagonality coefficient ($\mathcal{T}_H$ and $\mathcal{P}_H$) was calculated by dividing the number of hexagonal cells with six neighbors by the total number of cells. A high hexagonality coefficient indicates good corneal health, while a low value may indicate corneal damage or disease.

$$RE_H = |\frac{\mathcal{T}_H - \mathcal{P}_H}{\mathcal{T}_H}| \tag{3}$$

To measure the degree of correlation between the sizes (in pixels) of corresponding cells in the $\mathcal{T}$ and $\mathcal{P}$ images, the Pearson correlation coefficient (PCC) was utilized. A strong correlation is indicated by PCC values within the $[-1; -0.5] \bigcap [0.5; 1.0]$, while a perfect match is attained when the PCC value is equal to either 1 or –1.

The summary of the cell morphometry accuracy measures obtained for each considered version of the CNN-watershed for each testing fold is shown in Table 2. The best scores are shown in bold.

Table 2 Morphometric parameters accuracy measures for each testing fold (F1, F2, F3). PCC—the Pearson correlation coefficient, MAE_N—the mean absolute error between the number of neighbors, RE_H—the relative hexagonality error

Model	PCC			MAE_N			RE_H		
	F1	F2	F3	F1	F2	F3	F1	F2	F3
I2I attention U-Net	**0.898**	0.915	0.748	**0.136**	0.111	**0.239**	**0.035**	0.055	**0.048**
SW attention U-Net	0.765	0.909	0.729	0.319	0.123	0.480	0.085	0.071	0.251
I2I U-Net	0.885	**0.930**	**0.750**	0.156	**0.106**	0.263	0.053	**0.050**	0.094
SW U-Net	0.688	0.911	0.688	0.196	0.117	0.493	0.040	0.056	0.154

Table 3 Average prediction time in seconds for an image with the size of $384 \times 384 \times 1$

Model	Average time (s)
I2I attention U-Net	0.080
SW attention U-Net	3.190
I2I U-Net	**0.062**
SW U-Net	3.041

4.5 Prediction Time

Finally, the prediction time was measured for each considered CNN model and prediction setup. Table 3 shows the average time for a single image prediction for each considered model. Time measurements were performed on an NVIDIA GTX 1070 with 8 GB RAM DDR5 combined with an AMD Ryzen 5 5600X and 64 GB of RAM DDR4. All the experiments were conducted on a TensorFlow 2.9.1 with the Keras [13] library, and GPU was utilized for computations.

5 Discussion

Based on the visual assessment presented in the top panel of Fig. 6, it can be observed that the probability maps generated by the models used in the image-to-image (I2I) setup are more precise and less blurry compared to those produced by the sliding window (SW) setup. The probability for each class is higher for the I2I models, particularly for the cell centers, where the probabilities outputted by the SW models are notably weaker. This suggests that the sliding window models are less confident in their predictions, potentially due to the lack of global information related to the cells and their neighbors. This observation is further supported by the attention maps displayed in Fig. 7 for the Attention U-Net model, where the attention for the cell centers in the I2I setup is more concentrated compared to the SW counterparts. However, this shortcoming is primarily limited to the cell centers, as all model outputs exhibit similar levels of confidence in the case of the cell edges. As a result, this leads to comparable cell segmentation outcomes.

The numerical evaluation supports this finding, particularly when examining the accuracy measures for image segmentation, as shown in Table 1. Although the DICE scores vary by no more than 3.4% (and no less than 1.5%) between the worst and best-performing variants, the CNN models used in the I2I setup display slightly better results than their SW counterparts, with an average difference in segmentation scores of less than 0.5% for the Attention U-Net and around 2% for the U-Net. This small difference may be attributed to the application of the watershed transform to the edge probability maps, which resolves discontinuous edges resulting from the threshold-based post-processing that is typically applied to cell edge probability maps in state-of-the-art methods.

When the accuracy measures derived from cell morphometry are considered, the image-to-image setup still outperforms the sliding window approach for both variants of the U-Net model. However, the Attention U-Net is more accurate. The advantage of the model is, on average, 0.15% for the Pearson correlation coefficient of cell sizes, 6.7% for the mean absolute error of the number of neighboring cells, and almost 46% for the relative hexagonality error.

Finally, experiments confirmed that using a CNN model in a sliding window setup is an essential computational hurdle. Specifically, it increased the prediction time 40–50 times compared to the image-to-image setup (see Table 3).

Indirectly comparing our results to the corneal endothelium image segmentation results reported by selected authors is auspicious. In particular, our proposed I2I Attention U-Net scored the average cell-based DICE coefficient of 0.979, edge-based DICE of 0.876, and MHD of 0.400 on a challenging Rotterdam dataset. The corresponding scores for I2I U-Net were 0.980, 0.880, and 0.384. Vigueras-Guillen et al. [14] utilized a dataset of 50 images and corresponding masks, achieving an average DICE coefficient (for cells) of 0.981 and an MHD of 0.22, which is a comparable result in terms of cell-based DICE. In another study by [15], the authors proposed a method to segment the relatively simple Alizarine dataset [16]. With the best-fit algorithm [17], a DICE coefficient (for cells) on this dataset was equal to 0.94 and an MHD to 0.14, which is worse than we achieved in terms of the cell-based DICE. Lower MHD scores were the results of applied postprocessing using the best-fit method. Without the best-fit postprocessing, they achieved a DICE coefficient of 0.62 and an MHD of 1.26. our method avoids postprocessing, reducing the complexity and time required for segmentation while maintaining excellent results.

6 Conclusions

This study has shown that applying a marker-driven watershed transform to post-process the cell edges probability maps in the CNN-based corneal endothelium image segmentation can significantly improve the method's sensitivity to discontinuous edges. This approach allows replacing the sliding window setup with an image-to-image setup, decreasing computational overhead and shortening prediction times while maintaining similar or even better segmentation accuracy. Additionally, our results indicate that Attention U-Net outperforms the classical U-Net regarding segmentation cell quality measured in terms of cell morphometric parameters. These findings demonstrate the potential of our proposed method for efficient and accurate corneal endothelium image segmentation, which can have practical applications in diagnosing and treating various eye diseases.

References

1. Ronneberger, O., Fischer, P., Brox, T.: U-net: convolutional networks for biomedical image segmentation. In: Medical Image Computing and Computer-Assisted Intervention (MICCAI), vol. 9351 of LNCS, pp. 234–241. Springer (2015)
2. Fabijańska, A.: Segmentation of corneal endothelium images using a u-net-based convolutional neural network. Artif. Intell. Med. **88**, 1–13 (2018). https://doi.org/10.1016/j.artmed.2018.04.004
3. Daniel, M., Atzrodt, L., Bucher, F., Wacker, K., Böhringer, S., Reinhard, T., Böhringer, D.: Automated segmentation of the corneal endothelium in a large set of "real-world" specular microscopy images using the u-net architecture. Sci. Rep. **9**, 4752 (2019). https://doi.org/10.1038/s41598-019-41034-2
4. Vigueras-Guillén, J.P., Sari, B., Goes, S.F., Lemij, H.G., van Rooij, J., Vermeer, K.A., van Vliet, L.J.: Fully convolutional architecture versus sliding-window CNN for corneal endothelium cell segmentation. BMC Biomed. Eng. **1**, 4 (2019). https://doi.org/10.1186/s42490-019-0003-2
5. Kucharski, A., Fabijańska, A.: CNN-watershed: a watershed transform with predicted markers for corneal endothelium image segmentation. Biomed. Signal Process. Control **68**, 102805 (2021). https://doi.org/10.1016/j.bspc.2021.102805
6. Vigueras-Guillén, J.P., van Rooij, J., van Dooren, B.T.H., Lemij, H.G., Islamaj, E., van Vliet, L.J., Vermeer, K.A.: Denseunets with feedback non-local attention for the segmentation of specular microscopy images of the corneal endothelium with guttae (2022). https://doi.org/10.48550/ARXIV.2203.01882. arxiv:2203.01882
7. Zhang, Y., Higashita, R., Fu, H., Xu, Y., Zhang, Y., Liu, H., Zhang, J., Liu, J.: A multi-branch hybrid transformer network for corneal endothelial cell segmentation. In: de Bruijne, M, Cattin, P.C., Cotin, S., Padoy, N., Speidel, S., Zheng, Y., Essert, C. (eds.) Medical Image Computing and Computer Assisted Intervention—MICCAI 2021, pp. 99–108. Springer International Publishing, Cham (2021). https://doi.org/10.1007/978-3-030-87193-2_10
8. Selig, B., Vermeer, K.A., Rieger, B., Hillenaar, T., Hendriks, C.L.L.: Fully automatic evaluation of the corneal endothelium from in vivo confocal microscopy. BMC Med. Imaging **15**(1), 13 (2015). https://doi.org/10.1186/s12880-015-0054-3
9. Oktay, O., Schlemper, J., Folgoc, L.L., Lee, M., Heinrich, M., Misawa, K., Mori, K., McDonagh, Hammerla, N.Y., Kainz, B., Glocker, B., Rueckert, D.: Attention u-net: learning where to look for the pancreas (2018). https://doi.org/10.48550/ARXIV.1804.03999. arxiv:1804.03999

10. Li, C., Tam, P.: An iterative algorithm for minimum cross entropy thresholding. Pattern Recogn. Lett. **19**(8), 771–776 (1998). https://doi.org/10.1016/s0167-8655(98)00057-9
11. Dubuisson, M.-P., Jain, A.: A modified hausdorff distance for object matching. In: Proceedings of 12th International Conference on Pattern Recognition, vol. 1, pp. 566–568 (1994). https://doi.org/10.1109/ICPR.1994.576361
12. Freedman, D., Pisani, R., Purves, R.: Statistics (international student edition), Pisani, R. Purves, 4th edn. WW Norton & Company, New York
13. Sha, Y.: Keras-u-net-collection (2021). https://github.com/yingkaisha/keras-unet-collection. https://doi.org/10.5281/zenodo.5449801
14. Vigueras-Guillén, J.P., Sari, B., Goes, S.F., Lemij, H.G., van Rooij, J., Vermeer, K.A., van Vliet, L.J.: Fully convolutional architecture versus sliding-window CNN for corneal endothelium cell segmentation. BMC Biomed. Eng. **1**(1) (2019). https://doi.org/10.1186/s42490-019-0003-2
15. Nurzynska, K.: Deep learning as a tool for automatic segmentation of corneal endothelium images. Symmetry **10**(3). https://doi.org/10.3390/sym10030060. https://www.mdpi.com/2073-8994/10/3/60
16. Ruggeri, A., Scarpa, F., Luca, M.D., Meltendorf, C., Schroeter, J.: A system for the automatic estimation of morphometric parameters of corneal endothelium in Alizarine red-stained images. Br. J. Ophthalmol **94**(5), 643–647 (2010). arXiv:https://bjo.bmj.com/content/94/5/643.full.pdf, https://doi.org/10.1136/bjo.2009.166561. https://bjo.bmj.com/content/94/5/643
17. Piórkowski, A.: Best-fit segmentation created using flood-based iterative thinning. In: Advances in Intelligent Systems and Computing, pp. 61–68. Springer International Publishing (2016). https://doi.org/10.1007/978-3-319-47274-4_7

Tissue Pattern Classification with CNN in Histological Images

Krzysztof Siemion, Lukasz Roszkowiak, Jakub Zak, Antonina Pater, and Anna Korzynska

Abstract Tissue pattern is an important factor in morphological evaluation of tissue samples. It can be decisive in disease discrimination or for establishing disease subtypes. Tissue architecture can be generally described by pathologist as classical, hypocellular or hypercellular. This article presents a study used to establish classification convolutional neural network model for tissue compactness assessment. The VGG16 network was trained to classify image patches to create reliable heatmaps. The application of image augmentation, class-specific sampling, and hyperparameter tuning was used to prevent overfitting and increase the accuracy of the model. Based on the current results it can be concluded that the differentiation between hypo- and hypercellular tissue (compactness) is possible with application of deep learning classification model VGG16. We hope to find correlation of features related to tissue compactness with subtypes of the analysed disease that would become diagnostic markers or prognostic factors.

Keywords Biomedical engineering · Digital pathology · Deep learning · Image classification

1 Introduction

Tissue architecture is critical for cell homeostasis and physiological functions [1]. It can be also crucial factor in disease discrimination or establishing subtypes of the disease. Typically tissue architecture can be described by pathologist as hypocellular, hypercellular or classical. Such classification might aid the expert in evaluation of the

K. Siemion · L. Roszkowiak · J. Zak · A. Pater · A. Korzynska
Nalecz Institute of Biocybernetics and Biomedical Engineering, Polish Academy of Sciences, Ks. Trojdena 4 st., 02-109 Warsaw, Poland

K. Siemion (✉)
Medical Pathomorphology Department, Medical University of Bialystok, Bialystok, Poland
e-mail: ksiemion@ibib.waw.pl

P. Strumiłło et al. (eds.), *The Latest Developments and Challenges in Biomedical Engineering*, Lecture Notes in Networks and Systems 746,
https://doi.org/10.1007/978-3-031-38430-1_2

tissue and in cooperation with other features. It might also help the expert pathologist in disease differentiation, diagnosis and prognosis.

This article presents a study used to establish classification model for tissue compactness. The model was tested on the set of digital whole slide images from patients with inflammatory spindle cell lesions (ISCLs) [2], which are treated as a heterogenic group of diseases and the biology of many of them remains not fully understood. The histomorphology of these tumors seems to be diverse enough to examine if the correlation of features related to tissue compactness with subtypes of the analysed lesions exists. To achieve that goal, the convolutional neural network (CNN) was trained to classify compactness in the images of the tissue samples.

1.1 Related Works

Genetic and epigenetic factors have major influence on cancer cell phenotype and tumor architecture [3]. Other scientific teams presented different approaches to determining spatial contexts and tissue architecture, such as fast Fourier transform [4] or spatially resolved transcriptomics [5] and apply them to multiple tissues, i.e. muscle [6]. Also, some scientists provided insight on how tissue fixation protocol affects tissue architecture [7].

Up to date, there is no study focusing strictly on the automatic classification of tissue architecture. The method presented in this study could be used in the future to aid in differentiation of multiple diseases.

2 Materials and Methods

2.1 Disease

A group of "inflammatory myofibroblastic lesions" consists of neoplastic, e.g. inflammatory myofibroblastic tumor (IMT), inflammatory fibroid polyp, inflammatory liposarcoma, inflamed gastrointestinal stromal tumor, and reactive lesions, called inflammatory pseudotumor(IPT) [8]. The IMT is an intermediate-grade neoplasm, which often recurs after surgical excision and rarely gives metastases [9, 10]. Otherwise, IPT almost never recurs after resection and does not metastasize [2].

ISCLs can occur as a mass in every localization of the human body. IMTs occur most frequently in the abdominal cavity, less often in the lung, the genital tract, the head and neck region [11]. IPTs can occur in almost every anatomical location [12]. Both tumors consist of proliferating fibroblasts and/or myofibroblasts infiltrated by lymphocytes, plasma cells, eosinophils, and histiocytes [9].

Genetic test, such as gene sequencing or fluorescent in situ hybridization, are the most reliable methods, which enable to differentiate both disease. Unfortunately,

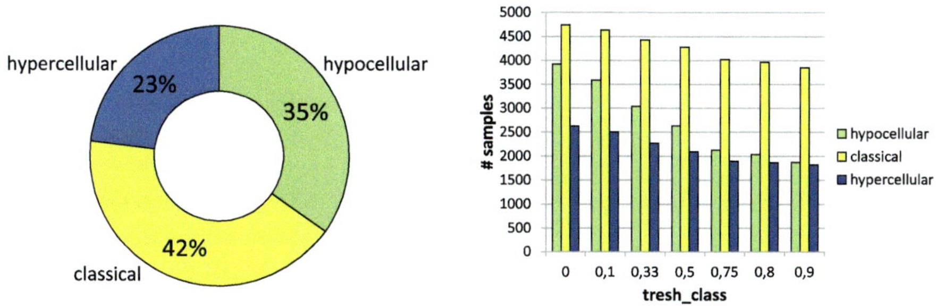

(a) Sample distribution per class

(b) Sample distribution per *class_tresh* parameter

Fig. 1 Dataset sample distribution annotated by expert; compactness classes: hypercellular in blue, hypocellular in green, classical in yellow for patch size of 256×256

these methods are expensive in money and resources. Finding correlation of tissue compactness features with subtypes of the analysed disease would provide efficient diagnostic markers or prognostic factors.

Three basic histologic patterns of ISCLs are distinguished: hypocellular (sclerotic, scar-like), classical (nodular fasciitis-like, myxoid) and hypercellular (compact, proliferating) [8, 13, 14]. The presence of giant cells, myxoid intercellular content, ganglion-like cells, lymphovascular invasion, necrosis, high mitotic activity and increased cellularity are considered as adverse factors that worsen the prognosis after tumor resection [8, 15, 16]. That is why assessment of the histologic pattern by the deep neural network may improve histopathologial diagnostics.

2.2 Image Data

This study dataset consisted of histological slides from patients with ISCLs collected from the archives of the Academic Center of Pathomorphological and Genetic-Molecular Diagnostics in Bialystok (Poland). The study obtained the consent of the bioethics committee at the Medical University of Bialystok—number APK.002. 339.2020. The hematoxylin and eosin stained microscopic slides were digitalized with Hamamatsu NanoZoomer SQ slide scanner. The 85 whole-slide images of 77 patients were collected in total. Then arbitrarily selected 3200×1800 pixels images were annotated by an expert pathologist with masks representing each histological tissue architecture type (example presented in Fig. 2). The annotated images were then subdivided into smaller fragments (patches) for deep learning (DL) model training. They were divided into patches of size 256×256, 128×128 and 64×64 pixels which resulted in about 9, 35 and 150 thousand images respectively. The distribution of samples with size 256×256 pixels that contained classes of hypocellular, hypercellular and classical tissue architecture are presented in Fig. 1.

2.3 Dataset Parameters

The available annotations in the dataset are relatively rough, due to uncertainty of the data. We established a crucial parameter, called *class_tresh* (class threshold), which value decides about the class assignment to the sample patch. Moreover, the labeled regions do not correspond directly to patch splitting as the annotations were made in bigger images (see Sect. 2.2). The set value of *class_tresh* is a minimal ratio of area in an image patch taken by annotation label. For example, when the *class_tresh* is set to 0.2 the label has to take up more than 20% of the area of the image patch to be included in the considered class.

Due to progressive manner of dataset creation during project development, at the first stage of our research the number of examples in our dataset was very limited and we wanted to try lowering the threshold to increase the number of examples in each class. Nevertheless, the logical assumption would be to set the *class_tresh* at 0.5, and that is the set value in current stage of the research.

The dataset comprised of image patches with different tissue architecture was quantitatively heterogeneous with the majority of classical architecture, as can be seen in Fig. 1. In case of strongly biased dataset if the validation set is too small there might occur a situation where all samples in the validation set will derivate from one class (the one in majority). This inevitably results in pushing the updated weights in the direction of assigning every sample to class in the majority of examples. To avoid this problem, instead of fully randomly, we assigned the samples with consideration of classes into each (train/validation/test) set. The other employed strategy was to limit the sets to the number of example in the least populated class. So, all subsets contained equally distributed classes with randomly selected examples.

Lastly, we arbitrarily opted for 50/10/40 (train/validation/test) dataset split. The test set was separated, while train and validation test were used for cross-validation.

2.4 The Deep Learning Model & Implementation

In these experiments, we used a well known VGG16 model [17] where its input and output layers were modified to fit our purposes. The model with about 14 million trainable parameters was initialized with random weights. Primarily stochastic gradient descent (SGD) optimizer was used with default parameters of learning rate and decay, and typical loss function—the "categorical crossentropy" was used. Batch size was set to 16 in all of the experiments.

The model, training and inference was implemented in Python with Keras Tensorflow framework. The model was obtained from the open-access Github repository.[1]

[1] https://github.com/qubvel/classification_models.

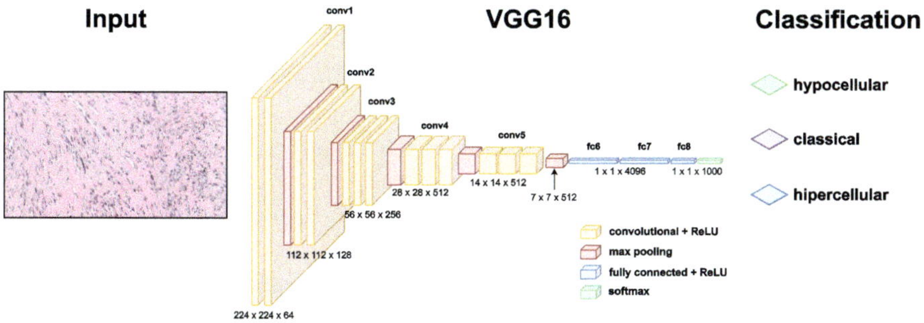

Fig. 2 Classification model with sample input image; classes: hypercellular, classical and hypocellular

Table 1 Comparison of tested augmentation schemes

Augmentation	Flips	Rotate	RandomCrop	Elastic transformations	ColorJitter	CLAHE	Brightness modification
Simple	v	v	v				
Med	v	v	v	v			
med+colorJitter	v	v	v	v	v		
med+CLAHE	v	v	v	v		v	
Heavy	v	v	v	v		v	v
no_deform	v	v	v		v	v	v
no_deform+no_colorJitter	v	v	v			v	v

2.5 *Data Augmentation*

In situations where there are very few examples, the data augmentation can improve the achieved results. In this study Albumentations[2] package was used to perform image modification. We proposed several augmentation strategies and called them: simple, med (medium) and heavy. We tested the influence of geometric deformations and ColorJitter on our data by implementing additional augmentation schemes: med+colorJitter, med+CLAHE, no_deform and no_deform+no_colorJitter. For clear comparison between tested augmentation schemes all of them are presented in Table 1.

Simple The simple augmentation consisted of only vertical and horizontal flips, image rotation and random cropping. This kind of augmentation has low impact on image content as it's not influential on values of pixels. Nevertheless it can improve the training capabilities of the model.

Medium (med) Along simple augmentation this strategy consisted of several additional distortions: grid distortion, optical distortion and elastic deformation. This

[2] https://github.com/albumentations-team/albumentations.

realizes multiple geometric transformations changing the location and shape of the objects in the images.

Med+colorJitter The *colorJitter* is simply realised by randomly changing the brightness, contrast, saturation and hue of an image. Adding this specific transform extends the scope of changes in images from geometry to intensity of pixels. The high variability of color and brightness in histopathology images is one of main issue with this type of data, hence adding this kind of transform should increase model generalization ability.

Med+CLAHE The algorithm of Contrast Limited Adaptive Histogram Equalization (CLAHE) is a variant of adaptive histogram equalization in which the contrast amplification is limited, so as to reduce this problem of noise amplification [18]. This particular modification tends to normalize the images hence improving the generalization.

Heavy The proposed heavy augmentation consist of multiple distortions in geometry and brightness of images along with Contrast Limited Adaptive Histogram Equalization (CLAHE). Such strong manipulation of data can sometimes have negative impact on training because of additional uncertainty of data.

no_deform The distortions in images caused by geometric transformations are sometimes very drastic and might hinder the process of learning. With this particular augmentation approach we wanted to test if the applied distortions really improve the training process and knowledge generalisation. This approach works as an ablation test for augmentation techniques.

no_deform+no_colorJitter Simple augmentation with added CLAHE and brightness modification. The geometric transformations and color modifications are omitted. This approach works as an ablation test for augmentation techniques.

2.6 Training Scenarios for Compactness Classification

We have organized the dataset so that all patches containing the annotation (with sufficient *class_tresh*) were treated as either classical, hypo- or hypercellular, while the rest of the samples were considered as "other" class. This resulted in a heavily biased dataset, with the majority of "other" class. We tackled this problem with the training in which the number of samples was limited per class (to the number of examples in the least populated class) to make the dataset more balanced.

Moreover, we tested two different approaches with and without inclusion of negative examples in the dataset. First, the 4-class approach included 3 positive classes and one negative. In the other approach the 3-class scenario was reduced to predict only classical, hypo- and hypercellular classes.

3 Results

First, we tested the capability of the VGG16 model to contain the knowledge sufficient for the task. In this experiment we did not use any data augmentation, since we wanted to try and overfit the model, see Fig. 3. The results in Table 2. confirm that the model is able to learn the subtle differences in tissue compactness, but regularization techniques are necessary to tackle the problem of overfitting.

Second, we concluded a comparative analysis for different *class_tresh* values. The evaluation was performed with 3-fold cross-validation and 3-class model. The resulting accuracy of models with different *class_tresh* parameter value showed the proportional increase as seen in Table 3.

Third, we compared the proposed augmentation strategies. The Table 4. shows that medium augmentation gives the best numerical results according to evaluation performed on test set. We tested two tile-sizes and achieved consistent results. For this evaluation the same 3-class model was used with 0.5 *class_tresh* parameter and same limit of samples.

Fourth, results of average compactness classification accuracy with 2 different training schemes are presented in Table 5. Distinct models with different number of classes explicitly treated are compared. For this evaluation the *class_tresh* parameter value was set to 0.5 as well as the same augmentation strategy, limit of samples, and train/val/test split was used to achieve comparable results.

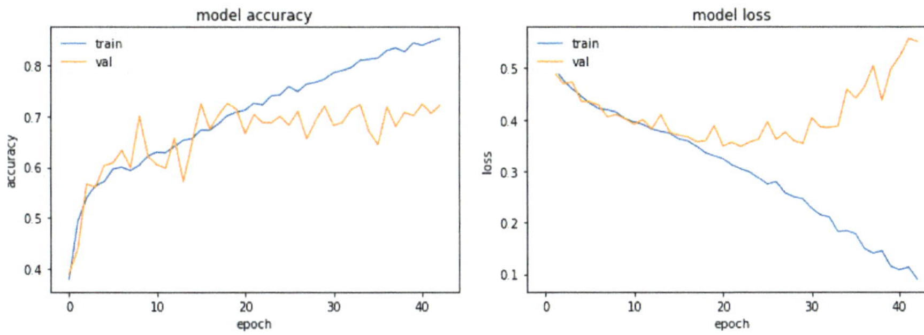

(a) Training model accuracy (b) Training model loss

Fig. 3 Example of VGG16 model overfitting to ISCL dataset, as presented with train and validation sets. Around 18th epoch the model overfits as the loss of validation set is no longer decreasing

Table 2 Overfit results for two tile-sizes (where, acc stands for accuracy)

	Train		Validation		Test			Evaluation on train set
tile_size	Loss	Acc	Loss	Acc	Loss	Acc	binary_acc	Acc
64	0,00	1,00	0,16	0,89	0,46	0,89	0,95	1,00
128	0,00	1,00	0,15	0,91	0,36	0,90	0,95	1,00

Table 3 Comparison of classification accuracy with different *class_tresh* parameter value. Given values are means acquired on test set with 3-fold cross-validation

class_tresh	Accuracy	Loss	Binary accuracy	Precision
0.90	97.07	6.05	98.05	97.09
0.75	96.84	6.88	97.90	96.85
0.50	95.57	9.23	97.05	95.60
0.33	94.83	10.65	96.42	95.55
0.10	93.47	11.37	95.26	95.83
0.00	82.35	14.41	88.27	70.91

Table 4 Comparison of augmentation schemes. Given values are means acquired on test set with 3-fold cross-validation. Headers description: epochs—number of epochs before the training was terminated; acc—accuracy on test set; bin_acc—binary accuracy o test set

tile_size	augmentation	Epochs	Acc	Loss	bin_acc	Precision
128	Simple	39	95.08	12.10	96.71	95.10
	Med	49	**95.39**	9.60	**96.92**	95.43
	med+colorJitter	69	93.89	12.70	95.93	93.95
	med+CLAHE	71	94.92	10.21	96.62	94.96
	Heavy	77	94.38	10.36	96.25	94.44
	no_deform	73	92.62	14.67	95.09	92.74
	no_deform+no_colorJitter	56	94.22	11.21	96.13	94.21
256	Simple	49	91.79	15.72	94.56	91.97
	Med	63	**95.83**	9.91	**97.24**	95.95
	med+colorJitter	104	89.29	18.13	92.87	89.43
	med+CLAHE	73	89.00	18.32	92.64	89.12
	Heavy	104	95.63	8.99	97.10	95.70
	no_deform	133	91.63	17.92	94.39	91.66
	no_deform+no_colorJitter	83	95.81	8.83	97.23	95.89

Table 5 Results of average compactness classification accuracy of models trained with different training scenarios. Randomly selected 4 images were used for F-score evaluation. Where: **3-class** model means discrimination of classical, hypo- and hypercellular compactness; **4-class** is the model with explicit "other" class included in the dataset; TP - true positive, FP—false positive, FN—false negative, TN—true negative. The best results marked in bold

Class	tile_size	epochs	train_acc	val_acc	test_acc	test_bin_acc	TP	FP	FN	TN	F-score
4	64	127	89.00	90.85	90.84	95.42	1929	601	671	9399	75.20
	128	111	92.19	91.86	91.98	96.01	504	177	148	2321	75.62
	256	153	89.82	93.14	**92.54**	**96.28**	123	28	20	585	**83.67**
3	64	53	92.15	93.39	93.59	95.74	2153	2024	447	7976	63.54
	128	72	95.27	95.83	94.30	96.20	565	406	87	2092	69.62
	256	63	91.11	97.04	**94.70**	**96.50**	131	106	12	507	68.95

4 Discussion

ISCLs can occur as masses in every localization of the human body including the inflammatory myofibroblastic tumors that are intermediate-grade neoplasms characterized by high recurrence rate after excision and low metastatic potential [9]. Three basic histologic patterns can be distinguished, i.e. hypocellular, classical and hypercellular [8, 13, 14]. Increased cellularity and presence of myxoid content in the classical morphology can worsen the prognosis of a patient [8]. That is why assessment of the histologic pattern by the deep neural network may improve histopathologial diagnostics.

We encountered two main problems with the automatic compactness classification. First, was the strong bias in number of examples in each class, with the classical tissue as far more common in our samples. It created a vast discrepancy in the training data. To avoid this problem, instead of fully randomly, we assigned the samples with consideration of classes into each (train,val,test) subset. All of the subsets contained equally distributed classes, and were limited to the number of example in the least populated class. This resulted in balanced datasets that provided consistent good classification results.

The second problem, compactness classification in locations where tissue with different patterns are mingled. This creates intertwined class labels (there were image patches with multiple types of compactness) and consequently that might confuse the learning model. There are two solutions that might work in this situation: (1) lowering the size of the image patch; (2) developing a model that gives continuous value, corresponding to the ratio of area taken by each class. Lowering the size of the image patch should give more precise results, and this was partially confirmed with the achieved results.

To cope with the mingled classes in patches, we established a crucial parameter, called *class_tresh*, which value decides about the class assignment to the sample patch. The set value of *class_tresh* is a minimal ratio of area in an image patch taken by annotation label. In general, the models accuracy increased with higher *class_tresh* parameter value (see Table 3). This confirmed that the model could achieve better results with fewer samples but consisting more reliable information.

The integration of augmentation during training of the model significantly improved the results. It provided for longer training without suffering from overfitting. The training should be sufficiently extended, gradually increasing the accuracy of the model.

According to the augmentation comparison, presented in Table 4, the medium (med) augmentation strategy gives best numerical results according to evaluation performed on test set. Applying different augmentation schemes we wanted to test the influence of different augmentation methods on our type of data. Based on the achieved results we have concluded that color manipulation (implemented as colorJitter) in histology images has negative influence on classification results. These might give too drastic changes to the images that contain very subtle information.

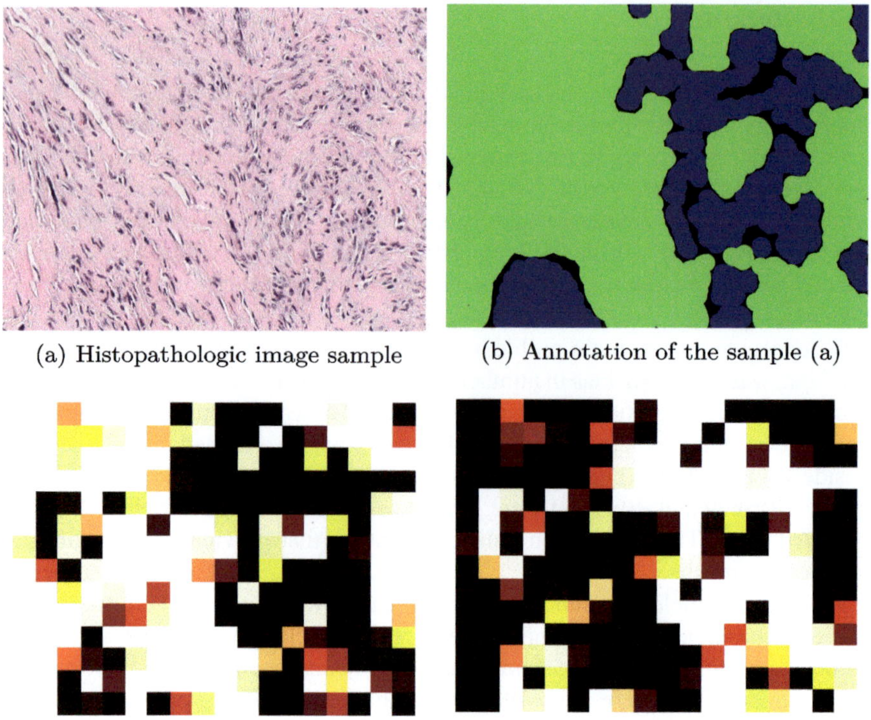

(a) Histopathologic image sample

(b) Annotation of the sample (a)

(c) Heatmap (soft predictions) of class hypocellular

(d) Heatmap (soft predictions) of class hypercellular

Fig. 4 Example of inference on sample image, generating separate heatmaps for each class. Relation of localization of hypercellular class could be clearly seen in the results image. Heatmaps (soft predictions) are presented with "hot" colormap, where white equals max value

This consequently makes it more difficult to classify tissue architecture, hence lower numerical results of these more complex augmentation strategies.

On the other hand, geometric deformations improve the results as they increase the generalisation of learning process. Also, the process of brightness and contrast manipulation have positive impact on learning process. To sum up, the **heavy** augmentation combines the augmentation techniques that are beneficial for tissue architecture classification. The numerical results are comparable between **med** and **heavy** augmentations as well as time taken per epoch. In the final experiments we decided to use **heavy** augmentation as more complex augmentation has better chance to prevent overfitting and because this method allowed for the longer effective training.

Most importantly we compared different training scenarios with different number of classes explicitly given to the model. The main difference was how the negative samples were provided and we tried to answer the question are they actually improving the classification accuracy. The model that was trained on patches with only relevant classes (3-class model) without negative examples achieved best accuracy. The example of inference in comparison to labels is shown in Fig. 4.

Our hypothesis was that introduction of the 4th class ("other") might increase the generality of the model, but the high variability of tissue in that class might be misleading for the trained model. A seen in Table 5 the 4-class model achieved comparable accuracy value to 3-class model. However, according to patch based evaluation of test set images the number of false positives significantly decreases.

The achieved results in automatic classification of tissue architecture were satisfactory, yet they were achieved only based on ISCL dataset. In the future we plan to further develop this method and test it with different kind of tissue.

5 Conclusions

To sum up, based on the current results, we can conclude that the differentiation between classical, hypo- and hypercellular tissue compactness is possible with application of deep learning classification model VGG16. The achieved accuracy was promising, although we hope to produce even better results by increasing the number of samples used in training the classification model. In the future we hope to find correlation of features related to tissue compactness with markers of prognostic or predictive factors of ISCLs.

Acknowledgements **Ethics approval** Consent of the Bioethics Committee at the Medical University of Bialystok—number APK.002.339.2020.

Funding This work has received support from Nalecz Institute of Biocybernetics and Biomedical Engineering Polish Academy of Sciences statutory financing. The research is partially funded by two subsidies from the Medical University of Bialystok: SUB/1/DN/22/002/1155 and SUB/1/DN/21/002/1194.

References

1. Nelson, C.M., Bissell, M.J.: Of extracellular matrix, scaffolds, and signaling: tissue architecture regulates development, homeostasis, and cancer. Ann. Rev. Cell Dev. Biol. **22**(1), 287–309 (2006)
2. Kutok, et al.: Inflammatory pseudotumor of lymph node and spleen: an entity biologically distinct from inflammatory myofibroblastic tumor. Hum. Pathol. **32**(12), 1382–1387 (2001)
3. Almagro, J., Messal, H.A., Elosegui-Artola, A., van Rheenen, J., Behrens, A.: Tissue architecture in tumor initiation and progression. Trends Cancer **8**(6), 494–505 (2022)
4. Zak, J., Siemion, K., Roszkowiak, L., Korzynska, A.: Fourier transform layer for fast foreground segmentation in samples' images of tissue biopsies. In: Biocybernetics and Biomedical Engineering—Current Trends and Challenges, pp. 118–125. Springer International Publishing (2021)

5. Chang, Y., He, F., Wang, J., Chen, S., Li, J., Liu, J., Yu, Y., Su, L., Ma, A., Allen, C., Lin, Y., Sun, S., Liu, B., Otero, J., Chung, D., Fu, H., Li, Z., Xu, D., Ma, Q.: Define and visualize pathological architectures of human tissues from spatially resolved transcriptomics using deep learning (2021)
6. Morris, T.A., Eldeen, S., Tran, R.D.H., Grosberg, A.: A comprehensive review of computational and image analysis techniques for quantitative evaluation of striated muscle tissue architecture. Biophys. Rev. **3**(4), 041302 (2022)
7. Singhal, P.: Evaluation of histomorphometric changes in tissue architecture in relation to alteration in fixation protocol—An invitro study. J. Clin. Diagn. Res. (2016)
8. Siemion, K., Reszec-Gielazyn, J., Kisluk, J., Roszkowiak, L., Zak, J., Korzynska, A.: What do we know about inflammatory myofibroblastic tumors?—A systematic review. Adv. Med. Sci. **67**(1), 129–138 (2022)
9. Antonescu, C.R., et al.: WHO classification of tumours. Soft tissue and bone tumours, International Agency for Research on Cancer (2020)
10. Gros, L., Tos, A.P.D., Jones, R.L., Digklia, A.: Inflammatory myofibroblastic tumour: state of the art. Cancers **14**(15), 3662 (2022)
11. Lindberg, M.R.: Diagnostic Pathology: soft Tissue Tumors E-Book. Elsevier Health Sciences (2019)
12. Zhu, et al.: Pulmonary inflammatory myofibroblastic tumor versus IgG4-related inflammatory pseudotumor: differential diagnosis based on a case series. J. Thorac. Disease **9**(3), 598–609 (2017)
13. Shenawi, H.A., Al-Shaibani, S.A., Saad, S.K.A., Al-Sindi, F., Al-Sindi, K., Shenawi, N.A., Naguib, Y., Yaghan, R.: An extremely rare case of malignant jejunal mesenteric inflammatory myofibroblastic tumor in a 61-year-old male patient: a case report and literature review. Front. Med. **9** (2022)
14. Khatri, A., Agrawal, A., Sikachi, R., Mehta, D., Sahni, S., Meena, N.: Inflammatory myofibroblastic tumor of the lung. Adv. Respir. Med. **86**(1), 27–35 (2018)
15. Coffin, C.M., Hornick, J.L., Fletcher, C.D.M.: Inflammatory myofibroblastic tumor. Am. J. Surg. Pathol. **31**(4), 509–520 (2007)
16. Bennett, J.A., Nardi, V., Rouzbahman, M., Morales-Oyarvide, V., Nielsen, G.P., Oliva, E.: Inflammatory myofibroblastic tumor of the uterus: a clinicopathological, immunohistochemical, and molecular analysis of 13 cases highlighting their broad morphologic spectrum. Modern Pathol. **30**(10), 1489–1503 (2017)
17. Simonyan, K., Zisserman, A.: Very deep convolutional networks for large-scale image recognition
18. Pizer, S.M., Amburn, E.P., Austin, J.D., Cromartie, R., Geselowitz, A., Greer, T., ter Haar Romeny, B., Zimmerman, J.B., Zuiderveld, K.: Adaptive histogram equalization and its variations. Comput. Vis. Graph. Image Process. **39**(3), 355–368 (1987)

Robust Multiresolution and Multistain Background Segmentation in Whole Slide Images

Artur Jurgas, Marek Wodzinski, Manfredo Atzori, and Henning Müller

Abstract Background segmentation is an important step in analysis of histopathological images. It allows one to remove irrelevant regions and focus on the tissue of interest. However, background segmentation is challenging due to the variability of stain colors and intensity levels across different images, modalities, and magnification levels. In this paper, we present a learning-based model for histopathology background segmentation based on convolutional neural networks. We compare two multiresolution approaches to deal with the variability of magnification in histopathology images: (i) model that uses upscaling of smaller patches of the image, and (ii) model simultaneously trained on multiple resolution levels. Our model is characterized by solid performance both in multiresolution and multistain dyes (H&E and IHC), achieving good performance on publicly available dataset. The quantitative scores are, in terms of the Dice score, close to 94.71. The qualitative analysis presents strong performance on previously unseen cases from different distributions and various dyes. We freely release the model, weights, and ground-truth annotations to promote the open science and reproducible research.

Keywords Computational pathology · Deep learning · Digital pathology Segmentation · Whole-slide images · WSI

A. Jurgas · M. Wodzinski · M. Atzori · H. Müller
University of Applied Sciences Western Switzerland (HES-SO Valais), Information Systems Institute, Sierre, Switzerland

A. Jurgas (✉) · M. Wodzinski
Faculty of Electrical Engineering, Automatics, Computer Science and Biomedical Engineering, AGH University of Science and Technology, Krakow, Poland
e-mail: arjurgas@student.agh.edu.pl

M. Atzori
Department of Neuroscience, University of Padova, Padova, Italy

H. Müller
Medical Faculty, University of Geneva, Geneva, Switzerland

P. Strumiłło et al. (eds.), *The Latest Developments and Challenges in Biomedical Engineering*, Lecture Notes in Networks and Systems 746,
https://doi.org/10.1007/978-3-031-38430-1_3

1 Introduction

Background segmentation is a basic step in most preprocessing tasks for whole slide images (WSIs), which are digital scans of tissue slides used for cancer diagnosis and prognosis. Segmentation aims to separate the foreground tissue regions from the background glass regions, which can reduce computational costs and improve accuracy for subsequent analysis such as classification, detection, grading, and registration [10]. However, background segmentation of WSIs is challenging due to variations in tissue appearance, staining quality, illumination conditions, and scanning artifacts. Existing methods for background segmentation of WSIs are either based on handcrafted features or deep learning models. Up to our knowledge, the currently existing methods are limited to H&E (hematoxylin and eosin) staining. Moreover, most existing methods for background segmentation are either not publicly available or require manual tuning of parameters for different datasets.

Many studies have explored histopathology segmentation, which is a technique to identify different regions in tissue images like [1, 5, 13] However, most of these studies focus on segmenting nuclei, which are small and distinct structures in the tissue. Segmenting the whole tissue is also challenging because it involves large areas that have low contrast and that are often similar to the background, especially when immunohistochemistry (IHC) staining is used. This type of staining is much less normalized and quality controlled [7, 17]. Even for H&E staining, the differences in dyes are problematic [16]. Therefore, existing methods for segmentation cannot be easily adapted to whole tissue segmentation.

Some previous studies [12] used a conventional method to segment tissue regions of interest (ROI) from histopathological images. However, this method works well only for images stained with H&E, and not for those stained with immunohistochemistry (IHC). Moreover, our goal is different from theirs. We want to segment the background from the tissue, not just the ROI within the tissue. This is important for some preprocessing techniques that require selecting the entire tissue area, including its folds and artifacts.

Recent studies have demonstrated that deep learning can achieve remarkable results in histopathology, enabling more accurate and detailed predictions for various diseases [6, 11, 13, 15] However, most of the existing work focuses on specific types of tissue, such as epidermal tissue [14]. While there is some research in background segmentation [2], the proposed models are not publicly available or accessible, limiting their reproducibility and applicability.

In this paper, we propose a deep learning-based pipeline for background segmentation of WSIs that can handle diverse types of tissues and stains without requiring any prior information or user intervention. Our framework consists of two main components: a patch-level segmentation network that predicts foreground probability maps for small patches extracted from WSIs, and a slide-level fusion during inference that combines the patch-level predictions into a final binary mask for the whole slide. We evaluate our framework on a public dataset from ACROBAT challenge, covering mainly breast cancer patients. We show that our framework achieves solid performance and high generalizability across different tissues. Additionally, we show solid performance in multiple resolutions of those images, as well as both H&E and IHC staining.

One of the applications of our background segmentation method is to improve the quality and efficiency of other algorithms that process histopathology images. Software tools like QuPath [3] and HistoQC [8] are widely used for various tasks such as tissue detection, annotation, classification, and quantification and are proven to benefit histopathology analysis [4]. However, these tools often rely on manual or semi-automatic methods to remove the background regions from the images, which can be time-consuming and inconsistent. Our background segmentation method can help streamline this process.

The source code, ground-truth segmentation masks, and model weights will be made publicly available [9].

2 Materials and Methods

2.1 Dataset

The ground-truth used for training is defined by manual segmentation of nine different tissues from the ACROBAT dataset [18]. The dataset is a collection of tissues in a pyramidal tiff format with varying resolutions at each level of the pyramid. Starting at 10x, authors provide 7 to 9 lower resolutions per image, with a downsampling factor of 2 in between. The second level which made up most of the training data has an average resolution of 12146 ± 2189 pixels along the X axis, and 24007 ± 4854 pixels along the Y axis. Each slide at this resolution can potentially generate up to 4449 unique, non-overlapping patches of size 256 by 256 pixels. We utilize a random sampling strategy. In each iteration, we sample 256 patches in random locations. This means over the course of training, there is some overlap of those patches. The final number of patches is not defined and depends on how long we train the network.

The dataset includes various artifacts such as markers, scratches, out-of-focus regions and coverslips. It also consists of two types of staining: H&E and IHC. We use three distinct IHC dyes during training and testing. We selected this dataset because it represents the real-world challenges of image analysis. We used eight images for training, one image for validation, and four images each in its two staining variants for qualitative evaluation without prior manual segmentations. We aimed to achieve high-quality segmentation by using a patch-based approach for both training and inference. This allows us to exploit the heterogeneity of the tissue structures across different slides and to generalize better with less data (Table 1).

We used Nvidia Tesla V100 graphics card configured with 300W TDP and 32 GB of memory hosted on PLGrid HPC cluster Prometheus. We did not utilize the tensor cores as of the time of writing, which means the inference times in Table 2 could be further improved.

Table 1 Models' performance during training

Dataset patches	Model	Dice
Training	Multiresolution	96.14
Validation	Multiresolution	92.27
Training	Upscaled	96.31
Validation	Upscaled	91.68

Table 2 Models' performance on validation image on different resolution levels

	Multiresolution model		*Upscaled model*	
Pyramid level	Dice	Latency [mm:ss]	Dice	Latency [mm:ss]
2nd	93.66	1:08	94.43	1:06
3rd	95.09	0:16	93.21	0:49
4th	95.61	0:03	92.85	0:44
5th	94.62	0:01	81.00	0:47
6th	94.57	0:00	11.16	0:45
Mean dice	94.71		74.53	

2.2 Model

We present an encoder-decoder Convolutional Neural Network (CNN) as our model for image segmentation in Fig. 2. The model architecture and the training pipeline are illustrated in Fig. 1. We trained our model using binary cross-entropy (BCE) loss and Adam optimizer, and we measured its performance on both training and validation datasets using soft Dice score.

We adopted a UNet-like architecture that consists of an encoder-decoder structure with both short and long skip connections. The short skip connections allow the network to preserve spatial information across different levels of abstraction, while the long skip connections enable the network to recover fine details from the encoder output. We present the architecture details of the model in Table 3.

Our training pipeline involves extracting image patches of size 256×256 from: (i) different resolutions of the image pyramid, or (ii) only the second level of the pyramid, and then inferring on other levels by extracting smaller patches proportional to the downsampling level and upscaling them using bilinear interpolation to 256×256 pixels.

Our random sampling strategy is dependent on previously generated probability maps. We apply gaussian blurring with a 3×3 kernel on segmentation maps and then subtract that from the original mask. This gives us a blurred edge of the segmentation mask. Then we transform this output so that the background and foreground of the tissue has a 25% sampling probability and the edge 50%.

Table 3 Model architecture with each convolution kernel size, the output shape of the layer, and the layer's number of parameters

Layer type	Kernel shape	Output shape	Param #
UNet	–	[1, 1, 256, 256]	–
+ Sequential	–	[1, 1, 128, 128]	–
+ Conv2d	[4, 4]	[1, 1, 128, 128]	17
+ GroupNorm	–	[1, 1, 128, 128]	2
+ LeakyReLU	–	[1, 1, 128, 128]	–
+ Sequential	–	[1, 32, 64, 64]	–
+ ResidualBlock	[3, 3]	[1, 32, 128, 128]	9,760
+ Conv2d	[4, 4]	[1, 32, 64, 64]	16,416
+ GroupNorm	–	[1, 32, 64, 64]	64
+ LeakyReLU	–	[1, 32, 64, 64]	–
+ Sequential	–	[1, 64, 32, 32]	–
+ ResidualBlock	[3, 3]	[1, 64, 64, 64]	57,792
+ ResidualBlock	[3, 3]	[1, 64, 64, 64]	78,272
+ Conv2d	[4, 4]	[1, 64, 32, 32]	65,600
+ GroupNorm	–	[1, 64, 32, 32]	128
+ LeakyReLU	–	[1, 64, 32, 32]	–
+ Sequential	–	[1, 128, 16, 16]	–
+ ResidualBlock	[3, 3]	[1, 128, 32, 32]	230,272
+ ResidualBlock	[3, 3]	[1, 128, 32, 32]	312,192
+ Conv2d	[4, 4]	[1, 128, 16, 16]	262,272
+ GroupNorm	–	[1, 128, 16, 16]	256
+ LeakyReLU	–	[1, 128, 16, 16]	–
+ Sequential	–	[1, 128, 32, 32]	–
+ ResidualBlock	[3, 3]	[1, 128, 16, 16]	312,192
+ ResidualBlock	[3, 3]	[1, 128, 16, 16]	312,192
+ ConvTranspose2d	[4, 4]	[1, 128, 32, 32]	262,272
+ GroupNorm	–	[1, 128, 32, 32]	256
+ LeakyReLU	–	[1, 128, 32, 32]	–
+ Sequential	–	[1, 64, 64, 64]	–
+ ResidualBlock	[3, 3]	[1, 64, 32, 32]	160,192
+ ConvTranspose2d	[4, 4]	[1, 64, 64, 64]	65,600
+ GroupNorm	–	[1, 64, 64, 64]	128
+ LeakyReLU	–	[1, 64, 64, 64]	–
+ Sequential	–	[1, 32, 128, 128]	–
+ ResidualBlock	[3, 3]	[1, 32, 64, 64]	40,160
+ ConvTranspose2d	[4, 4]	[1, 32, 128, 128]	16,416
+ GroupNorm	–	[1, 32, 128, 128]	64
+ LeakyReLU	–	[1, 32, 128, 128]	–
+ Sequential	–	[1, 1, 256, 256]	–
+ ResidualBlock	[3, 3]	[1, 1, 128, 128]	346
+ ConvTranspose2d	[4, 4]	[1, 1, 256, 256]	17
+ GroupNorm	–	[1, 1, 256, 256]	2
+ LeakyReLU	–	[1, 1, 256, 256]	–
+ Sequential	–	[1, 1, 256, 256]	–
+ Conv2d	[1, 1]	[1, 1, 256, 256]	2

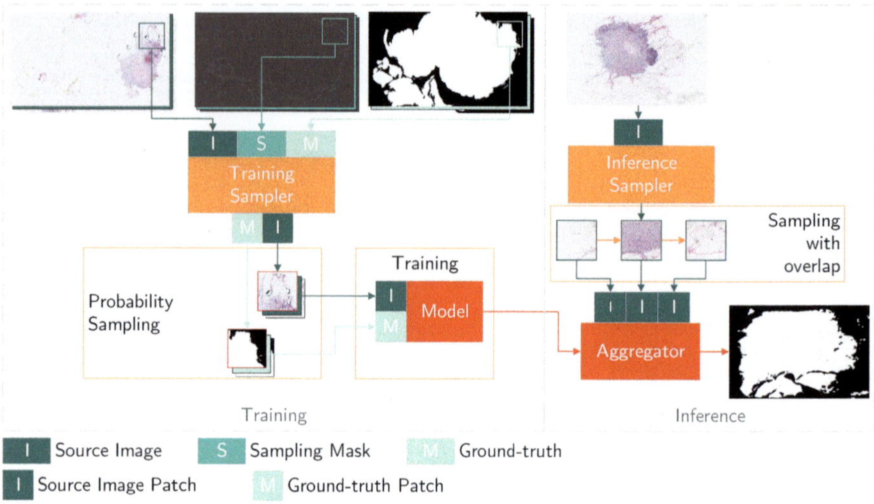

Fig. 1 A schematic diagram of the proposed deep learning pipeline. The pipeline consists of two stages: training and inference. Each block has marked inputs described in the legend

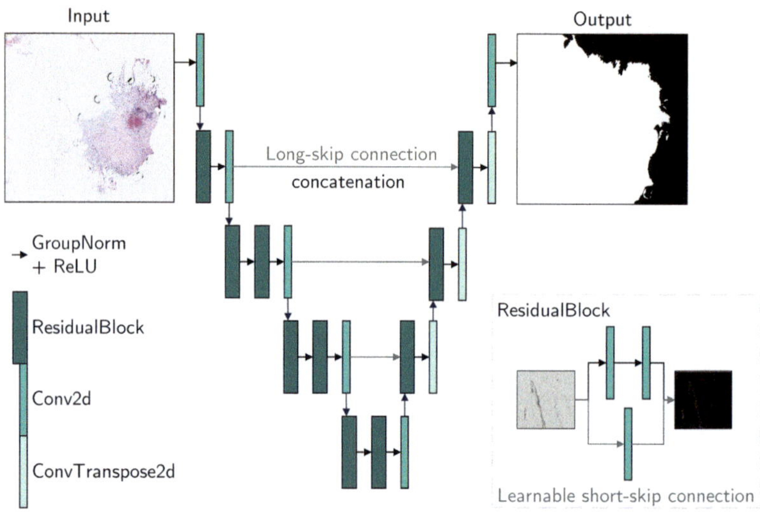

Fig. 2 A schematic illustration of the UNet-like model used in this study. The model consists of an encoder-decoder architecture with both short skip connections described in ResidualBlock and long skip connections based on concatenating feature maps

During the validation and inference steps, we employ an aggregator of individual patches by merging them together with a 20% overlap. In this overlap region, we average the value of pixels.

3 Results

The Figs. 3, 4 and 5 illustrate some examples of our method's output. We compared the results of the upscaled and multiresolution model in Tables 1 and 2. The scores reported in those tables are the average scores of all patches in each batch.

Results in Table 1 indicated that the multiresolution model had a similar performance on the training dataset but a slightly higher performance on the validation dataset than the single-resolution model. This indicates that the multiresolution model reduced overfitting and improved generalization. Figures 4 and 5 illustrate some examples of image reconstruction from both models on the test set Table 4 shows expanded descriptions.

4 Discussion and Conclusions

We proposed two methods for patch-based segmentation of histopathology images: (i) training and testing on a single pyramid level with patch resampling at inference time, and (ii) training and testing on multiple pyramid levels without patch resampling. We found that the second method achieved better performance across

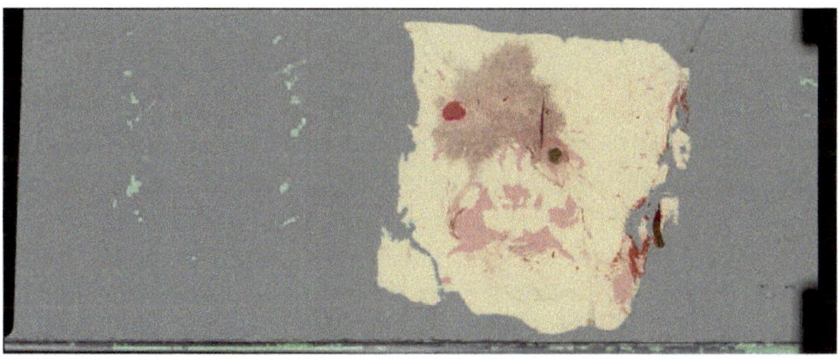

(a) Full image

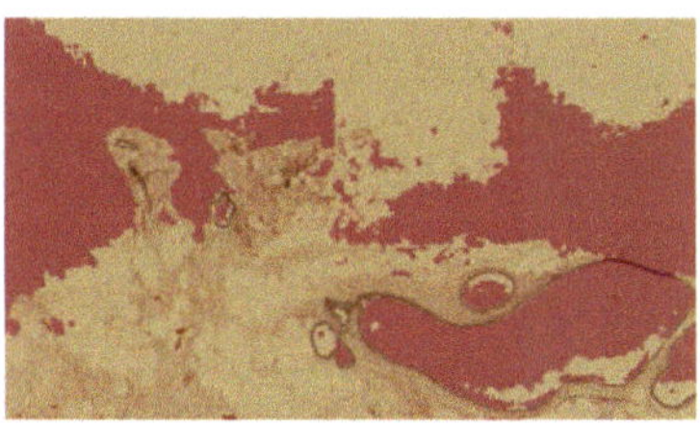

(b) Example of segmentation details

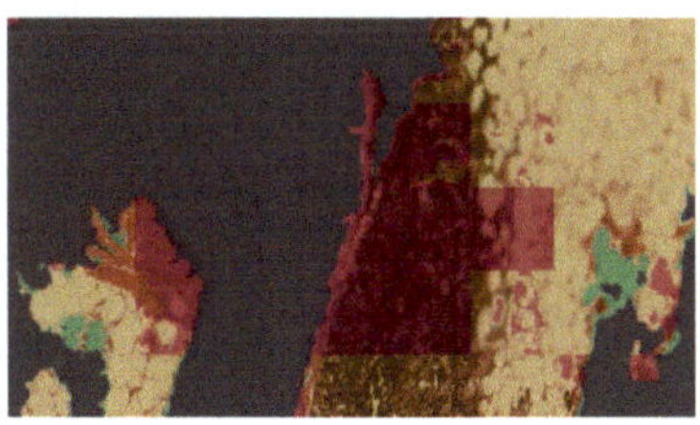

(c) Example of failure on out-of-focus regions

Fig. 3 Visualization of segmentation on the validation dataset. Segmentation output is marked with green color. Ground-truth masks are marked with red color. Overlap of the prediction with ground-truth is marked with orange color

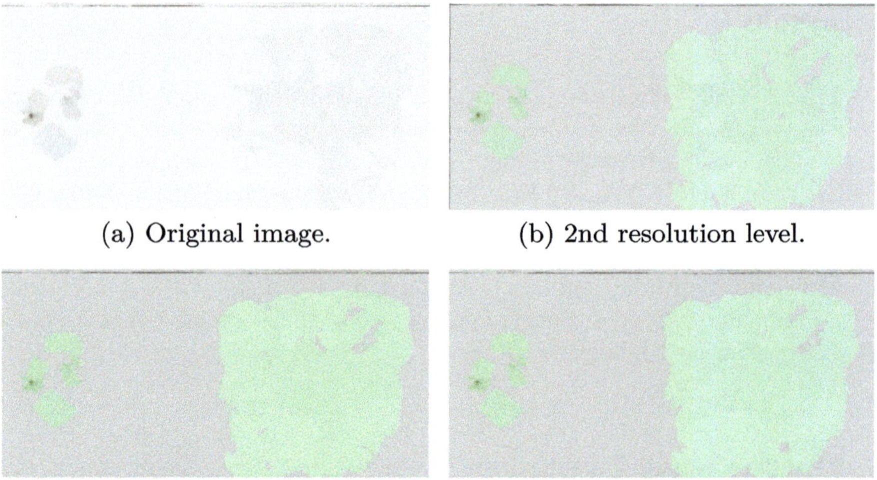

(a) Original image. (b) 2nd resolution level.

(c) 3rd resolution level. (d) 4th resolution level.

Fig. 4 Exemplary visualization from the test dataset (IHC staining). Segmentation output is marked with green color

different resolutions and significantly reduced the inference time. Our method is a general approach for histopathology slide analysis that does not target any specific regions of tissues. Instead, it aims to segment the background of the slide by inverting the segmentation of other structures, such as cells, nuclei, glands, vessels, etc.

According to Table 2 mutliresolution model achieved substantially better performance on lower magnification levels, which is supported by comparing visualizations in Figs. 5 and 6. The multiresolution model had also a faster inference time than the single-resolution model, which is beneficial for practical applications.

The second method required a larger model with a wider receptive field, but it is still feasible to deploy on even IoT class hardware such as Nvidia Jetson. This allows for deployment on modern WSIs digitization hardware or integration into most computational histopathology software.

Additionally, the first method incurred additional computational costs due to two interpolation steps. The first method suffered from information loss and artifact introduction when upscaling patches from lower resolutions. The difference in performance between the two methods can be intuitively explained by considering that the second method used more information from the original image and did not introduce any artificial structures due to patches interpolation.

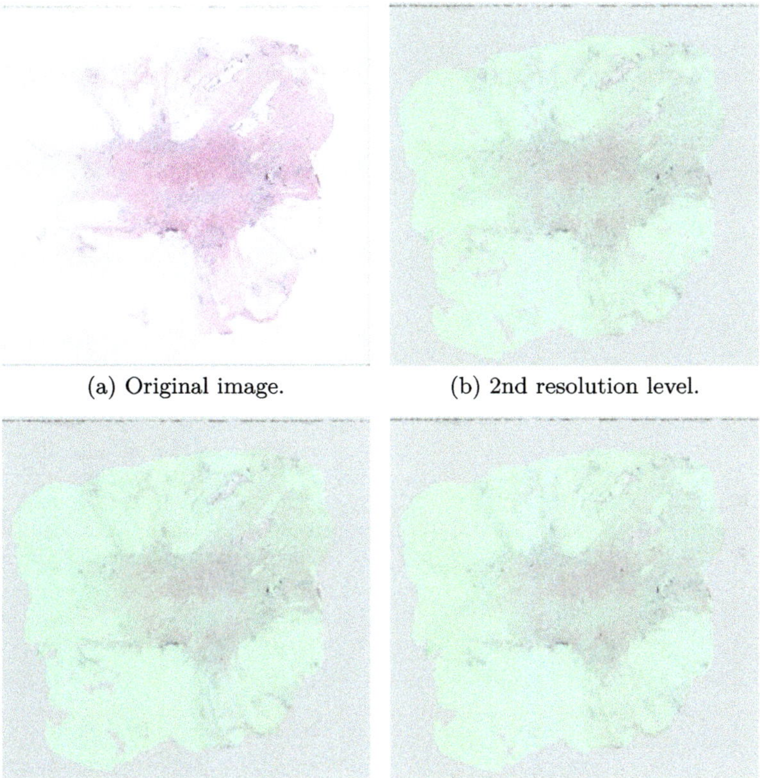

(a) Original image. (b) 2nd resolution level.

(c) 3rd resolution level. (d) 4th resolution level.

Fig. 5 Exemplary visualization from the test dataset (H&E staining). Segmentation output is marked with green color

A common problem in histopathology is the variability of staining techniques and dyes used to visualize different tissue structures and biomarkers. Different laboratories may use different protocols, reagents, and equipment to perform immunohistochemistry staining, which can result in inconsistent and non-reproducible results. This poses a challenge for developing computational models that can analyze histopathological images and extract meaningful features from them. Our model addresses this challenge by being stain-invariant, meaning that it can work on images stained with a wide range of different dyes without requiring any pre-processing or normalization steps. This makes our model more robust and generalizable to different datasets and applications. Segmentations of the same tissue dyed by IHC and HE staining is presented in Figs. 4 and 5.

Table 4 Qualitative analysis of the model's output in the test set. Each case's number matches the number in the original dataset. The images will be published as the supplementary material in the associated repository

Case from dataset	Output commentary
4	Small artifacts from aggregation in blurry regions in H&E variant. They disappear while going down the magnification levels. On the 3rd level, artifacts from scratches appear. IHC variant has no artifacts, even though they are present in the image
5	While correctly segmenting tissue, there is a larger artifact area in the HE variant. It disappears on lower resolution levels. IHC variant is segmented similarly to H&E one
8	Solid performance on both HE and IHC variants. A little noise in the largest IHC sample where coverslip is present in the original, though most of the coverslip is properly not segmented by the model. The artifact disappears at lower magnifications
14	Solid HE segmentation. Similar performance in IHC, although some noise is present on the highest resolution in less visible regions of the tissue
29	Solid HE segmentation. The coverslip on the IHC variant has been partly segmented
30	Solid HE and IHC segmentation. The coverslip on the IHC variant was properly not segmented

One of the limitations of our method is the sensitivity of the model to the resolution of the input images. Our model was trained on images with a fixed set of magnifications, and it may not perform well on images with a very different magnification level. Another limitation is the possibility of large or significantly different artifacts than in our dataset influencing the segmentation results. Particularly with higher resolutions, such artifacts could be segmented as part of the background, leading to inaccurate or incomplete segmentation.

In this paper, we proposed a novel method for fast and generalizable background segmentation in histopathological images. We tackled the problem from two perspectives: an upscaling of lower resolution images, and training the model on all magnification levels. We demonstrated that our method can achieve high accuracy and robustness on various tissues with different resolution levels and staining dyes.

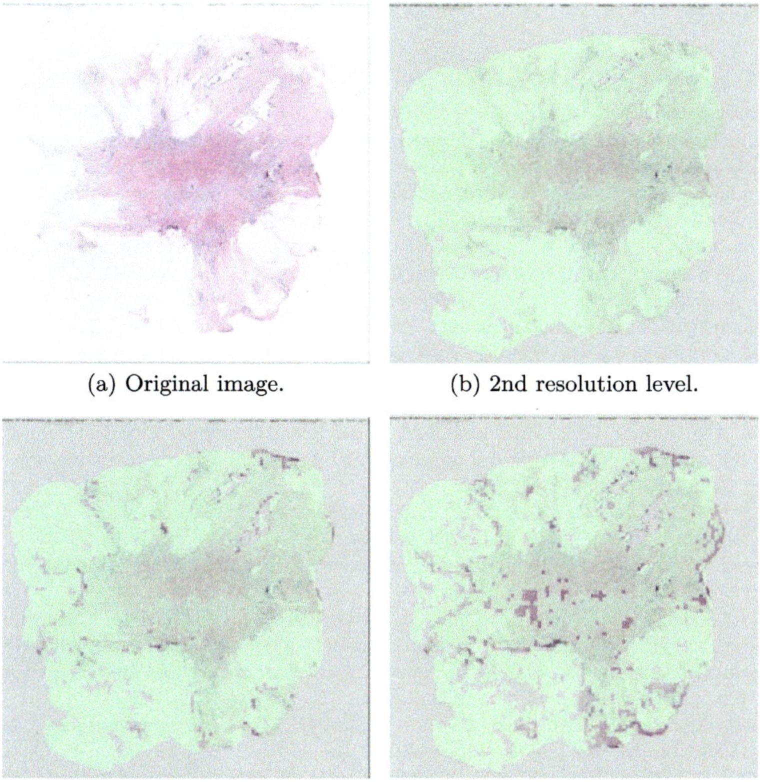

(a) Original image. (b) 2nd resolution level.

(c) 3rd resolution level. (d) 4th resolution level.

Fig. 6 Exemplary visualization from the test dataset (H&E staining) on the upscaling model. Segmentation output is marked with green color

Compared to other works like [12] our method does not require adjusting any hyperparameters. We also showed that our deep learning model can learn to segment tissues effectively even with a small amount of data, thanks to the patch-based approach. Our method can be useful for preprocessing histopathological images for further analysis and diagnosis.

Acknowledgements This work was done as a part of the IMI BigPicture project (IMI945358). We gratefully acknowledge Poland's high-performance computing infrastructure PLGrid (HPC Centers: ACK Cyfronet AGH) for providing computer facilities and support within computational grant no. PLG/ 2023/016239.

References

1. Al-Kofahi, Y., Lassoued, W., Lee, W., et al.: Improved automatic detection and segmentation of cell nuclei in histopathology images. IEEE Trans. Biomed. Eng. **57**(4), 841–852 (2010)
2. Bándi, P., Balkenhol, M., van Ginneken, B., et al.: Resolution-agnostic tissue segmentation in whole-slide histopathology images with convolutional neural networks. PeerJ **7**, e8242 (2019)
3. Bankhead, P., Loughrey, M.B., Fernández, J.A., et al.: QuPath: open source software for digital pathology image analysis. Sci. Rep. **7**(1), 16878 (2017)
4. Chen, Y., Zee, J., Smith, A., et al.: Assessment of a computerized quantitative quality control tool for whole slide images of kidney biopsies. J. Pathol. **253**(3), 268–278 (2021)
5. Cui, Y., Zhang, G., Liu, Z., et al.: A deep learning algorithm for one-step contour aware nuclei segmentation of histopathology images. Med. Biol. Eng. Comput. **57**(9), 2027–2043 (2019)
6. Ehteshami Bejnordi, B., Veta, M., Johannes van Diest, P., et al.: Diagnostic assessment of deep learning algorithms for detection of lymph node metastases in women with breast cancer. JAMA **318**(22), 2199–2210 (2017)
7. Elias, J.M., Gown, A.M., Nakamura, R.M., et al.: Special report: quality control in immunohistochemistry: report of a workshop sponsored by the biological stain commission. Am. J. Clin. Pathol. **92**(6), 836–843 (1989)
8. Janowczyk, A., Zuo, R., Gilmore, H., et al.: HistoQC: an open-source quality control tool for digital pathology slides. JCO Clin. Cancer Inform. **3**, 1–7 (2019)
9. Jurgas, A.: Jarartur/pcbbe23-histseg: multiresolution and multistain background segmentation in WSIs (2023)
10. Levy, J.J., Jackson, C.R., Haudenschild, C.C., et al.: PathFlow-MixMatch for whole slide image registration: an investigation of a segment-based scalable image registration method (2020)
11. Litjens, G., Sánchez, C.I., Timofeeva, N., et al.: Deep learning as a tool for increased accuracy and efficiency of histopathological diagnosis. Sci. Rep. **6**, 26286 (2016)
12. Muñoz-Aguirre, M., Ntasis, V.F., Rojas, S., et al.: PyHIST: a histological image segmentation tool. Plos Comput. Biol. **16**(10), e1008349 (2020)
13. Naylor, P., Laé, M., Reyal, F., et al.: Segmentation of nuclei in histopathology images by deep regression of the distance map. IEEE Trans. Med. Imaging **38**(2), 448–459 (2019)
14. Oskal, K.R.J., Risdal, M., Janssen, E.A.M., et al.: A U-net based approach to epidermal tissue segmentation in whole slide histopathological images. SN Appl. Sci. **1**(7), 672 (2019)
15. Shelhamer, E., Long, J., Darrell, T.: Fully convolutional networks for semantic segmentation. IEEE Trans. Pattern Anal. Mach. Intell. **39**(4), 640–651 (2017)
16. Tellez, D., Litjens, G., Bándi, P., et al.: Quantifying the effects of data augmentation and stain color normalization in convolutional neural networks for computational pathology. Med. Image Anal. **58**, 101544 (2019)
17. Tsutsumi, Y.: Pitfalls and caveats in applying chromogenic immunostaining to histopathological diagnosis. Cells **10**(6), 1501 (2021)
18. Weitz, P., Valkonen, M., Solorzano, L., et al.: ACROBAT—A multi-stain breast cancer histological whole-slide-image data set from routine diagnostics for computational pathology (2022). arxiv:abs/2211.13621

Impact of Visual Image Quality on Lymphocyte Detection Using YOLOv5 and RetinaNet Algorithms

A. Polejowska, M. Sobotka, M. Kalinowski, M. Kordowski, and T. Neumann

Abstract Lymphocytes, a type of leukocytes, play a vital role in the immune system. The precise quantification, spatial arrangement and phenotypic characterization of lymphocytes within haematological or histopathological images can serve as a diagnostic indicator of a particular lesion. Artificial neural networks, employed for the detection of lymphocytes, not only can provide support to the work of histopathologists but also enable better disease monitoring and faster analysis of the general immune system condition. In this study, the impact of visual quality on the performance of state-of-the-art algorithms for detecting lymphocytes in medical images was examined. Two datasets were used, and image modifications such as blur, sharpness, brightness, and contrast were applied to assess the performance of YOLOv5 and RetinaNet models. The study revealed that the visual quality of images exerts a substantial impact on the effectiveness of the deep learning methods in detecting lymphocytes accurately. These findings have significant implications for deep learning approaches used in digital pathology.

Keywords Digital pathology · Lymphocyte detection · YOLOv5 · RetinaNet · Image quality · Image degradation · Histopathology images · Tumor

1 Introduction

Histopathological images, which are images of tissue samples taken from a patient's body, are an invaluable tool in the diagnosis and treatment of diseases. Accurate detection of lymphocytes, which are a type of immune cells, in these images is

A. Polejowska (✉) · M. Sobotka · M. Kalinowski · M. Kordowski · T. Neumann
Department of Biomedical Engineering, Faculty of Electronics, Telecommunications and Informatics and BioTechMed Center, Gdańsk University of Technology, ul. Narutowicza 11/12, 80-233 Gdańsk, Poland
e-mail: s175588@student.pg.edu.pl

T. Neumann
e-mail: tomasz.neumann@pg.edu.pl

P. Strumiłło et al. (eds.), *The Latest Developments and Challenges in Biomedical Engineering*, Lecture Notes in Networks and Systems 746,
https://doi.org/10.1007/978-3-031-38430-1_4

critical for the diagnosis of certain conditions, particularly tumors [12]. However, the visual quality of histopathological images can affect the accuracy of algorithms in detecting lymphocytes.

In this research, the impact of visual quality on the performance of state-of-the-art algorithms on detecting lymphocytes in histopathological images was investigated. Two datasets were used and image modifications, including blur, sharpness, brightness and contrast, were systematically applied to evaluate the performance of YOLOv5 [14] and RetinaNet from Detectron2 [24] deep learning models. The effect of visual quality on the accuracy of these algorithms and any potential issues that may arise when using them to analyze histopathological images were determined.

Understanding the effect of visual quality on the performance of algorithms for detecting lymphocytes in histopathological images is crucial for advancing the application of computational methods in digital pathology. The findings of this study have significant implications for the implementation of these methods in pathology and underscore the importance of ensuring appropriate image quality for achieving reliable results.

2 Related Work

In the field of medical image analysis, the impact of image quality on the performance of deep learning algorithms has received increasing attention. While extensive research has been conducted on natural images, there is still limited knowledge of the effect of image quality on medical images, particularly within the domain of digital pathology in the task of detecting lymphocytes.

In recent years, various studies have examined the effect of image quality on deep neural networks in the natural image domain. One study that examined the effect of image quality on deep neural networks was published by Samuel Dodge and Lina Karam [5]. The authors investigated how blur, noise, and compression affect the performance of image classifiers. They found that blur and noise had a significantly negative impact on networks' performance, while the effect of contrast and compression was less significant.

Other researchers have focused on improving the performance of deep learning algorithms using image modification techniques. The literature has demonstrated the criticality of properly preparing histopathological images prior to utilizing neural network models, as the quality of the input data impacts the performance of a model [17, 20]. The work by Guerrero and Oliveira [9] proposed a preprocessing stage with stain normalization to enhance the accuracy of deep learning models in the lymphocytes detection task. The proposed method was found to be one of the best-ranked results in the state of the art. This research emphasizes how preprocessing histopathological images can boost the performance of deep learning models.

In digital pathology, researchers have also concentrated on developing specific deep learning models for lymphocyte detection. One such study (YOLLO) [22], proposed a model tailored for detection of lymphocytes in histopathology whole slide images (WSIs) stained with immunohistochemistry (IHC). The authors made modifications to the original YOLO architecture, including simplifying the architecture and implementing a guided sampling strategy. As a result, detection and processing times were sped up. This research highlights the potential of using object detection methods for lymphocyte detection in histopathology.

3 Experimental Setup

In the following, the experimental setup utilized to conduct experiments on the selected model architectures is described. Details regarding the dataset and the data annotation that was carried out to train models in an appropriate data format are presented. The neural network architectures implemented in the project are also outlined, and the results of their detection predictions on degraded images are compared.

Datasets. The first dataset used was the Leukocyte Images for Segmentation and Classification (LISC) [19], which contains hematological images taken from the peripheral blood of healthy individuals. These images were classified into five classes of leukocyte images, with a dimension of 720 $\times$ 576 pixels. The lymphocyte class consisted of 52 images, which were provided with ground truth for all images.

For the second dataset, a dataset from Andrew Janowczyk [13] was selected. This dataset comprises 100 images of estrogen receptor-positive (ER+) breast cancer (BC) images scanned at 20x, with each image measuring 100 $\times$ 100 pixels. An expert pathologist identified the lymphocyte centers.

The datasets were combined, and each image was resized to 256 $\times$ 256 pixels. Lymphocytes in the images were manually annotated using the Roboflow platform [6] based on the annotations prepared by experts. The final dataset contains 2858 lymphocyte determinations in the entire dataset consisting of 152 images. These images were split into: 53% training set, 26% test set and 21% validation set.

Data preprocessing. The raw ground truth images were manually annotated by an experienced pathologist. The images on which the model was trained were resized to 256 $\times$ 256 pixels and the original image scale was maintained. Subsequently, 152 lymphocyte images were manually annotated. A Reinhard normalization method was used to normalize the lymphocyte test dataset images. The Reinhard algorithm was used to match the color distribution of the source image to the color distribution of the target image, which was achieved through equalizing the mean and standard deviation of the pixel values in each channel [3].

Deep learning algorithms. Two one-stage machine learning detectors were considered (Fig. 1).

Model: YOLOv5. The architecture was developed to predict detected objects in real time. It consists of three modules: Backbone, Neck and Head. The Backbone mod-

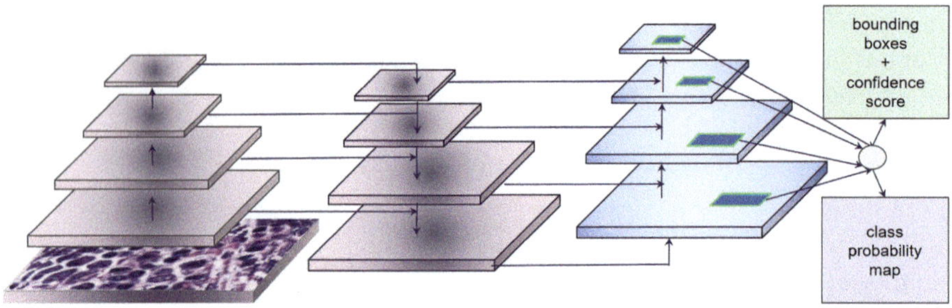

Fig. 1 The YOLOv5 model architecture [1]

ule is mainly used to extract features from input images. It aggregates and forms image features at different granularities [21]. The Model Neck is used to generate feature pyramids which help to identify the same object in different sizes and scales. It consists of a series of layers to mix and combine image features. The Head module performs final object detection. It predicts anchor boxes from features generating output vectors. Output vectors contain class probabilities, objectness scores and bounding box coordinates. YOLO family networks loss function is calculated on objectness score, class probability score and bounding box regression [23]. The YOLO neural network performs feature extraction and object classification methods at the same time during inference. In the YOLO network, images are divided into NxN grids. Candidate boxes are equally distributed on the x and y axis. The candidate boxes detect objects and predict the confidence of the existence of the object in each candidate box. Confidence scores reflect how confident the model is that the box contains an object and also how accurate is the prediction of the box. Authors defined confidence as Pr(Object) x IoU(truth prediction) [18]. If no object exists in that cell, the confidence score should be zero. In other cases, the confidence score should equal the Intersection over Union (IoU) between the predicted box and the ground truth. Each bounding box consists of five predictions: x, y, width, height, and confidence. The (x, y) coordinates represent the center of the box relative to the bounds of the grid cell. The width and height are predicted relative to the whole image [18]. Each grid cell predicts conditional class probabilities. These probabilities are conditioned on the grid cell containing an object. There is only one set of class probabilities predicted per grid cell, regardless of the number of boxes. Scores of probability of object occurrence and how well the predicted box fits the ground truth box are encoded into one result [18]. YOLO algorithm was used in the following application on lymphocyte detection [22]. There are five versions of YOLOv5 models. In this paper the small type is used to obtain maximum time performance during inference to compare the faster model with other approaches. For the training of the YOLOv5 model for lymphocyte detection, transfer learning with a setting of 500 epochs, a batch size of 16, an initial learning rate of 0.01 and a default hyperparameter configuration without augmentation (Fig. 2).

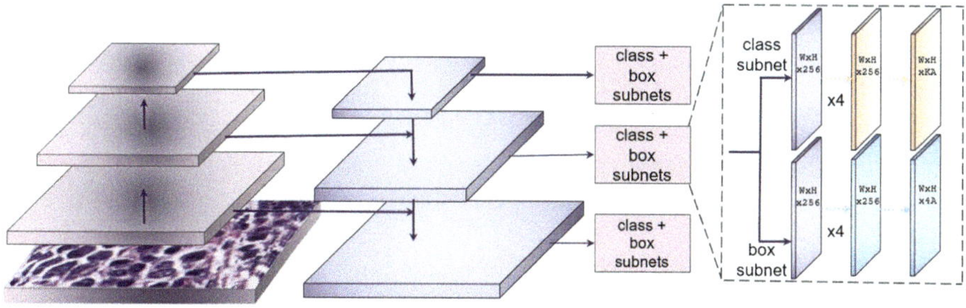

Fig. 2 The RetinaNet model architecture [15]

Model: RetinaNet. In order to compare the impact of image degradation on object detection between different networks the Detectron2 framework was used which offers various models. The RetinaNet model was selected [15]. It is an improvement of two existing structures: Feature Pyramid Networks (FPN) and Focal Loss. FPN networks are based on object detection at varying scales in an image. Featurized image pyramids are feature pyramids built upon image pyramids. Subsampled images have a lower resolution and a smaller size. This process is compute and memory intensive, which is why hand-engineered features were replaced by convolutional neural networks (CNNs). The feature maps in a CNN architecture decrease after every convolutional block, resulting in a pyramidal structure. The pyramidal structure is needed to get the most accurate results. The FPN combines low-resolution semantically strong features with high-resolution semantically weak features [15]. This is achieved by creating a top-down pathway with lateral connections to bottom-up convolutional layers. The RetinaNet consists of four elements: the bottom-up pathway, top-down pathway and lateral connections, classification subnetwork and regression subnetwork. The bottom-up pathway is the feed-forward computation of the backbone convolutional network, which computes a feature hierarchy consisting of feature maps at several scales [15]. The top-down pathway upsamples feature maps from higher pyramid levels. The lateral connections merge top-down layers with bottom-top layers with the same spatial size. At each stage, the classification subnetwork predicts the probability of every object detected. A regression subnetwork regresses offset for predicted box with ground truth object. Another key element of the RetinaNet network is the Focal Loss. The Focal Loss is a loss function that acts as a more effective alternative to previous approaches for dealing with class imbalance [16]. It assigns more weight to hard or easily misclassified examples. The Focal Loss is an extension of the cross entropy loss function that would down-weight easy examples and focus training on hard negatives. Using transfer learning, the RetinaNet model with a ResNet backbone depth of 50 was trained setting a maximum of 1500 solver steps with a batch size of 16 and a 0.001 base learning rate without defining augmentation in the training configuration.

Application. In order to systematically evaluate the impact of different image imperfections on the performance of lymphocyte detection algorithms, a web application was developed that facilitates modification of various image parameters and observation of the resulting detections. The application offers options to apply image degradation techniques, such as blur, noise, and contrast adjustments, and allows for the evaluation of the effect on the performance of the YOLOv5 and RetinaNet models.

In addition to providing a platform for conducting experiments, the application is also integrated with an experiment tracking platform [2]. This allows for tracking the performance of the models under various conditions and the identification of patterns and trends that may not be apparent from visual inspection alone. Through the use of this application, insights into the factors that influence the performance of lymphocyte detection algorithms were gained and strategies for improving their accuracy and robustness were identified.

Image degradation techniques. Different image degradation techniques were applied to the test dataset in order to evaluate the performance of the YOLOv5 and RetinaNet models in the presence of various types of image imperfections. These imperfections included blur, contrast, sharpness, and brightness modifications. In order to apply these modifications, various image processing techniques and algorithms were used. The results of these modifications were then used to evaluate the performance of the models.

Gaussian Blur. In digital pathology, blur can be particularly problematic for analyzing histopathological samples, as the camera focus may vary in natural settings. A study published by Arizona State University [5] found that blur has a negative impact on image classification, particularly when compared to other degradation methods. The authors of this study found that blur and noise had the greatest influence on the performance of convolutional neural networks for classification tasks. It is believed that blur interferes with the ability of convolutional layers to locate edges at early stages of feature abstraction, leading to inaccurate feature extraction at the beginning of the network training process.

To mitigate the negative effects of blur on object detection, researchers have proposed various approaches, such as augmenting the training dataset with blurred images [4]. By increasing the robustness of the model to blur, it is expected to improve the performance of object detection in real-world scenarios, where blur is a common issue.

To simulate blurring and examine its impact on lymphocyte detection, the Gaussian kernel was used, a method as described in Gwosdek et al. [10]. This method convolves the image with a Gaussian kernel, which is a matrix of weights generated based on a specified standard deviation. The size of the kernel is determined by the radius parameter, which specifies the standard deviation of the Gaussian distribution. A larger radius will result in a larger kernel and more smoothing, while a smaller radius will result in a smaller kernel and less smoothing. In experiments, the kernel standard deviation was varied. Figure 3 shows the same image with different lev-

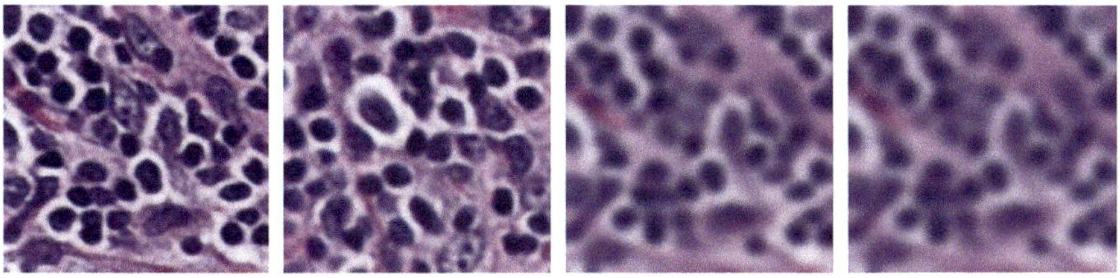

Fig. 3 Example images from the test dataset with different levels of Gaussian blur applied—radius size: 1.0, 2.0, 5.0, 6.0

els of blur applied, demonstrating the visual impact of this image degradation method.

Sharpness. Sharpness is inversely related to blur which is typically determined by the spread of edges in the spatial domain, and accordingly the attenuation of high frequency components [25]. Digital images have limitations of distinguishment lines of contrast with such clarity and sharpness as the human eye. Sensors of digital camera are limited by the number of pixels and their range of values. Sharpness is a combination of two factors: resolution and acutance. Resolution is straightforward and not subjective. It depends on the number of pixels in the image file. All other factors equal, the higher the resolution of the image, the sharper it can be. Acutance is a subjective measure of the contrast at an edge. Image sharpening is performed by applying image convolution with mask with specified parameters mentioned in [7] or by applying algorithm presented in [8]. To perform image sharpening the Python Pillow ImageEnhance module was used with specified sharpening factors. The factor value ranged from 1.0 to 12.0 (Fig. 4).

Brightness. Image brightness is one of the simplest pixel value modification to perform on. Modifying brightness can be accomplished by multiplication between pixels values with a specified ratio. Discrimination between different intensity levels on image is an important consideration. The range of light intensity levels to which the human visual system can adapt is wide from the scotopic threshold to the glare limit [8]. Computer colors representation is more limited. The proper settings of

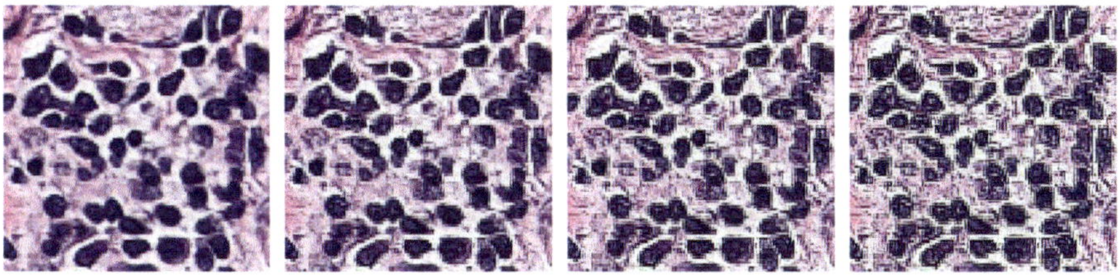

Fig. 4 Examples image from the test dataset with different levels of of sharpness applied—sharpness levels: 2, 5, 8 and 12

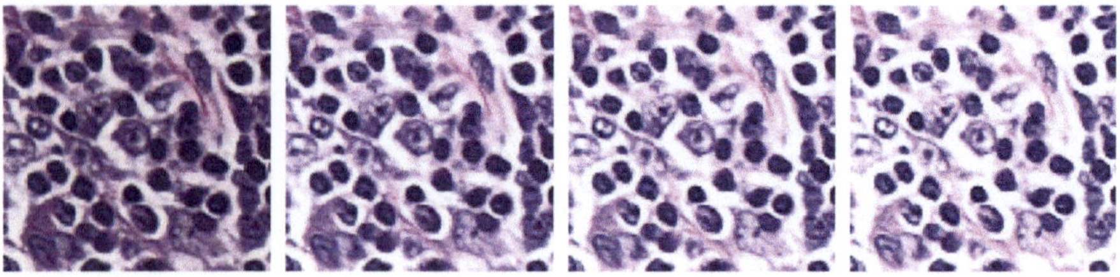

Fig. 5 Example images from the test dataset with different levels of brightness applied—brightness levels: 1.0, 1.3, 1.5, 1.7

image brightness modification can lead to increased model ability to generalization and brightness independence. Brightness manipulation on images allows us to generate new images samples in a dataset. This allows us to generate images with a range from black to white. Spot brightness modification can lead to add shadow or bright spots to simulate complicated light conditions in real situations. To perform brightness modification the Pillow module was used to manipulate brightness over dataset images to analyze how this parameter affects lymphocyte detection between two selected models. The factor value ranged from 1.0 to 1.7. As in the previous parameter, library component preserves the original image with factor value 1.0. The resulting images of the operations are presented in Fig. 5.

Contrast. Contrast is the electro-optical parameter of the image, which determines the ratio of the maximum luminance to the minimum luminance. When the contrast adjustment is raised, the image will have a higher percentage of dark pixels and whites (middle pixel values are eliminated).

Contrast reduction is obtained by blending the input image with a gray image [11]

$$output = (1 - alpha) \cdot image + alpha \cdot image. \quad (1)$$

This equation was used to change images contrast in Fig. 6. Alpha is a key factor, that has been changed in range from 0.1 to 3.0. The experiment was started with an alpha value 0.1, next from an alpha value of 0.2–2.0 in steps of 0.2. Then, the alpha value was increased to 2.5, where last step was 3.0. An alpha value of 1.0 is the same as in other image degradation experiments present the original image.

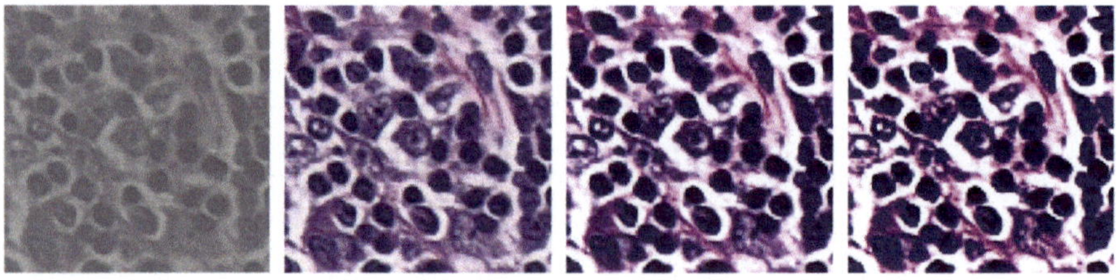

Fig. 6 Example image from the test dataset with different levels of contrast applied—contrast value: 0.2, 1.0, 2.0, 3.0

4 Experiments and Results

The impact of stain normalization on cell lymphocyte detection was quantified by comparing the testing metrics before and after the implementation of the Reinhard normalization algorithm on the test dataset (Table 1). The performance of the models was then measured using metrics such as Average Precision (AP) and F1-score (F1). On the test dataset, both models were predicted with a confidence threshold of 0.2. The average precision metric was calculated with the IoU parameter set to 0.7.

The results in Table 1 show that RetinaNet performs better than YOLOv5 on both original images and images with stain normalization. The performance of both models decreases significantly after applying stain normalization, indicating that the normalization process may not be suited for all types of images in the dataset, however, still RetinaNet performs better under this image modification. For further analysis, the original dataset without Reinhard normalization was selected.

The purpose of the next experiments in this study was to investigate the impact of various image imperfections on the performance of state-of-the-art object detection algorithms. To do this, a test dataset of 32 images was selected and various modifications were applied to them, including blur, sharpness, brightness and contrast.

Table 1 Comparing metrics on test dataset without and with Reinhard normalization

Image	YOLOv5 AP (%)	RetinaNet AP (%)	YOLOv5 F1 (%)	RetinaNet F1 (%)
Original	93.36	96.40	87.56	91.16
Reinhard norm	51.80	67.32	47.74	70.96

Gaussian Blur. The results of applying Gaussian blur on the test dataset are shown in Table 2.

Table 2 Metrics on test dataset with Gaussian Blur applied

Radius size	YOLOv5 AP (%)	RetinaNet AP (%)	YOLOv5 F1 (%)	RetinaNet F1 (%)
1	93.88	98.62	88.63	83.65
2	91.37	92.29	80.67	64.34
5	35.93	0	21.64	0
6	3.12	0	0.26	0

The blur radius size is varied from 1 to 6. Comparison of Table 2 metrics with Table 1 base results after visual inspection (Fig. 3) reveal that the models are not robust enough to detect lymphocytes on blurred images since the metrics values decrease below 36% for a radius size of 5. It can be also observed that YOLOv5 is more robust to blur with higher radius size modifications compared to RetinaNet.

Sharpness. The results of images sharpness modification are shown in Table 3. To perform image sharpening the sharpness factor value was ranged from 1.0 to 12.0. Both models lose their detection capability as the parameter increases. Based on the results, RetinaNet is more resilient to higher sharpness levels than YOLOv5 network.

Table 3 Metrics on test dataset with sharpness modification applied

Level	YOLOv5 AP (%)	RetinaNet AP (%)	YOLOv5 F1 (%)	RetinaNet F1 (%)
2	93.36	94.34	87.36	89.95
5	88.69	85.44	84.41	79.93
8	82.32	91.13	79.02	80.86
12	67.31	84.33	61.29	68.34

Brightness. The results of images brightness modification are shown in Table 4. During this test, the brightness factor 1.0, 1.3, 1.5, and 1.7 was selected. The factor value of 1.0 should preserve the original quality of the image. RetinaNet maintains AP value above 90% across all specified levels. F1 score results also suggest that RetinaNet is more robust to brightnesss changes than YOLOv5.

Table 4 Metrics on test dataset with brightness modification applied

Level	YOLOv5 AP (%)	RetinaNet AP (%)	YOLOv5 F1 (%)	RetinaNet F1 (%)
1.0	93.47	96.28	87.66	91.23
1.3	96.91	97.29	91.06	90.04
1.5	85.06	97.91	76.54	88.57
1.7	72.03	91.46	61.83	78.65

Contrast. The results of modifying image contrast on the test dataset are shown in Table 5. During this test, the Alpha parameter was selected from 0.2, 1.0, 2.0, and 3.0. The Alpha value of 1.0 should preserve the original quality of the image. It is worth noting that both models F1 scores remained at about the same level when the contrast was increased. The RetinaNet architecture is observed to be more robust to low levels of contrast compared to YOLOv5.

Table 5 Metrics on test dataset with diffrent contrast

Level	YOLOv5 AP (%)	RetinaNet AP (%)	YOLOv5 F1 (%)	RetinaNet F1 (%)
0.2	0	25	0	18.57
1.0	93.47	96.28	87.66	91.23
2.0	90.22	88.88	87.68	90.59
3.0	90.20	86.84	87.91	90.27

5 Discussion and Conclusion

In this study, the effect of image quality characteristics on lymphocyte detection in histopathological images using state-of-the-art algorithms was assessed. The results indicated that modifications such as image blur, contrast, brightness, and sharpness have a significant impact on the performance of these algorithms. The RetinaNet model detects objects with greater precision on the test dataset but with notably longer inference time that YOLOv5. The use of a specific model depends on the purpose of the application and the conditions under which it could be launched.

Furthermore, the results demonstrated that the performance of the algorithms varied depending on the specific method employed. For instance, the algorithm that performed optimally on images with high blur was not the same as the algorithm that performed optimally on images with low contrast. This highlights the importance of considering the specific visual quality characteristics of an image when selecting an algorithm for lymphocytes detection.

These findings have significant implications in the field of histopathological image analysis. They underscore the importance of maintaining optimal image quality during the digitization process of histopathological samples, such as by ensuring proper focus of the camera to minimize blur. Poor image quality may compromise the accuracy of object detection models, leading to unreliable results in lymphocyte detection tasks.

The further studies can be developed by the addition of another new public large-scale dataset of lymphocytes and set of the new operations such as analysis of noise, particularly adversarial noise. Moreover, a set-based object detector using a transformer architecture can be used in the new image analysis. This can offer a more comprehensive understanding of the strengths and limitations of various deep learning architectures for lymphocyte detection, which can aid in developing more reliable and effective models for computer-aided diagnosis in digital pathology.

References

1. Azam, M.A., Sampieri, C., Ioppi, A., et al.: Deep learning applied to white light and narrow band imaging videolaryngoscopy: toward real-time laryngeal cancer detection. Laryngoscope **132**(9), 1798–1806 (2022)
2. Biewald, L.: Experiment Tracking with Weights and Biases (2020)
3. Budginaitė, E., Morkūnas, M., Laurinavičius, A., Treigys, P.: Deep learning model for cell nuclei segmentation and lymphocyte identification in whole slide histology images. Informatica **32**(1), 23–40 (2021)
4. Chengji, L., Tao, Y.,Liang, J., et al.: Object detection based on yolo network. In: 2018 IEEE 4th Information Technology and Mechatronics Engineering Conference (ITOEC) (2018), pp. 799–803
5. Dodge, S.F., Karam, L.J.: Understanding how image quality affects deep neural networks, pp. 1–6 (2016). arxiv:1604.04004
6. Dwyer, B., Nelson, J., Solawetz, J., et al.: Roboflow (Version 1.0) (2022). [Software]
7. Fitzpatrick, J.: How digital image sharpening works, and why you should use it (2016)
8. Gonzalez, R.C., Woods, R.E., et al.: Image segmentation. Digital image processing **2**, 331–390 (2002)
9. Guerrero, R.E.D., Oliveira, J.L.: Improvements in lymphocytes detection using deep learning with a preprocessing stage. In: 2021 IEEE 34th International Symposium on Computer-Based Medical Systems (CBMS), pp. 178–182 (2021). https://doi.org/10.1109/CBMS52027.2021.00068
10. Gwosdek, P., Grewenig, S., Bruhn, A., Weickert, J.: Theoretical foundations of gaussian convolution by extended box filtering. Springer (2011)
11. Haeberli, P., Voorhies, D.: Image processing by linear interpolation and extrapolation. IRIS Univ. Mag. **28**, 8–9 (1994)
12. Hendry, S., Salgado, R., Gevaert, T., et al.: Assessing tumor infiltrating lymphocytes in solid tumors: a practical review for pathologists and proposal for a standardized method from the international immuno-oncology biomarkers working group: Part 2: Tils in melanoma, gastrointestinal tract carcinomas, non-small cell lung carcinoma and mesothelioma, endometrial and ovarian carcinomas, squamous cell carcinoma of the head and neck, genitourinary carcinomas, and primary brain tumors. NIH Public Access (2017)
13. Janowczyk, A.: Use case 4: lymphocyte detection. Deep learning, digital histology (2015)
14. Jocher, G., Chaurasia, A., Stoken, A., et al.: ultralytics/yolov5: v7.0 - YOLOv5 SOTA Realtime Instance Segmentation. Zenodo (2022)
15. Lin, T.Y., Dollár, P, Girshick, R., et al.: Feature pyramid networks for object detection. In: Proceedings of the IEEE Conference on Computer Vision and Pattern Recognition (CVPR), pp. 2117–2125 (2017a)
16. Lin. T.Y., Goyal, P., Girshick, R., et al.: Focal loss for dense object detection. In: Proceedings of the IEEE International Conference on Computer Vision (ICCV), pp. 2980–2988 (2017b)
17. Öztürk, Ş, Akdemir, B.: Effects of histopathological image pre-processing on convolutional neural networks. Procedia Comput. Sci. **132**, 396–403 (2018)
18. Redmon, J., Divvala, S., Girshick, R., Farhadi, A.: You only look once: unified, real-time object detection. In: Proceedings of the IEEE Conference on Computer Vision and Pattern Recognition (CVPR), pp. 779–788 (2016)
19. Rezatofighi, S.H., Soltanian-Zadeh, H.: Automatic recognition of five types of white blood cells in peripheral blood. Comput. Med. Imaging Graph. (2011)
20. Sengoz, N., Yigit, T., Ozmen, O., Isik, A.H.: Importance of preprocessing in histopathology image classification using deep convolutional neural network. Adv. Artif. Intell. Res. (2022)
21. Tan, M., Pang, R., Le, Q.V.: Efficientdet: scalable and efficient object detection (2019)
22. Van Rijthoven, M., Swiderska-Chadaj, Z., Seeliger, K., Van der Laak, J., et al.: You only look on lymphocytes once. In: 1st Conference on Medical Imaging with Deep Learning (MIDL) (2018)

23. Wang, C.Y., Liao, H.Y.M., et al.: CSPNet: a new backbone that can enhance learning capability of CNN. In: Proceedings of the IEEE/CVF Conference on Computer Vision and Pattern Recognition (CVPR) (2020)
24. Wu, Y., Kirillov, A., Massa, F., et al.: Detectron2 (2019)
25. Yu, S., Wu, S., Wang, L., et al.: A shallow convolutional neural network for blind image sharpness assessment. Public Library of Science San Francisco (2017)

Using Local Normalization and Local Thresholding in the Detection of Small Objects in MR Brain Images

Patrycja Kwiek and Elżbieta Pociask

Abstract A lot of time and effort has been put into finding an automatic segmentation system which could detect the region and position of white matter hypersensitivities in MRI brain scans. At the cellular level, changes in the white matter of the brain can be understood as a loss of myelin around axons. These changes might be detected by MRI due to local changes in water content. By establishing a method which could detect such irregularities, it would be possible to help diagnose some serious autoimmune diseases. In this work, a wide variety of local thresholding algorithms that are applied after local normalization is evaluated in terms of how accurately they segment these hypersensitivities. With the use of ImageJ, the following algorithms were tested: Bernsen, Contrast, Mean, Median, MidGrey, Otsu, Sauvola and SDA. The most accurate segmentation results are presented for all algorithms. For most of the ALT algorithms, the use of local normalization increased the achieved Dice value by as much as 100%. Contrary to what might be expected, the best results were achieved for various values of the $sigma1$ and $sigma2$ parameters of local normalization, both for single images and for a group of algorithms for a given image, thus it is difficult to apply the presented solutions to other examples.

Keywords Plaques · Segmentation · Local normalisation · ALT · SDA

1 Introduction

Detecting hypersensitivities in MRI brain scans could be important for detecting certain illnesses, such as multiple sclerosis or Alzheimer's disease (AD). One of the pathological signs of AD is deposition of amyloid beta protein in intra- and extracellular spaces. As such, detection of this protein could be valuable in the development of new treatments to prevent the progression of Alzheimer's disease. Amyloid beta

P. Kwiek · E. Pociask (✉)
Department of Biocybernetics and Biomedical Engineering, AGH University of Science and Technology, Mickiewicza 30 Av., 30–059 Krakow, Poland
e-mail: epociask@agh.edu.pl

P. Strumiłło et al. (eds.), *The Latest Developments and Challenges in Biomedical Engineering*, Lecture Notes in Networks and Systems 746,
https://doi.org/10.1007/978-3-031-38430-1_5

protein can be detected histochemically by using several amyloid binding dyes, but this is time-consuming and expensive. Changes in the white matter of the brain can be understood at the cellular level as a loss of myelin around axons. A plaque is formed when loss of myelin occurs not only for a single neuron but also for a larger group of neurons nearby. These changes can be detected by MRI because local changes in water content occur. By establishing a method which could detect these irregularities, it would be possible to help diagnose some serious autoimmune diseases. If plaques could be detected by means of image segmentation, this would reduce not only the cost but also the time needed to process the data, thus leading to quicker diagnosis and therapy [5, 6, 13]. There are also methods based on artificial intelligence, but they have limitations that do not allow efficient segmentation of small changes [22]. For many reasons, image processing is an important aspect of science that never stops developing. In the field of medicine, image processing allows several improvements, one of the most important of which is that it makes non-invasive diagnosis possible. More often than not, it also reduces diagnosis time and cost because it relieves doctors of the need to analyze images by "eye", which can be time-consuming. An algorithm which could (even preliminarily) select the ROI would greatly assist doctors' work. All these factors influence the reasoning behind the aim of this work, which was to evaluate local normalization and auto local thresholding methods in the detection of small objects in MRI brain images. Various methods of combining local normalization with local thresholding were tested and assessed based on the calculated Dice values [4, 6, 13]. The presented work consists of several parts. The first part was conducted in ImageJ and consisted of applying various local normalization and auto local thresholding algorithms (subsequently, also referred to as 'local thresholding') to three MRI images of the brain. In each case, the Dice value was calculated for each set of parameters. This data was then analysed using Python. In the first approach, the algorithms were used individually [13]. Subsequently, the calculations were performed using various local thresholding algorithms with the help of local normalization as it is theorized that local normalization improves the detection of small objects in images [6, 13]. For comparison, the results were also calculated with the use of local normalization with global thresholding. One measure of how much the achieved results improved was comparison with global thresholding results, as the defined threshold was expected to achieve worse results. The results were then summarized in the two most important graphs: distribution of maximum Dice values based on $sigma1$ and $sigma2$ values of the local normalization algorithm; and correlation of the maximum Dice value with and without applying the local normalization method before the local thresholding algorithms. The results of all of these methods were evaluated. The changes in parameters and the maximum Dice values are discussed, taking into consideration usage of the local normalization algorithm. Finally, the conclusions of this study are presented.

2 Materials and Methods

2.1 Data

3 sets of 12-bit DICOM images of brain MRI scans (Fig. 1) with different resolutions were used in this study. They were acquired using three different MRI scanners—

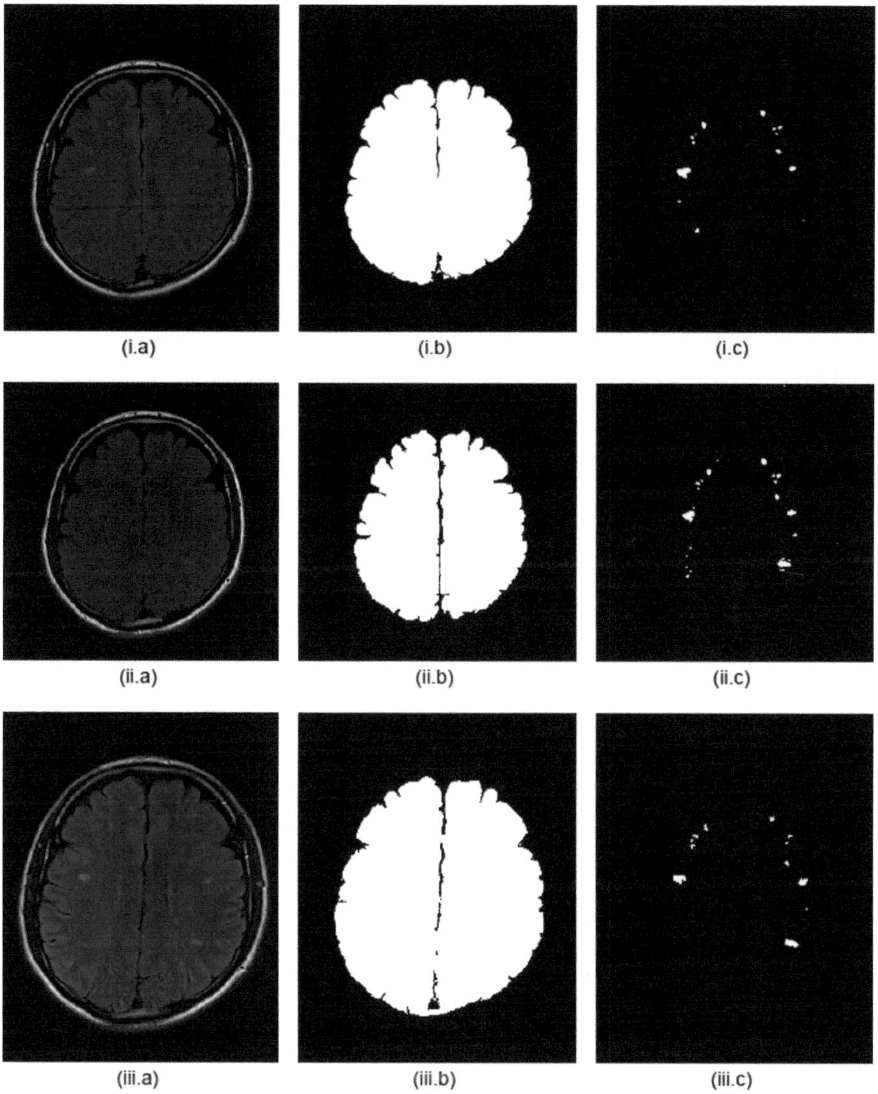

Fig. 1 **a** MRI brain scans. **b** ROI, **c** manually segmented plaques (i) A (ii) C and (iii) B

Magnetom Essenza, Avanto Fit and Skyra (3T)—and resulted in the following resolutions, respectively: 0.71875Ă-0.71875, 0.375Ă-0.375, and 0.859375Ă-0.859375. The ROI masks represent brain matter and were created with the use of a semi-automatic method. The plaques were manually selected on the pictures.

2.2 Normalization and Thresholding Methods

Local Normalization Local normalization is used to uniformize the variance and mean of a given image around a local neighborhood. This is useful when there are artifacts in the picture and when the illumination is uneven. In the used method, Gaussian filtering enables the rapid execution of the local normalization algorithm [6, 21].

The local normalization algorithm [13] is based on the formula:

$$g(x, y) = \frac{f(x, y) - m(x, y)}{\sigma(x, y)}$$

where $g(x, y)$ is the output image; $f(x, y)$ corresponds to the original image; $m(x, y)$ is an estimation of a local mean ($sigma1$) of $f(x, y)$; and $\sigma(x, y)$ is the local variance estimation ($sigma2$).

Global Thresholding Global thresholding uses a single threshold value for the whole image but ignores illumination, shades etc. It is based on the assumption that the processed image possesses a bimodal histogram, thus it is assumed that the object can be extracted from the background by comparing pixels from the image with a user-defined threshold value [21].

Local Thresholding Local adaptive thresholding makes it possible to convert a grayscale image to a black-and-white one. Usually in an image, 0 represents white pixels and 1 represents black. Unlike global thresholding, local adaptive thresholding uses different threshold values for every pixel based on its neighbors. There are many adaptive thresholding algorithms, but only the following were considered in this study: Bernsen, Contrast, Mean, MidGrey, Otsu, Sauvola and SDA [2, 13, 15, 21].

The Bernsen algorithm uses a user-provided contrast threshold. When the local threshold is higher or equal to the contrast threshold, the threshold is chosen to be at the local mid-gray value (which is equal to the mean value of the maximum and minimum grey values). However, when the contrast threshold is higher than the local threshold, the neighborhood is regarded as consisting of only one class and the pixel is classified as either background or object, depending on the mid-gray value [2, 6].

Contrast is a simple algorithm in that the only user-provided parameter it uses is kernel $radius$. It is based on a basic contrast toggle which sets the pixel value to one of two values (either black, i.e., 0, or white, i.e., 255). The value is found by establishing whether the current value of the pixel is closest to the local minimum or maximum [6].

In the case of the Mean algorithm, the user can choose the $radius$ value and the C value, whose default value is equal 0. This method selects a threshold equal to the mean of the local greyscale distribution minus the C value [6].

Similarly to the previous method, MidGrey algorithm users can choose the $radius$ and the C values; however, this algorithm chooses the threshold as a mid-gray, (max + min)/2, of the local greyscale distribution instead of the mean [6].

The Otsu algorithm takes the threshold that minimizes the intra-class variance. In this version of the algorithm, pixels are verified in regions instead of global Otsu. Users can only provide the $radius$ value [11].

Sauvola's thresholding method is a variation of Niblack's method. In addition to the $radius$ value, this algorithm requires the user to provide two extra parameters: k-value, which amplifies the contribution of standard deviation in an adaptive manner, with 0.5 as default; and r-value. It represents the dynamic range of the standard deviation, where 128 is used as default [6, 15].

The SDA algorithm, which stands for Statistical Dominance Algorithm, is intended for small object detection. There are two parameters: SDAr, which is the $radius$ of disc-shaped areas of interest; and SDAt, which is a threshold by which objects are contrasted against the background [9].

2.3 *Dice Coefficient*

The Dice coefficient is a similarity index used in statistics to determine the similarity of two samples.

Given two sets of data, X and Y, it is defined that:

$$DSC = \frac{2|X \cap Y|}{|X| + |Y|} \tag{1}$$

where |X| and |Y| represent the number of elements in sets.

Using the Boolean approach, DSC can be written as:

$$DSC = \frac{2TP}{2TP + FP + FN} \tag{2}$$

where TP means true positive, FP corresponds to false positive, and FN is false negative [3, 18].

3 Results

The Tables 1, 2 and 3 show that choosing an approach with additional local normalization at the beginning of the data processing produces better results. In 5 out of 6 situations, the best Dice was found to be generated with the use of the Sauvola method. The next best was the SDA approach; where, the results for image A were better for SDA with local normalization than for Sauvola with local normalization. The worst results in all cases were achieved with either the Otsu or Contrast methods, where the Dice reached values even lower than 0.1. When comparing the Dice value for a given method (ALT) with and without local normalization, the greatest difference (improvement) was noted for the Contrast method on image C. Both the Bernsen and the Mean methods are characterized by a high variety of $radius$ values between images. The lowest values of $sigma1$ and $sigma2$ for the Sauvola algorithm were found in image C. For some methods, the best Dice result was found with more than one set of parameters; this happened in the case of image B with LN+Bernsen and LN+SDA, and image C with LN+Bernsen. Out of all of the LN+ALT variations, three resulted in Dice higher than 0.5 for all images. The algorithms mentioned are LN+Mean, LN+Sauvola and LN+SDA. The best result (0.7357) was achieved for image C by using LN+Sauvola (Fig. 2).

Figure 3 shows that, before applying local threshold methods, the use of the local normalization algorithm allows for better maximum Dice values. The achieved result was worse in the case of LN+ALT in only one instance. The Contrast algorithm performed better without applying LN in the case of image A. LN+GT results were plotted as broken lines, which make it possible to decide whether the given LN+ALT

Table 1 Best DICE for thresholding for image A

Thresholding	Parameters	DICE	Parameters with LN	DICE with LN
GT	–	–	s1 = 17, s2 = 21, T = 161	0.5487
Mean	r = 47, C = –25	0.2583	s1 = 19, s2 = 21, r = 22, C = –41	0.5677
Median	r = 15, C = –21	0.3194	–	–
Bernsen	r = 78, C = 122	0.1603	s1 = 37, s2 = 3, r = 37, C = 214	0.4190
MidGrey	r = 65, C = –31	0.2567	s1 = 11 s2 = 22, r = 117, C = –51	0.446
Contrast	r = 62	0.1505	s1 = 11, s2 = 19, r = 19	0.07554
Otsu	r = 68	0.03645	s1 = 17, s2 = 40, r = 120	0.07553
Sauvola	r = 50, rv = 0.1, kv = 2.45	0.6158	s1 = 14, s2 = 31, r = 61, rv = 0.05, kv = 3.95	0.6371
SDA	SDAr = 25, SDAt = 14, binT = 198	0.4865	s1 = 24, s2 = 22, SDAr = 23, SDAt = 30, binT = 187	0.6468

Table 2 Best DICE for thresholding for image B

Thresholding	Parameters	DICE	Parameters with LN	DICE with LN
GT	–	–	s1 = 8, s2 = 10, T = 192	0.5232
Mean	r = 25, C = –34	0.1655	s1 = 7–8, s2 = 10, r = 12, C = –59	0.5407
Median	r = 22, C = –34	0.2897	–	–
Bernsen	r = 44, C = 163	0.2756	s1 = 17, s2 = 30, r = 4, C = 172–255	0.3812
MidGrey	r = 36, C = –36	0.3636	s1 = 6, s2 = 10, r = 110, C = –63	0.5249
Contrast	r = 37	0.04335	s1 = 40, s2 = 36, r = 39	0.2621
Otsu	r = 20	0.02843	s1 = 11, s2 = 40, r = 122	0.1676
Sauvola	r = 25, rv = 0.2, kv = 3.95	0.5185	s1 = 6, s2 = 23, r = 26, rv = 0.2, kv = 4.55	0.6926
SDA	SDAr = 14, SDAt = 17, binT = 217	0.4532	s1 = 8–9, s2 = 11, SDAr = 12, SDAt = 40–41, binT = 198 and binT = 201	0.6032

Table 3 Best DICE for thresholding for image C

Thresholding	Parameters	DICE	Parameters with LN	DICE with LN
GT	–	–	s1 = 28, s2 = 8, T = 184	0.5945
Mean	r = 22, C = –17	0.2091	s1 = 30, s2 = 8, r = 79, C = –44	0.6284
Median	r = 5, C = –9	0.3030	–	–
Bernsen	r = 32, C = 75	0.2484	s1 = 17, s2 = 21, r = 2, C = 152–255	0.4561
MidGrey	r = 28, C = –19	0.3904	s1 = 30, s2 = 9, r = 113, C = –55	0.6218
Contrast	r = 23	0.04588	s1 = 26, s2 = 40, r = 29	0.3647
Otsu	r = 34	0.06121	s1 = 25, s2 = 15, r = 101	0.4569
Sauvola	r = 14, rv = –0.1, kv = –1.15	0.6097	s1 = 5, s2 = 15, r = 24, rv = 0.05, kv = 3.6	0.7357
SDA	SDAr = 18, SDAt = 5, binT = 234	0.5566	s1 = 10, s2 = 8, SDAr = 11, SDAt = 25, binT = 200	0.699

methods have potential. In all instances, the Contrast and Otsu algorithms performed the worst. However, the Sauvola, SDA and Mean algorithms with the use of LN were better than the LN+GT method. LN+MidGrey showed very good prospects; however, in the case of image A, the maximum Dice value turned out to be lower than in the case of LN+GT.

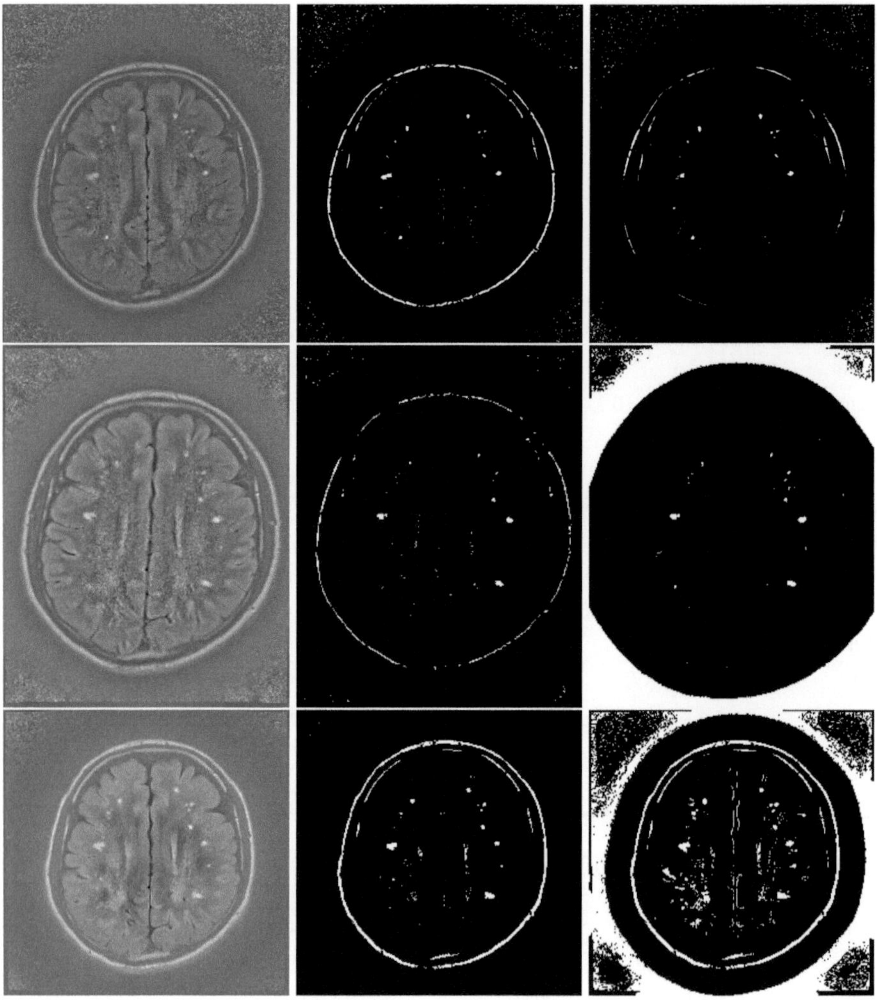

Fig. 2 Best results of LN+GT and LN+ALT for each image; rows: image A, image B, image C; columns: LN for GT, best result for LN+GT, best result for LN+ALT

The sigma distribution of local normalization shows which algorithms were more influenced by added ALT in terms of achieving better results (Fig. 4). It is interesting that LN+Mean for all images achieves similar results to LN+GT. Similarly, MidGrey parameters are located in the vicinity of the same position as GT. The biggest range of the $sigma1$ value can be noticed for image B, while for $sigma2$ it is image C. The biggest agglomeration of parameters was noted for image B with the SDA, GT, Sauvola and MidGrey algorithms. It should be noted that the ALTs that are mentioned as aggregated or being close to GT results are those which resulted in the highest maximum Dice values.

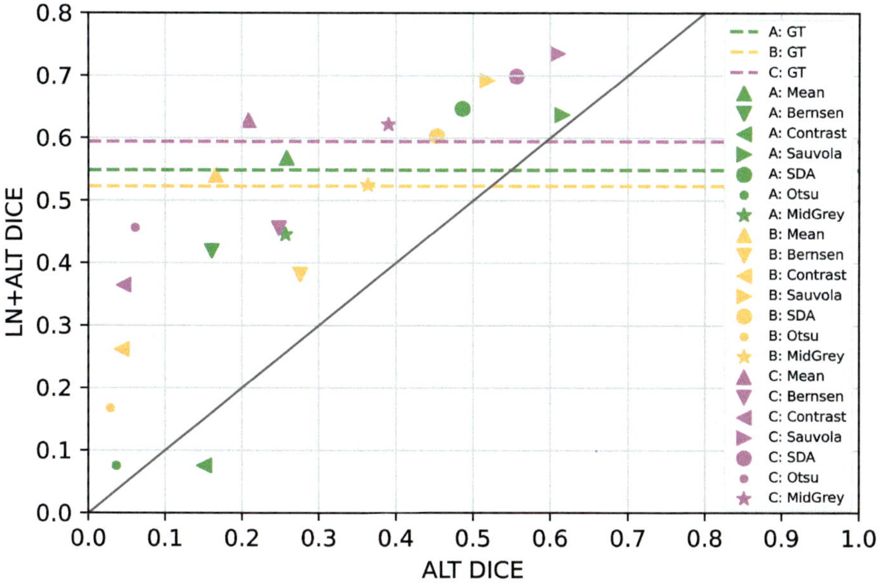

Fig. 3 DICE comparison with and without LN

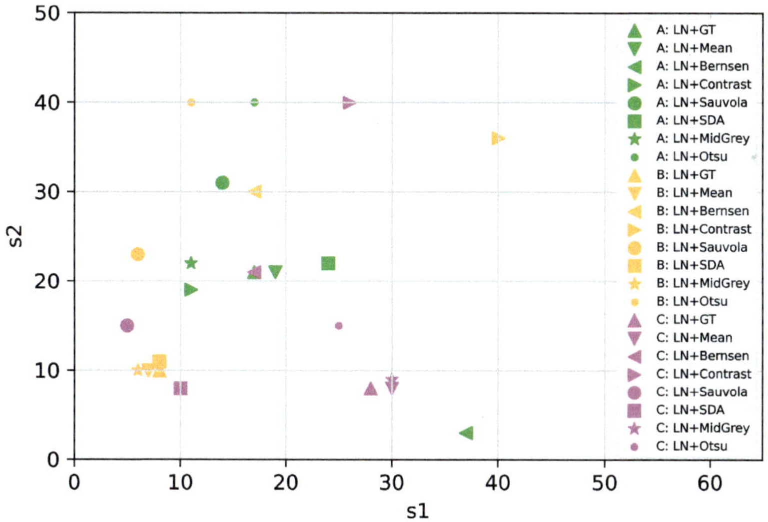

Fig. 4 Sigma parameters for best DICE values of each algorithm and image

The best result for each of the images after applying LN+GT and LN+ALT can be compared in Fig. 2, where the best results of applying LN+ALT for each given image are:

- Image A: LN+SDA (DICE = 0.6468)
- Image B: LN+Sauvola (DICE = 0.6926)
- Image C: LN+Sauvola (DICE = 0.7357).

4 Conclusions

It can be seen that applying local normalization before the use of ALT improves detection of small objects in MRI images of the brain for most of the algorithms, with a percentage difference of up to 100% of the Dice value. The results for algorithms such as LN+Sauvola, LN+SDA and LN+Mean are very promising. Most noteworthy is the LN, even though it was designed to unify the mean and variance of the image clearly enhanced such changes, what distinguishes the described approach from the analyzed algorithms. However, as could be seen during the course of this work, the best results are achieved for various values of $sigma1$ and $sigma2$, both for single images and for a group of algorithms for a given image, which could raise concerns about possible generalization of the method for other images. As the study was conducted on a limited number of images, it would be best to repeat the calculations with different images for the parameters which achieved the best Dice values. Further work could consist of the creation of different segmentation techniques for plaque detection, taking into consideration algorithms based on artificial intelligence.

Acknowledgements This publication was funded by AGH University of Science and Technology, Faculty of Electrical Engineering, Automatics, Computer Science and Biomedical Engineering, KBIB no 16.16.120.773.

References

1. Balakrishnan, R., Valdés Hernández MDC, F.A.: Automatic segmentation of white matter hyperintensities from brain magnetic resonance images in the era of deep learning and big data—A systematic review. Comput. Med. Imaging Graph. 101–867 (2021)
2. Bernsen, J.: Dynamic thresholding of gray-level images. In: Proceedings of the Eighth International Conference on Pattern Recognition, Paris, 1986 (1986)
3. Dice, L.R.: Measures of the amount of ecologic association between species. Ecology **5**(3), 297–302 (1945)
4. Frey, B.M., Petersen, M., Mayer, C., Schulz, M., Cheng, B., Thomalla, G.: Characterization of white matter hyperintensities in large-scale MRI-studies. Front. Neurol. **10** (2019). https://doi.org/10.3389/fneur.2019.00238
5. Maiti, P., Plemmons, A.: Labeling and imaging of amyloid plaques in brain tissue using the natural polyphenol curcumin. J. Vis. Exp. **153** (2019)
6. Milewska, K., Obuchowicz, R., Piórkowski, A.: A preliminary approach to plaque detection in MRI brain images. In: Piaseczna, N., Gorczowska, M., Łach, A. (eds.) Innovations and Developments of Technologies in Medicine, Biology and Healthcare, pp. 94–105. Springer International Publishing, Cham (2022)
7. Niblack, W.: An Introduction to Digital Image Processing, 115–116 prentice hall. Englewood Cliffs, New Jersey (1986)

8. Nichele, L., Persichetti, V., Lucidi, M., Cincotti, G.: Quantitative evaluation of imagej thresholding algorithms for microbial cell counting. OSA Contin. **3**(6), 1417–1427 (2020). https://doi.org/10.1364/OSAC.393971. Jun
9. Nurzynska, K., Mikhalkin, A., Piorkowski, A.: CAS: cell annotation software—Research on neuronal tissue has never been so transparent. Neuroinformatics **15**, 365–382 (2017)
10. Oksuz, I.: Brain MRI artefact detection and correction using convolutional neural networks. Comput. Methods Prog. Biomed. **199** (2020)
11. Otsu, N.: A threshold selection method from gray-level histograms. IEEE Trans. Syst. Man Cybern. **SMC-9**(1), 62–66 (1979)
12. Phansalkar, N., More, S., Sabale, A., Joshi, M.: Adaptive local thresholding for detection of nuclei in diversity stained cytology images. In: 2011 International Conference on Communications and Signal Processing, pp. 218–220. IEEE (2011)
13. Piórkowski, A., Lasek, J.: Evaluation of local thresholding algorithms for segmentation of white matter hyperintensities in magnetic resonance images of the brain, pp. 331–345. Springer (10 2021). https://doi.org/10.1007/978-3-030-89654-6_24
14. Sage, D., Unser, M.: Easy Java programming for teaching image-processing. In: Proceedings of 2001 International Conference on Image Processing. vol. 3, pp. 298–301. IEEE (2001)
15. Sauvola, J., Pietikäinen, M.: Adaptive document image binarization. Pattern Recogn. **33**(2), 225–236 (2000)
16. Schneider: Nih image to imagej: 25 years of image analysis. Nature Methods **9** (2012)
17. Soille, P.: Morphological Image Analysis. Springer (2004)
18. Sorensen, T.: A method of estabilishing groups of equal amplitude in plant sociology based on similarity of species content and its application to analyses of the vegetation on Danish commons. Kobenhavn **5**(4), 1–46 (1948)
19. Sorysz, J., Sorysz, D.: Efficiency of local binarization methods in segmentation of selected objects in echocardiographic images, pp. 179–192 (07 2022). https://doi.org/10.1007/978-3-031-10467-1_10
20. Sorysz, J., Sorysz, D., Piórkowski, A.: Segmentation of a first generation agent bubbles in the B-mode echocardiographic images, pp. 127–135 (01 2022). https://doi.org/10.1007/978-3-030-88976-0_17
21. Tadeusiewicz, R., Korohoda, P.: Komputerowa analiza i przetwarzanie obrazów. Wydawnictwo Fundacja Postępu Telekomunikacji, Kraków (1997)
22. Zhang, Y., Duan, Y., Wang, X., Zhuo, Z., Haller, S., Barkhof, F., Liu, Y.: A deep learning algorithm for white matter hyperintensity lesion detection and segmentation. Neuroradiology **64** (2022). https://doi.org/10.1007/s00234-021-02820-w

Using Histogram Skewness and Kurtosis Features for Detection of White Matter Hyperintensities in MRI Images

Anna Baran and Adam Piórkowski

Abstract White matter hyperintensities are regions of hyperintensive signal in white brain matter that appear in T2 or FLAIR MRI imaging. They show demyelination, cerebral oedema, tumors, or angiogenesis. Their presence is associated with a number of neurological diseases such as dementia, cognitive impairment, depression, schizophrenia, or even cerebral small vessel disease and multiple sclerosis. WMH areas are thought to appear on MRI images even a few years before the clinical manifestation of certain neurological diseases. Automatic segmentation algorithms can help better understand white matter lesions as part of large-scale studies. It is believed that changes in lesions' appearance can be monitored over time, thus their correlation with certain diseases can be better understood. This work describes the use of histogram features for the detection of white matter hyperintensities, with a focus on skewness and kurtosis. The first aim of this study is to find features that can be used to identify white matter hyperintensities; then, the authors propose a fully automatic detection algorithm based on a sliding window method to obtain local histograms, and a Support Vector Machine algorithm to perform binary classification. The authors explored the effects of different combinations of characteristics and the relationship between feature values and hyperintensity percentage in a disc-shaped window. The described study is preliminary and contains an initial examination of the effectiveness of using histogram features for the detection of white matter hyperintensities in MRI images.

Keywords Skewness · Kurtosis · Histogram features · White matter hyperintensities · Detection

A. Baran (✉) · A. Piórkowski
Department of Biocybernetics and Biomedical Engineering, AGH University of Science and Technology, Mickiewicza 30 Av., 30–059 Krakow, Poland
e-mail: annabaran@student.agh.edu.pl

A. Piórkowski
e-mail: pioro@agh.edu.pl

P. Strumiłło et al. (eds.), *The Latest Developments and Challenges in Biomedical Engineering*, Lecture Notes in Networks and Systems 746,
https://doi.org/10.1007/978-3-031-38430-1_6

1 Introduction

White matter hyperintensities (WMH) can be found in places where small vessels are compressed and blood flow to brain structures is restricted. Regional blood flow restriction frequently leads to demyelination [7]. Hyperintensive areas are also associated with depression, cognitive impairment, and a decrease in control of fine motor skills and balance; they are thought to play an important role in neurodegradation [3]. An MRI review can be extremely important for diagnosing patients with non-typical symptoms because it shows abnormalities in white matter even when there are no clinical manifestations of disease, therefore recognition of these hyperintensities improves diagnostic accuracy [6]. However, there are no standardized approaches to WMH segmentation [3, 12]. Both detection and segmentation of white matter lesions are challenging because their properties vary and they do not have sharp edges that can be easily detected. It is crucial to find features that can effectively distinguish these areas if this method is to support the diagnosis and prognosis of certain diseases.

Local thresholding methods of MRI have been proven to be unsuitable for this task as they do not give promising results for similar problems [8, 10]. However, the use of histogram features for detection of white matter hyperintensities has scarcely been described in the literature. The most common approach is to use fully automated Neural Networks. This method is complex, time consuming, and uses large, high-quality datasets that can be difficult to obtain [1, 4].

One article describes work on a histogram-based segmentation algorithm that analyzes pre-processed images that have been corrected by mean and anisotropic filters and from which the skull, scalp, and flow-related artifacts have been manually removed. Statistical features of the central peak, including mean, skewness, standard deviation, kurtosis, and total number of pixels, were used for segmentation [5].

The method described in this work uses original images scaled to 8-bit. No elements were manually removed or filtered. The resolution of these images and the number, size, and arrangement of white matter hyperintensities within them are different. This should make this method universal, meaning it can be used to analyze images—regardless of size, resolution, number and arrangement of lesions—from different types of MRI machines. The proposed method does not use Neural Networks and is therefore less computationally complex. Paragraph 2 presents the data and methods that were used in this work; paragraphs 3 and 4 contain results and conclusions.

2 Materials and Methods

2.1 Data

The dataset used in this work contains 3 DICOM MRI FLAIR images of white matter with hyperintensities that are scaled linearly to 8 bit from min-max to 0–255. The images are 2D, meaning only one slice of the 3D volume was chosen for examination. White matter (also referred to as the region of interest (ROI)) and white matter hyperintensities were manually segmented in cooperation with radiologists (Fig. 1). Their properties are presented in Table 1.

2.2 Histogram Features

It was observed that histograms of the border of WHMs are uniquely asymmetrical, which can be used for detection if the correct approach is used. First, all images were analyzed using qMaZda [11]. After performing map computation in this software, it was observed that hyperintensive areas are clearly visible after computing histogram skewness and kurtosis.

Skewness is a measure of asymmetry. It describes how far from the normal distribution a histogram is. A skewed histogram is longer on one side of its peak than on the other [2].

Kurtosis is a measure of the "tailedness" of the histogram; this means it describes the tail that appears on asymmetrical distributions.

Minimum is the smallest value in a distribution. The background of MRI images has the lowest value.

Maximum is the highest value in a distribution; in this case, either white matter hyperintensities or the cranium will have the highest maximum, meaning the highest brightness. How bright a hyperintensity is depends on the type of disease and its stage.

Mean is the average value of a distribution. It is calculated by summing up the values and dividing this sum by their number.

Standard deviation describes whether data are clustered around the mean or dispersed. Windows with hyperintensities have a higher standard deviation than other elements.

Table 1 Image properties

Image ID	Pixel spacing	Size [px]
H 16	0.71875 × 0.71875	252 × 320
XY 27	0.375 × 0.375	540 × 640
Z 15	0.859375 × 0.859375	256 × 256

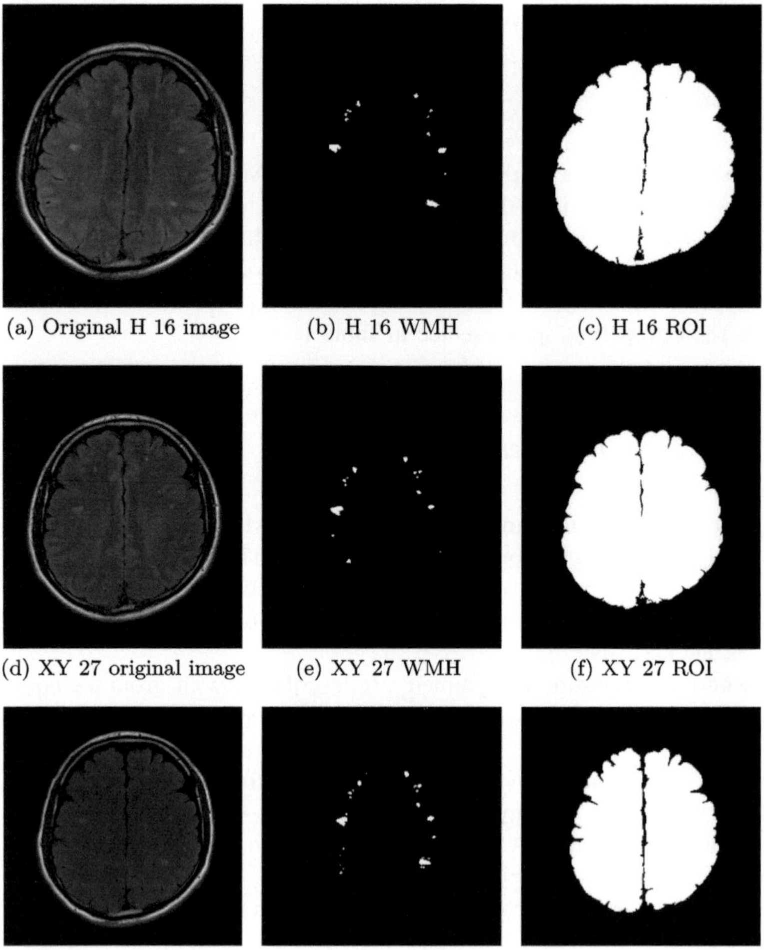

(a) Original H 16 image (b) H 16 WMH (c) H 16 ROI

(d) XY 27 original image (e) XY 27 WMH (f) XY 27 ROI

(g) Z 15 original image (h) Z 15 WMH (i) Z 15 ROI

Fig. 1 Dataset with manual segmentation of hyperintensities (WMH) and white matter (ROI)

2.3 Histogram Analysis

ImageJ software was used to analyze histograms of certain areas of MRI images. It was observed that the histogram of an area containing WHM is elongated on one side of the peak, whereas healthy tissue has a more symmetrical histogram. This observation steered this work in the direction of focusing on histogram features that describe asymmetry. Some hyperintensive areas can be as bright as the cranium, while others are only slightly brighter than the surrounding white matter. Their borders are not sharp or easy to detect, and they vary in shape and size (Figs. 2, 3 and 4).

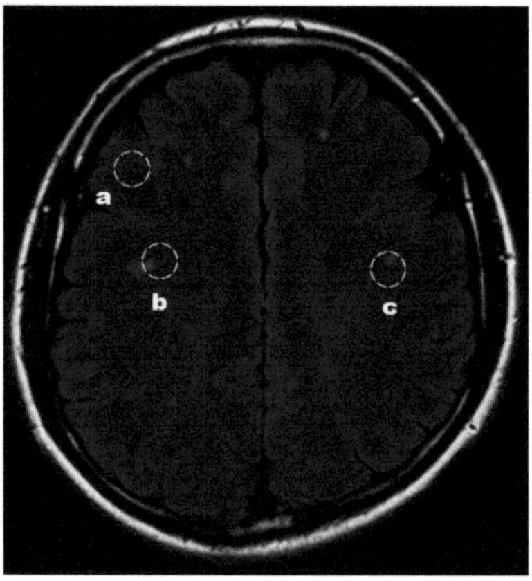

Fig. 2 Areas chosen for histogram analysis

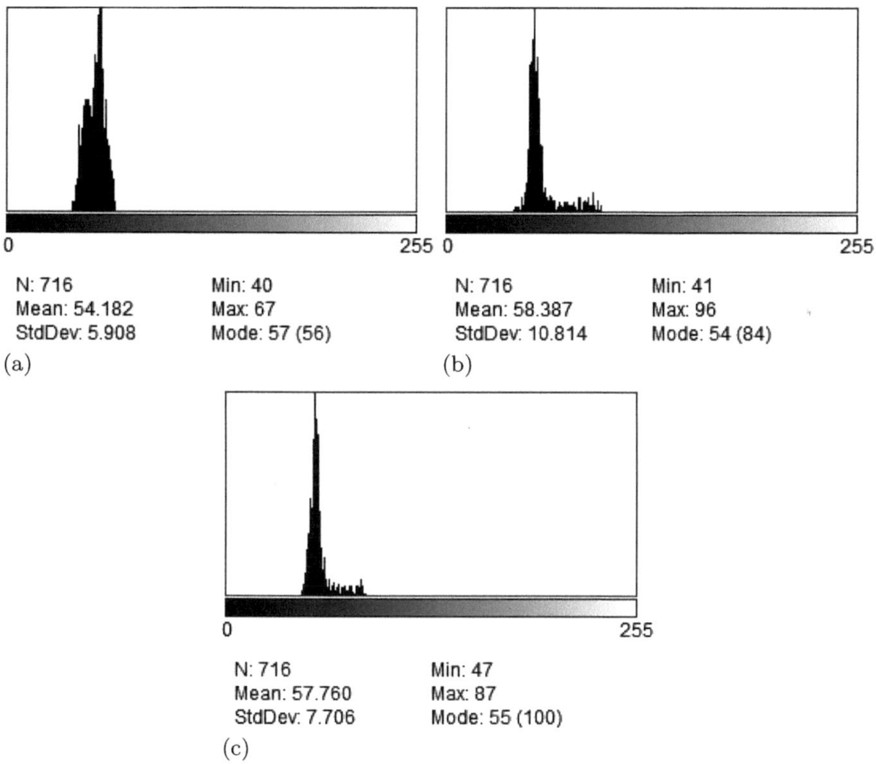

Fig. 3 Histograms of chosen areas. **a** Healthy tissue, **b**, **c** white matter hyperintensity

2.4 MaZda Map Computation

MaZda software was used to analyze the MRI images. The map computation option made it possible to calculate a large number of features and observe whether any features made hyperintensive areas stand out. As it turned out, histogram skewness and kurtosis provided images that contained WMH objects that were easily distinguishable from other elements of the image. Figure 4 shows skewness and kurtosis computed by MaZda software. After computing skewness, it can be clearly seen a darker border around WMHs; this means skewness has a specific value on the edge of these areas. A similar phenomenon can be observed on the kurtosis histogram, but this time the WMHs have a light border around them. None of the other computed features proved to be unique to white matter hyperintensities. Maps calculated using histogram features such as mean, minimum, and maximum do not provide a clear distinction between WHM objects and other elements of the image.

Dark edges on skewness images and light edges on kurtosis images also appear for the cranium, white matter, and even for artifacts on the background. This is why this work also proposes using other statistical features for detection and segmentation. The background has the lowest minimum and maximum, and in many cases the cranium is the brightest element. On average, windows containing white matter hyperintensities also have higher standard deviation than other elements.

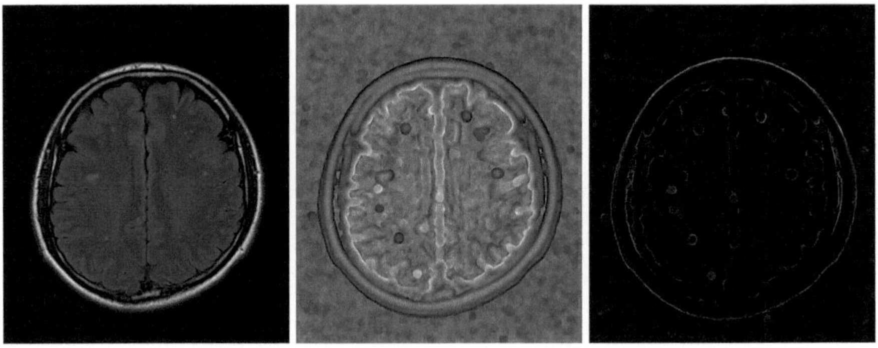

Fig. 4 Original image and MaZda computed skewness and kurtosis

2.5 Skewness and Kurtosis Formulas

The next step of the experiment was to implement known skewness and kurtosis formulas to determine which ones can be used to detect white matter hyperintensities.

Pearson's First Skewness Coefficient (PFSC)

$$PFSC = \frac{\mu - D}{\sigma} \tag{1}$$

where:
μ—mean,
D—mode,
σ—standard deviation.

Pearson's Second Skewness Coefficient (PSSC)

$$PSSC = \frac{3 \cdot (\mu - Me)}{\sigma} \tag{2}$$

where:
μ—mean,
Me—median,
σ—standard deviation.

Skewness formula used in MaZda

$$Skewness = Variance^{-\frac{3}{2}} \left(\sum_{k=0}^{\theta} (k - \mu)^3 p(k) \right) \tag{3}$$

where:

$$Variance = \sum_{k=0}^{\theta} (k - \mu)^2 p(k) \tag{4}$$

$$p(k) = \frac{1}{area} \sum_{(x,y)\epsilon ROI} \begin{cases} 1 : I(x, y) = k \\ 0 : I(x, y) \neq k \end{cases} \tag{5}$$

μ—mean,
area—total pixel count.
This formula can also be found in literature [9].

Kurtosis
Kurtosis is the fourth central moment divided by the square of variance or standard deviation to the power of 4.

$$Kurtosis = \frac{\mu_4}{\sigma^4} \tag{6}$$

where:
μ_4—fourth central moment,
σ—standard deviation.

Kurtosis formula used in MaZda

$$Kurtosis = Variance^{-2}(\sum_{k=0}^{\theta}(k-\mu)^3 p(k)) - 3 \tag{7}$$

where:

$$Variance = \sum_{k=0}^{\theta}(k-\mu)^2 p(k) \tag{8}$$

$$p(k) = \frac{1}{area} \sum_{(x,y)\epsilon ROI} \begin{cases} 1 : I(x,y) = k \\ 0 : I(x,y) \neq k \end{cases} \tag{9}$$

μ—mean,
area—total pixel count.
This formula can also be found in literature [9].

2.6 Feature Maps

Feature maps were generated using five formulas presented in paragraph 2.5 and the sliding window method. An optimal radius of 7 pixels was chosen experimentally. A larger radius provided less distinction between structures, meaning images were blurry and hard to read. Not all skewness and kurtosis formulas provide sufficient distinction of white matter hyperintensities. Only the formulas used in MaZda give promising results (Fig. 5).

2.7 Image Analysis

Each image was divided into circular windows with a radius ranging between 5 and 30 pixels. The window of a chosen radius was moved through an image with a step equal to 1 pixel with the aim of analyzing the relationship between the size of the window, feature values, and detection efficiency. It is known that specific skewness values only appear on the edge of WMH objects, but to obtain this the window has to fit a good proportion of healthy white matter to a hyperintensive area. A radius ranging from 5 to 30 pixels is believed to provide a good variety of window sizes for analysis.

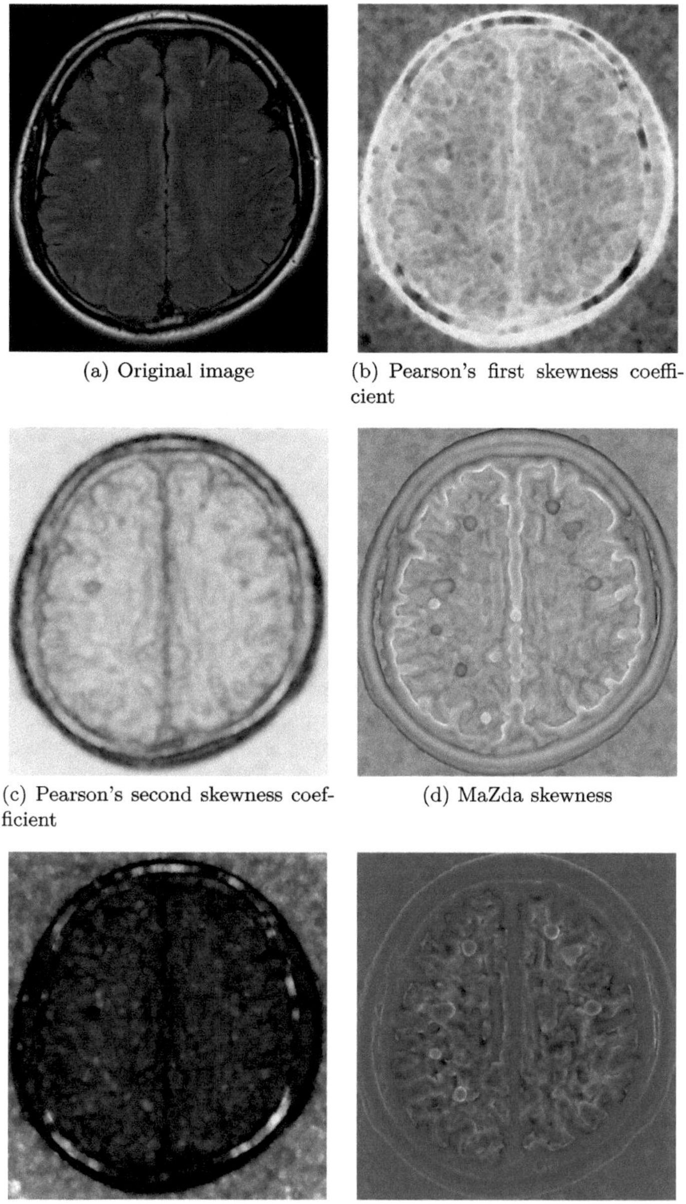

(a) Original image

(b) Pearson's first skewness coefficient

(c) Pearson's second skewness coefficient

(d) MaZda skewness

(e) Kurtosis

(f) MaZda kurtosis

Fig. 5 Feature maps

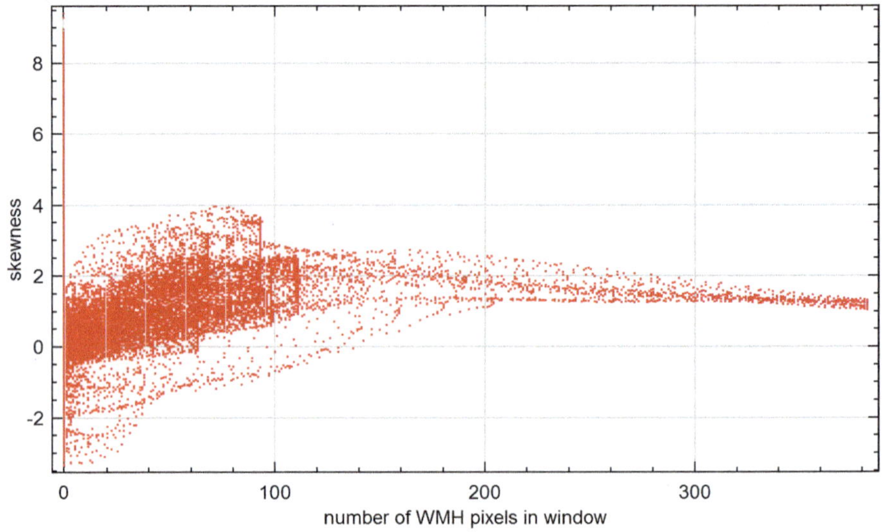

Fig. 6 Skewness values in relation to number of WMH pixels in a circular window with a radius equal to 20 pixels

A set of data was created for every image and radius. It contained information about the center position, how many pixels are in the WMH area, and histogram values. The chosen statistical features were computed for every histogram. Based on this computation, a graph showing the relationship between skewness values and the number of WMH pixels in the window was generated and is shown (Fig. 6).

The horizontal axis represents the number of pixels containing white matter hyperintensities in a circular window. In this case, the window radius is 20 pixels, which means it has 1257 pixels in total. If the WMH pixel number is above approx. 200 pixels (16% of window volume), the range of skewness values becomes narrower.

In all analyzed cases, if the percentage was above 16–26%, skewness had values in a narrower range. This means that if the right radius is chosen based on the size of the hyperintensive area, then skewness will have a specific range of values, whereas the rest of the image has a wide range of skewness values. This information can be used for detection.

2.8 Classification Algorithm

The Support Vector Machine algorithm was used to verify whether the chosen histogram features can be used to detect WMHs. In this case, the SVM algorithm is used to perform binary classification.

First, the model is trained to classify datapoints based on given features by finding an optimal hyperplane to separate data of different classes. Second, a testing dataset is used to obtain the evaluation metric of said model. The single dataset that was examined in one round contained around 38,416–333,900 datapoints, depending on the radius and image resolution. All datapoints contained data collected from circular windows of the same image but from different randomly chosen parts of it. The algorithm does not classify the whole image as an image of either a healthy or a diseased patient; instead, it determines whether a certain window contains a white matter hyperintensity or not. Because the circular window slides with a step of 1 pixel, every pixel is classified as either a white matter hyperintensity or not after combining the information from all windows. The training set contained 60% of the data, and the testing set contained 40% of the data. No datapoints belonged to both sets.

In this work, different combinations of minimum, maximum, standard deviation, Pearson's first skewness coefficient, Pearson's second skewness coefficient, skewness from MaZda, kurtosis, and kurtosis from MaZda were used along with different kernels (linear, polynomial and radial based functions) to obtain the best evaluation metrics.

3 Results

The support Vector Machine algorithm can classify hyperintensive areas with a F1-score of 87%. These results confirm that histogram skewness and kurtosis can be used to detect white matter hyperintensities. Evaluation metrics tend to change depending on the order in which features are arranged when creating a model.

Best results are presented in Table 2.

These results are promising. The Support Vector Machine algorithm was used to prove the hypothesis that histogram features would be useful in the detection of white matter lesions in MRI images. The algorithm is not perfect yet, but more combinations will be experimented with to achieve better outcomes.

Table 2 Best results

Features	F1-score (%)
Min, Max, PFSC, Kurtosis, SD	86.83
Min, Max, MaZda skewness, Kurtosis, SD	87.03
Min, Max, MaZda skewness, MaZda kurtosis, SD	87.30
Min, Max, MaZda skewness, Mean	87.02

Where: PFSC—Pearson's First Skewness Coefficient; SD—standard deviation

4 Conclusions

This preliminary study confirmed that using the sliding window method to obtain local histograms of an image and calculate their features can be effective for detection of white matter hyperintensities. The formulas for kurtosis and skewness that are used in MaZda have been proven to correlate most with white matter lesions, but WMH objects must take up at least 16–26% of the circular window volume in order to be detected. To achieve this, the right radius must be selected.

Feature maps show that a radius equal to 7 pixels provides the least blurry images and makes white matter hyperintensities most visible. However not all skewness and kurtosis formulas help differentiate WMH objects in MRI images. This is why features calculated from MaZda formulas appear in Table 2.

Further work will involve computing standard deviation for a higher radius than skewness, while maintaining the same center. It is currently believed that this might provide appealing results, especially in cases where the optimal window for calculating skewness only fits part of the lesion. Standard deviation is more unique for WMH areas when a circular window fits a whole volume of white matter lesion along with some of the surrounding tissues. Also, different combinations of features and kernels will be experimented with. The aim is to create a fully automatic diagnostic tool that can detect and segment white matter hyperintensities without the use of neural networks. The obtained results are good, but the achieved knowledge and experience will be used in further work to improve the described method.

Acknowledgements This work was financed by the AGH University of Science and Technology, Faculty of EAIIB, KBIB no 16.16.120.773.

References

1. Balakrishnan, R., Hernández, M.d.C.V., Farrall, A.J.: Automatic segmentation of white matter hyperintensities from brain magnetic resonance images in the era of deep learning and big data—A systematic review. Comput. Med. Imaging Graph. **88**, 101867 (2021)
2. Doane, D.P., Seward, L.E.: Measuring skewness: a forgotten statistic? J. Stat. Educ. **19**(2) (2011)
3. Frey, B.M., Petersen, M., Mayer, C., Schulz, M., Cheng, B., Thomalla, G.: Characterization of white matter hyperintensities in large-scale MRI-studies. Front. Neurol. **10**, 238 (2019)
4. Heinen, R., Steenwijk, M.D., Barkhof, F., Biesbroek, J.M., van der Flier, W.M., Kuijf, H.J., Prins, N.D., Vrenken, H., Biessels, G.J., de Bresser, J.: Performance of five automated white matter hyperintensity segmentation methods in a multicenter dataset. Sci. Rep. **9**(1), 1–12 (2019)
5. Jack, C.R., Jr., O'Brien, P.C., Rettman, D.W., Shiung, M.M., Xu, Y., Muthupillai, R., Manduca, A., Avula, R., Erickson, B.J.: Flair histogram segmentation for measurement of leukoaraiosis volume. J. Magn. Reson. Imaging: Off. J. Int. Soc. Magn. Reson. Med. **14**(6), 668–676 (2001)
6. Lebrun, C., Cohen, M., Chaussenot, A., Mondot, L., Chanalet, S.: A prospective study of patients with brain MRI showing incidental t2 hyperintensities addressed as multiple sclerosis: a lot of work to do before treating. Neurol. Therapy **3**(2), 123–132 (2014)

7. Milewska, K., Obuchowicz, R., Piórkowski, A.: A preliminary approach to plaque detection in mri brain images. In: Innovations and Developments of Technologies in Medicine, Biology and Healthcare: proceedings of the IEEE EMBS International Student Conference (ISC), pp. 94–105. Springer (2022)
8. Piórkowski, A., Lasek, J.: Evaluation of local thresholding algorithms for segmentation of white matter hyperintensities in magnetic resonance images of the brain. In: Applied Informatics: Fourth International Conference, ICAI 2021, Buenos Aires, Argentina, Proceedings, vol. 4. pp. 331–345. Springer (2021)
9. Pratt, W.K.: Digital Image Processing. Wiley (1991)
10. Sorysz, J., Sorysz, D.: Efficiency of local binarization methods in segmentation of selected objects in echocardiographic images. In: Intelligent Computing: Proceedings of the 2022 Computing Conference, vol. 3, pp. 179–192. Springer (2022)
11. Szczypiński, P.M., Strzelecki, M., Materka, A., Klepaczko, A.: Mazda—A software package for image texture analysis. Comput. Methods Programs Biomed. **94**(1), 66–76 (2009)
12. Zhang, Y., Duan, Y., Wang, X., Zhuo, Z., Haller, S., Barkhof, F., Liu, Y.: A deep learning algorithm for white matter hyperintensity lesion detection and segmentation. Neuroradiology 1–8 (2022)

Texture Analysis Versus Deep Learning in MRI-based Classification of Renal Failure

Artur Klepaczko, Marcin Majos, Ludomir Stefańczyk, Katarzyna Szychowska, and Ilona Kurnatowska

Abstract In this paper, we compare two approaches to automatically classify MR images of kidney lesions. The first approach involves the extraction of texture features for manually delineated kidney regions of interest (ROI) and then the classification of feature vectors with a model trained in a supervised manner. This classic machine learning approach is then challenged by a convolutional neural network-based method, which performs image classification in the learned latent feature space. In both cases, We aim to verify the hypothesis that it is possible to differentiate the state of renal failure between three classes: control, active inflammation, and chronic malformations based on the information content of the T1-weighted non-contrast enhanced MRI. The experiments performed on a sample of 25 showed superior performance of the convolutional neural network, for which we obtained the accuracy score at the level of 94% against 80% for the texture-based classification.

Keywords Kidney disease · Texture analysis · Deep learning · MR imaging

---

A. Klepaczko (✉)
Institute of Electronics, Lodz University of Technology, Łódź, Poland
e-mail: artur.klepaczko@p.lodz.pl

M. Majos · L. Stefańczyk
Department of Radiology—Diagnostic Imaging, Medical University of Lodz, Łódź, Poland
e-mail: marcin.majos@umed.lodz.pl

L. Stefańczyk
e-mail: ludomir.stefanczyk@umed.lodz.pl

K. Szychowska · I. Kurnatowska
Department of Internal Medicine and Transplantic Nephrology, Medical University of Lodz, Łódź, Poland
e-mail: ilona.kurnatowska@umed.lodz.pl

I. Kurnatowska
e-mail: katarzyna.szychowska@umed.lodz.pl

P. Strumiłło et al. (eds.), *The Latest Developments and Challenges in Biomedical Engineering*, Lecture Notes in Networks and Systems 746,
https://doi.org/10.1007/978-3-031-38430-1_7

1 Introduction

Renal failure is a growing medical, diagnostic, and social challenge. It is estimated that 12% of the American and European population [16, 17] is suffering from chronic kidney disease (CKD), and even though it is vastly surpassed by the number of patients suffering from cardio-vascular or neoplastic diseases, its therapy is thought to generate greater costs—even 1% now and probably 2% by 2030 CKD patients must be treated with hemodialysis or renal transplants, as well as their lifespan is reduced considerably [17].

CKD is a state when the glomerular filtration rate (GFR) is lower than 60 ml/min/1.73 m^2 for longer than 3 months or when the GFR level is higher than 60 ml/min/1.73 m^2 but it is associated with the destruction of renal microscopic or macroscopic structure [21]. CKD can be reversible in specific conditions but most frequently its natural evolution leads to the constant, graduate, and the inevitable loss of renal function ended by end-stage kidney failure. Sometimes CKD progresses faster as it is accompanied by acute kidney injuries (AKI) which further damage renal morphological structure [2, 15].

Monitoring of CKD is based on verifying the blood level of creatinine. The achieved value combined with patients' age, mass, and sex, is then used to estimate GFR [16]. Unfortunately, a decrease of GFR, and therefore renal function, is secondary to renal parenchyma damage, so a low level of GFR means that the best moment for beginning more aggressive treatment has already passed. GFR is the founding stone of renal function diagnostics but in recent monitoring algorithms, there are several other diagnostic tools enabling the prediction of increased risk of CKD progression including examining urine for proteinuria and other sediment abnormalities, verifying electrolyte levels, and performing kidney biopsy or diagnostic imaging [9, 14, 24].

Up to now, the role of radiological examination is still marginal in the management of CKD. In the vast majority of cases, the diagnostic imaging is restricted to ultrasound (US) both in M-mode to determine organ size and morphological features as well as in doppler and spectral-doppler modes to evaluate the vascular distribution and the character of blood flow. Other modalities, like computed tomography (CT) and magnetic resonance imaging (MRI), are rather restricted to emergency situations or are conducted for diagnosing additional, non-kidney-related purposes. All these methods—radiological and laboratory—present already established processes of renal function loss at a different moment of development but far later than at its beginning. The only examination, included in a scientifically approved diagnostic scheme, presenting an up-to-date kidney state is the outcome of histopathological biopsy which is undertaken quite rarely due to its invasive nature.

On the other hand, usage of magnetic resonance imaging (MRI) is limited due to lower spatial resolution, longer imaging time, insensitivity in renal calcifications detection, or the problem of motion artifact. Nevertheless, the role of MRI increases as it provides excellent soft-tissue contrast (enabling, e.g., differentiation between renal cortex and medulla) and is safer for patients who cannot receive iodinated tracer agents or for whom exposure to ionizing radiation is unacceptable.

The progression of CKD is supposed to be related to tubule-interstitial injury followed by parenchymal atrophy and fibrosis. Recent studies suggest that these pathologies are caused by hypoxia associated with arteriolar damage due to fibrosis and glomerular damage leading to a reduction in peritubular perfusion. Some researchers suspect that both states can be visualized by two functional MRI sequences—diffusion-weighted images (DWI) and blood oxygen level-dependent MRI (BOLD-MRI). The first sequence measures the Brown movement of water particles which presents information about the effects of capillary perfusion and water diffusion in extracellular spaces. The second one uses deoxyhemoglobin as a marker of tissue oxygenation. Functional MRI techniques are not yet validated—they give, however, promise to be the most reliable tools for monitoring the actual state of CKD enabling implying most fitted treatment [25, 29].

On the other hand, some studies suggest that not only advanced MRI sequences can visualize the damage to renal parenchyma on a microscopic level, but CKD may give detectable differences in signal on even most basic MRI sequences, such as T1- and T2-weighted images [5, 23].

In this paper, we aim to verify the hypothesis that it is possible to differentiate the state of renal failure between three classes: control, active inflammation, and chronic malformations based on the information content of the T1-weighted non-contrast enhanced MRI. For that purpose, we first employ a classic machine learning scheme involving image texture analysis, selection of discriminative texture features, and feature vector classification. We then challenge this approach by deep learning methodology to evaluate whether manually engineered features can compete with latent representations extracted by a convolutional neural network. The below-reported experiments are the results of a preliminary study performed on a relatively limited sample of 25 subjects.

2 Related Work

Computer methods for aiding image-based diagnosis of kidney disease are relatively widely described in the literature. A large group of contributions concern kidney segmentation for the need to calculate the total kidney and renal cortex volumes [3, 19, 32]. Both metrics are simple measures allowing assessment of tissue atrophy and hence can also serve as an indirect indicator of disease progression. The majority of these works utilize the CT modality [8, 18, 28, 30, 31], which, as said, is not recommended for patients with declined renal performance. Therefore, attempts were made to develop segmentation models dedicated specifically to MR images, both contrast- and non-contrast enhanced.

As you might expect, the vast majority of the most recent publications present results obtained using deep neural networks, especially models based on the U-Net architecture. These methods, compared to classical image processing techniques, give competitive segmentation results, because the convolutional filters used in them are characterized by excellent properties of encoding texture and shape features of the objects of interest. For the same reason, examples of accurate performance of convolutional neural networks in various medical image classification tasks are abundant. These examples include lesion classification in endoscopy images, brain tumor classification in MRI, COVID recognition in CT chest images, and thyroid nodules in ultrasound data, to name just a few.

Apparently, the usage of deep learning models in application to renal tissue state discrimination is less frequent, giving precedence to texture analysis and other feature extraction methods. For example, in [7] the usefulness of texture analysis was evaluated in regard to renal dysfunction classification into three stages: healthy control, moderate and severe state. The study was conducted on a cohort of 166 CKD patients divided into 3 groups based on the GFR level with 2 cutoff thresholds (30 and 60 ml/min/1.73 m^2). The patients underwent several different MR imaging examinations (T1-weighted, DWI, and BOLD). The extracted texture features embraced first- and second-order statistics such as histogram, grey-level co-occurrence matrix, run-length matrix, and grey-level dependence matrix. The model construction pipeline consisted of the sequential selection of the most significant features and training of a series of classifiers, including decision trees, support vector machines, and linear discriminant analysis. The best-achieved accuracy was reported for the SVM algorithm with the radial basis function kernel and was equal to 82.8% (macro average).

Deep learning-based approaches, on the other hand, were proposed in [13] in relation to US modality, in [20] for CT images, and in [6] for electron microscopy. In any of the cited studies, the classification metrics surpassed those reported earlier for texture-based analysis of MRI data. For example, in the case of CKD detection in US scans, the average accuracy was 85.6%, although the classification problem was reduced to a binary choice between positive and negative CKD status.

Apart from image-based approaches, there appeared attempts to develop deep learning CKD prediction models which operated on clinical features, such as blood pressure, albumin, serum creatinine, blood urea, glucose level, etc., e.g. [11, 26] According to our survey, however, there are no published results comparing the performance of deep learning models with machine learning predictors trained on hand-crafted texture features, both applied to MRI images of the kidney.

3 Materials and Methods

3.1 MRI Data

The current study was performed on a collection of 25 datasets obtained for 5 control volunteers and 20 patients with clear indications to undergo kidney state diagnostics. All examinations were performed in Barlicki University Hospital, Lodz, Poland, and embraced taking a biopsy in the left kidney and afterward, MR scanning using

Dixon's T1-weighted method [1]. Dixon's method was implemented on the Siemens Magnetom Vida scanner using a 3D spoiled gradient echo sequence and the VIBE protocol (volumetric interpolated breath-hold examination).

Let us recall, that in Dixon's method, two images are acquired—a so-called in-phase (denoted as *IP*) and an opposed-phase (*OP*) image. While in the prior image, the orientation of magnetization vectors of water and fat isochromats are aligned at the central row of the k-space, their phase is shifted by 180 °C in the latter case. Hence, by subtracting or aggregating IP and OP images, one obtains a water-only (*WO*) and a fat-only (*FO*) image, where the contributions from fat or water isochromats are effectively suppressed. In our experiments, we used all four image types. Although WO and FO data are correlated with IP and OP images, they exhibit various texture and contrast characteristics, which may affect the performance of machine learning algorithms.

The in-plane resolution of the images was 260×320 with the pixel size 1.4×1.4 mm^2 and 2 mm slice thickness. The images were acquired in a *quasi* coronal plane, as the exact oblique slice orientation was adjusted for each subject to reflect individual kidney rotation angles with respect to anatomical planes.

The biopsy material was used to determine the character of the renal tissue lesion. The result of the histopathology allowed us to label each subject as belonging either to the active inflammation or chronic fibrosis class. Hence, with the addition of the control subjects, the machine learning problem undertaken in this study is a three-category classification task. Example image cross-sections for each of the three classes and four image types are illustrated in Fig. 1.

All patients and volunteers gave their written informed consent for partici-pation in the study which was approved by the Bioethics Committee at the Medical University of Lodz, Poland.

3.2 Texture Analysis

Although the image acquisition protocol was 3-dimensional, we decided to perform the analysis in 2D. The primary reason for this action was the relatively low number of subjects enrolled in this initial feasibility study and hence small variation presented to a convolutional neural network. In contrast to classical machine learning strategies, deep learning models require very large training databases to be able to extract relevant image features. By reducing the problem to 2D, we gain a lot more training examples representing subsequent slices of the kidney from a single subject. In order to ensure comparative conditions for deep learning and texture-based analysis, also the latter is conducted for 2D cross-sections.

Also, it is important to note, that for each patient only the right kidney was included in the investigation. As said before, the left kidney was used for biopsy and hence its appearance in the MR image might have been affected by temporal inflammation caused by the puncture.

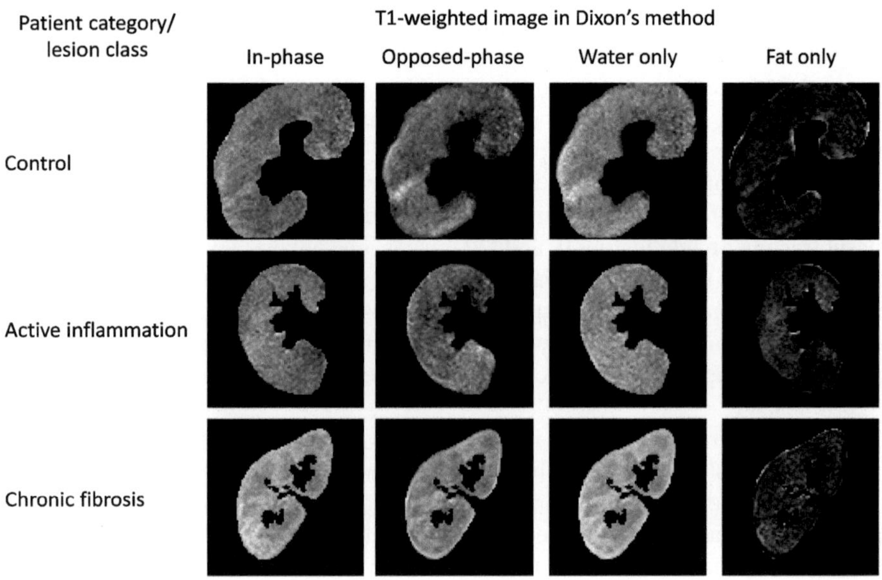

Fig. 1 Examples of T1-weighted MR images acquired by Dixon's method for three patient categories examined in the study.

The texture features for each of the four image types were extracted using the grey-level cooccurrence matrix. We chose this texture model due to its capability to encapsulate short- and long-range image intensity patterns in various directions. From the technical perspective, GLCM is also widely available in many popular software packages, such as SciKit-Learn [22], Matlab [10], or MaZda [27], allowing other researchers to reproduce the results presented in this paper. Specifically, we calculated 6 features of the GLCM matrix, i.e. energy, homogeneity, dissimilarity, contrast, correlation, and the angular second moment, at 3 distances (1, 3, and 5 pixels) in 4 directions ($0, \pi/4, \pi/2, 3\pi/4$), giving in total 144-feature vector for each analyzed kidney cross-section. The reader is referred to Scikit-Learn documentation for the definition of specific GLCM-derived parameters [22].

Moreover, feature extraction was performed only within regions of interest (ROI) covering exclusively the renal tissue. For that purpose, in each image we manually segmented kidneys. The annotation was performed by one radiologist and one expert in medical image processing. The segmentations were then cross-checked and corrected when necessary. Eventually, square-shaped image patches of size 96×96 pixels centered over the delineated ROIs were cropped-off from the original cross-section images for further processing.

3.3 Kidney Texture Classification

As a result of 2D image slice selection and texture feature extraction, a dataset composed of 1000 feature vectors was constructed. This number divides into 500 vectors representing the active inflammation class, 335 vectors of the chronic fibrosis category, and 165 control samples. Since for each image type we extract 144 features, data vectors are determined in a 288-dimensional feature space. Compared to the data set size, such a complex, and potentially redundant domain clearly calls for dimensionality reduction. It can be achieved either through feature selection or feature transformation. Since we are interested in whether any of the images acquired or determined in the Dixon method convey more lesion-discriminating information than the others, we chose the first option.

Prior to feature selection, all attributes were standardized to obtain zero mean and unit standard deviation and thus balance the impact of various features with significantly different ranges on data vector similarity metrics. Then, we selected the subset of the 20 most discriminative attributes using the *wrapper* technique. The feature space was explored by the floating forward selection method, as implemented in Weka software [4]. The wrapped estimator used to evaluate feature subsets was the support vector machines (SVM) classifier, i.e., the same algorithm later utilized for final data discrimination.

In our experiments, we applied the SVM algorithm with the radial basis function kernel

$$k(\mathbf{x}_i, \mathbf{x}_j) = \exp\left(-\gamma\|\mathbf{x}_i - \mathbf{x}_j\|^2\right), \tag{1}$$

where $\|\cdot\|$ denotes the ℓ^2-norm. An additional method parameter usually designated with the letter C allows a certain number of data points to violate the decision boundary. The γ and C parameters contribute conversely to the complexity of the SVM model. Therefore, we tuned their values using the exhaustive grid search algorithm in a five-fold cross-validation experiment. The best results were reported for $\gamma = 0.05$ and $C = 5$.

3.4 Deep Learning Classification

The major challenge for deep learning model development is the availability of training examples. As noted above, the size of our dataset is equal to 1000 image patches, with the least frequent control group of 165 samples. Therefore, in order to increase the data variability and simultaneously prevent a neural network from overfitting, we performed augmentation of the input data set. Each cropped kidney cross-section was randomly rotated (by an angle in the range $\pm 15°$), shifted (within a limit of 10 pixels), and horizontally or vertically flipped. The last augmentation method was image brightness scaling (up to the 20% limit), however without any contrast alteration. The number of augmentations was varied for each of the subjects

Table 1 Architecture of the AlexNet model used for kidney image classification

Layer type	Layer specification
Input	$96 \times 96 \times 1$
2D convolution	Size = 11×11, stride = 4×4, channels = 64, ReLU activation
2D Max pooling	Size = 3×3, stride = 2×2
2D convolution	Size = 5×5, stride = 1, channels = 192, ReLU activation
2D Max pooling	Size = 3×3, stride = 2×2
2D convolution	Size = 3×3, stride = 1, channels = 384, ReLU activation
2D convolution	Size = 3×3, stride = 1, channels = 256, ReLU activation
2D convolution	Size = 3×3, stride = 1, channels = 256, ReLU activation
2D Max pooling	Size = 3×3, stride = 2×2
Average pooling	Output size = 6×6
Dropout	Probability = 0.5
Dense	Input size = $256 * 6 * 6$, output size = 4096, ReLU activation
Dense	Input size = 4096, output size = 4096, ReLU activation
Dense	Input size = 4096, output size = 3

categories in order to obtain balanced class distribution in the final data set. Hence, the active inflammation class images were augmented 7 times (resulting in 3500 augmented instances), chronic fibrosis—10 times (3350 instances), and the control group—20 times (3300 instances).

The classification itself was performed using the AlexNet architecture [12]. The structure of the network layers, as implemented in the Torchvision framework utilized in our study, is recalled in Table 1. The differences with the original model consist in the size of the input layer (adjusted to width, height, and the number of channels of the cropped image patches), the number of filters of convolutional layers, and the final dense layer, which must be adjusted to the number of recognized classes (in our case 3). We decided not to use any pre-trained weights, hence the training was launched from a randomly initiated network state. The model optimization was accomplished with the help of the Adam algorithm with the learning rate = 0.001, and the number of epochs = 50.

4 Results

The experiments for texture-based and deep-learning classification paths were performed using different testing strategies. In the prior case, the 5-fold cross-validation technique was employed, both during the final classifier training as well as in the feature selection stage. However, feature selection was performed only once, and cross-validation was used to evaluate subsequent feature subset candidates. After feature selection was completed, the whole dataset was reduced to the most significant attributes and submitted to SVM classifier training and validation. The division into folds was performed in such a way, that none of the folds contained data extracted from patients included in other buckets.

The list of selected attributes is shown in Table 2. Note that the wrapper approach identifies a subset of relevant attributes which show their predictive performance only if they occur together, and does not assess their individual significance. Therefore, the presented list cannot be interpreted as a ranking, where the first feature is anyway better than the last one. Apparently, according to our research, it turned out that the most informative features were extracted for the water-only and in-phase images. As regards specific statistics derived from the GLCM matrix, it is the dissimilarity metric followed by correlation and energy in various combinations of distances and orientations which appear most frequently in the selected features subset.

In the case of the deep learning approach, since the overall number of image slices obtained after the augmentation procedure was relatively large, instead of cross-

Table 2 Selected most significant texture features.

Feature type	Image type	
	WO	IP
Dissimilarity	$d = 1, \theta = 0°$	$d = 1, \theta = 90°$
	$d = 3, \theta = 0°$	$d = 1, \theta = 135°$
	$d = 5, \theta = 45°$	$d = 5, \theta = 135°$
Correlation	$d = 5, \theta = 45°$	$d = 1, \theta = 0°$
	$d = 5, \theta = 135°$	$d = 3, \theta = 90°$
		$d = 5, \theta = 135°$
ASM	$d = 3, \theta = 135°$	
	$d = 5, \theta = 0°$	
Energy		$d = 1, \theta = 90°$
		$d = 3, \theta = 135°$
		$d = 5, \theta = 0°$
		$d = 5, \theta = 45°$
		$d = 5, \theta = 125°$
Homogeneity		$d = 1, \theta = 90°$
Contrast		$d = 1, \theta = 90°$

d—pixel offset, θ—direction of the GLCM method

Table 3 Classification performance in texture and latent features domains

Feature domain/classifier	Accuracy	Precision*	Recall*	F-measure*
Texture analysis/SVM	0.794	0.793	0.794	0.793
WO/AlexNet	0.940	0.941	0.940	0.940
FO/AlexNet	0.542	0.542	0.542	0.542
IP/AlexNet	0.916	0.918	0.916	0.916
OP/AlexNet	0.918	0.921	0.917	0.918

*Weighted average

validation we decided to divide the dataset into the training and test subsets in proportions of 80% to 20%, respectively. While partitioning the dataset, we made sure to maintain the class balance and that the test set contains image slices only from patients not included in the training cohort. Then, 20% of the training data was set apart for validation. The confusion matrices shown in Fig. 2 present normalized results for SVM and AlexNet classifiers, whereas Table 3 collects other performance metrics, such as accuracy, recall, and precision. All metrics were calculated as weighted averages to account for class imbalance in texture-based inference. Although in deep learning experiments, class representations are approximately equal, we left the same metrics for consistency and for the sake of objective comparisons.

5 Discussion

Our experiments confirm that it is feasible to differentiate renal tissue lesions using texture analysis, although the classification performance is rather moderate. There is especially high ambiguity in properly distinguishing chronic fibrosis from active inflammation (recall = 71%). It is instructive to note that we obtained an accuracy score (79.4%) similar to that reported in [7], where the same imaging modality and texture-based image processing approach were utilized. In the corresponding experiment setting, with the SVM classifier and RBF kernel, and features selected from all T1-weighted images acquired in Dixon's method, they achieved an accuracy of 82.8%. One must note, however, that the classification problem was defined differently, as the target therein was to differentiate between moderate and severe CKD states.

On the other hand, it emerges that the applied convolutional neural network outperforms texture analysis in extracting relevant image descriptors. The best-achieved result in our experiments was obtained for the water-only image type, where the accuracy reached the value of 94%. Other image types seem to be less informative, however, still they ensure better recognition rates than the SVM trained on manually-engineered features, giving an accuracy of 92% in the case of in-phase and

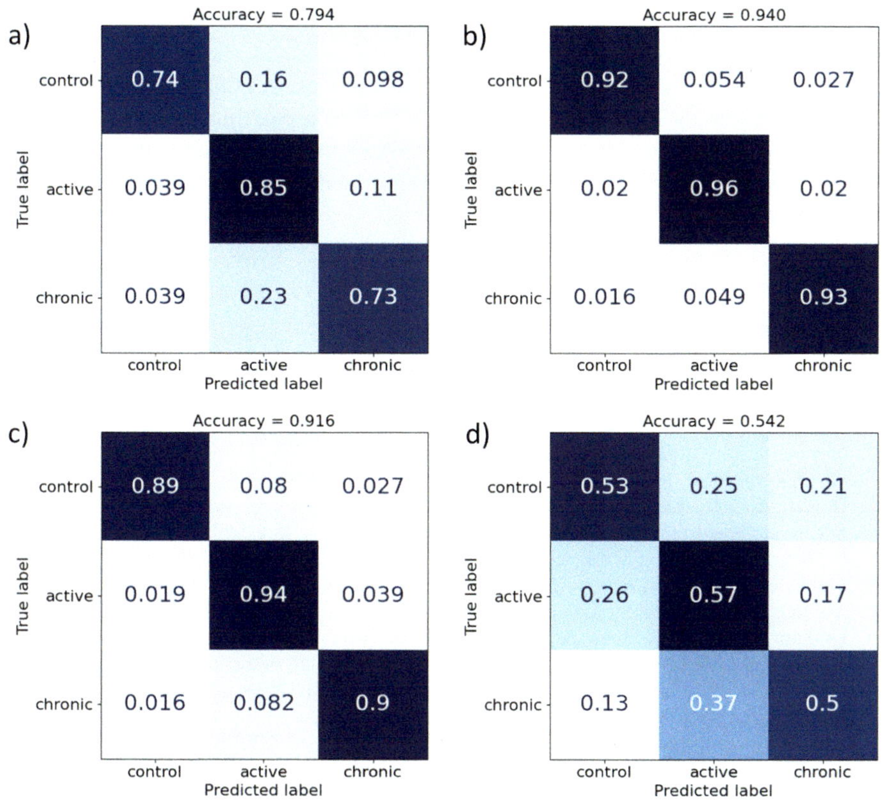

Fig. 2 Confusion matrices obtained for **a** SVM classifier tested on texture feature vectors, **b–d** AlexNet tested on WO, IP, and FO images.

89% for opposed-phase images. The least informative are the fat-only images, which in fact do not lead to a reasonable prediction performance (accuracy = 54%). Our recommendation is therefore to focus on deep learning in the future development of automatic kidney disease classification systems, although these findings must be confirmed on a larger database of training and testing examples.

Eventually, it may be argued that the AlexNet architecture is obsolete at the time of publishing this paper. However, in our application, it ensured higher classification accuracy than other tested models, i.e. ResNet-50, ResNet-101, and EfficientNetV2. For example, the accuracy scores for the WO images obtained by these 3 classifiers were equal to 73.5, 75.2%, and 63.2% respectively. Although more effort would be required to fine-tune the hyperparameters of these models, they apparently exhibit significantly lower performance in the studied case. One possible remedy here could be the use of transfer learning. An initial experiment using weights pre-trained on the ImageNet dataset improved the accuracy of ResNet-101 to 91.5%, which is still worse than AlexNet trained from scratch.

The future work, apart from further validation of the observed superior performance of deep learning in comparison with the classical texture analysis-based approach, will focus on the fusion of both techniques. It could be accomplished either by presenting a multi-channel input image to a classifier network, where a gray-scale channel is accompanied by manually-engineered texture feature maps, or, alternatively, by extending texture feature vectors with the latent features extracted in a network backbone.

References

1. Bernstein, M.A., King, K.F., Zhou, X.J.: Handbook of MRI Pulse Sequences. Academic Press, Amsterdam; Oxford (2004)
2. Chawla, L.S., Kimmel, P.L.: Acute kidney injury and chronic kidney disease: an integrated clinical syndrome. Kidney Int. **82**(5), 516–524 (2012)
3. Daniel, A.J., Buchanan, C.E., Allcock, T., Scerri, D., Cox, E.F., Prestwich, B.L., Francis, S.T.: Automated renal segmentation in healthy and chronic kidney disease subjects using a convolutional neural network. Magn. Reson. Med. **86**(2), 1125–1136 (2021)
4. Frank, E., Hall, M.A., Witten, I.H.: The WEKA workbench. Online appendix for. In: Data Mining: Practical Machine Learning Tools and Techniques. Morgan Kaufmann (2016)
5. Graham-Brown, M.P., Singh, A., Wormleighton, J., Brunskill, N.J., McCann, G.P., Barratt, J., Burton, J.O., Xu, G.: Association between native T1 mapping of the kidney and renal fibrosis in patients with IgA nephropathy. BMC Nephrol. **20**(1), 256 (2019)
6. Hacking, S., Bijol, V.: Deep learning for the classification of medical kidney disease: a pilot study for electron microscopy. Ultrastruct. Pathol. **45**(2), 118–127 (2021)
7. Hara, Y., Nagawa, K., Yamamoto, Y., Inoue, K., Funakoshi, K., Inoue, T., Okada, H., Ishikawa, M., Kobayashi, N., Kozawa, E.: The utility of texture analysis of kidney MRI for evaluating renal dysfunction with multiclass classification model. Sci. Rep. **12**(1), 14776 (2022)
8. Heller, N., Isensee, F., Maier-Hein, K.H., Hou, X., Xie, C., Li, F., Nan, Y., Mu, G., Lin, Z., Han, M., Yao, G., Gao, Y., Zhang, Y., Wang, Y., Hou, F., Yang, J., Xiong, G., Tian, J., Zhong, C., Ma, J., Rickman, J., Dean, J., Stai, B., Tejpaul, R., Oestreich, M., Blake, P., Kaluzniak, H., Raza, S., Rosenberg, J., Moore, K., Walczak, E., Rengel, Z., Edgerton, Z., Vasdev, R., Peterson, M., McSweeney, S., Peterson, S., Kalapara, A., Sathianathen, N., Papanikolopoulos, N., Weight, C.: The state of the art in kidney and kidney tumor segmentation in contrast-enhanced CT imaging: results of the kits19 challenge. Med. Image Anal. **67**, 101821 (2021)
9. Hemmelgarn, B.R., Manns, B.J., Lloyd, A., James, M.T., Klarenbach, S., Quinn, R.R., Wiebe, N., Tonelli, M.: Alberta Kidney Disease Network: Relation between kidney function, proteinuria, and adverse outcomes. JAMA **303**(5), 423–429 (2010)
10. Inc., T.M.: Matlab version: 9.13.0 (r2022b) (2022). https://www.mathworks.com
11. Islam, M.A., Majumder, M.Z.H., Hussein, M.A.: Chronic kidney disease prediction based on machine learning algorithms. J. Pathol. Inform. **14**(100189), 100189 (2023)
12. Krizhevsky, A., Sutskever, I., Hinton, G.E.: Imagenet classification with deep convolutional neural networks. In: Pereira, F., Burges, C., Bottou, L., Weinberger, K. (eds.) Advances in Neural Information Processing Systems, vol. 25. Curran Associates, Inc. (2012)
13. Kuo, C.C., Chang, C.M., Liu, K.T., Lin, W.K., Chiang, H.Y., Chung, C.W., Ho, M.R., Sun, P.R., Yang, R.L., Chen, K.T.: Automation of the kidney function prediction and classification through ultrasound-based kidney imaging using deep learning. NPJ Digit. Med. **2**(1), 29 (2019)
14. Lamb, E., Price, C.: Tietz textbook of clinical chemistry and molecular diagnostics. In: Kidney Function Tests, pp. 669–708. Elsevier (2012)
15. Leung, K.C.W., Tonelli, M., James, M.T.: Chronic kidney disease following acute kidney injury-risk and outcomes. Nat. Rev. Nephrol. **9**(2), 77–85 (2013)

16. Levey, A.S., Stevens, L.A., Schmid, C.H., Castro, A.F., Feldman, H.I., Kusek, J.W., Van Lente, F., Greene, T., Coresh, J.: A new equation to estimate glomerular filtration rate. Ann. Intern. Med. **150**(9), 604–612 (2009)
17. Li, P.K.T., Garcia-Garcia, G., Lui, S.F., Andreoli, S., Fung, W.W.S., Hradsky, A., Kumaraswami, L., Liakopoulos, V., Rakhimova, Z., Saadi, G., Strani, L., Ulasi, I., Kalantar-Zadeh, K.: Kidney health for everyone everywhere—from prevention to detection and equitable access to care. Clin. Nephrol. **93**(3), 111–122 (2020)
18. Lin, Z., Cui, Y., Liu, J., Sun, Z., Ma, S., Zhang, X., Wang, X.: Automated segmentation of kidney and renal mass and automated detection of renal mass in CT urography using 3D U-Net-based deep convolutional neural network. Eur. Radiol. **31**, 5021–5031 (2021)
19. Lundervold, A.S., Rorvik, J., Lundervold, A.: Fast semi-supervised segmentation of the kidneys in DCE-MRI using convolutional neural networks and transfer learning. In: 2nd International Scientific Symposium, pp. 79–81 (2017)
20. Ma, F., Sun, T., Liu, L., Jing, H.: Detection and diagnosis of chronic kidney disease using deep learning-based heterogeneous modified artificial neural network. Futur. Gener. Comput. Syst. **111**, 17–26 (2020)
21. National Kidney Foundation: K/DOQI clinical practice guidelines for chronic kidney disease: evaluation, classification, and stratification. Am. J. Kidney Dis. **39**(2 Suppl 1), S1-266 (2002)
22. Pedregosa, F., Varoquaux, G., Gramfort, A., Michel, V., Thirion, B., Grisel, O., Blondel, M., Prettenhofer, P., Weiss, R., Dubourg, V., Vanderplas, J., Passos, A., Cournapeau, D., Brucher, M., Perrot, M., Duchesnay, E.: Scikit-learn: machine learning in Python. J. Mach. Learn. Res. **12**, 2825–2830 (2011)
23. Peperhove, M., Vo Chieu, V.D., Jang, M.S., Gutberlet, M., Hartung, D., Tewes, S., Warnecke, G., Fegbeutel, C., Haverich, A., Gwinner, W., Lehner, F., Bräsen, J.H., Haller, H., Wacker, F., Gueler, F., Hueper, K.: Assessment of acute kidney injury with T1 mapping MRI following solid organ transplantation. Eur. Radiol. **28**(1), 44–50 (2018)
24. Rahn, K.H., Heidenreich, S., Brückner, D.: How to assess glomerular function and damage in humans. J. Hypertens. **17**(3), 309–317 (1999)
25. Schachinger, H., Klarhöfer, M., Linder, L., Drewe, J., Scheffler, K.: Angiotensin II decreases the renal MRI blood oxygenation level-dependent signal. Hypertension **47**(6), 1062–1066 (2006)
26. Singh, V., Asari, V.K., Rajasekaran, R.: A deep neural network for early detection and prediction of chronic kidney disease. Diagnostics (Basel) **12**(1), 116 (2022)
27. Strzelecki, M., Szczypinski, P., Materka, A., Klepaczko, A.: A software tool for automatic classification and segmentation of 2D/3D medical images. Nucl. Instrum. Methods Phys. Res. Sect. A: Accel. Spectrometers Detect. Assoc. Equip. **702**, 137–140 (2013)
28. Türk, F., Lüy, M., Barışç, N.: Kidney and renal tumor segmentation using a hybrid V-Net-based model. Mathematics **8**(10), 1772 (2020). https://doi.org/10.3390/math8101772
29. Xu, X., Fang, W., Ling, H., Chai, W., Chen, K.: Diffusion-weighted MR imaging of kidneys in patients with chronic kidney disease: initial study. Eur. Radiol. **20**(4), 978–983 (2010)
30. Yin, K., Liu, C., Bardis, M., Martin, J., Liu, H., Ushinsky, A., Glavis-Bloom, J., Chantaduly, C., Chow, D.S., Houshyar, R., Chang, P.: Deep learning segmentation of kidneys with renal cell carcinoma. J. Clin. Oncol. **37**(15_suppl), e16098–e16098 (2019)
31. Zhao, W., Jiang, D., Queralta, J.P., Westerlund, T.: MSS U-Net: 3D segmentation of kidneys and tumors from CT images with a multi-scale supervised U-Net. Inform. Med. Unlocked **19** (2020)
32. Zöllner, F.G., Kociński, M., Hansen, L., Golla, A.K., Trbalić, A.S., Lundervold, A., Materka, A., Rogelj, P.: Kidney segmentation in renal magnetic resonance imaging—current status and prospects. IEEE Access **9**, 71577–71605 (2021)

Mobile Application for Learning Polish Sign Language

Anna Slian, Joanna Czajkowska, and Monika Bugdol

Abstract The increasing awareness and popularization of sign language in many countries have contributed to the development of mobile educational applications. This also applies to Polish Sign Language, used by the Deaf community in Poland. However, among the available solutions, there is a need for one which would engage the user in independent signing, i.e., forming the necessary finger motor skills and dexterity. To address this challenge, the paper focuses on developing a system for recognizing Polish Sign Language symbols that could work inside a mobile application. The publicly available MediaPipe library is used to determine the hand's characteristic points and their coordinates. The detected points are then normalized and stored. Three machine learning techniques are verified at the character recognition stage: support vector machine, k-nearest neighbors, and random forest. Each was taught on the coordinate values obtained from a set containing 29,945 images of hands, divided into 29 classes (considering selected letters, supporting characters, and numbers). The obtained accuracies of all the methods are equal to about 98%.

Keywords Polish sign language · Recognition · Random forest · SVM · k-nearest neighbors

1 Introduction

According to WHO estimates, over 1.5 billion people live with hearing loss (meaning hearing thresholds <20 dB in one or two ears), which can rise to 2.5 billion within 25 years. Due to recreational exposure to loud sounds, more than 1 billion young

A. Slian · J. Czajkowska · M. Bugdol (✉)
Faculty of Biomedical Engineering, Silesian University of Technology, Roosevelta 40, 41-800 Zabrze, Poland
e-mail: joanna.czajkowska@polsl.pl

P. Strumiłło et al. (eds.), *The Latest Developments and Challenges in Biomedical Engineering*, Lecture Notes in Networks and Systems 746,
https://doi.org/10.1007/978-3-031-38430-1_8

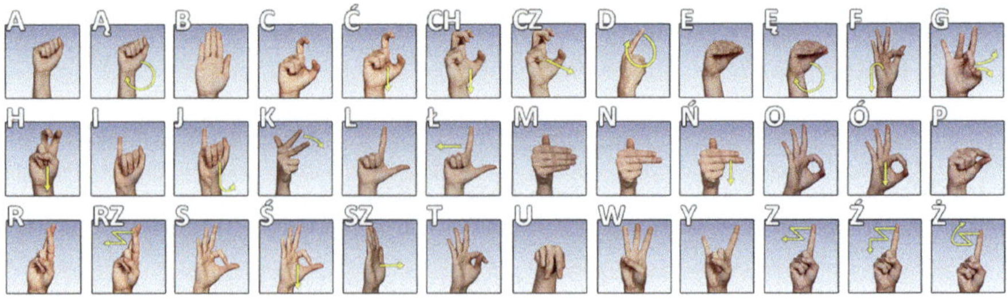

Fig. 1 Letters of the alphabet in PSJ [4]

people are at risk for hearing loss. In Europe, there are about 34.4 million deaf and hard-of-hearing adults (35 dB or greater) [1], and this number in Poland is estimated to be above 500 thousand. However, the reported values are rather underestimated. Over 100 people lose hearing yearly due to accidents, diseases, or aging. Difficult reception of auditory stimuli may translate into obstacles in communication with the environment, reduced or insufficient knowledge of the native phonic language, and a varied level of reading and understanding of speech.

The primary language of the Deaf, which is a kind of indicator of their cultural identity, is the sign language used in a given country [2]. Sign language refers to visual-spatial language that uses the sense of sight used by deaf parents to contact deaf children [3]. This type of communication evolved spontaneously wherever deaf communities existed. So it can be classified as a natural language—subject to constant changes and gradual shaping. When signing, the people do not use their voice but can move their lips soundlessly to make it easier for the interlocutor to read the message. PSL (Polish sign language) is an entirely separate entity, having different grammar than the Polish language. The basis are ideographic signs, equivalents of words, and idioms (short phrases, untranslatable with a single word). They are complemented by dactylographic signs (letters of the alphabet (see Fig. 1) and numerals) and supplementary ones, such as punctuation marks [3]. Studying PSL should include all these elements.

1.1 Mobile Applications for Learning Sign Language

Due to the growing popularity of mobile applications, different tools for learning sign language appeared, mostly American [5, 6]. In 2017, the first mobile application for PSL was provided—"Migaj z nami: PJM" [7].

It was developed by the educational platform Migam.eu, and contains over 500 videos, thematic dictionaries, and quizzes checking the user's knowledge. The application works on Android 5.1 and above and on iOS. Currently, the application is entirely free. However, it uses internet access, Bluetooth, location, adversely affecting the life of the phone's battery [7].

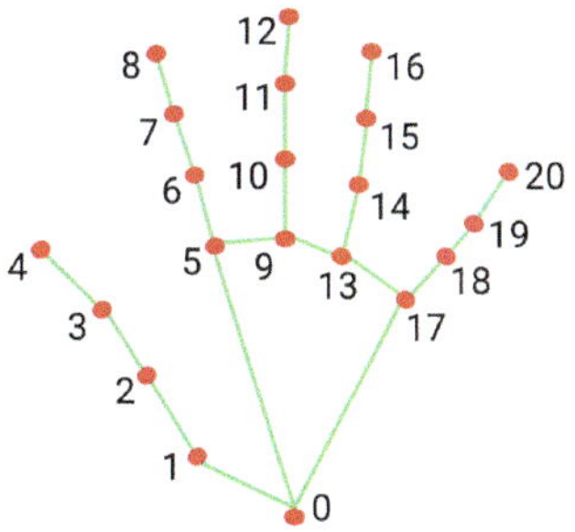

Fig. 2 21 landmarks recognized by the MediaPipe algorithm [8]

An interesting solution for real-time hand detection and tracking is the MediaPipe library. It allows users to infer 21 hand landmarks, as shown in Fig. 2. It utilizes several cooperating ML models to detect the palm area, define the region of interest, find the relevant points, and return their 3D coordinates. According to the developers, the hand detection algorithm has an average precision of 95.7% [8].

Due to the popularity of American sign language (ASL) worldwide, more apps are available on the market. An example of a learning app is "The ASL App", accessible for Android and iOS devices. It contains over 2,500 characters and phrases, allowing users to save favorite phrases, change the playback speed, rewind, and pause the video. The wide range of material to learn, the pleasant format, and the ability to navigate through the watched video make this one of the best apps available for learning ASL [5].

To sum up, mobile apps on the market mainly focus on presenting lessons in video form. However, their quizzes only require the user to translate sign language content by selecting the correct answer. They do not test the user's ability to show signs, i.e. they do not engage the user in practicing signing on their own. Adding such a feature would require implementing a module in the app to recognize sign language characters shown to the phone's camera.

The MediaPipe library combined with common machine learning algorithms is described in [9]. For the algorithm selection, the authors consider different sign language fingerspelling datasets: American, Indian, and Italian for the alphabet and American and Turkish for numbers. Among all the solutions, the best classification accuracy—over 99% for all the datasets was obtained for Support Vector Machine (SVM).

1.2 *The Aims of the Study*

The study aims to develop a solution to recognize Polish sign language symbols in an image shown to a mobile phone camera. The algorithm should be resistant to background and lighting changes. In addition, the target device on which the algorithm is to run is the user's cell phone, which imposes limitations on memory usage and computing capabilities. After analyzing similar solutions used for American sign language, we decided to use the MediaPipe library and assess whether a classification based on the obtained locations of feature points would be possible as an alternative to the computationally demanding and unreliable analysis of whole hand images.

Based on the experiments described in [9], we consider the three most promising algorithms described by the authors: SVM, K-nearest neighbors (KNN), and random forest (RF).

2 Materials and Methods

2.1 Dataset

The dataset used to generate training and testing data consists of 29,945 images containing various signs used in PSL words and phrases: numbers, letters (see Fig. 1), and additional characters. The photos were recorded with cell phone cameras of people aged 21–80 with varying degrees of experience with PSL (but all experienced). The image set was divided into 29 groups corresponding to the meaning in sign language. Each class was assigned a numerical label (0–28), according to Table 1. Characters where the fingers are placed similarly (e.g., the letter "O" and the number "0") have been combined into one class.

The signs that have been selected have the following characteristics:

- require the use of only one hand,
- do not require movement, or it is insignificant,
- are often used in PSL words and phrases.

2.2 Data Pre-processing

The data pre-processing scheme is presented in Fig. 3. The first step utilizes the MediaPipe library to determine the 21 hand landmarks. To minimize the probability of incorrectly detected hands and, as a result, incorrect data sent to the training set, the minimum detection confidence for the MediaPipe hand detector was set to 0.7. It resulted in reducing the dataset, yet the cardinality of individual sets still ranged from 800 to 1200 cases. The MediaPipe library pre-normalizes the point coordinates to the dimensions of the input image—its width and height [8]. However, the model may predict a specific landmark outside the area of the analyzed image. Therefore, it was necessary to normalize them. For this, the detected landmarks' coordinates (x,y) are re-scaled to the range (0,1).

Exemplary detection results are shown in Figs. 4 and 5.

Table 1 Selected character classes with numerical labels

Sign from PSL	Class label	Sign from PSL	Class label
Number 0, letter O	0	Letter M	15
Number 1	1	Letter N	16
Number 2	2	Letter P	17
Number 3	3	Letter R	18
Number 4	4	Letter S	19
Number 5	5	Letter U	20
Letter A	6	Letter W	21
Letter B	7	Letter Y	22
Letter C	8	Additional character Aw	23
Letter D	9	Additional character Bk	24
Letter E	10	Additional character Cm	25
Letter F and T	11	Additional character Ik	26
Letter H	12	Additional character Om	27
Letter I and J	13	Additional character Um	28
Letter L	14		

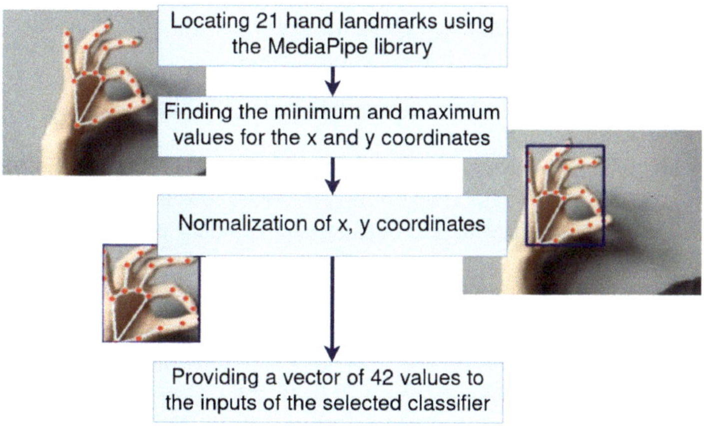

Fig. 3 Input data processing scheme

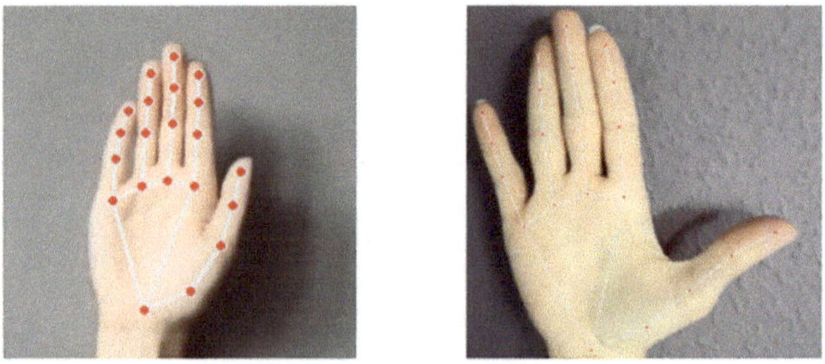

The "B" character in PSL The "Bk" character in PSL

Fig. 4 An example of the letter B and the additional character Bk with marked landmarks

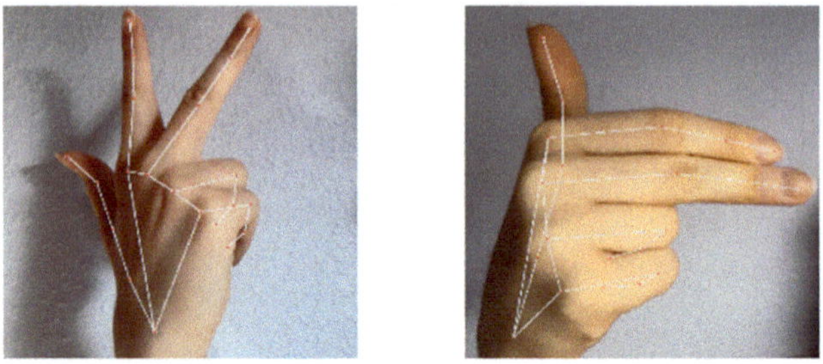

The "3" character in PSL The "N" character in PSL

Fig. 5 An example of the number 3 and the letter N with marked landmarks

2.3 Selected Classification Algorithms

Three widely described in literature classifiers were considered the best solution for the main classification step.

Support vector machine targets to find the best hyperplane separating classes in multidimensional feature space. The optimal hyperplane is defined by the maximum margin—the largest distance to the nearest training-data point of any class [10]. For the non-linearly separable data, the authors introduced two modifications of the fundamental optimization problem. The first one utilizes the kernel trick to shift the analysis into a feature space with higher dimensionality, whereas the second one defines an additional penalty term for classifying the data with errors. SVM classifier is widely described in the literature for sign language recognition [9, 11, 12]. Ali et al. [12] utilize SIFT features and SVM for Arabic sign language recognition. Reventh et al. [11] apply SVM for Indian sign language recognition, whereas Halder et al. [9] test its accuracy in different sign language analyses.

The K-nearest neighbor algorithm assumes that given instances in a dataset will usually occur near objects with similar properties [13, 14]. The class of a new object can therefore be predicted based on the classes of that object's k-nearest neighbors by voting. It was first used in ASL character recognition in 2012, in a paper by Pansare et al. [15].

The random forest algorithm can be used to solve classification and regression problems. It is based on combining multiple decision trees into different subsets of a given dataset, taking predictions from each tree instead of relying on a single one. The outcome is predicted based on the majority vote of the predictions. The algorithm has already been used in the analysis of Indian sign language by Potluri et al. [16], ASL by Bajaj et al. [17], or Chinese sign language by Yuan et al. [18], among others.

2.4 Parameters and Implementation

The normalized data provided by the MediaPipe library and their assigned labels are then saved to a .csv file, greatly facilitating their subsequent use for learning models. All the components of the algorithm were implemented in Python language, in environments Spyder and Google Colab.

For the input of each classifier, 42 values are given (coordinates x, y from 21 characteristic points, visible in Fig. 2). Each method classification accuracy was assessed in k-fold cross-validation (k=10). In addition, we observed which classes were most often misclassified and measured the classification time.

Based on a literature review, a set of the most promising parameters was defined for each of the considered methods. The radial basis function ($K(\mathbf{x}, \mathbf{x}') = exp(-\gamma||\mathbf{x} - \mathbf{x}'||^2)$) was selected as kernel function in SVM classifier with $\gamma = \{0.01, 0.1, 0.5, 0.8, 1\}$ and penalty C regulating the error term, equal to $C = \{1, 10, 100, 1000\}$. For the kNN classifier, the parameter k, relating to the number of neighbors, was chosen from the set $k = \{3, 5, 7\}$. All the parameters were evaluated, and the next section summarizes the best-obtained results.

3 Results and Discussion

The best classification results (accuracy, precision, recall, f1-score), provided by the following parameters: $\gamma = 0.8$ and $C = 10$ for SVM, and $k = 3$ for kNN algorithm, as well as classification times for each machine learning algorithm are presented in Table 2.

The highest values were obtained for the SVM classifier (accuracy equal to 98.62%). A bit worse results can be observed for the kNN and RF algorithms (97.66% and 97.74%, respectively). In each method, only single cases were classified incorrectly.

The multiple proportions test has been computed to verify whether the obtained accuracies difference is statistically significant. The obtained p value ($p<0.0001$) shows a significant effect. However, the Crammers V $= 0.03107(0.02416 - 0.03799)$

Table 2 Comparison of selected classifiers

ML algorithm	Accuracy [%]	Precision [%]	Recall [%]	f1-score [%]	time [s]
SVM	98.62	98.64	98.62	98.62	1.66
kNN classifier	97.66	97.69	97.66	97.66	0.63
Random Forest classifier	97.74	97.77	97.74	97.74	0.09

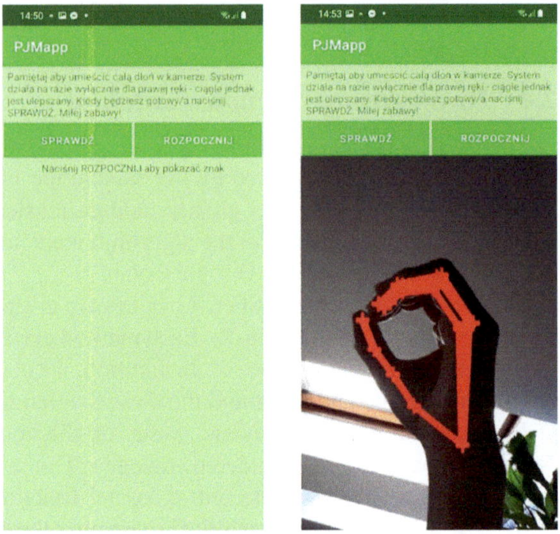

Fig. 6 Screenshot of the application with the PSL recognition system

indicates that the effect size is small. Therefore it should not be considered practically important.

Due to the high accuracy and the shortest classification time, the random forest classifier was considered the best for implementation. It became part of a mobile application created in the Android Studio environment to enable recognition of characters shown to a cell phone camera (in real-time), as shown in Fig. 6.

4 Conclusion

The study proved that using common machine learning techniques utilizing the locations of hand landmarks could provide a convenient solution for recognizing Polish sign language characters in the mobile application. Since the classification was based on a much smaller number of features than in the case of analyzing whole images, the proposed solution is low computationally demanding. The normalization of coordinate values reduces the influence of margins or hand positioning on the image (described in [9]), and using the MediaPipe library ultimately decreases the problem of background color or lighting problem.

All the considered algorithms provide accurate classification results, and the statistical analysis indicates that the differences should not be reported as having practical importance due to the small effect size.

The MediaPipe library used in the project is available for Python, JavaScript, and Java. In addition, the developers offer a preview of sample solutions running on the Android Studio platform. The supported configuration options offer, among others, point detection on a static (photo) or dynamic image. Locating points works much better when the model can support itself on predictions from the previous frame. The implementation, therefore, took advantage of the ability to operate on a live image captured from the user's camera. Testing of the algorithm showed that locating hands in the image and finding landmarks using the MediaPipe library is highly reliable.

While evaluating selected classifiers, additional attention was paid to ensure that they had adequate runtime and accuracy and could be easily implemented inside the mobile application. The selected random forest classifier meets all the requirements, proving that it is possible to create a sign language recognition system based only on information about the location of hand landmarks obtained from the MediaPipe library.

References

1. Hear-it (2023). https://www.hear-it.org/hearing-loss-in-different-countries
2. Rutkowski, P., Łozińska, S.: Linguistics of space and movement. In: Sign Communication and Corpus Methods, Warsaw (2014) (in Polish)
3. Kosiba, O., Grenda, P.: Sign language lexicon. In: Silentium, Bogatynia (2011) (in Polish)
4. West Pomeranian branch of the Polish Association of the Deaf (2023). https://www.pzg.szczecin.pl/kurs-jezyka-migowego/polski-alfabet-palcowy/
5. The ASL App (2021). https://theaslapp.com/
6. Spread Signs (2021). https://play.google.com/store/apps/details?id=com.spreadthesign.androidapp_paid&hl=pl&gl=US
7. Migaj z nami (2021). https://play.google.com/store/apps/details?id=mobi.qrtag.migajznami&hl=pl&gl=US
8. Hands Mediapipe (2021). https://google.github.io/mediapipe/solutions/hands.html
9. Halder, A., Akshit, T.: Real-time vernacular sign language recognition using MediaPipe and machine learning. Journal homepage: www. ijrpr.com (2021) ISSN 2582-7421
10. Cortes, C., Vapnik, V.: Support-vector networks. Mach. Learn. **20**, 273–297 (1995). https://doi.org/10.1007/BF00994018
11. Revanth, K., Raja, N.S.M.: Comprehensive SVM based Indian sign language recognition. In: 2019 IEEE International Conference on System, Computation, Automation and Networking (ICSCAN), Pondicherry, India, 2019, pp. 1–4. https://doi.org/10.1109/ICSCAN.2019.8878787
12. Ali, M.A., Ewis, M.R., Mohamed, G.E., Ali, H.H., Moftah, H.M.: Arabic sign language recognition (ArSL) approach using support vector machine. In: 2017 27th International Conference on Computer Theory and Applications (ICCTA), Alexandria, Egypt, 2017, pp. 17–21. https://doi.org/10.1109/ICCTA43079.2017.9497164
13. Adeyanju, I.A., Bello, O.O., Adegboye, M.A.: Machine learning methods for sign language recognition: a critical review and analysis. Intell. Syst. Appl. **12**, 200056 (2021). https://doi.org/10.1016/j.iswa.2021.200056

14. Fix, E., Hodges, J.L.: Discriminatory analysis. In: Consistency Properties, Nonparametric Discrimination (1951)
15. Pansare, J.R., Gawande, S.H., Ingle, M.: Real-time static hand gesture recognition for American sign language (ASL) in complex background. J. Signal Inf. Process. **3**(03), 364–367. https://doi.org/10.4236/jsip.2012.33047
16. Potluri, A. S. A., George, S.M., Gaurav, R., Anusri, S.: Indian sign language recognition using random forest classifier. In: 2021 IEEE International Conference on Electronics, Computing and Communication Technologies (CONECCT), Bangalore, India, 2021, pp. 1–6. https://doi.org/10.1109/CONECCT52877.2021.9622672
17. Bajaj, Y., Malhotra, P.: American sign language identification using hand trackpoint analysis (2020). https://doi.org/10.48550/arxiv.2010.10590
18. Yuan, S., Wang, Y., Wang, X., Deng, H., Sun, S., Wang, H., Huang, P., Li, G.: Chinese sign language alphabet recognition based on random forest algorithm. In: IEEE International Workshop on Metrology for Industry 4.0 & IoT, Roma, Italy, 2020, pp. 340–344 (2020). https://doi.org/10.1109/MetroInd4.0IoT48571.2020.9138285
19. Taskiran, M., Killioglu, M., Kahraman, N.: A real-time system for recognition of American sign language by using deep learning. In: 2018 41st International Conference on Telecommunications and Signal Processing (TSP), Athens, Greece, pp. 1–5 (2018). https://doi.org/10.1109/TSP.2018.8441304
20. Tolentino, L.K.S., Juan, S.R., Thio-ac, A.C., Pamahoy, M.A.B., Forteza, J.R.R., Garcia X.J.O.: A static sign language recognition using deep learning. Int. J. Mach. Learn. Comput. **9**, 821–827 (2019). https://doi.org/10.18178/ijmlc.2019.9.6.879
21. Kishore, P.V.V., Prasad, M.V.D., Prasad, C.R., Rahul, R.: 4-camera model for sign language recognition using elliptical Fourier descriptors and ANN. In: Proceedings of the International Conference on Signal Processing and Communication Engineering Systems, SPACES 2015, in Association with IEEE, pp. 34–38. https://doi.org/10.1109/SPACES.2015.7058288
22. Islam, M.M., Siddiqua, S., Afnan, J.: Real time hand gesture recognition using different algorithms based on American sign language. In: Proceedings of the IEEE International Conference on Imaging, Vision and Pattern Recognition, IcIVPR 2017: 13, pp. 1031–1036 (2017). https://doi.org/10.1109/ICIVPR.2017.7890854

Colour Clustering and Deep Transfer Learning Techniques for Breast Cancer Detection Using Mammography Images

Hosameldin O. A. Ahmed and Asoke K. Nandi

Abstract Breast cancer is a major global health concern affecting millions of women each year. Computer-aided diagnosis (CAD) systems have the potential to contribute significantly to early detection and reducing the mortality rate of breast cancer. This paper proposes a new methodology for breast cancer detection utilising data analytics, artificial intelligence, and mammograms. The approach is a mixed methodology based on colour clustering and deep transfer learning techniques to extract features from mammogram images. The proposed method was validated using the mini-DDSM mammogram images dataset, and its effectiveness was evaluated using various metrics such as accuracy, specificity, precision, recall, and F1 score. The results showed that all networks had high detection accuracy, with GoogleNet achieving the highest (99.58%) and ShuffleNet the lowest (97.08%). The proposed method achieved 100% detection accuracy using ResNet18, VGG16, ShuffleNet, DarkNet, and NasnetLarge, while Inception-ResNet-v2 had a detection accuracy of 98.33% with LRC and 99.17% with SVM. The proposed method has demonstrated the potential to improve the performance of CAD systems.

Keywords Breast cancer detection · Mammography images · Image clustering · Deep transfer learning

1 Introduction

Breast cancer is a major global public health issue that affects millions of women annually. In 2020, there were 2.3 million new cases and 685,000 deaths from breast cancer worldwide [1, 2]. Breast cancer is the most prevalent cancer among women,

H. O. A. Ahmed · A. K. Nandi (✉)
Department of Electronic and Electrical Engineering, Brunel University London, London UB8 3PH, UK
e-mail: asoke.nandi@brunel.ac.uk

H. O. A. Ahmed
OpenAITech LTD, Old Marylebone, London NW1 5RA, UK

P. Strumiłło et al. (eds.), *The Latest Developments and Challenges in Biomedical Engineering*, Lecture Notes in Networks and Systems 746,
https://doi.org/10.1007/978-3-031-38430-1_9

with 2.26 million cases in 2020, emphasizing the importance of continuous efforts to raise awareness, improve early detection, and develop effective treatment and care for breast cancer patients globally. Early detection is critical in improving treatment outcomes and preventing the spread of cancer and potentially life-threatening complications [3]. Various imaging techniques, including mammograms, breast thermography, magnetic resonance imaging, ultrasound, positron emission tomography, histopathology, and computed tomography [4–10], have been developed for breast cancer diagnosis, with mammograms being one of the most widely used methods for detecting breast cancer at an early stage [11].

Breast cancer diagnosis is a challenging process that often relies on manual techniques, leading to potential errors, particularly in detecting early signs of the disease on mammograms [12]. Screening programs may miss up to 35% of cancers due to factors such as errors or undetectable cancer [13]. AI-powered Computer-Aided Diagnostic (CAD) systems have shown promise in improving breast cancer detection rates and reducing mortality rates [14–16]. However, further research and development are needed to improve CAD system performance for early detection of breast cancer. In this paper, a new methodology for breast cancer diagnosis utilising data analytics, artificial intelligence, and mammograms is proposed. The methodology is based on a mixed methodology of colour clustering and deep transfer learning techniques to learn features from mammogram images. These features can provide valuable information on the anatomy, physiology, and biology of the disease or condition, leading to improved accuracy and precision of diagnosis, prediction of clinical outcomes and treatment response, and identification of treatment options for individual patients.

This study aims to investigate whether a pre-processing step of clustering and colourizing clusters followed by transfer learning using a pre-trained network can enhance the accuracy of breast cancer diagnosis through classification algorithms. The study addresses a common challenge in medical image analysis: the requirement for a large amount of labeled data. Clustering and colourizing can help minimise the amount of labeled data necessary, while transfer learning can leverage pre-existing knowledge from pre-trained networks to improve the model's performance. The use of colourized clusters in medical imaging may have originated in part from the need to adapt pre-existing CNNs designed for 3-channel inputs, but its key function is to provide a more informative and visually distinctive representation of the image data for medical interpretation. This approach could potentially improve the precision and efficiency of breast cancer diagnosis, offering significant clinical and societal advantages.

The reminder of this paper is structured as follows: Section 2 provides an overview of the related work. Section 3 explains the proposed method in detail. Section 4 outlines the experimental study conducted to validate the proposed framework. Section 5 presents a discussion on the experimental results. Finally, Section 6 provides concluding remarks.

2 Related Work

Conventional approaches to diagnosing breast cancer involve manual examination by expert radiologists or pathologists, which have limitations. In recent years, AI-powered tools have been developed to detect breast cancer through image recognition, and they have demonstrated superior performance compared to conventional CAD systems and manual examinations [17]. Various computer-aided support systems have been developed to improve the performance of CAD using mammograms. Some of these systems involve image preprocessing and classification techniques for breast cancer diagnosis [18]. Other studies have explored the use of machine learning, deep learning, and Deep Reinforcement Learning for the detection of breast cancer in various medical imaging modalities [11]. One study proposed a CAD system that uses an artificial neural network (ANN) to assist in the diagnosis of masses, classifying them as benign or malignant [19]. The system achieved a sensitivity of 92.6% and specificity of 93.3%, with an overall classification accuracy of 92.9%. A study analyzed nine techniques for classifying breast masses in digital mammography. The best feature performance was shown by the Gray-Level Run-Length Matrices (GLRLM) descriptor [20]. Mavroforakis et al. explored textural features and fractal dimension analysis for mammogram images. Linear, neural, and SVM classifiers were used to classify masses, with promising results found in the dataset's quality and information content [21]. An automated confirmatory system (ACS) was proposed to support radiologists in mammography screening. Pre-processing and segmentation techniques generate a high-quality filtered image with texture-based diagnosis. The system employs rough-set approach and ANN to classify mammograms as normal, benign, or malignant, with the ANN classifier showing high performance (sensitivity, specificity, and accuracy of 0.98, 0.89, and 0.96) [22]. A deep convolutional neural network and extreme learning machine technique achieved high accuracy for detecting normal/abnormal and benign/malignant cases in mammogram datasets. The improved crow-search optimized extreme learning machine (ICS-ELM2) classifier using ResNet18 features provided the highest classification accuracy [23]. The proposed Breast Mass Classification (BMC) technique utilises a combination of k-means clustering, recurrent neural network (RNN), convolutional neural network (CNN), random forest, and boosting techniques to classify breast mass into benign, malignant, and normal, achieving high sensitivity, specificity, F-measure, and accuracy rates [24]. Sun et al. proposed a modified CNN architecture called multi-view CNN for mammographic image classification, which uses dilated convolutional layers to extract features from different views of mammography and multi-scale features. The proposed method is compared to state-of-the-art methods on the DDSM and MIAS datasets [25]. Another technique presented in the literature is a conditional Generative Adversarial Network (cGAN) for breast tumor segmentation in mammograms, which achieved an overall accuracy of 80% using a CNN-based shape descriptor to classify generated masks into four tumor shapes [26]. Moreover, an ensemble of classifiers for breast cancer recognition was presented, using advanced techniques for image representation, and applying a weighted majority voting approach for ensemble integration, achieving superior accuracy in detecting abnormal cases from a database of over 10,000 mammograms [27]. In [28], a technique based on CNN for classifying mammogram images into benign, malignant, and normal was proposed, achieving high accuracy rates.

One common approach for developing deep learning-based CAD systems is transfer learning, where pre-trained convolutional neural network (CNN) models are used as feature extractors. Several studies have utilised transfer learning-based approaches for detecting and classifying breast cancer in medical images. For instance, GoogLeNet, VGGNet, and ResNet were used as pre-trained CNN models to extract features from breast cytology images for the classification of malignant and benign cells [29]. Falconí et al. proposed a transfer learning-based approach for classifying mammogram images using the BI-RADS scale in the INbreast dataset, achieving an accuracy of 90.9% [30]. Zhang et al. utilized transfer learning and data augmentation for mammogram and tomosynthesis classification, achieving promising results with multiple CNN models [31]. Bagchi et al. compared the performance of ResNet18, ResNet50, and ResNet101 for classifying pre-processed mammograms into normal, benign, and malignant categories, achieving an accuracy of 79.27% with the ResNet50 model [32]. Arevalo et al. presented a supervised framework for classifying mass lesions in mammography images using CNNs, achieving an AUC performance of 0.822 [33]. A CNN-SVM hybrid framework was proposed for image classification, achieving an accuracy of 93.35% [34]. In [35], a customized deep CNN was proposed for breast cancer detection and classification, achieving an accuracy of 88.7%. Muduli et al. proposed a CNN model for breast cancer classification using mammograms and ultrasound images, achieving high accuracy on multiple datasets with data augmentation techniques [36]. Furthermore, five deep hybrid CNN-based frameworks were presented for breast cancer detection in [37], achieving high accuracy rates for abnormality and malignancy detection with mini-DDSM (mammogram), BUSI, and BUS2 (ultrasound) datasets.

3 Methodology

The methodological approach taken in this study is a mixed methodology based on colour clustering and deep transfer learning techniques to learn features from mammogram images. These features can provide important information about the anatomy, physiology, and biology of a disease or condition and consequently may improve the accuracy and precision of diagnosis, to predict clinical outcomes and treatment response. By performing pre-processing and colour clustering procedures on mammogram images we produce clustered and coloured images that may offer distinct regions that make it easier to differentiate between different regions in mammogram images and accordingly may produce different features on mammograms for each health condition of the breast. In deep transfer learning, a pre-trained model is typically a deep neural network that has been trained on a large dataset such as ImageNet for image recognition tasks [38]. The pre-trained network has already learned to recognise a wide range of features from the data it was trained on, and these learned features can be very useful for recognising similar features in a new dataset. To use the pre-trained model for feature learning in a new task, the last layer or layers of the pre-trained model are removed, and the remaining layers are used to extract features from the new dataset [39]. In our case the new dataset contains the produced clustered and coloured images. These extracted features are used as inputs to train a new model on the specific task to be performed, which is the detection of the breast cancer using the mammogram images dataset in our case study. The benefit of using deep transfer learning for feature learning is that the pre-trained model has already learned to recognise a wide range of features that are useful for many different tasks [40]. This means that the new model can be trained more quickly and effectively, using fewer data, than if it were trained from scratch. In our proposed method, we used deep transfer learning to extract features from the produced clustered and coloured images of the original mammogram images. Figure 1 presents the proposed method. In the first stage [see Fig. 1], a two steps technique is utilised to learn features from the mammograms images as follows: (a) Mammogram image pre-processing and colour clustering to produce clustered and coloured images from the original mammogram images, and (b) the use of deep transfer learning to learn features from the obtained clustered and coloured images. The following subsections describe these two steps for feature learning in detail.

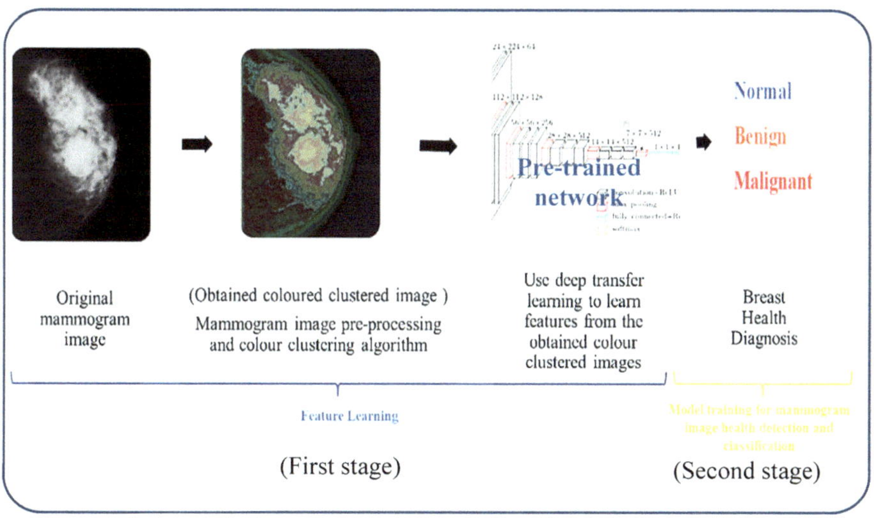

Fig. 1 The proposed method

3.1 Mammogram Image Pre-Processing and Colour Clustering Algorithm

In our proposed technique a pre-processing and colour clustering process is performed [see Fig. 2] on mammogram images. The process begins by inputting the original mammogram image, which is then converted to grayscale to simplify the image and makes it easier to perform further processing. Next, the image is flattened into a 1-dimensional array, and k-means clustering [41] is performed on this flattened image by grouping similar pixels together based on their RGB values. The k-means clustering algorithm divides the image into distinct regions, making it easier to visualise and analyse. The clusters are then assigned different colours, and these coloured pixels are overlaid onto the original image, which can effectively highlight colour variations in the image. The colour assignment makes it easier to distinguish between different regions. The resulting image shows a simplified version of the tissue colour, which can help in identifying areas of interest and potentially aiding in the diagnosis. The final output is a clustered and coloured image, where each cluster is represented by a different colour. Nevertheless, the current state-of-the-art deep learning methods for mammography image analysis focus on finding regions of interest (ROIs) with lesions and categorising them as malignant or benign, our approach utilises the k-means clustering procedure to divide the image into different regions, which can help in identifying areas of interest. The proposed method focuses on identifying patterns and structures within the image and provides a simplified, clustered, and coloured version of the mammography image. This method can be useful in medical image analysis and useful in supporting the diagnosis process, as it can help to identify patterns and structures within the image that may not be easily visible in the original grayscale image.

3.2 Use of Deep Transfer Learning to Learn Features from the Obtained Clustered and Coloured Images

To learn features from the obtained clustered and coloured images. pre-trained models can be used as a starting point for a new image recognition task, allowing the transfer of knowledge gained from the large dataset to the new task. The idea behind using a pre-trained model is that the lower layers of the network have learned general features, such as edges and simple shapes, which can be used to recognise objects in a wide range of images. By reusing these learned features, our trained model can focus on learning task-specific features in the higher layers of the network, which can lead to faster and more accurate training of the model. Using a pre-trained model for transfer learning in image recognition tasks has several benefits, including, reduced training time, improved accuracy, and increased generalisation. There are several popular pre-trained models available for use in image recognition tasks such as VGG16 and

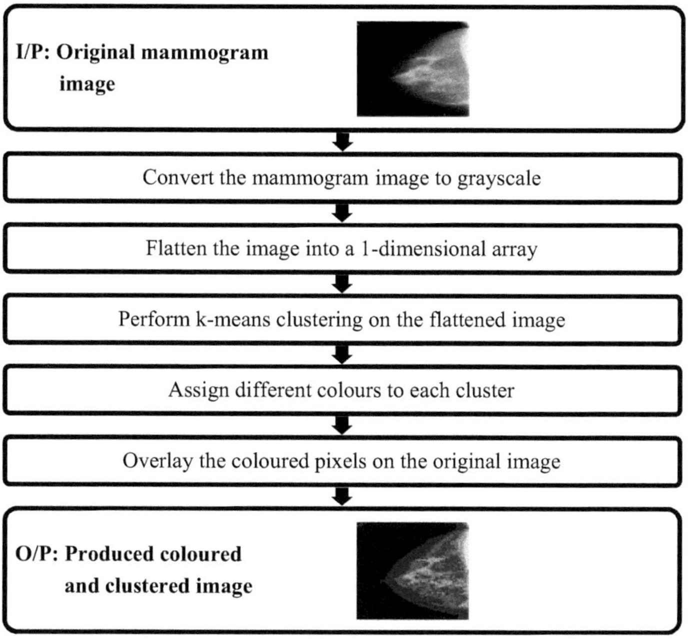

Fig. 2 The colour clustering process

ResNet18. These models can be fine-tuned by adding new layers to the network and training them on a smaller dataset specific to the new task.

3.3 Breast Health Diagnosis

In the second stage [see Fig. 1], the obtained features from the process in subsection 3.2 above are used to train a detection model using a classification algorithm. Then, the trained model can be used to detect and identify breast health conditions.

3.4 Dataset

For this case study, the mini-DDSM dataset [42] that contains 2620 cases of women's breasts and 9684 mammogram images in 16-bit "png" format is used to validate our proposed method. It includes CC and MLO views of both the right and left breast, with ground truth data curated by experts and classified as malignant, benign, or normal. In this study, the following subsets are used:

1. **Dataset A**: A subset comprising 1200 mammogram images (Normal—600, Abnormal—600, which includes malignant—300, and benign—300) is extracted randomly from the mini-DDSM dataset to examine the performance of the proposed method in detecting Abnormality condition of the mammogram image.
2. **Dataset B**: A subset containing 600 mammogram images (Malignant—300 and Benign—300) is extracted randomly to validate the performance of the proposed method in detecting the malignancy condition of the mammogram image.

4 Experiments and Results

In this study, the performance of a proposed method for detecting abnormality and malignancy conditions in mammogram images was examined. We conducted several experiments to evaluate the method's effectiveness. The dataset used in the study was randomly divided into 80% for training and 20% for testing the models. We began by converting the original mammograms to grayscale images to simplify the images and facilitate processing. The grayscale images were flattened into one-dimensional arrays to serve as input for the k-means clustering algorithm. The k-means clustering algorithm was used to group similar pixels based on their RGB values. The number of clusters used in the clustering process was set to 30, which was found to be appropriate for achieving high detection accuracy. The clustering process involved assigning each pixel to the cluster whose mean RGB values were closest to it. The clustering process was repeated until the mean RGB values of each cluster no longer changed significantly. The clustering process resulted in a clustered and coloured mammogram image as the final output. The colours assigned to each cluster were generated using the "viridis" colourmap from the Matplotlib library [43]. Examples of the obtained coloured and clustered images for images with normal, malignant, and benign conditions are shown in Fig. 3.

The obtained clustered and coloured images were then used to learn features using deep transfer learning techniques. Eight pre-trained networks, including ResNet18, Inception-ResNet-v2, AlexNet, VGG16, GoogleNet, ShuffleNet, DarkNet19, and NasNet Large, were examined in this study [44–48]. To carry out the detection tasks, the authors employed two different classifiers: logistic regression classifier (LRC) and support vector machine (SVM) [49, 50]. Sixty trials were conducted for each classifier, and the results were averaged to obtain the detection accuracy levels for each experiment. The performance of the proposed method is assessed through validation measures including detection and classification accuracy, precision, recall, specificity, and F1 score. The subsequent subsections provide a detailed description of the results obtained.

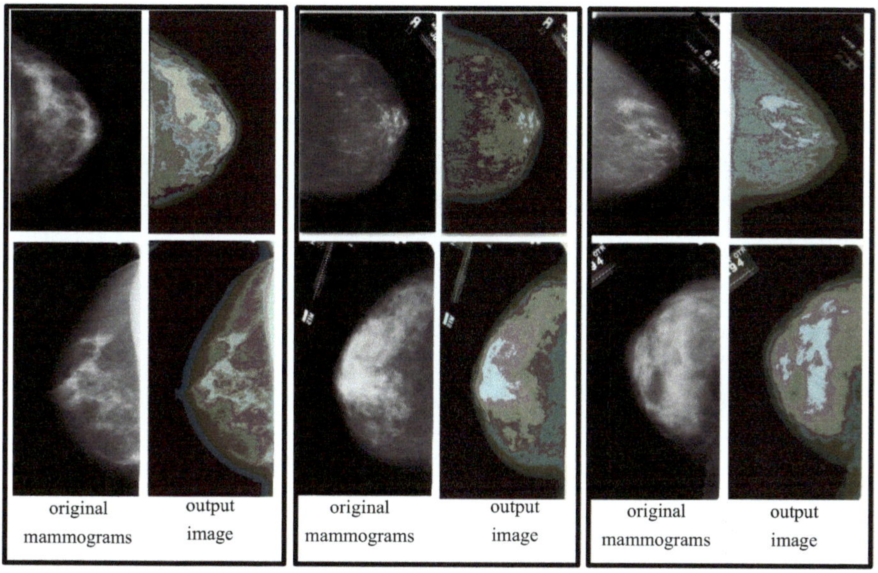

(1) Normal condition) (2) Malignant condition (3) Benign condition

Fig. 3 Examples of the obtained coloured and clustered images using the feature learning technique of our proposed method

4.1 Abnormality Detection

We conducted multiple experiments to evaluate the performance of the proposed method with LRC and SVM in detecting abnormality in mammogram images using our proposed method and Dataset A. The results of these experiments are presented in Tables 1 and 2, respectively.

Table 1 The results of detecting abnormality in mammograms using our proposed method with LRC and Dataset A

Pre-trained networks	Detection accuracy (%)	Precision	Recall	Specificity	F1 score
ResNet18	99.14 ± 0.6	99.14	99.15	100	99.14
Inception-ResNet-v2	97.92 ± 1.6	97.92	97.92	100	97.92
AlexNet	97.50 ± 2.5	97.50	97.62	100	97.56
VGG16	98.35 ± 1.6	98.35	98.41	100	98.39
GoogleNet	99.58 ± 0.4	99.58	99.59	100	99.59
ShuffleNet	97.08 ± 2.1	97.08	97.17	100	97.12
DarkNet	97.50 ± 2.3	97.5	97.51	100	97.51
NasnetLarge	98.33 ± 1.6	98.33	98.33	100	98.33

Table 2 The results of detecting abnormality in mammograms using our proposed method with SVM and Dataset A

Pre-trained networks	Detection accuracy (%)	Precision	Recall	Specificity	F1 score
ResNet18	99.17 ± 0.8	99.17	99.18	100	99.17
Inception-ResNet-v2	97.50 ± 2.0	97.50	97.62	100	97.56
AlexNet	95.83 ± 4.1	95.83	96.15	100	95.99
VGG16	97.52 ± 2.5	97.52	97.53	100	97.53
GoogleNet	97.50 ± 2.5	97.50	97.62	100	97.56
ShuffleNet	98.75 ± 1.2	98.75	98.75	100	98.75
DarkNet	99.15 ± 0.7	99.15	99.16	100	99.16
NasnetLarge	98.75 ± 0.9	98.75	98.78	100	98.77

Table 3 The results of detecting malignancy in mammograms using our proposed method with LRC and Dataset B

Pre-trained networks	Detection accuracy (%)	Precision	Recall	Specificity	F1 score
ResNet18	100.00 ± 0.0	100.00	100.00	100.00	100.00
Inception-ResNet-v2	98.33 ± 1.61	98.33	98.39	100.00	98.36
AlexNet	96.67 ± 3.3	96.67	96.72	100.00	96.69
VGG16	100.00 ± 0.0	100.00	100.00	100.00	100.00
GoogleNet	95.00 ± 0.5	95.00	95.05	100.00	95.03
ShuffleNet	100.00 ± 0.0	100.00	100.00	100.00	100.00
DarkNet	100.00 ± 0.0	100.00	100.00	100.00	100.00
NasnetLarge	100.00 ± 0.0	100.00	100.00	100.00	100.00

4.2 *Malignancy Detection*

Multiple experiments were carried out to assess the effectiveness of the proposed approach, LRC and SVM, in detecting malignancy in mammogram images utilising Dataset B. The outcomes of these experiments are displayed in Tables 3 and 4, respectively.

5 Discussion

Tables 1 and 2 validate the effectiveness of our proposed method with LRC and SVM in detecting abnormalities in mammograms as proposed in this study. Table 1 shows high detection accuracy for all networks, with GoogleNet having the highest (99.58%) and ShuffleNet the lowest (97.08%). Specificity is perfect (100%) for all networks. Precision, recall, and F1 scores vary, with ResNet18 achieving the highest

Table 4 The results of detecting malignancy in mammograms using our proposed method with SVM and Dataset B

Pre-trained networks	Detection accuracy (%)	Precision	Recall	Specificity	F1 score
ResNet18	100.00 ± 0.0	100.00	100.00	100.00	100.00
Inception-ResNet-v2	99.17 ± 0.8	99.17	99.18	100.00	99.17
AlexNet	93.33 ± 0.6	93.33	93.38	100.00	93.36
VGG16	97.50 ± 2.3	97.50	97.51	100.00	97.51
GoogleNet	96.67 ± 3.3	96.67	96.72	100.00	96.69
ShuffleNet	99.58 ± 0.4	99.58	99.59	100.00	99.59
DarkNet	100.00 ± 0.0	100.00	100.00	100.00	100.00
NasnetLarge	100.00 ± 0.0	100.00	100.00	100.00	100.00

accuracy and F1 score in Table 2 (99.17%). All models have high precision, recall, and F1 score, indicating good performance in identifying positive samples and avoiding false positives.

Tables 3 and 4 show the results of detecting malignancy in mammograms using a proposed method with various pre-trained networks and SVM on Dataset B. The proposed method achieved 100% detection accuracy using ResNet18, VGG16, ShuffleNet, DarkNet, and NasnetLarge with both LRC and SVM. Inception-ResNet-v2 had a detection accuracy of 98.33% with LRC and 99.17% with SVM. AlexNet had a detection accuracy of 96.67% with LRC and 93.33% with SVM. GoogleNet had a detection accuracy of 95% with LRC and 96.67% with SVM. The precision, recall, specificity, and F1 score were also reported.

Table 5 compares the proposed method with recently published results [37] to verify its efficiency. Results show that our proposed method with ResNet18 and LRC achieved the highest accuracy of 99.58% for abnormality detection, but slightly lower for malignancy detection compared to other methods. On the other hand, the proposed method with GoogleNet and SVM achieved 100% accuracy for malignancy detection and 99.17% for abnormality detection, which is similar to the reported results in reference [37]. These results demonstrate the effectiveness of combining deep learning and traditional machine learning approaches for medical image analysis.

Table 5 A comparison with the results from literature mini-DDSM dataset of breast cases of women's breasts

Ref	Method (%)	Abnormality (%)	Malignancy (%)
[37]	AlexNet	98.33	96.25
	VGG16	98.33	97.5
	ResNet18	99.17	97.5
	ShuffleNet	99.17	97.5
	Hybrid1	97.5	97.5
	Hybrid2	98.33	97.5
	Hybrid3	99.17	97.5
	Hybrid4	99.17	97.5
	Hybrid5	99.17	98.75
*	The proposed method with GoogleNet and LRC	99.58 ± 0.4	95.00 ± 0.5
	The proposed method with ResNet18 and SVM	99.17 ± 0.8	100.00 ± 0.0

6 Conclusion

This paper proposed a new methodology for the detection of breast cancer using data analytics, artificial intelligence, and mammograms. The proposed method employs a mixed methodology based on colour clustering and deep transfer learning techniques to learn features from mammogram images. Eight pre-trained networks are examined in this study, and the results show that ResNet18, Googlenet, and AlexNet achieved the highest accuracy in the test set. The proposed method achieved 100% detection accuracy using ResNet18, VGG16, ShuffleNet, DarkNet, and NasnetLarge with both LRC and SVM. The precision, recall, specificity, and F1 score were also reported. In general, the proposed method achieved high accuracy and F1 scores for detecting abnormality and malignancy in mammograms. The pre-trained networks used in the experiments showed high precision and recall for both LRC and SVM, with ResNet18 and GoogleNet achieving the best results. These findings suggest that the proposed method with pre-trained deep neural networks and SVM classifiers can be an effective approach for the automatic detection of abnormality and malignancy in mammograms.

References

1. WHO webpage, https://www.who.int/news-room/fact-sheets/detail/breast-cancer, last accessed 2023/03/04
2. WHO webpage, https://www.who.int/news-room/fact-sheets/detail/cancer, last accessed 2023/03/04
3. Youlden, D.R., Cramb, S.M., Dunn, N.A., Muller, J.M., Pyke, C.M., Baade, P.D.: The descriptive epidemiology of female breast cancer: An international comparison of screening, incidence, survival and mortality. Cancer Epidemiol. **36**(3), 237–248 (2012)

4. Heath, M., Bowyer, K., Kopans, D., Kegelmeyer, P., Moore, R., Chang, K., Munishkumaran, S.: Current status of the digital database for screening mammography. Digit. Mammogr.: Nijmegen **1998**, 457–460 (1998)
5. Moghbel, M., Mashohor, S.: A review of computer assisted detection/diagnosis (CAD) in breast thermography for breast cancer detection. Artif. Intell. Rev. **39**, 305–313 (2013)
6. Kuhl, C.K., Schrading, S., Strobel, K., Schild, H.H., Hilgers, R.D., Bieling, H.B.: Abbreviated breast magnetic resonance imaging (MRI): First postcontrast subtracted images and maximum-intensity projection—a novel approach to breast cancer screening with MRI. J. Clin. Oncol. 1 **32**(22), 2304–2310 (2014)
7. Lee, C.H., Dershaw, D.D., Kopans, D., Evans, P., Monsees, B., Monticciolo, D., Brenner, R.J., Bassett, L., Berg, W., Feig, S., Hendrick, E.: Breast cancer screening with imaging: Recommendations from the Society of Breast Imaging and the ACR on the use of mammography, breast MRI, breast ultrasound, and other technologies for the detection of clinically occult breast cancer. J. Am. Coll. Radiol. 1 **7**(1), 18–27 (2010)
8. Lavayssière, R., Cabée, A.E., Filmont, J.E.: Positron emission tomography (PET) and breast cancer in clinical practice. Eur. J. Radiol. 1 **69**(1), 50–58 (2009)
9. Veta, M., Pluim, J.P., Van Diest, P.J., Viergever, M.A.: Breast cancer histopathology image analysis: A review. IEEE Trans. Biomed. Eng. 30 **61**(5), 1400–1411 (2014)
10. Chang, C.H., Sibala, J.L., Fritz, S.L., Dwyer, S.J., 3rd., Templeton, A.W., Lin, F., Jewell, W.R.: Computed tomography in detection and diagnosis of breast cancer. Cancer **46**(4 Suppl), 939–946 (1980)
11. Dar, R.A., Rasool, M., Assad, A.: Breast cancer detection using deep learning: Datasets, methods, and challenges ahead. Comput. Biol. Med. 106073 (2022)
12. Madani, M., Behzadi, M.M., Nabavi, S.: The role of deep learning in advancing breast cancer detection using different imaging modalities: A systematic review. Cancers 29 **14**(21), 5334 (2022)
13. Freeman, K., Geppert, J., Stinton, C., Todkill, D., Johnson, S., Clarke, A., Taylor-Phillips, S.: Use of artificial intelligence for image analysis in breast cancer screening programmes: Systematic review of test accuracy. Bmj 374 (2021)
14. Topol, E.J.: High-performance medicine: the convergence of human and artificial intelligence. Nat. Med. **25**(1), 44–56 (2019)
15. Tang, J., Rangayyan, R.M., Xu, J., El Naqa, I., Yang, Y.: Computer-aided detection and diagnosis of breast cancer with mammography: Recent advances. IEEE Trans. Inf Technol. Biomed. **13**(2), 236–251 (2009)
16. Yassin, N.I., Omran, S., El Houby, E.M., Allam, H.: Machine learning techniques for breast cancer computer aided diagnosis using different image modalities: A systematic review. Comput. Methods Programs Biomed. **156**, 25–45 (2018)
17. McKinney, S.M., Sieniek, M., Godbole, V., Godwin, J., Antropova, N., Ashrafian, H., Back, T., Chesus, M., Corrado, G.S., Darzi, A., Etemadi, M.: International evaluation of an AI system for breast cancer screening. Nature **577**(7788), 89–94 (2020)
18. Jalalian, A., Mashohor, S.B., Mahmud, H.R., Saripan, M.I., Ramli, A.R., Karasfi, B.: Computer-aided detection/diagnosis of breast cancer in mammography and ultrasound: A review. Clin. Imaging **37**(3), 420–426 (2013)
19. Singh, L., Jaffery, Z.A.: Computer-aided diagnosis of breast cancer in digital mammograms. Int. J. Biomed. Eng. Technol. **27**(3), 233–246 (2018)
20. Chaieb, R., Kalti, K.: Feature subset selection for classification of malignant and benign breast masses in digital mammography. Pattern Anal. Appl. **22**, 803–829 (2019)
21. Mavroforakis, M.E., Georgiou, H.V., Dimitropoulos, N., Cavouras, D., Theodoridis, S.: Mammographic masses characterization based on localized texture and dataset fractal analysis using linear, neural and support vector machine classifiers. Artif. Intell. Med. **37**(2), 145–162 (2006)
22. Peng, W., Mayorga, R.V., Hussein, E.M.: An automated confirmatory system for analysis of mammograms. Comput. Methods Programs Biomed. **125**, 134–44 (2016)

23. Chakravarthy, S.S., Rajaguru, H.: Automatic detection and classification of mammograms using improved extreme learning machine with deep learning. Irbm. **43**(1), 49–61 (2022)
24. Malebary, S.J., Hashmi, A.: Automated breast mass classification system using deep learning and ensemble learning in digital mammogram. IEEE Access. **9**, 55312–55328 (2021)
25. Sun, L., Wang, J., Hu, Z., Xu, Y., Cui, Z.: Multi-view convolutional neural networks for mammographic image classification. IEEE Access. **7**, 126273–126282 (2019)
26. Singh, V.K., Rashwan, H.A., Romani, S., Akram, F., Pandey, N., Sarker, M.M., Saleh, A., Arenas, M., Arquez, M., Puig, D., Torrents-Barrena, J.: Breast tumor segmentation and shape classification in mammograms using generative adversarial and convolutional neural network. Expert Syst. Appl. **139**, 112855 (2020)
27. Swiderski, B., Osowski, S., Kurek, J., Kruk, M., Lugowska, I., Rutkowski, P., Barhoumi, W.: Novel methods of image description and ensemble of classifiers in application to mammogram analysis. Expert Syst. Appl. **81**, 67–78 (2017)
28. Gnanasekaran, V.S., Joypaul, S., Meenakshi Sundaram, P., Chairman, D.D.: Deep learning algorithm for breast masses classification in mammograms. IET Image Proc. **14**(12), 2860–2868 (2020)
29. Khan, S., Islam, N., Jan, Z., Din, I.U., Rodrigues, J.J.: A novel deep learning based framework for the detection and classification of breast cancer using transfer learning. Pattern Recogn. Lett. **125**, 1–6 (2019)
30. Falconí, L., Pérez, M., Aguilar, W., Conci, A.: Transfer learning and fine tuning in mammogram bi-rads classification. In: 2020 IEEE 33rd International Symposium on Computer-Based Medical Systems (CBMS), pp. 475–480 (2020)
31. Zhang, X., Zhang, Y., Han, E.Y., Jacobs, N., Han, Q., Wang, X., Liu, J.: Classification of whole mammogram and tomosynthesis images using deep convolutional neural networks. IEEE Trans. Nanobiosci. **17**(3), 237–242 (2018)
32. Bagchi, S., Mohd, M.N., Debnath, S.K., Nafea, M., Suriani, N.S., Nizam, Y.: Performance comparison of pre-trained residual networks for classification of the whole mammograms with smaller dataset. In: 2020 IEEE Student Conference on Research and Development (SCOReD), pp. 368–373 (2020)
33. Arevalo, J., González, F.A., Ramos-Pollán, R., Oliveira, J.L., Lopez, M.A.: Representation learning for mammography mass lesion classification with convolutional neural networks. Comput. Methods Programs Biomed. **127**, 248–257 (2016)
34. Jaffar, M.A.: Deep learning-based computer aided diagnosis system for breast mammograms. Int. J. Adv. Comput. Sci. Appl. **8**(7), 286–290 (2017)
35. Shakeel, S., Raja, G.: Classification of breast cancer from mammogram images using deep convolution neural networks. In: 2021 International Bhurban Conference on Applied Sciences and Technologies (IBCAST), IEEE, pp. 595–599 (2021)
36. Muduli, D., Dash, R., Majhi, B.: Automated diagnosis of breast cancer using multi-modal datasets: A deep convolution neural network-based approach. Biomed. Signal Process. Control **71**, 102825 (2022)
37. Sahu, A., Das, P.K., Meher, S.: High accuracy hybrid CNN classifiers for breast cancer detection using mammogram and ultrasound datasets. Biomed. Signal Process. Control **80**, 104292 (2023)
38. Kim, H.E., Cosa-Linan, A., Santhanam, N., Jannesari, M., Maros, M.E., Ganslandt, T.: Transfer learning for medical image classification: A literature review. BMC Med. Imaging **22**(1), 69 (2022)
39. Belhi, A., Ahmed, H.O., Alfaqheri, T., Bouras, A., Sadka, A.H., Foufou, S.: Study and evaluation of pre-trained CNN networks for cultural heritage image classification. Data Anal. Cult. Herit.: Curr. Trends Concepts, 47–69 (2021)
40. Pan, S.J., Yang, Q.: A survey on transfer learning. IEEE Trans. Knowl. Data Eng. **22**(10), 1345–1359 (2010)
41. Hartigan, J.A., Wong, M.A.: Algorithm AS 136: A k-means clustering algorithm. J. R. Stat. Soc. Ser. C (Appl. Stat.) **28**(1), 100–8 (1979)

42. Lekamlage, C.D., Afzal, F., Westerberg, E., Cheddad, A.: Mini-DDSM: Mammography-based automatic age estimation. In: 2020 3rd International Conference on Digital Medicine and Image Processing, pp. 1–6 (2020)
43. Matplotlib webpage, https://matplotlib.org/stable/tutorials/colors/colormap-manipulation.html, last accessed 2023/03/05
44. Szegedy, C., Ioffe, S., Vanhoucke, V., Alemi, A.: Inception-v4, inception-resnet and the impact of residual connections on learning. Proc. AAAI Conf. Artif. Intell. **31**(1), 4278–4284 (2017)
45. Krizhevsky, A., Sutskever, I., Hinton, G.E.: Imagenet classification with deep convolutional neural networks. Commun. ACM **60**(6), 84–90 (2017)
46. Szegedy, C., Liu, W., Jia, Y., Sermanet, P., Reed, S., Anguelov, D., Erhan, D., Vanhoucke, V., Rabinovich, A.: Going deeper with convolutions. Proc. IEEE Conf. Comput. Vis. Pattern Recognit. 1–9 (2015)
47. Zhang, X., Zhou, X., Lin, M., Sun, J.: Shufflenet: An extremely efficient convolutional neural network for mobile devices. Proc. IEEE Conf. Comput. Vis. Pattern Recognit. 6848–6856 (2018)
48. Redmon, J., Farhadi, A.: YOLO9000: better, faster, stronger. Proc. IEEE Conf. Comput. Vis. Pattern Recognit. 7263–7271 (2017)
49. Hosmer, Jr. D.W., Lemeshow, S., Sturdivant, R.X.: Applied logistic regression. John Wiley & Sons (2013)
50. Brereton, R.G., Lloyd, G.R.: Support vector machines for classification and regression. Analyst. **135**(2), 230–267 (2010)

Constructing a Panoramic Radiograph Image Based on Magnetic Resonance Imaging Data

Piotr Cenda, Adam Cieślak, Elżbieta Pociask, Rafał Obuchowicz, and Adam Piórkowski

Abstract Panoramic radiograph images are one of the basic examinations used during dental and orthodontic treatment. Their creation is based on ionizing radiation, to which the patient is therefore exposed. The adverse effects of this radiation on the patient's health is a topic that has been increasingly raised in numerous publications and journals. Therefore, the motivation for this study was to develop a mechanism that reconstructs panoramic radiograph images from MRI data, which is a harmless method. To achieve this, an algorithm described in a previous article is used to create a three-dimensional bone and skin tissue model from MRI data derived from T1-weighted and T2-weighted sequences. Subsequently, the craniofacial part of this model is projected onto a plane to form a panoramic radiograph-like image. Different approaches to model projection are presented. Despite the noise which comes from the high degree of interpolation in the model used, the generated results are the first step in the generation of panoramic craniofacial images without the use of ionizing radiation.

Keywords Reconstruction · Panoramic radiograph · Magnetic resonance

1 Objective

Panoramic radiography is one of the most common examinations used during dental and orthodontic treatment and diagnosis, despite its known adverse effects on health [7, 9, 10, 12, 13]. Ionizing radiation is often used in dental and orthodontic treatment, especially during routine and periodic examinations; thus, the body is unnecessarily exposed to this radiation, which can increase the risk of cancer. Panoramic images can be useful in a variety of medical applications. For example, they can be used to

P. Cenda (✉) · A. Cieślak · E. Pociask · A. Piórkowski
AGH University of Science and Technology, Mickiewicza 30 Av., 30–059 Kraków, Poland
e-mail: cenda@student.agh.edu.pl

R. Obuchowicz
Jagiellonian University Medical College, Kopernika 19, 31–501 Kraków, Poland

P. Strumiłło et al. (eds.), *The Latest Developments and Challenges in Biomedical Engineering*, Lecture Notes in Networks and Systems 746,
https://doi.org/10.1007/978-3-031-38430-1_10

visualize the curvature of blood vessels or the shape of the inner ear. Additionally, panoramic images can help to identify abnormal or diseased areas in the body that may not be visible on traditional cross-sectional images. Therefore, panoramic images provide clinicians with an additional tool for diagnosing and treating patients.

Scientific papers [4, 5, 11] indicate that there is a correlation between examinations performed with ionizing radiation and future cases of meningiomas in children as young as 10 years old (an almost fivefold increase in risk). This is just one example of the studied long-term adverse effects of ionizing radiation in medical treatment and imaging methods. Emerging techniques to reduce this risk are based solely on reducing the dose of radioactive radiation rather than the use of non-ionizing imaging methods such as MRI.

A previous article [2] described a method by which a 3D model of bone and soft tissues was obtained from a series of MRI-derived images. Then, a cephalometric-like image was reconstructed from that 3D model. The data were derived from standard T1 and T2 sequences, as opposed to approaches that use unusual and hard-to-obtain sequences. For example, "black-bone" MRI [8, 15] is a currently studied magnetic resonance imaging method that makes bone tissue very easy to segment. These techniques show satisfactory results but involve the use of non-standard MRI machines, which precludes their use from most hospitals. In the aforementioned article, there was a slight problem with poor contrast between bone structures and calcified tissues, soft tissues and air in the body; therefore, in the tissue model, air inside the body (such as in the sinuses and esophagus) was segmented as bone tissue. However, despite these errors, the quality of the resulting model was good enough to conduct further work on it.

The goal of the project is to develop a method whereby an image similar to a panoramic radiograph-like image will be generated using a 3D model of bone and soft tissues derived from MRI data. This includes mapping the model to a panoramic image so that the visibility and accuracy of tissue outlines is preserved, which is important in the orthodontic treatment process.

2 State of the Science

Despite the large number of research papers describing segmentation methods from MRI imaging, most are concerned with segmentation of brain and bone tissue (without the craniofacial region) and mostly use non-standard MRI imaging sequences, which are not widely available. Additionally, panoramic image reconstruction itself from such data does not appear in the literature.

The main factor which explains why there have been so few similar segmentation and reconstruction attempts using MRI data is the short relaxation time of bone tissue. Approaches that attempt to circumvent this difficulty are based on the use of "black bone" sequences ("Black Bone" MRI), sometimes combining these results with gradient echo (GRE) and short echo (TE) sequences. These techniques provide fairly good results in which bone tissue is well distinguished from soft tissue and sinus air. However, these studies have focused only on bone segmentation for 3D models, so other tissue types are ignored; moreover, they require special equipment because so many different MRI sequences are needed.

Attempts have also been made to segment tissues using only T1 sequences, which are easily obtainable. The method mentioned in [6] involves only simple morphological operations, but the result has a relatively low level of position shift error relative to CT (within 3mm). The limitation of this method is that it can only be applied to the part of the skull containing the brain (the craniofacial region), so this precludes the use of this method for dental and orthodontic treatment.

Taking this into account, as described in a previous article [2], a method was developed based solely on T1 and T2 sequences that covers the entire head and the facial region. This allows the use of a segmented tissue model for dental and orthodontic treatment by reconstructing panoramic radiograph-like images from a previously generated tissue model.

3 Data

The magnetic resonance imaging (MRI) data used in this project include two distinct series: T1-weighted and T2-weighted images. These images were obtained from a sample of human subjects. They contain head and neck area, and are intended for use in a comprehensive analysis of the craniofacial, cerebrocranial and cervical spine areas. Specifically, each series contains 105 cross-sectional images (example in Fig. 1) that are oriented in the transverse plane. These T1-weighted images provide a clear depiction of the anatomical structures within the brain and surrounding areas, while the T2-weighted images highlight the presence of fluid and other soft tissues.

The resolution of each image is 448×448 pixels, thus providing a high level of detail and precision in the captured data in the transverse plane. Additionally, the 16-bit depth of each image ensures that it is able to accurately represent the range of intensities present in the original data. Overall, the T1-weighted and T2-weighted MRI data utilized in this study provide a comprehensive view of the craniofacial, cerebrocranial, and cervical spine regions.

4 Algorithm

The presented algorithm can be divided into three stages: data preparation, tissue segmentation, and construction of a panoramic radiograph-like image.

4.1 Data Preprocessing

Data preparation includes data loading, normalization, and matching of T1-weighted and T2-weighted image series. The algorithm takes into account the fact that the T1 and T2 sequences are not perfectly aligned (this is due to the patient's shift relative to the camera during examination and can be seen in Fig. 1), adjusting the sequences accordingly. This was achieved by applying appropriate geometric transformations, such as translation, rotation and scaling, to minimize any differences between the sequences. The parameters of these transformations are adjusted automatically by the optimization algorithm. The algorithm considers the distinct characteristics of T1 and T2 sequences and utilizes a Broyden-Fletcher-Goldfarb-Shanno (BFGS) optimization method [1] that finds proper parameter matrices in order to align the sequences. BFGS is an optimization method for finding the minimum of a function. It updates an estimate of the Hessian matrix using information from the function's gradients and is well-suited for problems with many variables, such as selecting multiple transformation parameters. In this application, it minimizes the differences between two sequences, which are described as outlines of shapes obtained using the Canny algorithm. That ensures that the data obtained from each sequence is compatible and can be utilized together in further analysis.

After alignment, processing includes filtering, thresholding, and de-noising of the prepared data according to diagram shown in Fig. 2.

This allows the extraction of relevant data describing bone and soft tissues, from which 3D models of these tissues are then created. In the resulting series of masks of specific tissues, bicubic interpolation in the vertical axis is applied, thus increasing the number of cross-sections in the transverse plane. Bicubic interpolation was chosen because it introduces the least distortion in MRI images [3]. This makes it possible to create a 3D model in which the points that make up the model are evenly spaced. This process ensures that the geometry of the tissue is preserved to the greatest extent

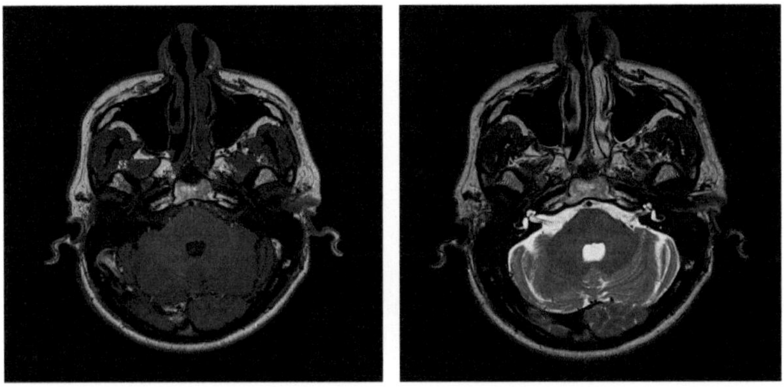

(a) Cross-section 50 (b) Cross-section 50

Fig. 1 Selected cross-section of the data used, T1 (**a**) and T2 (**b**) sequences

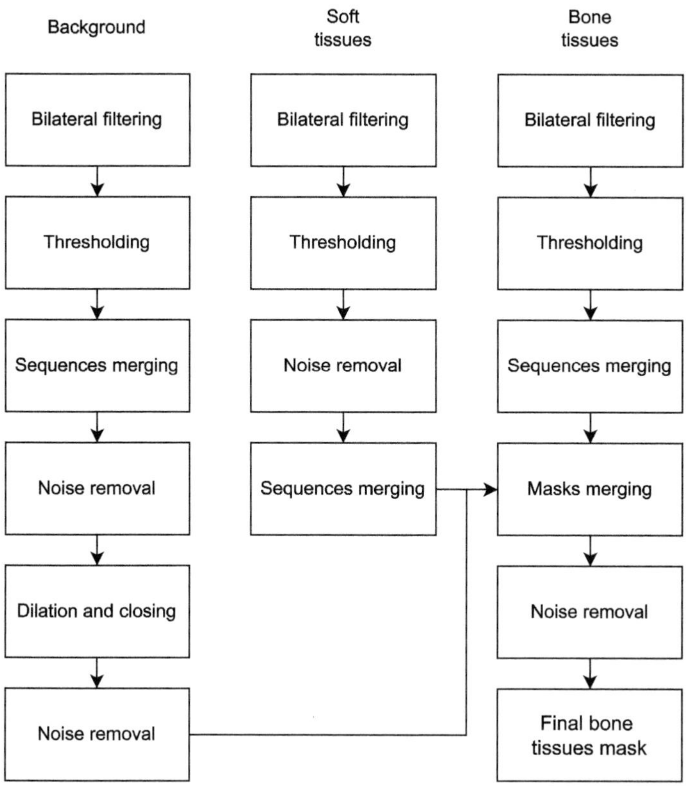

Fig. 2 Diagram showing the tissue segmentation process

possible, and the resulting distortion is minimized as much as possible by keeping the spacing of the model points identical in all planes. An example of a model generated in this way can be seen in Fig. 3.

4.2 Reconstruction

A CT scan captures multiple X-ray images of a body part from different angles, which are then processed by a computer to create detailed cross-sectional images. A panoramic image, is a 3D image created from CT data by projecting the maximum intensity value of the voxels along a certain axis. A diagram showing the symbolic mechanism of a panoramic CT scan is shown in Fig. 4. To generate a panoramic image, the CT data is first reconstructed into a 3D volume using algorithms such as filtered back projection [14]. The volume data is then resampled and projected onto a 2D plane along a specified axis, which can be horizontal, vertical, or diagonal. The intensity value of each pixel in the projected image is determined by finding the

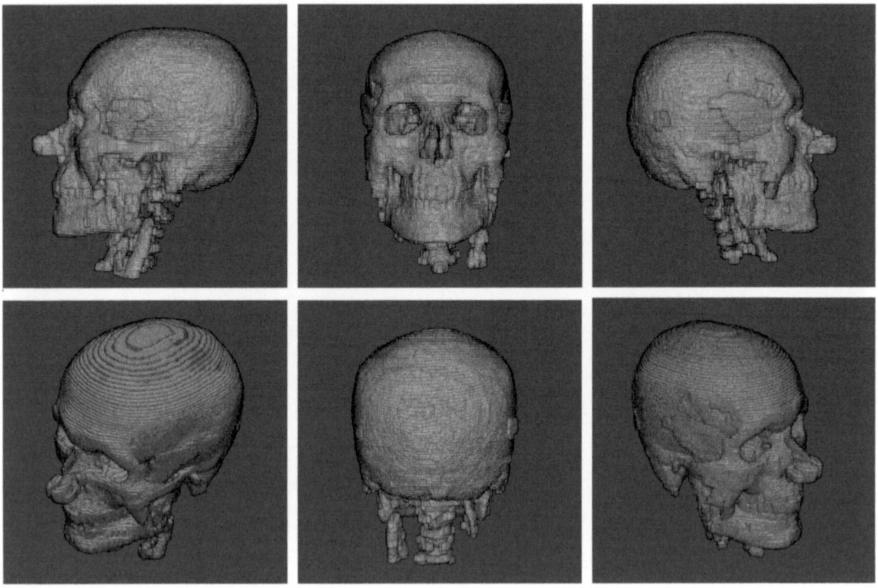

Fig. 3 Generated model from interpolated bone tissue masks

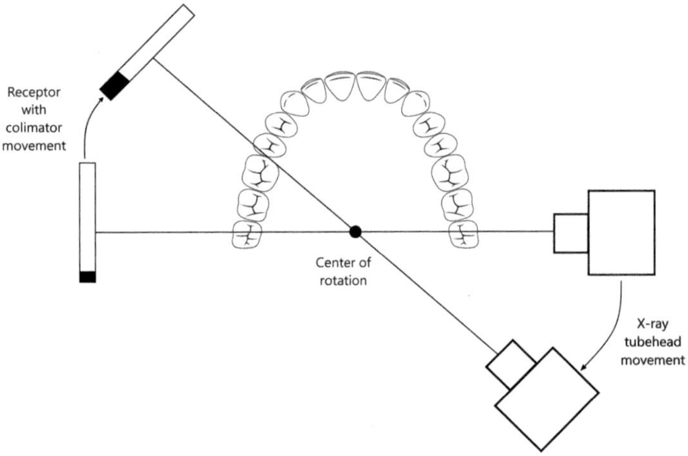

Fig. 4 Diagram showing mechanism behind CT panoramic scan

maximum intensity value of all the voxels along the projection axis that intersects the pixel.

In order to simulate the real mechanism of CT-based panoramic image creation, the construction part of the algorithm, which produces a panoramic radiograph-like image, consisted of two main steps: sampling points in the space of the prepared model, and assembling these points into a single pixel in the constructed image. The

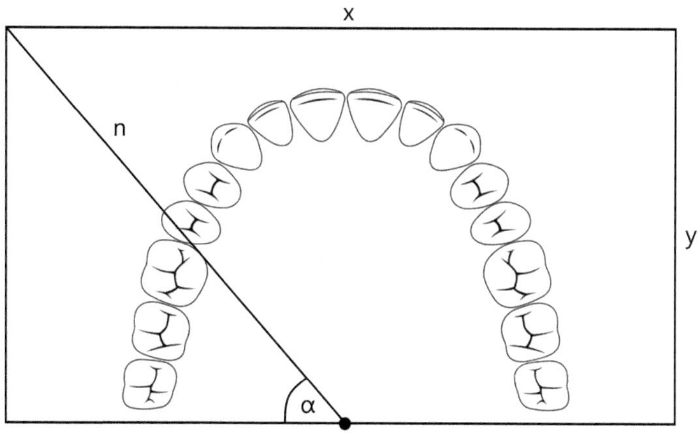

Fig. 5 Example slice with geometrical explanation of the algorithm's first part

sampling of points in the model is responsible for simulating the CT mechanism, which uses data on a certain axis when generating a panoramic image. Similarly to this mechanism, these data are then projected as a single pixel onto the 2d image. A previously prepared 3D tissue model is the basis of this part of the algorithm.

The first part of the algorithm creates a virtual beam that passes through the tissue model, according to which pixels are projected onto the resulting panoramic image; this image is created by sampling 1,000 points in space, which are arranged at equal distances from each other on a single straight axis. The model's values at these selected points are then extracted and projected into the next part of the algorithm.

The straight lines (described in Fig. 5 as n), each of which is made up of sampled points, are arranged radially from the center of the model along the intersection of the sagittal and coronal planes. This causes the arc formed by the jaw to be projected onto the plane that is the resulting panoramic photo. The line shift along the intersection of the sagittal and coronal planes is parallel and occurs for every pixel of the 3D tissue model, which corresponds to the resulting resolution in the z-axis (height). On the other hand, the offset between points on the x-axis in the resulting image (width) comes from the rotation of the line n on which the sampled points are located. This rotation is performed with a fixed angle (α in the Fig. 5), which depends on the number of pixels in the x-axis, which depends on the number of pixels in the x-axis, from 0 °C to 180 °C. Thus, a line with an offset angle of zero starts in front of the molars on one side, then with increasing angle follows the arc on which the teeth lie, and ends with an offset angle of 180 °C behind the molars on the other side.

The second part of the proposed algorithm is designed to map sampled points on the same axis (line n in the Fig. 5) to a single pixel in the resulting image (denoted as I in Eq. 1, which describes the process), thereby constructing an image similar to a panoramic photo. The resulting image has dimensions x by z, where the size of x is equal to $\frac{180}{\alpha}$, and z corresponds to the number of cross-sections of the model in the transverse plane. To achieve this, a mapping process is employed which involves calculating the mean value from a list of sampled points. Thus, by using the mean value, the algorithm effectively captures the underlying features of the 3D tissue model and projects them onto the 2D image plane.

$$I = \frac{1}{n} \sum_{i=0}^{n=\frac{180}{\alpha}} m_i \tag{1}$$

Furthermore, the mapping process plays an important role in determining the quality of the resulting panoramic image. A key factor that affects the quality of the image is the number of sampled points used in the mapping process and the number of axes that create the arc (which is equivalent to the value of the angle). The more points that are sampled, the higher the resolution of the resulting image (up to a certain point, of course). However, increasing the number of sampled points also increases the computational complexity of the algorithm, which may lead to longer processing times. Therefore, a trade-off between computational efficiency and image quality needs to be considered when selecting the number of sampled points. The parameters that were chosen as optimal were 1000 sampled points in one line and 0.2 °C of rotation of the line which determines the direction of direction the points are sampled around the vertical axis.

It is also worth noting that the proposed algorithm is not limited to the use of mean and median values for mapping the sampled points. Other statistical measures can also be used to construct the resulting image; thus, adjusting the contrast and brightness of the image and making different structures more visible. That would create a similar effect to changing the Hounsfield range in a real CT scan visualization.

5 Results

The algorithm described above generates an image resembling the panoramic radiograph shown in Fig. 6a) next to the true panoramic image shown in Fig. 6b). It covers the lower craniofacial region, including the maxilla, mandible, zygomatic bone and situs bone. In addition, parts of the cervical spine are visible below the mandible, and parts of cartilaginous tissue are also visible at the sides of the image. The resulting image has less contrast and brightness than the originals, but this is a matter of normalizing the image at the discretion of the algorithm user. The incisors, and canines are well delineated but the molars are slightly blended, which is due to the number of cross sections in the data.

The technique presented in this study provides a novel approach to imaging that may serve as a promising alternative to conventional methods that utilize ionizing radiation. This approach aims to mitigate the potential risks associated with the use

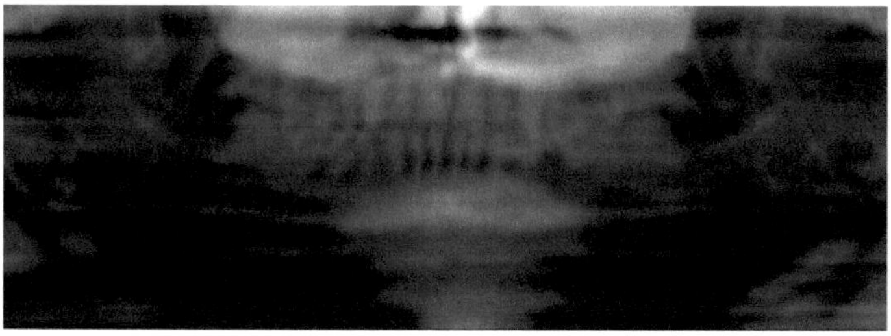

(a) Generated image

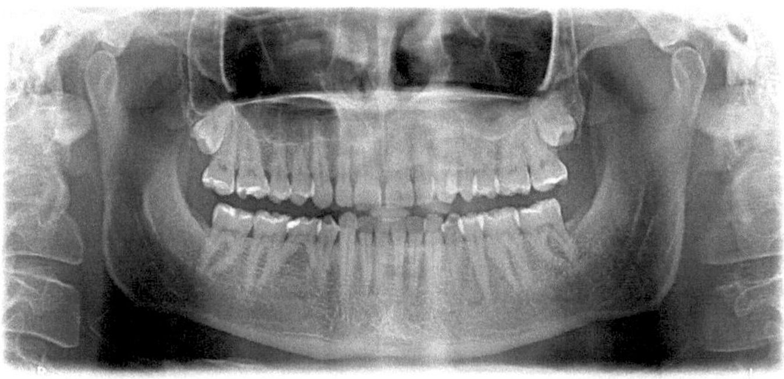

(b) Original image

Fig. 6 Constructed panoramic radiograph-like image and original image

of ionizing radiation in imaging. Although the resolution of the generated images in this study may appear low, this serves as a proof-of-concept for the proposed method. The main objective of this study was to demonstrate the possibility of avoiding the use of ionizing radiation by utilizing the presented approach. The image quality is largely attributed to the resolution of the data utilized in this study, which resulted in a high degree of interpolation during model generation. It is expected that increasing the number of sections in the data sequences used will improve the accuracy and readability of the reconstructed images. This will minimize the impact of interpolation, which, despite having the lowest level of noise among all methods, can still potentially compromise the integrity of the image data. Ultimately, this would improve the performance of the overall algorithm. Further research is warranted to assess the potential clinical applications of this alternative imaging technique and its ability to provide diagnostically useful images.

Acknowledgements This work was financed by the AGH University of Science and Technology, Faculty of EAIIB, KBIB on 16.16.120.773 and the grant "Studenckie koła tworzą innowacje"—II edition, project no. SKN/SP/535131/2022.

References

1. Broyden, C.G.: The convergence of a class of double-rank minimization algorithms 1. General considerations. IMA J. Appl. Math. **6**(1), 76–90 (1970). https://doi.org/10.1093/imamat/6.1.76
2. Cenda, P., Obuchowicz, R., Piórkowski, A.: Construction of a cephalometric image based on magnetic resonance imaging data. In: Pietka, E., Badura, P., Kawa, J., Wieclawek, W. (eds.) Information Technology in Biomedicine, pp. 143–154. Springer International Publishing, Cham (2022)
3. Cieślak, A., Piórkowski, A., Obuchowicz, R.: Comparison of interpolation methods for MRI Images Acquired with Different Matrix Sizes. In: Pietka, E., Badura, P., Kawa, J., Wieclawek, W. (eds.) Information Technology in Biomedicine, pp. 119–131. Springer International Publishing, Cham (2022)
4. Claus, E.B., Calvocoressi, L., Bondy, M.L., Schildkraut, J.M., Wiemels, J.L., Wrensch, M.: Dental x-rays and risk of meningioma. Cancer **118**(18), 4530–4537 (2012)
5. Cung, W., Freedman, L.A., Khan, N.E., Romberg, E., Gardner, P.J., Bassim, C.W., Baldwin, A.M., Widemann, B.C., Stewart, D.R.: Cephalometry in adults and children with neurofibromatosis type 1: implications for the pathogenesis of sphenoid wing dysplasia and the "NF1 facies". Eur. J. Med. Genet. **58**(11), 584–590 (2015)
6. Dogdas, B., Shattuck, D.W., Leahy, R.M.: Segmentation of skull and scalp in 3-D human MRI using mathematical morphology. Hum. Brain Mapp. **26**(4), 273–285 (2005)
7. Domeshek, L.F., Mukundan, S., Yoshizumi, T., Marcus, J.R.: Increasing concern regarding computed tomography irradiation in craniofacial surgery. Plast. Reconstr. Surg. **123**(4), 1313–1320 (2009)
8. Eley, K.A., Delso, G.: Automated 3D MRI rendering of the craniofacial skeleton: using ZTE to drive the segmentation of black bone and FIESTA-C images. Neuroradiology **63**(1), 91–98 (2021)
9. Hwang, S.Y., Choi, E.S., Kim, Y.S., Gim, B.E., Ha, M., Kim, H.Y.: Health effects from exposure to dental diagnostic X-ray. Environ. Health Toxicol. **33**(4), e2018,017 (2018)
10. Krille, L., Dreger, S., Schindel, R., Albrecht, T., Asmussen, M., Barkhausen, J., Berthold, J.D., Chavan, A., Claussen, C., Forsting, M., Gianicolo, E.A.L., Jablonka, K., Jahnen, A., Langer, M., Laniado, M., Lotz, J., Mentzel, H.J., Queißer-Wahrendorf, A., Rompel, O., Schlick, I., Schneider, K., Schumacher, M., Seidenbusch, M., Spix, C., Spors, B., Staatz, G., Vogl, T., Wagner, J., Weisser, G., Zeeb, H., Blettner, M.: Risk of cancer incidence before the age of 15 years after exposure to ionising radiation from computed tomography: results from a German cohort study. Radiat. Environ. Biophys. **54**(1), 1–12 (2015)
11. Maillie, H.D., Gilda, J.E.: Radiation-induced cancer risk in radiographic cephalometry. Oral Surg. Oral Med. Oral Pathol. **75**(5), 631–637 (1993)
12. Pflugbeil, S., Pflugbeil, C., Schmitz-Feuerhake, I.: Risk estimates for meningiomas and other late effects after diagnostic X-ray exposure of the skull. Radiat. Prot. Dosim. **147**(1–2), 305–309 (2011)
13. Smith-Bindman, R., Lipson, J., Marcus, R., Kim, K.P., Mahesh, M., Gould, R., Berrington de González, A., Miglioretti, D.L.: Radiation dose associated with common computed tomography examinations and the associated lifetime attributable risk of cancer. Arch. Intern. Med. **169**(22), 2078–2086 (2009)
14. Willemink, M.J., Noël, P.B.: The evolution of image reconstruction for CT-from filtered back projection to artificial intelligence. Eur. Radiol. **29**, 2185–2195 (2019)
15. Zhang, R., Lee, H., Zhao, X., Song, H.K., Vossough, A., Wehrli, F.W., Bartlett, S.P.: Bone-Selective MRI as a Nonradiative Alternative to CT for Craniofacial Imaging. Acad. Radiol. **27**(11), 1515–1522 (2020)

Optimization of the BOLD Hemodynamic Response Function for EEG-FMRI Studies in Epilepsy

Nikodem Hryniewicz, Rafał Rola, Kamil Lipiński, Ewa Piątkowska-Janko, and Piotr Bogorodzki

Abstract Simultaneous EEG-fMRI measurements are used to study epileptic patients in order to localize brain regions associated with interictal epileptiform discharges (IEDs). The common approach to this analysis uses the generalized linear model (GLM) with a canonical hemodynamic response function (HRF) to relate EEG activity to predicted BOLD responses. The study aimed to determine whether the canonical HRF is optimal for IED-induced hotspot fMRI analysis and develop optimized HRFs. The optimization of four HRF models was performed on BOLD responses derived from SPM activation clusters from 36 epileptic patients. The results showed that the optimized HRF models improved the sensitivity of the EEG-fMRI analysis of IED events in epilepsy patients compared to standard HRFs. This was evidenced by an increase in the size, number, and maximum *t*-score values of activated areas.

Keywords EEG-fMRI · Hemodynamics · Epilepsy · Hemodynamic response function · Optimization · GLM

N. Hryniewicz (✉) · E. Piątkowska-Janko · P. Bogorodzki
Nalecz Institute of Biocybernetics and Biomedical Engineering PAS, Warsaw, Poland
e-mail: nhryniewicz@ibib.waw.pl

R. Rola
Military Institute of Aviation Medicine, Warsaw, Poland

K. Lipiński · P. Bogorodzki
Warsaw University of Technology, Warsaw, Poland

R. Rola
National Institute of Oncology, Warsaw, Poland

P. Strumiłło et al. (eds.), *The Latest Developments and Challenges in Biomedical Engineering*, Lecture Notes in Networks and Systems 746,
https://doi.org/10.1007/978-3-031-38430-1_11

1 Introduction

Simultaneous recordings of electroencephalography (EEG) and functional magnetic resonance imaging (fMRI) are crucial for investigating patients with epilepsy. fMRI tracks brain activity non-invasively by observing changes in hemodynamics, which are related to deoxyhemoglobin concentration. The recorded signal is called blood oxygenation level dependent (BOLD) [1]. The EEG method consists in tracking changes in summed local electrical potentials generated in brain cortex recorded by the electrodes placed on the scalp. The measured electrical activity includes summed short action potentials and slower postsynaptic potentials [2].

Both of these methods, while providing information about brain activity, have their limitations. The fMRI has good spatial resolution, covering the entire volume of the brain with a voxel size of a few to several tens of cubic millimeters, but is limited by a temporal resolution of seconds. EEG provides high time resolution with measurements in the range of milliseconds. However, its spatial resolution is constrained by the number of electrodes placed on the surface of the head [3]. The use of the EEG-fMRI technique makes it possible to overcome temporal and spatial limitations, allowing for the complementary use of EEG and fMRI.

The first EEG measurements during the recording of echo planar imaging (EPI) sequences took place in 1993 [4]. Nowadays EEG-fMRI is widely used in the studies of patients with epilepsy. There is a relationship between the hemodynamics of the brain and its neural activity associated with interictal epileptiform discharges (IEDs). Therefore, the observation of hemodynamic changes and the calculation of maps providing areas associated with those discharges is possible. EEG-fMRI makes it possible to identify the onset times of IEDs and incorporate them into statistical analysis of fMRI data through the generalized linear model (GLM). Thanks to this, presurgical localization of epileptic focus is possible [5, 6].

As previously mentioned, the analysis of fMRI data involves utilizing a GLM which requires the preparation of a regressor by convolving the stimuli function with the hemodynamic response function (HRF—Fig. 1.). The HRF is an assumed response of the BOLD signal to a single stimulus. In the classical approach to fMRI analysis, for example using the SPM12 (statistical parametric mapping [7]) toolbox, the Canonical model of HRF is used. However, this is a kind of simplification, assuming that the hemodynamic response is stationary in time, for each voxel in the brain and for each subject. It was shown that there are inter-subject differences in hemodynamic response [8]. Many studies have been devoted to HRF estimation, including taking into account asynchronous event-related paradigms, different types of trials and taking into consideration several fMRI sessions to estimate the response [9]. The EEG-fMRI is also used to model the hemodynamic response in resting state, task-based or IED-driven data [10].

IEDs are a unique type of the stimuli function, thus suggesting that the hemodynamic response may differ from that assumed in "classical" approaches to analysis. In contrast to classical fMRI paradigms e.g. finger tapping, the IED is a pathological activity of the brain which is generated spontaneously without any external

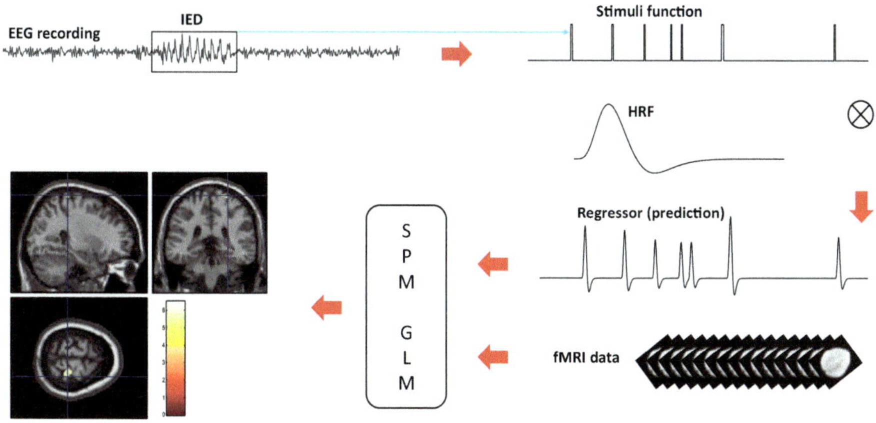

Fig. 1 Statistical analysis of simultaneous, time aligned fMRI and EEG data with GLM modelling. First, a neurologist manually identifies the onset and duration of IED events, which are presented as a boxcar function. Stimuli convolved with the HRF generates the effect size regressor for the GLM model. Finally, activation voxel clusters are detected and displayed on anatomical template at a significance level of $p < 0.05$ (FWE corrected)

stimuli. Moreover, the bioelectrical activity of IEDs are characterized by pathological hypersynchrony of the cortical networks, while physiological activity of motor areas during finger tapping is characterized by bioelectrical desynchrony. These differences in bioelectrical activity may result in different hemodynamic response. This is why we were interested in the optimization of HRF for IEDs as regressors in fMRI analysis. As early as 2003, Kang et. al. noted that using the same HRF for epilepsy data as for other experiments may not be optimal. They proposed a subject-specific approach to HRF modeling. The aim of their study was to compare activation areas (activation voxel clusters) and *t*-scores from the results of statistical analysis using patient-specific HRF in comparison with the standard HRF model. The estimation of the HRF was performed within the region exhibiting the highest *t*-score value from the standard analysis. The results of 8 patients were compared, and it turned out that in the case of using patient-specific HRF the activation areas were similar or larger. For 5 patients the analysis showed additional activation clusters. It was therefore concluded that the use of patient-specific HRF increases the sensitivity of the statistical analysis in epileptic patients' data [11].

Some studies have utilized models that allow for the HRF to vary across different voxels [12, 13]. In 2006, Lu et al. proposed the use of voxel-specific HRF in the analysis of EEG-fMRI data of patients with epilepsy. They proposed a deconvolution method to determine the voxel-specific HRF, which they used for analysis. The fixed three HRF approaches were compared to results obtained using the deconvolution method. The latter method yielded a larger volume of activated brain regions, and in some cases, additional activation clusters were detected [13]. This study concluded that utilizing modeled HRF instead of a fixed ones leads to more accurate statistical analysis.

The study aimed to develop optimal HRF models for IED-driven statistical fMRI analysis, in terms of: number, spatial extent and maximal *t*-score of detected activation clusters. Optimization process covered four HRF BOLD models:

(1) Canonical
(2) Gamma
(3) Glover, and for the first time in epilepsy studies to our knowledge:
(4) Physiological Balloon model.

Proposed models were evaluated within standard SPM analysis on group of 36 epileptic patients to assess whether the utilization of optimized HRFs improves statistical analysis outcomes.

2 Methods

2.1 MRI Acquisition

In this study the GE MRI Discovery MR750w magnetic resonance scanner with a magnetic field induction of 3 T was used. For each patient, a structural T1-weighted image was acquired (repetition time TR = 6.936 ms, echo time TE = 2.968 ms, resolution 1 mm × 1 mm × 1.2 mm) and from one to three 10-min rs-fMRI (resting state fMRI) sessions using EPI sequences (repetition time TR = 2.5 s, echo time TE = 25 ms, resolution 3 mm × 3 mm × 3 mm) with simultaneous acquisition of the EEG. The EEG and MR system were synchronized. The fMRI data was processed using SPM and the process included realignment, coregistration, segmentation, normalization and smoothing.

2.2 EEG Acquisition

The 64-channel Neuroscan Synamps RT EEG system with the QuickCap EEG cap were used to record the EEG signal during fMRI acquisition (sampling rate fs = 1 kHz). The transmission of 'triggers' from the scanner to the EEG system informing about the moments of collecting images of the full volume of the brain was provided. Data was processed in the CURRY7 software and included removal of MR gradient coil artifacts and filtering (1–30 Hz).

Drawing upon expertise and familiarity with the patient's condition, the neurologist manually identified relevant IED events (their start and duration [14]) by utilizing specific EEG montages and leads. They were later used as a stimuli function in building a regressor for the GLM model.

2.3 Patients

Through the cooperation of the Nalecz Institute of Biocybernetics and Biomedical Engineering PAS and the Military Institute of Aviation Medicine, data of patients with epilepsy was collected using the method of EEG-fMRI. It should be noted that each of examined persons did not show external manifestations of the epileptic disease, such as severe convulsions or uncontrollable movements. However IEDs were seen on the EEG registration. These discharges can last up to a few seconds, causing brief absence seizures, myoclonus of the eyelids or limbs with minimal motor manifestation.

Datasets of patients who did not exhibit any IED events during any of the EEG-fMRI sessions, as well as those whose EEG records were burdened by high skin–electrode impedance or artifacts that hindered the detection of such events, were excluded. Taking into account above criteria, from 54 registered EEG-fMRI datasets, 36 (14 male patients aged 16 Y.O. to 55 Y.O. with average age 33.9 Y.O. and 22 females aged 17 Y.O. to 66 Y.O. with average age 31.5 Y.O.) were selected. The study was conducted under the supervision of a neurologist and approved by the bioethics committee, with participation from patients who provided their consent.

2.4 Single Dataset Analysis

For each validated EEG-fMRI dataset, a statistical analysis was performed, to calculate statistical parametric maps in the SPM toolbox [7]. The analysis uses the GLM model, where the Canonical hemodynamic response model with default parameter settings was used to construct the GLM regressor.

Figure 1. shows the scheme of the experiment with an exemplary result of the analysis.

2.5 Hemodynamic Response Function Models

In order to obtain optimal HRFs, four models have been optimized. We extended Lu's optimization approach [13] where he used 3 models (Canonical, Gamma and Glover) by the physiological HRF model originally proposed by Buxton et.al. called the Balloon model [15, 16]. A modified version of this model is incorporated into the hemodynamics section of SPM toolbox [7, 17]. The following part of this section describes the HRF models used in this work.

2.5.1 Canonical Model

The Canonical HRF model is one of the most frequently used in the analysis of fMRI data utilizing the GLM. This model is set as the default HRF in the SPM toolbox. The shape of this model is based on two gamma functions and is described by seven parameters. In the optimization process we used SPM implementation (*spm_hrf* function) of Canonical model. Default values of Canonical HRF parameters are listed in Table 1.

2.5.2 Balloon Model

The Balloon model is our Simulink (MATLAB) implementation of the full model proposed by Buxton et al. and describes the causal sequence from the neuronal response to the final BOLD signal response [15, 16]. First the neural response to an input stimuli function is described with Eq. (1).

$$N(t) = S(t) - I(t) \tag{1}$$

where: $N(t)$—neural response,
$S(t)$—stimuli function,
$I(t)$—inhibitory input.
Inhibitory input is defined by the differential Eq. (2).

$$\frac{dI}{dt} = \frac{\kappa \cdot N(t) - I(t)}{\tau_i} \tag{2}$$

where: κ—inhibitory feedback gain factor,
τ_i—inhibitory feedback time constant.

Knowing the course of the neuronal response, it is possible to describe the cerebral blood flow and cerebral metabolic rate of oxygen responses with Eq. (3) and (4).

$$f(t) = 1 + n \cdot h\left(t - \delta t_f\right) * N(t) \tag{3}$$

$$m(t) = 1 + g(t - \delta t_m) * N(t) \tag{4}$$

where: $f(t)$—cerebral blood flow (CBF) response,
$m(t)$—cerebral metabolic rate of oxygen ($CMRO_2$) response,
δt_m—delay between stimuli function and $CMRO_2$ response,
δt_f—delay between stimuli function and CBF response,
n—flow/metabolism relation,
g, h—gamma functions.

The Balloon model is described using two dynamic variables q (total deoxyhemoglobin) and v (volume of the balloon), defined with Eqs. (5), (6) and (7).

$$\frac{dq}{dt} = \frac{1}{\tau_{MTT}}\left[m(t) - \frac{q(t)}{v(t)} f_{out}(v, t)\right] \tag{5}$$

$$\frac{dv}{dt} = \frac{1}{\tau_{MTT}}[f(t) - f_{out}(v, t)] \tag{6}$$

$$f_{out}(v) = v^{\frac{1}{\alpha}} + \tau \frac{dv}{dt} \tag{7}$$

where: $q(t)$—total deoxyhemoglobin,
$v(t)$—cerebral blood volume (CBV),
f_{out}—outflow of the balloon,
τ_{MTT}—balloon transit time,
τ—viscoelastic time constant,
α—flow/volume relation.
Finally, the BOLD response is described with Eq. (8).

$$\frac{\Delta S}{S} = V_0[a_1(1 - q) - a_2(1 - V)] \tag{8}$$

where: S—BOLD signal,
V_0—baseline blood volume,
a_1, a_2—model gain parameters (estimated in 2004 by Obata et. al. as 3.4 and 1.0, respectively [18]).

Our Simulink implementation is described with 9 parameters from Eqs. (1–8) which are listed in Table 1. The rest of the model parameters were considered constant and along with default Balloon HRF model parameters were selected on the basis of literature [15, 16, 18].

2.5.3 Gamma Model

The Gamma HRF is a simple model based on one gamma function. The HRF is defined by the Eq. (9), where the parameters determine the shape of the curve [13].

$$h_{gamma}(t) = \left(\frac{t}{b \cdot c}\right)^{b} \cdot e^{\left(b - \frac{t}{c}\right)} \tag{9}$$

Model is described with two parameters, which values are listed in Table 1. Default parameters were calculated on the basis of the visual stimulation experiment [19, 20], $b \cdot c$ refers to time to peak value (4.7 s), FWHM (full width at half-maximum) = 3.8 s.

2.5.4 Glover Model

The HRF Glover model, similarly to the Canonical model, is described with two gamma functions. The Glover HRF model is defined by the Eq. (10), where the parameters determine the shape of the curve [13].

$$h_{glover}(t) = \left(\frac{t}{d_1}\right)^{a_1} \cdot e^{-\left(\frac{t-d_1}{b_1}\right)} - c \cdot \left(\frac{t}{d_2}\right)^{d_2} \cdot e^{-\left(\frac{t-d_2}{b_2}\right)} \tag{10}$$

Model is described with five parameters, which values are listed in Table 1. Default parameters were calculated on the basis of the auditory stimulation experiment [21, 22]. The time to peak of the first gamma equals 5.4 s with FWHM = 5.2 s. The second gamma function is characterized by a time to peak of 10.8 s and FWHM = 7.35 s, $d_1 = a_1 \cdot b_1$; $d_2 = a_2 \cdot b_2$.

2.6 *HRF Parameters Optimization*

Optimization was performed with self-written MATLAB scripts, with use of SPM12 functions. For each patient BOLD signal from three largest clusters per session identified from the SPM analysis using the Canonical HRF model after FWE correction ($p < 0.05$) were extracted. Corresponding time-intensity curves will be called mean regional response (MRR) [23].

Optimization was employed with the *fmincon* MATLAB function with the following objective function (11):

$$MSE(P) = \sum_{i=1}^{n} \left(MRR(i) - \overline{MRR}(i, P)\right)^2 \tag{11}$$

where: MSE—mean squared error between MRR and its prediction (as described in Section 2.4.),

MRR—BOLD mean regional response,

$\overline{MRR}$—MRR prediction (regressor),

P—HRF model parameters that depend on the specific model utilized (as listed in Table 1),

n—length of the MRR curve.

The minimum value of objective function was found starting for the initial point set as default model parameters (Table 1.). The bound constrains of the optimized parameters were set as 0.5 and 1.5 of their default values.

The optimization scheme of HRF model parameters for single MRR is shown in Fig. 2.

Following optimization, the resultant parameters were averaged based on the number of clusters and then by the number of sessions. To determine the optimized

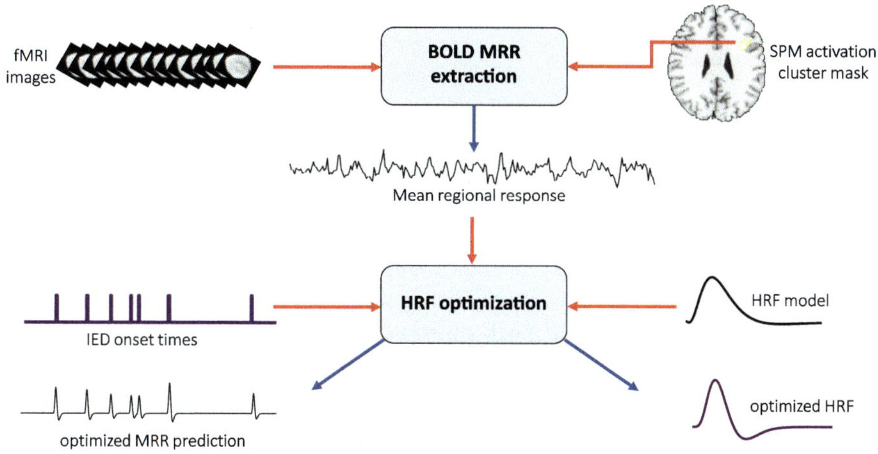

Fig. 2 The diagram illustrates the HRF optimization process. Initially, the MRR of the BOLD signal is calculated from the fMRI data, using the activation cluster identified through standard statistical analysis. Next, the predicted response with given either stimuli function or HRF model is prepared. Optimization with *fmincon* function minimalizes the mean square error (MSE) between the predicted response and the MRR

HRF model, the parameters were averaged based on the number of subjects for each of the proposed hemodynamic response models.

2.7 Statistical Analysis

For each patient, statistical analysis were performed (as described in Section 2.4) using the Canonical, Balloon, Gamma and Glover HRF models in GLM. Hemodynamic response parameters were selected based on the literature. After optimization (described in 2.6), SPM statistical analysis was again performed for each patient for the four HRF models using the new optimized HRF parameters. The parameters have been rounded to two decimal places. The analysis results were reported, containing information about the number and size of activation clusters, as well as the maximum t-score value. All of the results have been FWE-corrected at a significance level of $p < 0.05$.

3 Results

Optimization led to the acquisition of 35 novel sets of parameters for each HRF (analysis of one patient revealed no statistically significant results for any session) which after averaging gave new optimized parameters for four models of HRF. Table 1 present the optimized values alongside their default counterparts. Figure 3. shows the shapes of four HRF models, both before and after optimization of its parameters.

A significant differences between optimal and default models are observed. Delay of response was reduced from 6 s to 5.08 s for canonical model, CBF delay decreased from 5 s to 3.63 s and the flow metabolism ratio dropped from 2.5 to 1.73 for the Balloon model. Similarly HRF peak position changed for the Gamma model (b decreased from 8.6 to 4.65) and Glover model (a1 decreased from 6 to 3.43).

Figure 4. shows the improvement of fitting BOLD signal prediction (made with Gamma HRF) to the example MRR obtained from the Patient 1 activation cluster.

Employing various models with their respective default or optimized parameters can lead to dissimilar statistical analysis outcomes.

To aid in comprehending these results, we present the whole-brain SPM analysis of the data illustrated in Fig. 4, taking into account all four HRF models. The results are displayed in the axial view, showcasing the slice that yielded the highest *t*-score value in the analysis conducted with the default Canonical HRF model (Fig. 5).

Comparing these results, it is apparent that using the optimized Canonical HRF has led to an increase in the activation cluster size, which is 273% of its initial size. Moreover, with the use of optimized HRFs a previously undiscovered clusters has emerged.

By comparing the analysis results of all patients between the optimized and default HRF models, we found that the optimized Canonical HRF resulted in an increased number of clusters for 24 patients, increased number of voxels per max. *t*-score cluster for 28 patients, and increased max. *t*-score for 28 patients. Similarly, utilizing the optimized Balloon HRF model resulted in an increased number of clusters for 28 patients, increased number of voxels per max. *t*-score cluster for 28 patients, and increased max. *t*-score for 29 patients. In the case of the optimized Gamma HRF, we observed an increased number of clusters for 28 patients, increased number of voxels per max. *t*-score cluster for 28 patients, and increased max. *t*-score for 26 patients. Finally, when comparing the results of the optimized Glover HRF with the default Glover HRF, we noted an increased number of clusters for 28 patients, increased number of voxels per max. *t*-score cluster for 31 patients, and increased max. *t*-score for 28 patients. Table 2. shows the average values of the statistical analysis parameters mentioned above. Comparison of each optimized and default models showed an increase in all mean statistical parameters of interest for optimized HRFs. In addition, comparing the optimized Balloon, Gamma and Glover models with the default Canonical HRF model, one can notice higher mean statistical parameters for all of the optimized models (excluding the max. *t*-score for the Gamma model).

Table 1 Default and optimized parameters of the Canonical HRF model (SPM), Balloon HRF model (Eqs. 1–8), Gamma HRF model (Eq. 9) and Glover HRF model (Eq. 10)

Canonical HRF model

	Delay of response	Delay of undershoot	Dispersion of response	Dispersion of undershoot	Response/ undershoot	Onset	Length of kernel
Default	6	16	1	1	6	0	32
Optimized	5.08	15.74	1.16	0.59	3.56	0	32

Balloon HRF model

	Inhibitory feedback gain	Inhibitory feedback t constant	CBF delay	$CMRO_2$ delay	Flow/ metabolism	Flow/volume	Balloon transit time	Viscoelastic t. constant	Baseline blood volume
Default	2	3	5	4	2.5	0.38	3	20	0.03
Optimized	1.79	2.97	3.73	4.1	1.73	0.34	3.32	16.61	0.03

Gamma HRF model

	b	c
Default	8.6	0.547
Optimized	4.65	0.65

Glover HRF model

	a1	a2	b1	b2	c
Default	6	12	0.9	0.9	0.35
Optimized	3.43	13.01	0.89	0.84	0.37

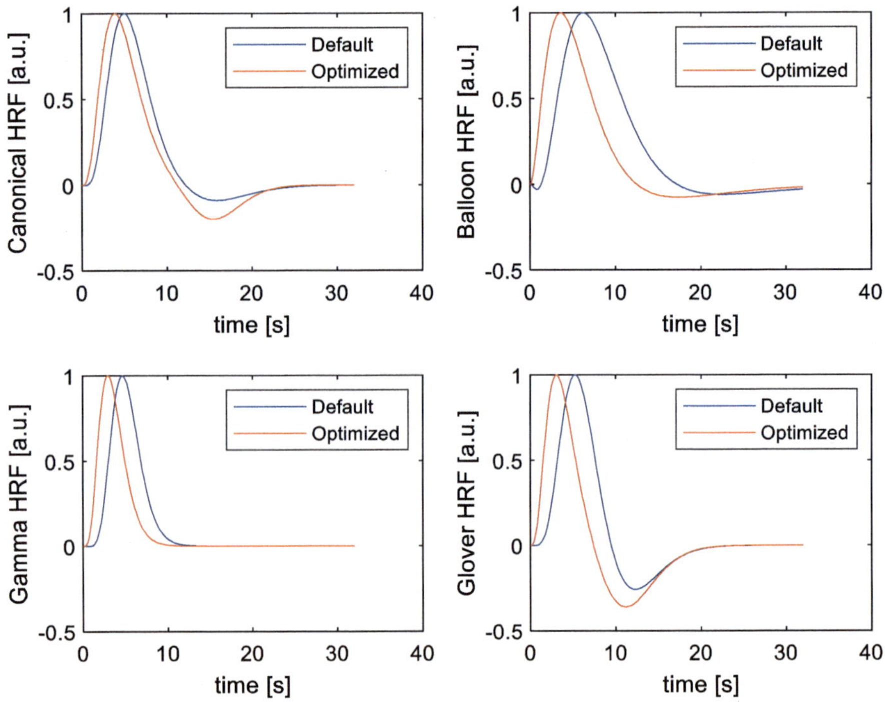

Fig. 3 Comparison of the shapes of the HRF models using the default (blue curves) and optimized (orange curves) parameters. HRFs are normalized to maximum value of 1

4 Discussion

Our study compares four different HRF models in order to answer the question of whether HRF optimization for IED events improves fMRI statistical analysis. Better statistical outcomes (in terms of: activation cluster size, max. *t*-score in activated areas and the number of detected activation clusters) obtained with optimized HRFs supports our proposition. Ito et al. recently explored increasing the sensitivity of fMRI analysis for epileptic patients data. They proposed SSWAS (spike-and-slow-wave-activation-summary) methodology which increased the number of patients with significant hemodynamic responses from 36% to 91% [24]. In our study we observed that optimal HRF increased number of patients with activations from 86% to 92% for the Canonical HRF, from 72% to 86% for the Balloon HRF, from 75% to 92% for the Gamma HRF, and from 69% to 92% for the Glover HRF. Significantly, the main innovation of our work lies in the successful implementation of the optimized Balloon HRF model into the SPM analysis. This may give insight into the physiological phenomena that are driven by IEDs. The lack of differentiation between the epilepsy types due to the small number of individual groups, may limits

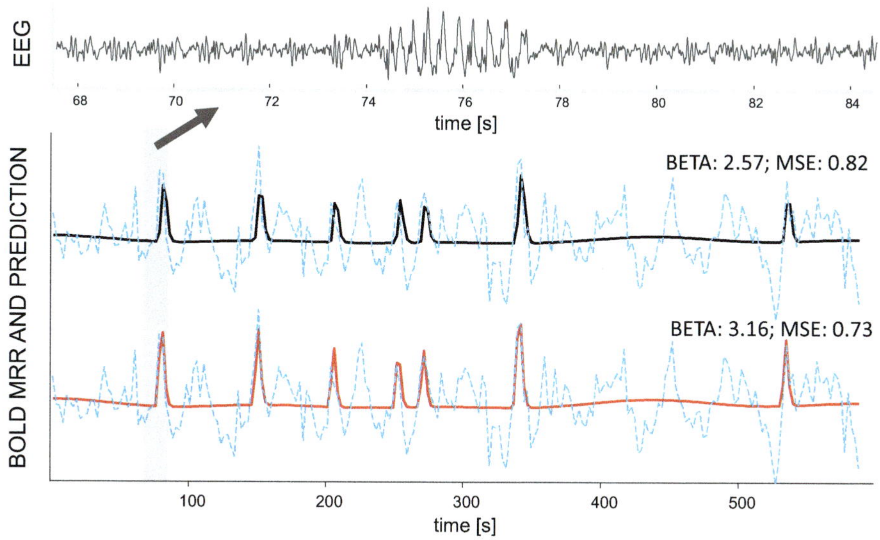

Fig. 4 The figure shows an exemplary 1st patient's (2nd session) IED (upper section) and the results of fitting gamma HRF prediction (lower section). The MRR is presented by a blue dashed curve, and the black curve represents the prediction before optimization. The lower red curve represents the prediction after optimization. Corresponding beta values and mean squared error between the BOLD signal prediction and MRR, both before and after optimization, are shown in the upper right corners of the graphs. The grey rectangle highlights the time range corresponding to the EEG

validity of the study. Introducing a new model would be reasonable to provide an even more accurate explanation of the hemodynamics associated with IED events.

5 Conclusions

We compared the results of the statistical analysis using four HRF models—with their default set of parameters and with the optimized. The results suggest that the use of models with standard parameter settings (including the Canonical model, widely used in the analysis of fMRI data), may not be optimal for the analysis of fMRI data of epileptic patients. Obtained results showed that the use of the optimized HRF models improves the sensitivity of the analysis and may be a better choice for the analysis of interictal epileptiform discharge events. Additionally, shapes of new HRFs revealed several differences in relation to default HRFs like e.g. location of the maximum response. This finding suggests that the hemodynamics associated with an IED differ significantly from the typical hemodynamic responses of the human brain.

By taking into account this difference when performing an analysis of the EEG-fMRI dataset, the sensitivity of the statistics has significantly improved. For example,

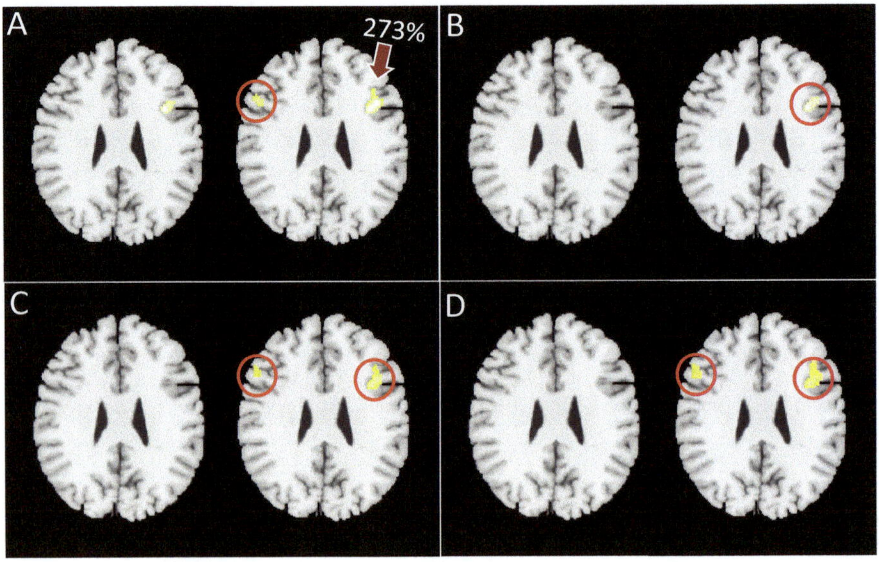

Fig. 5 Comparison of Patient 1 (2nd session) analysis results using default (left) and optimized (right) HRFs (**a** Canonical, **b** Balloon, **c** Gamma, **d** Glover). The results have been FWE-corrected at a significance level of $p < 0.05$. The red circles show the clusters that appeared after analysis with the optimized HRF, the arrow points to the cluster that has increased in volume

Table 2 Mean statistical results obtained with default and optimized HRFs

MODEL	Mean number of clusters		Mean number of voxels per max. *t*-score cluster		Mean max. *t*-score	
	Default	Optimized	Default	Optimized	Default	Optimized
Canonical	12.21	16.85	291.31	932.24	7.24	7.88
Balloon	4.15	15.63	59.2	389.84	6.03	7.43
Gamma	5.41	14.39	51.83	370.24	6.23	7.09
Glover	8.64	16.88	89.32	1259.21	6.61	7.81

for the canonical model, which is one of the most commonly used in fMRI data analysis, the optimized HRF allowed for a 38% increase in the average number of obtained activation clusters, a 227% increase in the average number of voxels in a cluster with max. *t*-score, and an average increase in maximum *t*-score by 9%.

References

1. Ogawa, S., Lee, T.M., Kay, A.R., Tank, D.W.: Brain magnetic resonance imaging with contrast dependent on blood oxygenation. Proc. Natl. Acad. Sci. **87**(24), 9868–9872 (1990)
2. Binnie, C.D., Prior, P.F.: Electroencephalography. J. Neurol. Neurosurg. Psychiatry **57**(11), 1308–1319 (1994)
3. Ritter, P., Villringer, A.: Simultaneous EEG–fMRI. Neurosci. Biobehav. Rev. **30**(6), 823–838 (2006)
4. Ives, J.R., Warach, S., Schmitt, F., Edelman, R.R., Schomer, D.L.: Monitoring the patient's EEG during echo planar MRI. Electroencephalogr. Clin. Neurophysiol. **87**(6), 417–420 (1993)
5. Sadjadi, S.M., Ebrahimzadeh, E., Shams, M., Seraji, M., Soltanian-Zadeh, H.: Localization of epileptic foci based on simultaneous EEG-fMRI data. Front. Neurol. **12** (2021)
6. Kowalczyk, M.A., Omidvarnia, A., Abbott, D.F., Tailby, C., Vaughan, D.N., Jackson, G.D.: Clinical benefit of presurgical EEG-fMRI in difficult-to-localize focal epilepsy: A single-institution retrospective review. Epilepsia **61**(1), 49–60 (2019)
7. Penny, W., Friston, K., Ashburner, J., Kiebel, S., Nichols, T.: Statistical parametric mapping: The analysis of functional brain images. 1st edn., Elsevier (2007)
8. Aguirre, G.K., Zarahn, E., D'esposito, M.: The variability of human bold hemodynamic responses. NeuroImage **8**(4), 360–369 (1998)
9. Ciuciu, P., Poline, J.B., Marrelec, G., Idier, J., Pallier, C., Benali, H.: Unsupervised robust nonparametric estimation of the hemodynamic response function for any fMRI experiment. IEEE Trans. Med. Imaging **22**(1), 1235–1251 (2003)
10. Prokopiou, P.C., Xifra-Porxas, A., Kassinopoulos, M., Boudrias, M.-H., Mitsis, G.D.: Modeling the hemodynamic response function using EEG-fMRI data during eyes-open resting-state conditions and motor task execution. Brain Topogr. **35**(3), 302–321 (2022)
11. Kang, J.K., Benar, C., Al-Asmi, A., Khani, Y.A., Pike, G.B., Dubeau, F., Gotman, J.: Using patient-specific hemodynamic response functions in combined EEG-fMRI studies in epilepsy. Neuroimage **20**(2), 1162–1170 (2003)
12. Pedregosa, F., Eickenberg, M., Ciuciu, P., Thirion, B., Gramfort, A.: Data-driven HRF estimation for encoding and decoding models. Neuroimage **104**, 209–220 (2015)
13. Lu, Y., Bagshaw, A.P., Grova, C., Kobayashi, E., Dubeau, F., Gotman, J.: Using voxel-specific hemodynamic response function in EEG-fMRI data analysis. Neuroimage **32**(1), 238–247 (2006)
14. Bagshaw, A.P., Hawco, C., Benar, C.G., Kobayashi, E., Aghakhani, Y., Dubeau, F., Pike, G.B., Gotman, J.: Analysis of the EEG-fMRI response to prolonged bursts of interictal epileptiform activity. Neuroimage **24**(4), 1099–1112 (2005)
15. Buxton, R.B., Wong, E.C., Frank, L.R.: Dynamics of blood flow and oxygenation changes during brain activation: The balloon model. Magn. Reson. Med. **39**(6), 855–864 (1998)
16. Buxton, R.B., Uludag, K., Dubowitz, D.J., Liu, T.T.: Modeling the hemodynamic response to brain activation. Neuroimage **23**, 220–233 (2004)
17. Adaszewski, S., Slater, D., Melie-Garcia, L., Draganski, B., Bogorodzki, P.: Simultaneous estimation of population receptive field and hemodynamic parameters from single point BOLD responses using Metropolis-Hastings sampling. NeuroImage **172**, 175–193 (2018)
18. Obata, T., Liu, T.T., Miller, K.L., Luh, W.M., Wong, E.C., Frank, L.R., Buxton, R.B.: Discrepancies between BOLD and flow dynamics in primary and supplementary motor areas: Application of the balloon model to the interpretation of BOLD transients. Neuroimage **21**(1), 144–153 (2004)
19. Savoy, R.L., O'Craven, K.M., Weisskoff, R.M., Davis, T.L., Baker, J., Rosen, B.: Exploring the temporal boundaries of fMRI: measuring responses to very brief visual stimuli. Society for Neuroscience 24th Annual Meeting, Miami Beach, FL, p. 518 (1994)
20. Cohen, M.S.: Parametric analysis of fMRI data using linear systems methods. Neuroimage **6**(2), 93–103 (1997)
21. Glover, G.H.: Deconvolution of impulse response in event-related BOLD fMRI. Neuroimage **9**(4), 416–429 (1999)

22. Worsley, K.J., Liao, C.H., Aston, J., Petre, V., Duncan, G.H., Morales, F., Evans, A.C.: A general statistical analysis for fMRI data. Neuroimage **15**(1), 1–15 (2002)
23. Bogorodzki, P., Rogowska, J., Yurgelun-Todd, D.A.: Structural group classification technique based on regional fMRI BOLD responses. IEEE Trans Med Imaging **24**(3), 389–398 (2005)
24. Ito, Y., Maesawa, S., Bagarinao, E., Okai, Y., Nakatsubo, D., Yamamoto, H., Kidokoro, H., Usui, N., Natsume, J., Hoshiyama, M., Wakabayashi, T., Sobue, G., Ozaki, N.: Subsecond EEG-fMRI analysis for presurgical evaluation in focal epilepsy. J Neurosurg. **134**(3), 1027–1036 (2020)

Improving the Resolution and SNR of Diffusion Magnetic Resonance Images From a Low-Field Scanner

Jakub Jurek, Kamil Ludwisiak, Andrzej Materka, and Filip Szczepankiewicz

Abstract Spatial resolution, signal-to-noise ratio (SNR) and acquisition time are interconnected in magnetic resonance imaging (MRI). Trade-offs are made to keep the SNR at the acceptable level, maximizing the resolution, minimizing the acquisition time and maintaining radiologically useful images. In low-field MRI scanners and especially in diffusion imaging, these trade-offs are even more crucial due to a generally lower image quality. Image post-processing is necessary in such cases to improve image quality. In this work, we alleviate the challenges of low SNR in dMRI at low magnetic fields by performing super-resolution reconstruction (SRR). Our approach combines multiple low-resolution images acquired at different image slice rotations and employs a convolutional neural network to perform the SRR. Training is performed on noisy images. The network learns to extract and compose complementary image details into a super-resolution output image. Because of the properties of noise and the training process, the super-resolution images are less noisy than the directly acquired high-resolution ones, contain more high-resolution details than the input low-resolution images and the total acquisition time is decreased.

Keywords Super-resolution reconstruction · Low-field MRI · Diffusion MRI · Neural networks

---

J. Jurek (✉) · A. Materka
Institute of Electronics, Lodz University of Technology, Aleja Politechniki 10, 93590 Lodz, Poland
e-mail: jakub.jurek@p.lodz.pl

K. Ludwisiak
Department of Diagnostic Imaging, Independent Public Health Care, Central Clinical Hospital, Medical University of Lodz, Pomorska 251, 92213 Lodz, Poland

F. Szczepankiewicz
Medical Radiation Physics, Lund University, Barngatan 4, 22185 Lund, Sweden

P. Strumiłło et al. (eds.), *The Latest Developments and Challenges in Biomedical Engineering*, Lecture Notes in Networks and Systems 746,
https://doi.org/10.1007/978-3-031-38430-1_12

1 Introduction

1.1 Motivation

Resolution, signal-to-noise ratio (SNR) and acquisition time are mutually limiting factors in magnetic resonance imaging (MRI), crucial for image quality and hence the utility for diagnostic purposes. There are limits to image resolution increase, for example, since it is accompanied by lower SNR and/or longer acquisition time. By contrast, SNR can be improved by reducing resolution and with a time benefit. It was first demonstrated by Peled et al. [10] that super-resolution reconstruction (SRR)—a concept of overcoming the limitations of image acquisition systems by combining several poor quality measurements into a better quality SR image—is feasible for MRI. Following that paper, which was focused on diffusion tensor imaging, SRR was demonstrated to be useful in different MRI domains, for example in [3, 7, 13, 14]. This emphasizes that SRR can enable vastly improved image quality in scenarios where imaging would otherwise be impossible, or have prohibitively long acquisition times.

Recent studies have considered super-resolution of ultra low-field neurological MR images [1, 5]. Machines with lower magnetic field strength generally offer lower image quality. Low-field scanners have significantly limited resolution, although they are much more accessible due to lower price, better portability and smaller size, for example.

Diffusion MRI (dMRI), a tool of growing importance [12], especially in studying the brain, has generally lower resolution and lower SNR than morphological images like T2- or T1-weighted. In combination with low magnetic field strengths, imaging diffusion becomes a formidable challenge, especially if isotropic resolution is warranted. The SNR of such images may be prohibitive for radiological interpretation and other analyses.

The present work is a proof of concept for applying SRR techniques to reconstruct isotropic brain diffusion MRI data based on low-resolution scans from a 0.4 T scanner. This field strength is considerably higher than in [1, 5], but still offers significantly worse image quality than routinely used field strengths (1.5–3 T). Inspired by our previous work [13], we apply a similar image acquisition scheme, that involved collecting multiple thick-slice low-resolution images with slices lying in different planes. In contrast to the mentioned work, which used a linear SR reconstruction scheme, we exploit the richness of the dataset using a nonlinear convolutional neural network (CNN). Similarly as in [7], we use a multichannel extension of the Super-Resolution CNN (SRCNN) [2] that extracts complementary high-resolution features from its low-resolution inputs to estimate the SR image. But in contrast to [7], we use directly acquired isotropic dMRI for CNN training. We explore the influence of background masking, training on dMRI of different b-values and slice thickness of low-resolution images on SRR results. We verify the methods on dMRI scans of a diffusion phantom and a volunteer's brain at the diffusion encoding strengths of $b = 600\ \mathrm{s/mm}^2$.

1.2 Super-Resolution Reconstruction by Neural Networks

In the past, super-resolution reconstruction was accomplished both using linear and nonlinear methods [2, 11]. The linear paradigm is based on attempts to invert the low resolution image formation model, approximated by:

$$y = Ax + \epsilon \tag{1}$$

where A is a linear operator involving both blurring and downsampling. This model relates the high-resolution image x to a low-resolution image (or images) y, and includes additive noise ϵ. Commonly, the solution is sought by finding the optimal solution in the least-squares sense, which leads to the following [13] or similar equations:

$$\hat{x} = (A^\top A + \lambda I)^{-1} A^\top y \tag{2}$$

Finding the solution involves regularization, in [13] represented by λI, which reduces noise amplification at the cost of blurring. $A^\top y$ reflects averaging all the low-resolution images resampled to a common high-resolution grid.

Nonlinear methods based on neural networks are model free. Estimation of $\hat{x}$ is done via a neural network:

$$\hat{x} = NN(y) \tag{3}$$

where, NN is an operator representing the neural network, whose form is implicit and depends on the neural network parameters. In the case of convolutional networks, these parameters are weights of linear filters that are convolved with the images and then nonlinearly transformed by neural network activation functions, the rectified linear unit for instance. Optimal weights are found by optimization of a cost function calculated between network output (SR images) and target high-resolution images, and parametrized by these weights.

Using a convolutional neural network for the SRR problem has several benefits. It changes the problem of formulating image formation models like (1), finding their explicit solutions and designing regularization terms, to the problem of acquiring target high-resolution images. Secondly, it encompasses the actual, hidden or complex relationship between the low-resolution and high-resolution images, for example nonlinear slice excitation profiles. It avoids immediate averaging of low-resolution images as in $A^\top y$, which on one hand reduces noise, but on the other blurs the complementary high-resolution details present in these images. And importantly, it can deal with nonlinear effects of the linear sampling operation, such as aliasing artifacts [6].

In this paper, the neural network is trained with the following assumptions. The inputs are low-resolution images with rotated slice planes, co-registered to the high-resolution grid and resampled using linear interpolation. They are fed to the neural network as individual channels and are combined optimally by the network. Targets are directly acquired high-resolution images. This implies that both input and target images are noisy. Since we are dealing with modulus images, the noise distribution is Rician [4] and has non-zero mean value. It was shown in Noise to Noise theory [9], that zero-mean noise does not alter the optimum of the mean squared error cost function, making it possible to learn clean images based on independent noisy pairs. We notice that when noise has non-zero mean, the network is expected to estimate its mean value and add it to output images in the form of a bias. Leaving theoretical proofs to another work, we demonstrate empirical evidence for this reasoning. Rician bias is expected to be greater in high-resolution target images than in the low-resolution inputs.

We also notice that noisy background regions may have negative influence on training, so we performed background masking.

In its full form, the developed SRR algorithm involves registration of low-resolution images to the high-resolution image grid, background masking and SR reconstruction by means of a convolutional neural network.

2 Materials

dMR images at b-values of 0 and 600 s/mm^2 (b0 and b600 in short) were acquired in a healthy volunteer and in a diffusion phantom [8] on a Hitachi Aperto Lucent scanner with 0.4 T magnetic field. Low resolution imaging was performed at two slice thicknesses with the following settings (values in the brackets are for 5 mm slice thickness, others are common): 2D EPI sequence, TR = 5214 ms (6162 ms), TE = 116 ms, single diffusion gradient direction, number of signal averages = 4, field of view (FOV) = 250 × 250 mm^2, 128 × 128 reconstruction matrix, slice thickness = 6.2 mm (5 mm), 1.95 × 1.95 mm^2 in-plane resolution, 22 (26) contiguous slices, bandwidth = 142.9 kHz, distortion correction on. To ensure that data can support the super-resolution reconstruction, multiple rotations of the FOV must be used such that the low resolution direction is evenly distributed on a circle. We used the rotation scheme from [13], resulting in five rotations about the anterior-posterior axis for both slice thicknesses. The rotation angles were sampled equally from the 180° range, yielding 0° (axial slices), 36°, 72°, 108° and 144°. Acquisition time was 475 s for volumes with 6.2 mm slices (95 s per rotation angle) and 560 s for volumes with 5 mm slices (112 s per rotation angle). The rotation scheme is visualized in Fig. 1. For super-resolution reconstruction, only one low resolution dataset is necessary: either with 5 mm or 6.2 mm slices. We compared the results of these options.

High resolution imaging was performed with the following settings: 2D EPI sequence, TR = 8769 ms, TE = 116 ms, single diffusion gradient direction, number of signal averages = 4 and 16 (for b = 0 and b = 600 s/mm^2), FOV = 250 × 250 mm^2, 128 x 128 reconstruction matrix, slice thickness = 2 mm, 1.95 × 1.95 mm^2 in-plane resolution, no gap between slices, bandwidth = 142.9 kHz, distortion correction on, b-values = 0 and 600 s/mm^2. 2 mm slices were acquired in the axial plane. 37 slices were acquired for the phantom. Due to scanner limitations, for the brain scan, the brain was covered by two 'slabs', each of 37 slices, incorporating the superior and

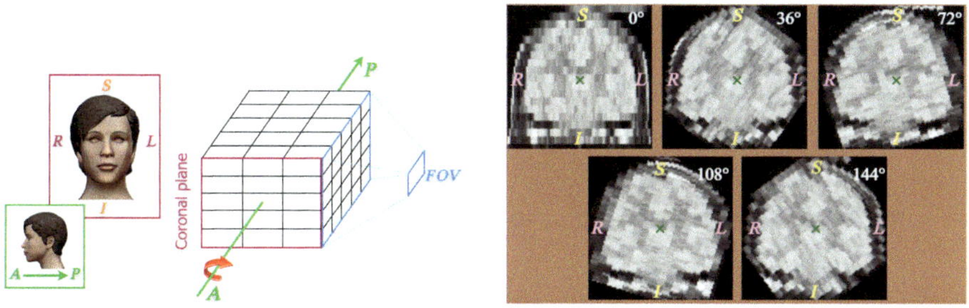

(a) The FOV is rotated about the fixed antero-posterior axis. A case of a 90° rotation is shown

(b) Coronal plane view of the LR images obtained for five rotation angles

Fig. 1 The FOV rotation scheme used in the experiment. The orientation of thick slices is different for each LR image, but the rotation axis is fixed (AP)

inferior brain, respectively. The fields of view for the slabs were positioned manually by the scanner operator so that there was no gap between them. Scan time was 74 s for the b0 with NEX = 4 and 290 s for the b600 with NEX = 16 (per slab), yielding 728 s in total.

3 Methods

3.1 Image Registration

SRR usually requires resampling low-resolution images to a common, high-resolution grid. If global image coordinates are known, this transform can be calculated, but since object motion between scans is possible, it is better to optimize the transformation using image similarity metrics. Thus, we co-registered the low-resolution images to the high-resolution ones using the *affine_registration* function from the DIPY library (Diffusion Imaging in Python), that matches images based on the mutual information metric. Registration was performed separately for b0 and b600 images, and, due to possible motion, separately for the two 'slabs' of the brain. Linear interpolation was used for resampling.

3.2 Background Masking

Background is the area of largest bias and contains no tissue, hence the utility of background examples in training a SRR neural network is minimal.

We thus performed background masking based on simple thresholding. The threshold was determined manually in low-resolution images. The resulting region was masked in both low- and high-resolution examples by setting the intensity to 0.

3.3 Neural Network Architecture and Training

For SR reconstruction, we used the SRCNN architecture [2, 7]. Compared to its original implementation, we reduced the number of filters in the first two layers to 8 and 64, which was based on our previous experience with this network. Thus, the final architecture was composed of three layers, with 8 filters of size 9×9 in the input layer, 64 filters of size 1×1 in the middle layer and 1 filter of size 5×5 in the output layer. The input to the network was organised similarly as in [7]. Interpolated low-resolution slices from the five rotated scans were set up as input channels. Training was performed based on the mean squared error as the cost function:

$$J(w) = \frac{1}{M}\sum_{m=1}^{M}(x - \hat{x})^2 \tag{4}$$

where w are network parameters, the summation is over all training examples, and x, $\hat{x}$ are defined as in Eqs. (1) and (3). The examples were 16×16 image patches, with M, the number of training examples, equal to 720. At this point, we considered the 2D reconstruction scheme only, involving individual slices and not 3D volumes, which corresponds to the applied FOV rotation pattern. $J(w)$ was optimized using the Adam algorithm in the mini-batch training approach with randomly shuffled mini-batches of size 16, with the learning rate of 0.001. We used an early stopping algorithm as in [7] to identify the epoch at which training should stop due to little progress or overfitting. The validation data used for early stopping were slices cropped to the size of 74×74 (valid region of overlapping FOVs) and there were five of them.

4 Results

We conducted several SRR experiments described in the following subsections: on the type of diffusion data used for training, on the influence of background masking, and on the influence of slice thickness. We then selected these steps that improved the SRR result and combined them to reconstruct b600 dMRI of the brain.

4.1 Training on b600 Versus b0 Images

It is a natural and theoretically well-grounded approach to train a neural network on data from the same distribution as the data on which the trained network is to be deployed. Thus, to reconstruct b600 dMRI of the brain, we used low-resolution (6.2 mm)—high-resolution pairs from b600 images for training. For comparison, we also trained the neural network with pairs from b0 images. Results are visible in Fig. 2. It follows from Fig. 2 that training on targets with low SNR and Rician noise, as in the case of b600 images, imposes a bias on the SR output. This bias is most likely equal to the Rician bias—the mean value of Rician noise at a particular voxel. By contrast, the SR output from a network trained on b0 data has much smaller bias due to a higher SNR of the targets [4]. The SR output for the b600-trained network seems to be less noisy and more blurry than the b0-trained network SR output, but both have a better definition of high-resolution details than the low-resolution slice shown in Fig. 2b. In subsequent experiments we continued training the SRR network on b0 images.

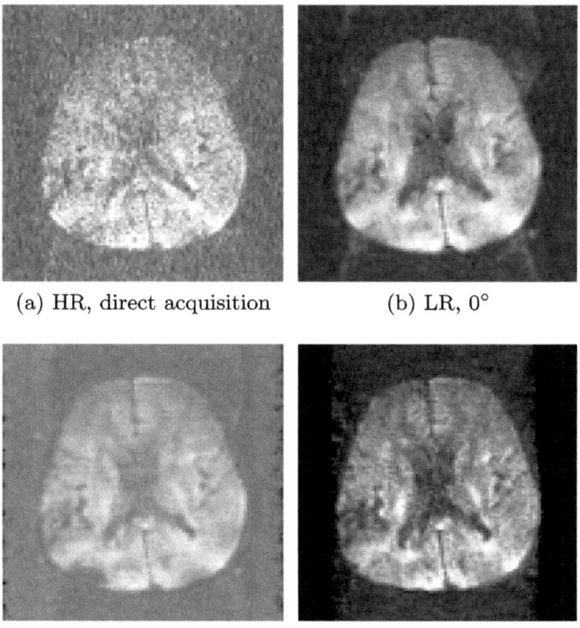

(a) HR, direct acquisition (b) LR, 0°

(c) SR, b600 training (d) SR, b0 training

Fig. 2 Comparison of super-resolution reconstruction of a b600 image slice by neural networks trained on b0 and b600 images. SR images inherit the bias, but not the noise variance, from the high-resolution targets. Hence, training on b600 data is infeasible. **a**—acquired image with 2 mm slice thickness, **b**—acquired image with 6.2 mm slice thickness, **c**, **d**—reconstructed images with 2 mm slice thickness

4.2 Influence of Background Masking

In Fig. 2d, bias is still visible. Figure 3b shows an intensity profile drawn from a slice of the b600 SR image. The blue line concerns the network trained on b0 images. A groove is visible in the intensity profile at a coordinate of approximately 42. This is an area filled with the cerebrospinal fluid that has a high diffusion coefficient and thus should have very low intensity in strongly diffusion-weighted images. We identified the reason for this residual bias in the relatively large number of examples containing the background. Bias is maximal in the background region since there is no tissue both in low- and high-resolution images. Thus, it is likely that the neural network will learn to keep or even increase the background bias, which might also concern other low-intensity regions.

Training was performed on b0 pairs with masked background, 6.2 mm slice thickness for low-resolution images. Figure 3 shows the SR result and the corresponding intensity profile. It is visible that the contrast between the cerebrospinal fluid and brain parenchyma was improved after application of background masking to the training data.

Background was not masked for SRR reconstruction using the trained networks. It follows from Fig. 3 that background masking imposes on the network a property of detecting and zeroing the background. Interestingly, this effect is observed for background and not for the cerebrospinal fluid region, which should have non-zero intensity indeed.

4.3 SRR Reconstruction of Phantom and Brain dMRI

We combined training on b0 images with background masking to train neural network for SRR of diffusion phantom and brain b600 dMRI, from low-resolution images with 6.2 mm thick slices. The results are shown in Figs. 4 and 5. For comparison with the neural-network-based SRR, we also show the trivial reconstruction obtained by averaging interpolated, low-resolution images. We refer to this method as fusion. It is apparent from Figs. 4 and 5 that fusion yields blurry results. For the phantom, we notice that automatic registration of the input low-resolution images may be inaccurate (Fig. 4d). This causes fusion not only to be blurry, but also shaken. Despite the slight misregistration, the SR image is sharp. It has lower bias than the directly acquired HR image, and lower noise. Both effects are also visible in the profile drawn in Fig. 4i. All phantom images have visible ghosting artifacts, also sustained by SRR.

Results obtained for the brain show superior quality of the SR results. The output is visibly less noisy and significantly less biased than the HR image. It is however as noisy as the input low-resolution images.

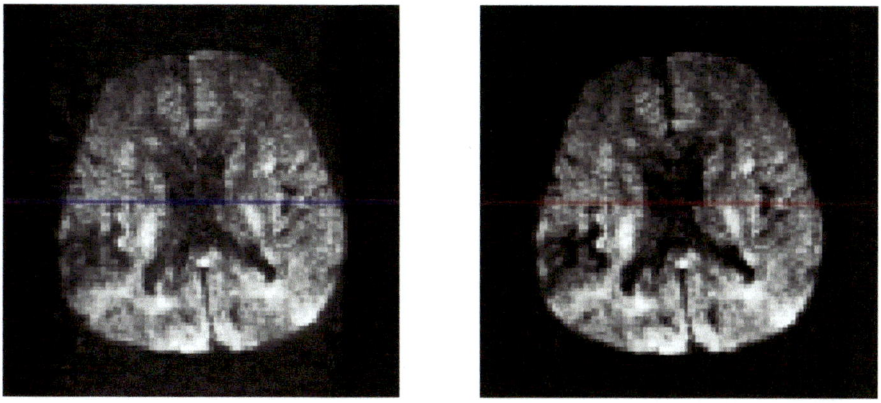

(a) Without (left) and with (right) background masking

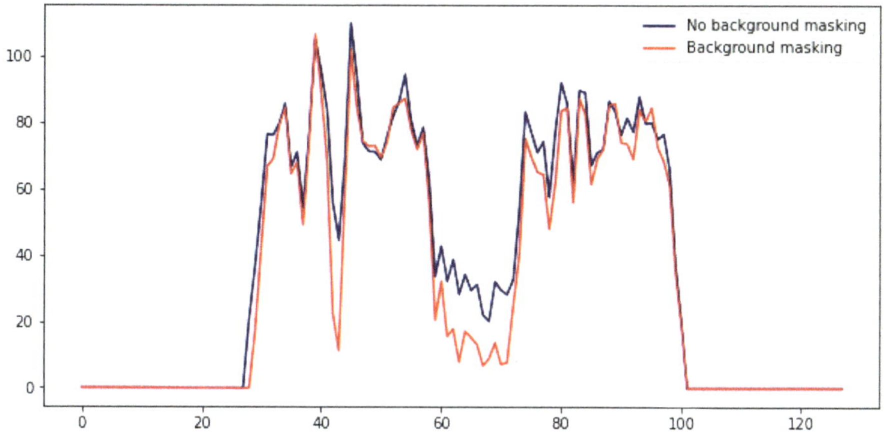

(b) Intensity profile

Fig. 3 Comparison of SRR results between a network trained without background masking and with it. The profile (**b**) drawn along the blue and red lines in (**a**) and a visual examination of (**a**) reveal that contrast is improved between the low-intensity areas and brain parenchyma

4.4 Impact of Decreasing the Slice Thickness of Acquired Low-Resolution Images

The rule for the number of rotated low-resolution images N_R necessary for SRR, as given by [13], states $N_R \geq \frac{\pi}{2} \cdot \alpha$, where α is the ratio of slice thickness to the in-plane voxel side length (the upsampling factor). For the acquired images with 1.95×1.95 mm^2 in-plane voxel size and $N_R = 5$, the maximum slice thicknesses is 6.2 mm according to the formula.

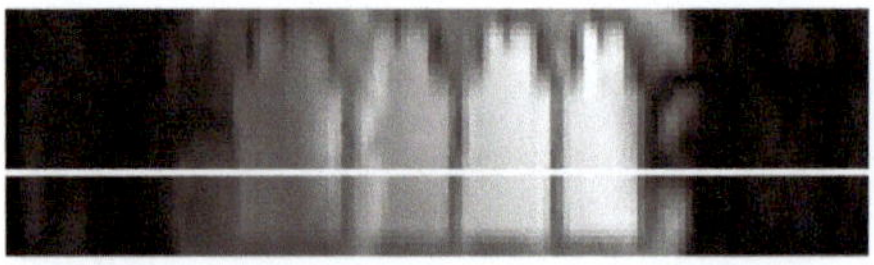

(a) 0°, LR, NEX=4

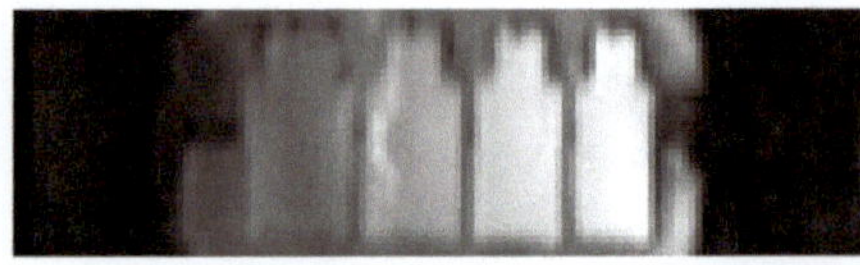

(b) 36°, LR, NEX=4

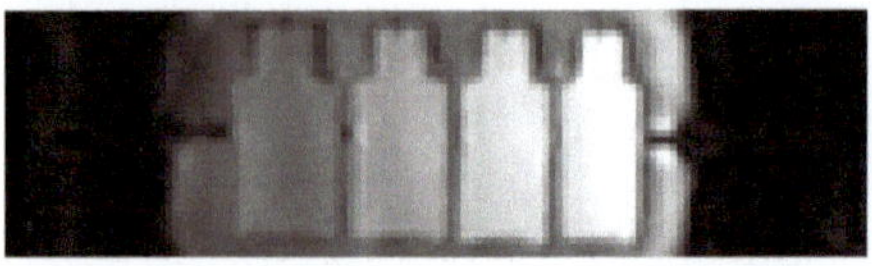

(c) 72°, LR, NEX=4

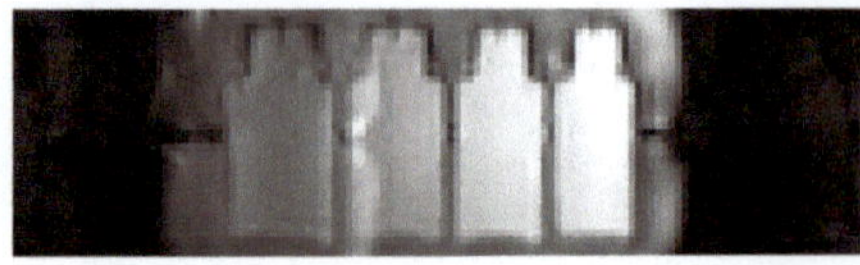

(d) 108°, LR, NEX=4

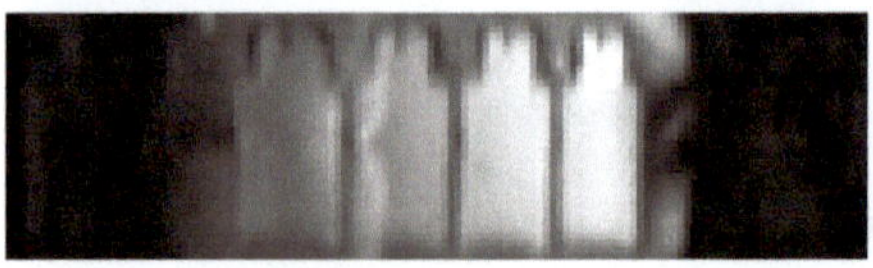

(e) 144°, LR, NEX=4

(f) Fusion, NEX=4

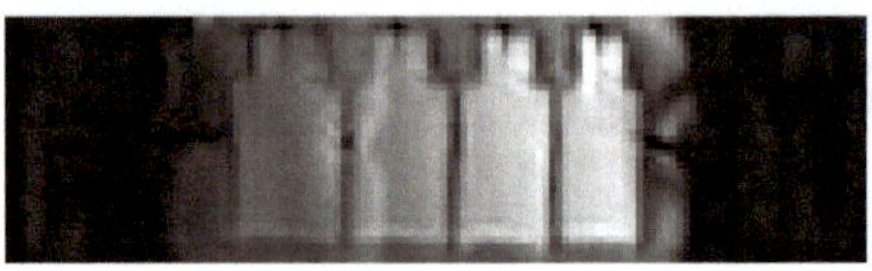

(g) SR, NEX=4

(h) HR, NEX=16

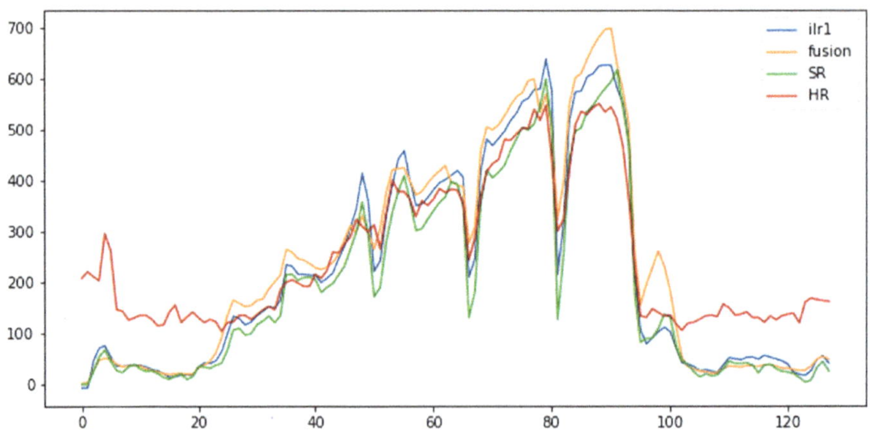

(i) Intensity profile along the line drawn in (a)

Fig. 4 b = 600 s/mm^2 dMRI of the diffusion phantom from the 0.4 T scanner. A slice from rotated low-resolution scans, their fusion, the SR image and from a high-resolution image are shown. Considerable Rician bias and poor SNR is visible for the high-resolution image. The low-resolution image in (**d**) is visibly misregistered even after automatic registration, which causes the fusion result to be shaken. This is not observed for the SR result, which is less noisy and less biased then the HR image

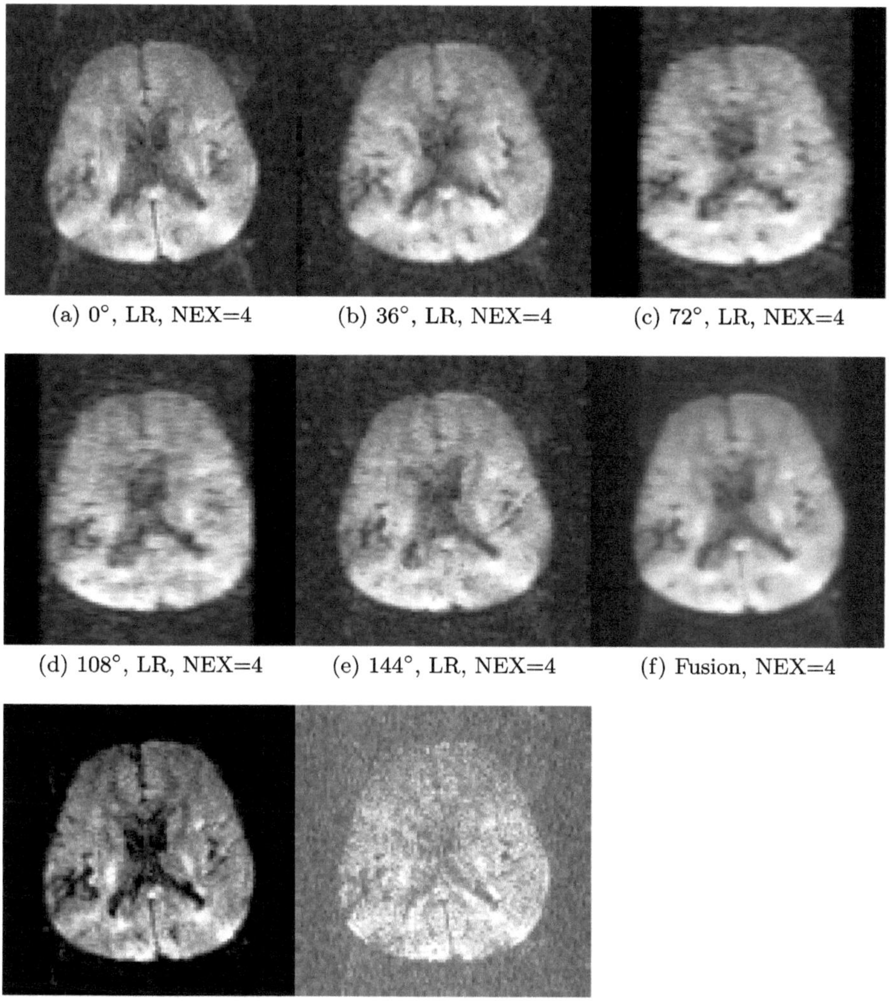

(a) 0°, LR, NEX=4 (b) 36°, LR, NEX=4 (c) 72°, LR, NEX=4

(d) 108°, LR, NEX=4 (e) 144°, LR, NEX=4 (f) Fusion, NEX=4

(g) SR, NEX=4 (h) HR, NEX=16

Fig. 5 b = 600 s/mm^2 dMRI of the brain from the 0.4 T scanner. A slice from rotated low-resolution scans, their fusion, the SR image and from a high-resolution are shown. Considerable Rician bias and poor SNR is visible for the high-resolution image. The SR image seems to preserve the details of the HR images, but is less noisy and less biased

Figure 6 shows a comparison of SR and fusion results for low-resolution images acquired with 5 and 6.2 mm slice thickness. To better visualise the effects of decreasing the slice thickness, differences were computed between these images and their fusion and SRR, after co-registration and resampling to the same HR grid. The intensity of 5 and 6.2 mm images were normalized before computing the errors.

Difference images computed for low-resolution axial (0°) images reveal high resolution details and noise. The difference is visibly smaller for fusion images and

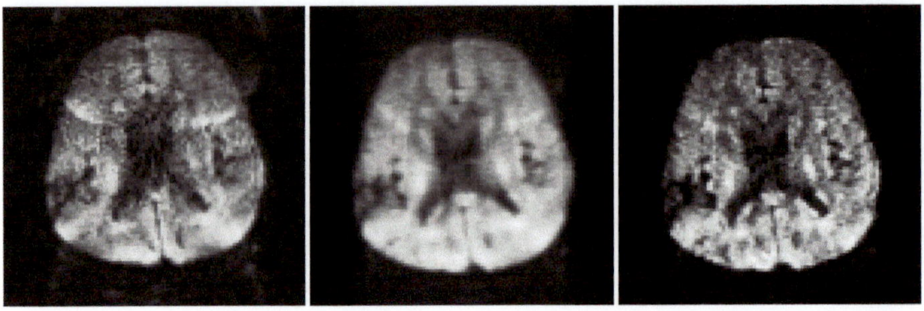

(a) 0°, LR, 5 mm (b) Fusion, 5 mm (c) SR, 5 mm

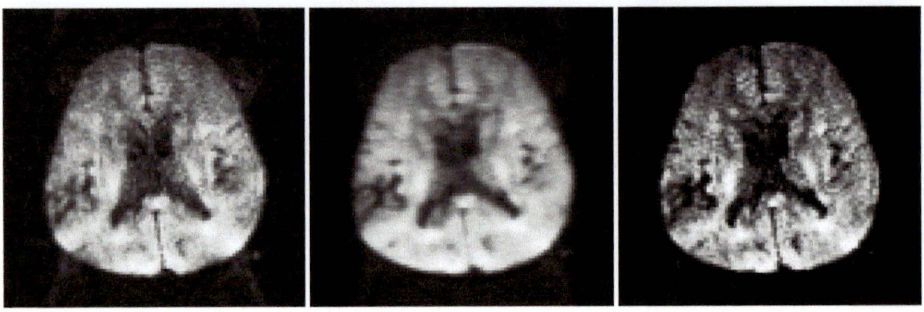

(d) 0°, LR, 6.2 mm (e) Fusion, 6.2 mm (f) SR, 6.2 mm

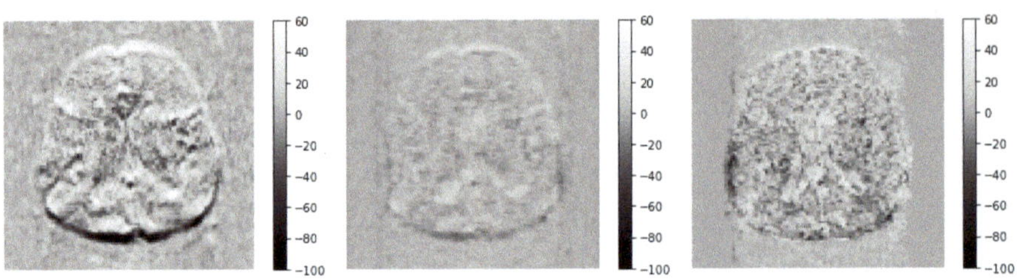

(g) Diff. (a)-(d) (h) Diff. (b)-(e) (i) Diff. (c)-(f)

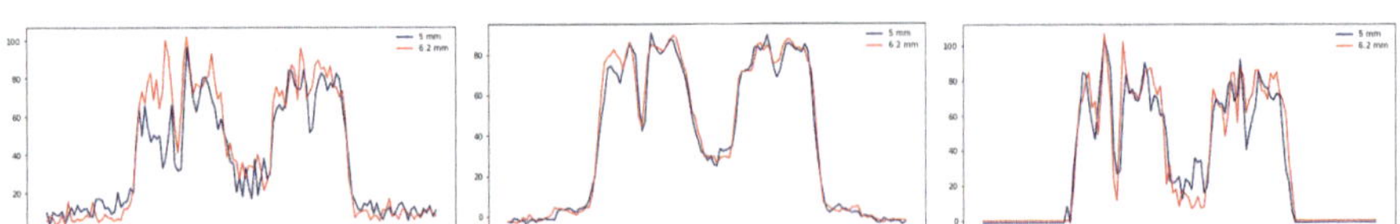

(j) Intensity profiles for (a, d) (k) Intensity profiles (b, e) (l) Intensity profiles (c, f)

Fig. 6 Comparison of b = 600 s/mm^2 images of the brain, acquired as 5 and 6.2 mm slices. The difference maps and the intensity profiles, drawn along the same line as in Fig. 3a, suggest that the difference between the fusion and SR images, as evaluated by visible residual details, is reduced compared to the LR images. Slice thickness seems to have limited influence on the SRR for the considered number of rotated low-resolution images and upsampling factor, but smaller thickness implies lower SNR and thus greater bias the SR ones, although noise is higher in the latter. Comparison of the SR images and their intensity profiles leads to the conclusion that SR images obtained from 6.2 mm images have smaller bias and lower noise. This is in accordance with the theory, since images with thinner slices are expected to have lower SNR and thus higher bias. Both factors influence the SR results adversely. Overall, the level of details retained in both SR images is similar, but images with thicker slices are acquired faster, have lower noise and bias.

5 Discussion and Conclusions

Isotropic brain dMRI images at b = 600 s/mm^2, 0.4 T field strength and 1.95 $\times$ 1.95 $\times$ 2 mm^3 resolution were successfully reconstructed from five low-resolution ones with rotated FOV and 6.2 mm slices thickness. SR images were visually compared to their high-resolution counterparts. Comparison reveals that SR images have a similar level of detail and considerably lower noise and reduced Rician bias. One should note the SR images were obtained with four repeated acquisitions (NEX = 4), while sixteen repetitions were used to acquire the high-resolution images. In total, image acquisition that allowed SRR reconstruction from volumes with 6.2 mm slices was 254 s faster than acquisition of a HR image directly. We investigated the b-value of 600 s/mm^2 and not higher ones since that was the maximum value allowed by the used scanner model. We did not use quantitative measures at the current stage of the study due to the lack of an appropriate reference image with high resolution and high SNR. This will be alleviated in the future by using digital phantoms.

Our experiments showed that SRR was not successful when the neural network was trained on b600 data. SR images inherit the bias present in the noisy high-resolution targets. However, training on b0 data gave satisfying visual results. This is in accordance with our previous work, where a SRR neural network trained on synthetic brain images was deployed successfully on real MRI from a different domain without additional training [7]. Interestingly, the SR images are robust to the variance of noise in the targets, which is in accordance with Noise2Noise theory [9]. The results suggest that low-resolution images should be acquired with maximal possible slice thickness for a given number of FOV rotations [13], which minimizes the bias and maximizes the SNR. Background should be masked to further reduce the bias.

In the future, we will study algorithms for denoising the input images and debiasing the target images, which is expected to further improve reconstruction quality. We will also investigate diffusion parameter mapping from 0.4T dMRI.

To conclude, neural network super-resolution reconstruction allowed to obtain an isotropic diffusion image of the brain from five low-resolution input images. The required scanning time was shorter than direct acquisition, and the resulting images were less noisy.

References

1. de Leeuw den Bouter, M.L., Ippolito, G., O'Reilly, T.P.A., Remis, R.F., van Gijzen, M.B., Webb, A.G.: Deep learning-based single image super-resolution for low-field MR brain images. Sci Rep **12**(1) (2022). https://doi.org/10.1038/s41598-022-10298-6
2. Dong, C., Loy, C.C., He, K., Tang, X.: Image super-resolution using deep convolutional networks. IEEE Trans. Pattern Anal. Mach. Intell. **38**, 295–307 (2016). https://doi.org/10.1109/TPAMI.2015.2439281
3. Gu, J., Li, Z., Wang, Y., Yang, H., Qiao, Z., Yu, J.: Deep generative adversarial networks for thin-section infant MR image reconstruction. IEEE Access **7**, 68290–68304 (2019). https://doi.org/10.1109/access.2019.2918926
4. Gudbjartsson, H., Patz, S.: The Rician distribution of noisy MRI data. Magn. Reson. Med. **34**, 910–914 (1995). https://doi.org/10.1002/mrm.1910340618
5. Iglesias, J.E., Schleicher, R., Laguna, S., Billot, B., Schaefer, P., McKaig, B., Goldstein, J.N., Sheth, K.N., Rosen, M.S., Kimberly, W.T.: Accurate Super-Resolution Low-Field Brain MRI (2022). https://doi.org/10.48550/ARXIV.2202.03564
6. Jurek, J.: Super-Resolution Reconstruction of Three-dimensional Magnetic Resonance Images Using Deep and Transfer Learning (2020). https://doi.org/10.13140/RG.2.2.23612.46723
7. Jurek, J., Kociński, M., Materka, A., Elgalal, M., Majos, A.: CNN-based superresolution reconstruction of 3D MR images using thick-slice scans. Biocybern. Biomed. Eng. **40**(1), 111–125 (2020). https://doi.org/10.1016/j.bbe.2019.10.003
8. Jurek, J., Materka, A., Ludwisiak, K., Majos, A., Gorczewski, K., Cepuch, K., Zawadzka, A.: Supervised denoising of diffusion-weighted magnetic resonance images using a convolutional neural network and transfer learning. Biocybern. Biomed. Eng. **43**(1), 206–232 (2023). https://doi.org/10.1016/j.bbe.2022.12.006
9. Lehtinen, J., Munkberg, J., Hasselgren, J., Laine, S., Karras, T., Aittala, M., Aila, T.: Noise2noise: learning image restoration without clean data. In: Dy, J.G., Krause, A. (eds.) Proceedings of the 35th International Conference on Machine Learning, ICML 2018, Stockholmsmässan, Stockholm, Sweden. Proceedings of Machine Learning Research, vol. 80, pp. 2971–2980. PMLR (2018). http://proceedings.mlr.press/v80/lehtinen18a.html
10. Peled, S., Yeshurun, Y.: Superresolution in MRI: Application to human white matter fiber tract visualization by diffusion tensor imaging. Magn. Reson. Med. **45**(1), 29–35 (2001). https://doi.org/10.1002/1522-2594(200101)45:1<29::aid-mrm1005>3.0.co;2-z
11. Reeth, E.V., Tham, I.W.K., Tan, C.H., Poh, C.L.: Super-resolution in magnetic resonance imaging: a review. Concepts Magn Reson Part **40A**(6), 306–325 (2012). https://doi.org/10.1002/cmr.a.21249
12. Tax, C.M., Bastiani, M., Veraart, J., Garyfallidis, E., Irfanoglu, M.O.: What's new and what's next in diffusion MRI preprocessing. NeuroImage **249**, 118830 (2022). https://doi.org/10.1016/j.neuroimage.2021.118830
13. Vis, G., Nilsson, M., Westin, C.F., Szczepankiewicz, F.: Accuracy and precision in super-resolution MRI: Enabling spherical tensor diffusion encoding at ultra-high b-values and high resolution. NeuroImage **245**, 118673 (2021). https://doi.org/10.1016/j.neuroimage.2021.118673
14. Woo, J., Murano, E.Z., Stone, M., Prince, J.L.: Reconstruction of high-resolution tongue volumes from MRI. IEEE Trans. Biomed. Eng. **59**(12), 3511–3524 (2012). https://doi.org/10.1109/tbme.2012.2218246

Modeling and Machine Learning

Improving the Predictive Ability of Radiomics-Based Regression Survival Models Through Incorporating Multiple Regions of Interest

Agata Małgorzata Wilk, Emilia Kozłowska, Damian Borys, Andrea D'Amico, Izabela Gorczewska, Iwona Debosz-Suwińska, Seweryn Gałecki, Krzysztof Fujarewicz, Rafał Suwiński, and Andrzej Świerniak

Abstract Radiomic features, numeric values extracted from a region of interest (ROI) in medical images, can be used to train prognostic models for various types of cancer. However, in locally advanced diseases, more than one lesion may be present. Using the information contained in multiple regions increases the complexity and necessitates additional processing. Here, we tested seven strategies of handling multiple regions in radiomic-based regularized Cox regression for predicting metastasis-free survival using a cohort of 115 non-small cell lung cancer patients. We have found that using all ROIs to fit the model allowed for better results than using only the largest ROI, achieving c-indexes of 0.617 and 0.581, respectively.

Keywords Lung cancer · Radiomics · ROI · Metastasis free survival · Regularized Cox regression

---

A. M. Wilk (✉) · E. Kozłowska · D. Borys · S. Gałecki · K. Fujarewicz · A. Świerniak
Department of Systems Biology and Engineering, Silesian University of Technology, 44-100 Gliwice, Poland
e-mail: andrzej.swierniak@polsl.pl

A. M. Wilk
Department of Biostatistics and Bioinformatics, Maria Sklodowska-Curie National Research Institute of Oncology, Gliwice Branch, 44-102 Gliwice, Poland

D. Borys · A. D'Amico · I. Gorczewska
Department of Nuclear Medicine and Endocrine Oncology, PET Diagnostics Unit, Maria Sklodowska-Curie National Research Institute of Oncology, Gliwice Branch, 44-102 Gliwice, Poland

I. Debosz-Suwińska
Department of Radiotherapy, Maria Sklodowska-Curie National Research Institute of Oncology, Gliwice Branch, 44-102 Gliwice, Poland

R. Suwiński
II-nd Radiotherapy and Chemotherapy Clinic and Teaching Hospital, Maria Sklodowska-Curie National Research Institute of Oncology, Gliwice Branch, 44-102 Gliwice, Poland

P. Strumiłło et al. (eds.), *The Latest Developments and Challenges in Biomedical Engineering*, Lecture Notes in Networks and Systems 746,
https://doi.org/10.1007/978-3-031-38430-1_13

1 Introduction

Medical imaging is a standard procedure used in cancer both as part of diagnostics, and disease management, for example in radiotherapy planning [5] The high-depth images generated during PET/CT, MRI, or other types of scans contain an abundance of information, invisible to the human eye, but usable in machine learning models. One of the ways of using this information is radiomics.

In radiomics, quantitative features are extracted from an expert-defined region of interest (ROI) in the image, describing various characteristics related to, among others, voxel intensity distributions, shape, or texture. Following successful standardization efforts [11], radiomic biomarkers are widely researched for use in cancer prognostics [4, 6, 10]. Usually, the modeling focuses on the primary tumor as a source of features and thus incorporates only one region of interest per patient. However, especially for locally advanced disease, the image may include multiple lesions, lymph nodes, or other affected regions, disregarded in the one patient—one ROI approach.

One example of cancer that is often already locally disseminated at diagnosis, is lung cancer, with most patients diagnosed at an advanced clinical stage of the disease. Combined with high metastatic potential, the late diagnosis contributes to low survival rates, with an estimated five-year survival of below 20% [3].

In our earlier work, we demonstrated the potential of radiomic features for predicting metastasis-free survival in non-small cell lung cancer (NSCLC) patients [2, 9]. Building upon this research, we investigate whether incorporating multiple ROIs from one patient in the model can improve its predictive ability. In a cross-validation procedure modified to account for the dependence of samples from one patient, we proposed and evaluated seven different methods of handling multiple ROIs. We show that although introducing multiple ROIs is associated with the higher noisiness of the data, using an appropriate strategy allows for more accurate prediction than using only one lesion.

2 Materials and Methods

2.1 Patient Characteristics

The cohort consists of 115 NSCLC patients treated in the Maria Sklodowska-Curie National Research Institute of Oncology, Gliwice Branch between 2009 and 2017. The patients received combination treatment, with between one and six cycles of platinum-based chemotherapy followed by a 60–70 Gy total dose of radiotherapy. The primary endpoint of interest was metastasis-free survival (MFS), defined as the time between cancer diagnosis and detection of distant metastases or time to last follow-up otherwise. Therefore, only patients without distant tumor spread detectable at diagnosis were included in the study group.

Table 1 Patient characteristics. For age, median and quartiles are given

Sex	Male	83 (72.2%)
	Female	32 (27.8%)
Age		61 (57–67)
Histopathology	Squamous	77(67.0%)
	Large cell	8 (7.0%)
	Adenocarcinoma	28 (24.3%)
	Other	2 (1.7%)
Location	Left	65 (56.5%)
	Right	50 (43.5%)
T	1	4 (3.5%)
	2	37 (32.2%)
	3	37 (32.2%)
	4	37 (32.2%)
N	0	19 (16.5%)
	1	6 (5.2%)
	2	83 (72.2%)
	3	7 (6.1%)
M	0	115 (100%)
	1	0 (0%)
Zubrod score	0	34 (29.6%)
	1	80 (69.6%)
	2	1 (0.9%)

Typically for lung cancer, the majority of patients were male. The median age at diagnosis was 61 years. Two-thirds of the cohort had squamous cell carcinoma, other represented subtypes were adenocarcinoma (24.3%) and large cell (7%). The left lung was slightly more prevalent as the tumor location. Most patients had advanced disease, with over 60% classified as T3 or T4 in the TNM staging, and over 80% exhibiting various degrees of lymph node spread. However, the general condition of the patients was good, as all but two patients were categorized as 0 or 1 on the Zubrod performance scale. Detailed information about the study group is presented in Table 1.

2.2 Image Acquisition

For 24 patients the planning PET/CT images were acquired using Philips GeminiGXL 16 (Philips, Amsterdam, Netherlands) scanner, and for 88 patients using Siemens Biograph mCT 128 (Siemens AG, Munich, Germany) scanner. Images were contoured by an experienced nuclear medicine specialist utilizing MIM 7.0.1 soft-

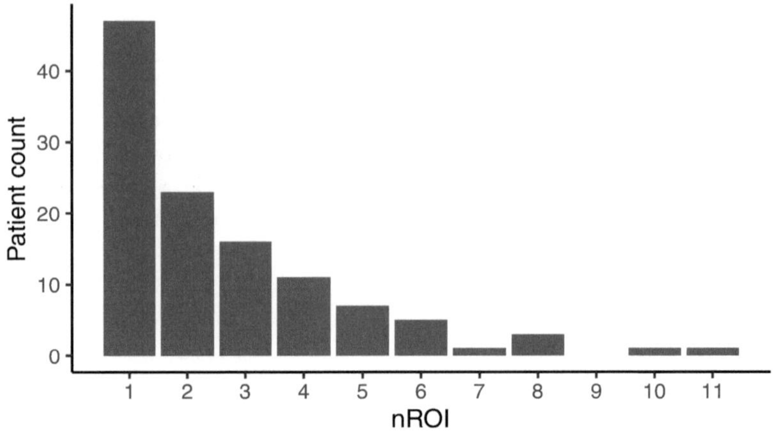

Fig. 1 Number of contoured ROIs per patient

ware and the PET Edge™ tool (both MIM Software Inc., OH, USA). In addition to the primary tumor, the ROIs included other tumor lesions in the lung and lymph nodes. Out of the entire cohort, 47 patients had only one ROI contoured, the remaining 68 had between two and eleven ROIs (Fig. 1).

2.3 Radiomic Feature Extraction

For each of the contoured ROIs, we extracted 105 radiomic features describing their shape, texture etc., using PyRadiomics version 3.0.1 [8]. The features can be divided into seven categories: first-order statistics (18 features), shape (14 features), Gray Level Co-occurrence Matrix (GLCM, 22 features), Gray Level Dependence Matrix (GLDM, 14 features), Gray Level Run Length Matrix (GLRLM, 16 features), Gray Level Size Zone Matrix (GLSZM, 16 features), and Neighboring-Gray Tone Difference Matrix (NGTDM, 5 features).

2.4 Survival Analysis—General Framework

To test the constructed models, we employed a 1000-iteration Monte Carlo Cross-Validation scheme with fixed subsets. Since the dataset contained multiple observations (ROIs) per patient, the subsets had to be generated in a way preventing information leakage. Therefore, in each iteration, the patients were divided into a training (two-thirds) and test (one-third) groups, stratified according to the metastasis status. The resulting training and test sets contained all ROIs from the assigned patients.

While such partitioning ensured that all ROIs of a given patient were used either for training or for testing, and the number of patients in the subsets was always the same, the set cardinalities could be different in each iteration (depending on method) due to varying number of contoured lesions per patient.

To explore the association between radiomic features and the metastasis-free surival we applied the Regularized Cox Regression model (CoxNet) implemented in the glmnet R package, version 4.1–6 [1, 7]. The optimal λ value was found using a cross-validation procedure. We assessed the predictive ability of the fitted models using the Harrell's c-index calculated for patients in the test set, adopting the median across all iterations as the final quality index.

2.5 Handling Multiple ROIs

We focused on two main approaches to handling multiple ROIs:

1. selecting (or constructing) a single ROI per patient and using it to fit the model, or
2. fitting the model on a full dataset (containing all ROIs) and reconstructing patient risk from the risks obtained for the ROIs.

The advantage of the second idea is having a larger dataset to build the model. However, because of the substantial heterogeneity, it may also be more subject to data noise.

Between the two approaches, we tested a total of seven distinct methods of handling multiple ROIs. The overview of the tested strategies is presented in Fig. 2.

RandomROI In this approach, one ROI is randomly selected to represent a patient. It requires no defined order existing for the ROIs, but the results are largely dependent on chance and change with each realization of the method. Considering the pronounced differences between ROIs from the same patient, the instability makes it largely unsuitable for potential clinical use. Here it will be regarded as the baseline model for comparison.

LargestROI Again, one ROI is selected to represent each patient, taking advantage of the intuitive order according to size (here defined as the voxel number of the mask). Selecting the largest ROI ensures that the result is identical each time the method is called. However, it does not utilize any information from the remaining ROIs. Indeed, since the largest of the lesions is usually the primary tumor, this method corresponds to the classic approach adopted in radiomics studies.

ArithmeticMeanROI A representative ROI is constructed, where each feature is an arithmetic average calculated across all the patient's ROIs. Although this method incorporates information from all the contoured lesions, it puts into question the interpretability of obtained radiomic features. For example, shape features like major axis length have fairly straightforward interpretation when applied to one ROI, but are no longer meaningful when averaged.

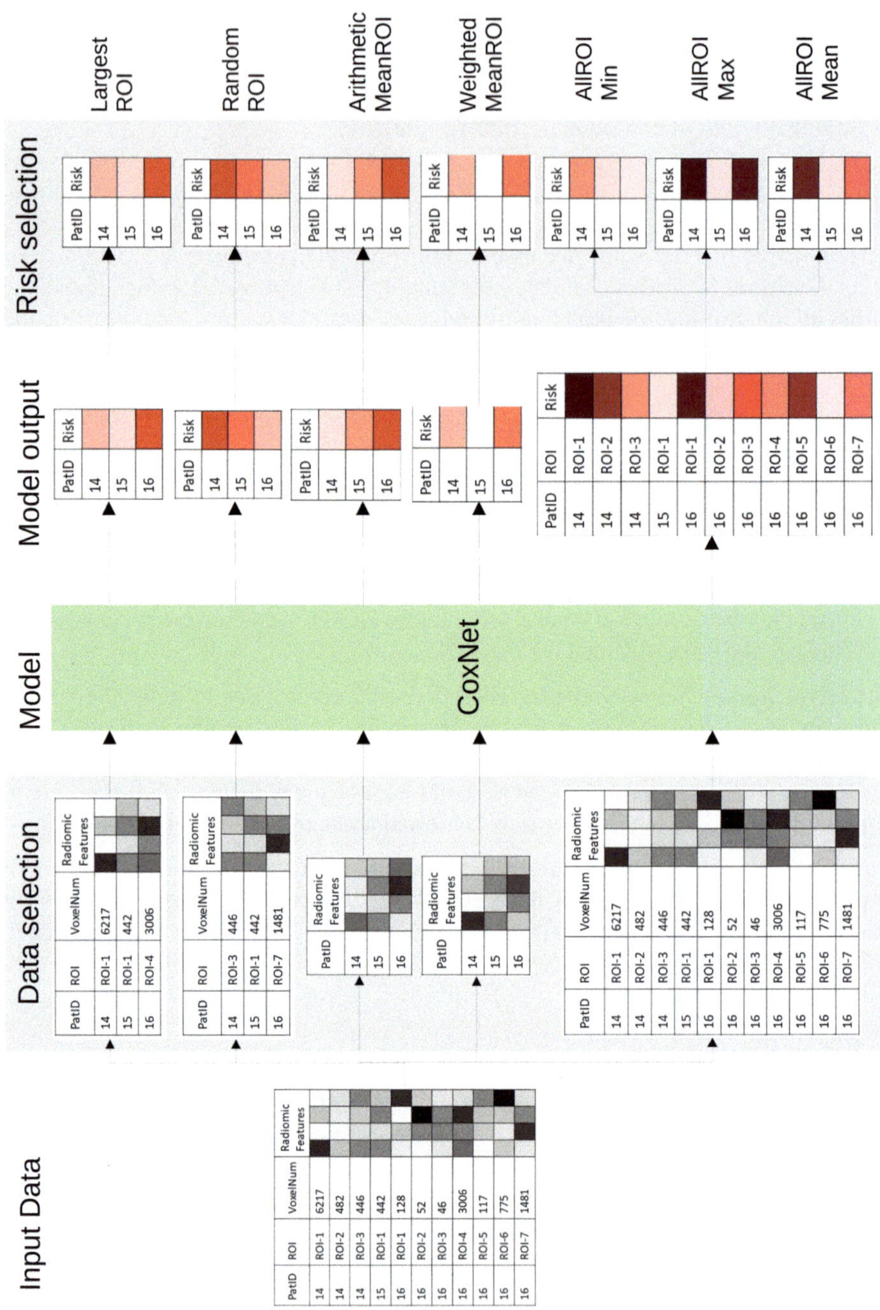

Fig. 2 Methods of handling multiple ROIs

WeightedMeanROI Analogous to the previous method. The representative ROI is calculated as a weighted mean of all ROIs with their sizes (voxel numbers) as weights.

AllROIMin In the subsequent three strategies, no selection takes place before training. The model is fitted using all available ROIs, resulting in a risk value assigned to each ROI. Thus, the methods require post-processing, in which the risk for the patient is derived. In the first version, the smallest risk is selected. As a consequence of such an approach, patients with multiple ROIs are more likely to be assigned smaller risks.

AllROIMax Opposite to the previous method, the highest risk is chosen for the patient, resulting in patients with multiple ROIs generally displaying higher risks. Notably, the highest risk is not necessarily associated with the largest ROI.

AllROIMean The risk for the patient is calculated as the mean of the risks of all the ROIs. Here, the risk is not directly dependent on the initial ROI number.

3 Results

3.1 ROI Characteristics

As expected, there is high intra-patient heterogeneity—the contoured ROIs had varying shapes, sizes, and even origin (lung tissue vs lymph nodes), which was reflected in the extracted radiomic features.

Perhaps surprisingly, there was no direct link between number of ROIs and MFS (Fig. 3a). The p-value in the log-rank test between groups with one, two to three, and more contoured lesions was 0.85. However, some differences could be observed in the radiomic profiles between patients with long and short MFS, even with the same number of ROIs, as seen in Fig. 3b, c. For clarity, features (depicted in columns) were not clustered.

3.2 Prediction of Metastasis Free Survival

Although the values of the obtained c-indexes were highly dependent on iteration, which can be attributed to the small sample size, distinct trends emerged in performance of the different strategies. Predictably, the worst result (median c-index 0.534) was obtained by the *randomROI* method. The "classic" approach using only the largest ROI yielded better results (median c-index 0.581), but was still outperformed by the *weightedMeanROI* method incorporating information from multiple ROI (c-index 0.592). The *arithmeticMeanROI* worked worse (c-index 0.557), confirming that a good strategy for incorporating multiple samples is crucial as the model is susceptible to noise.

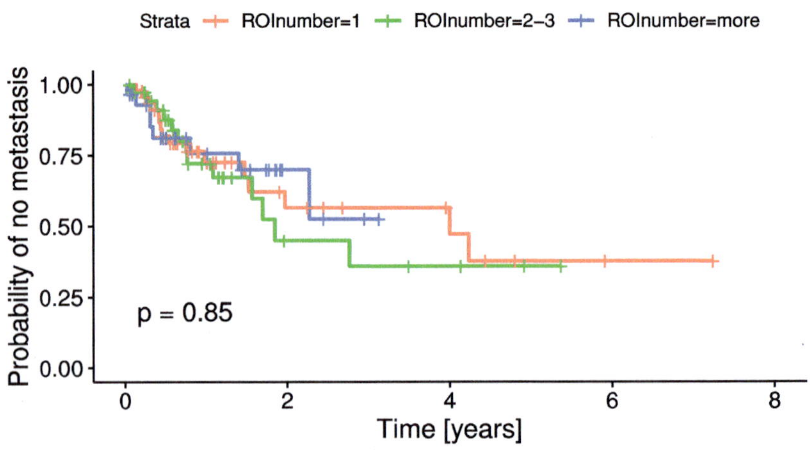

(a) Kaplan-Meier plot with patients stratified according to ROI number

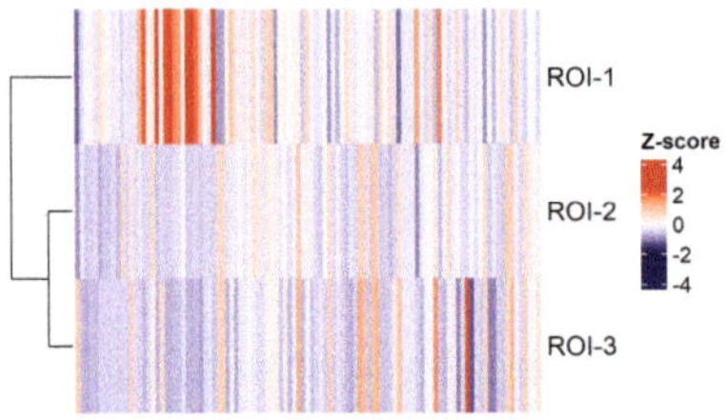

(b) Radiomic features for a sample patient with long MFS (over four years) and three contoured ROIs

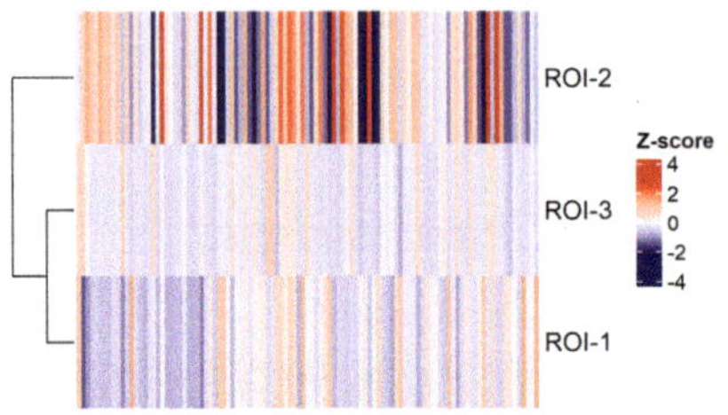

(c) Radiomic features for a sample patient with short MFS (below three months) and three contoured ROIs

Fig. 3 Characterization of the dataset with respect to ROIs

Among the methods using all available ROIs to fit the model, *allROIMin* had the poorest performance, giving a median c-index of 0.566. The other two methods, *allROIMax*, and *allROIMean* achieved the best overall performance, yielding median c-indexes of 0.617 and 0.616, respectively. Notably, the *allROIMean* approach displayed the lowest variability between subsets, emphasizing the added advantage of a larger training set associated with these approaches (Fig. 4).

4 Discussion

Medical images are gaining popularity as a source of prognostic biomarkers in cancer due to the relative simplicity and low invasiveness of their acquisition. With great emphasis and effort toward reproducibility and standardization, radiomics provides

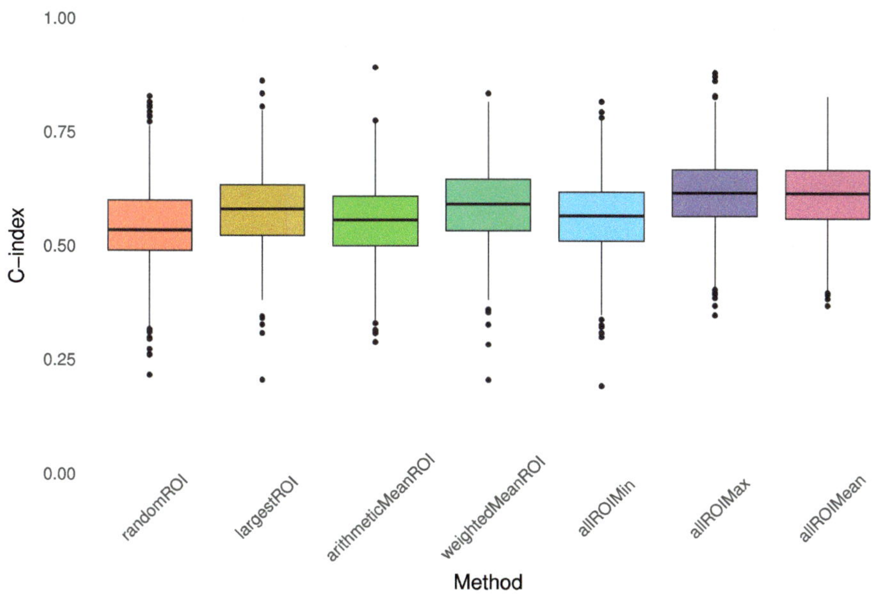

Fig. 4 Benchmark results. Boxplots depict c-indexes obtained in 1000 cross-validation iterations

a way to extract quantitative features from regions of interest in the image. The extracted variables, incorporated into predictive models such as Cox regression, can be used, for example, to assess the risk of distant metastasis. However, in malignancies like lung cancer, the disease is often locally disseminated at diagnosis. Hence, apart from the primary tumor, the regions of interest can include other tumor lesions or lymph nodes, resulting in an additional layer of information often overlooked in radiomic studies.

While the idea of utilizing the additional information in modeling is tempting, the exact strategy of handling multiple ROIs is a non-trivial task. It presents a set of unique challenges, from dataset structure, through inference of risk for a particular patient, to validation preventing information leakage. In an attempt to tackle these challenges, we performed a benchmark of seven different approaches, evaluating their performance in a tailored Monte Carlo Cross-Validation scheme.

On a cohort of 115 non-small cell lung cancer patients, we found that training the model on all available ROIs allowed for better prediction than using only the largest one. Setting the patient's risk to the maximum over all ROIs the median c-index in all iterations was 0.617, which was the highest reached quality. Also a representative ROI constructed as a weighted average outperformed the largestROI approach.

There are some limitations to the presented study, the most important being the limited cohort size. In every iteration, the training set contained only 77 patients, further reduced in the internal cross-validation within CoxNet. It resulted in a certain instability that could be observed in the large dispersion between iterations. Furthermore, we tested only one type of survival model, a limitation we plan to address in the future.

5 Conclusion

In radiomics-based survival modeling, incorporating multiple ROIs per patient can improve the predictive ability, provided that an appropriate strategy is used. For our dataset, the best results were achieved when all ROIs were used for prediction and the risk for a particular patient was derived subsequently.

Author Contributions A. M. W. designed the computational study and methodology, performed the analysis and visualizations, and wrote the initial draft of the manuscript. D. B. extracted the radiomic features and edited the manuscript. E. K. participated in the study design, conceptualization and methodology, prepared the data for analysis and edited the manuscript. A. D'A. and I. G. assembled and prepared the imaging data. I. D. S. and R. S. provided and assembled the patient database. S.G. participated in the analysis and manuscript writing. K. F. provided administrative support. A. S. participated in the study design and conceptualization and supervised the project. All authors have read and approved the final version of the manuscript.

Acknowledgements This work was supported by the Polish National Science Centre, grant number: UMO-2020/37/B/ST6/01959, and Silesian University of Technology statutory research funds. Calculations were performed on the Ziemowit computer cluster in the Laboratory of Bioinformatics and Computational Biology created in the EU Innovative Economy Programme POIG.02.01.00-00-166/08 and expanded in the POIG.02.03.01-00-040/13 project.

References

1. Friedman, J., Hastie, T., Tibshirani, R.: Regularization paths for generalized linear models via coordinate descent. J. Stat. Softw. **33**(1), 1–22 (2010)
2. Fujarewicz, K., Wilk, A., Borys, D., d'Amico, A., Suwiński, R., Świerniak, A.: Machine learning approach to predict metastasis in lung cancer based on radiomic features. In: Nguyen, N.T., Tran, T.K., Tukayev, U., Hong, T.P., Trawiński, B., Szczerbicki, E. (eds.) Intelligent Information and Database Systems, pp. 40–50. Cham, Springer Nature Switzerland (2022)
3. Lu, T., Yang, X., Huang, Y., Zhao, M., Li, M., Ma, K., Yin, J., Zhan, C., Wang, Q.: Trends in the incidence, treatment, and survival of patients with lung cancer in the last four decades. Cancer Manag. Res. **11**, 943–953 (2019). January
4. Reginelli, A., Nardone, V., Giacobbe, G., Belfiore, M.P., Grassi, R., Schettino, F., Del Canto, M., Grassi, R., Cappabianca, S.: Radiomics as a new frontier of imaging for cancer prognosis: a narrative review. Diagnostics **11**(10) (2021)
5. Saif, W., Tzannou, I., Makrilia, N., Syrigos, K.: Role and cost effectiveness of pet/ct in management of patients with cancer. Yale J. Biol. Med. **83**, 53–65 (2010)

6. Shen, C., Liu, Z., Guan, M., Song, J., Lian, Y., Wang, S., Tang, Z., Dong, D., Kong, L., Wang, M., Shi, D., Tian, J.: 2d and 3d ct radiomics features prognostic performance comparison in non-small cell lung cancer. Transl. Oncol. **10**(6), 886–894 (2017)
7. Simon, N., Friedman, J., Hastie, T., Tibshirani, R.: Regularization paths for cox's proportional hazards model via coordinate descent. J. Stat. Softw. **39**(5), 1–13 (2011)
8. van Griethuysen, J.J., Fedorov, A., Parmar, C., Hosny, A., Aucoin, N., Narayan, V., Beets-Tan, R.G., Fillion-Robin, J.C., Pieper, S., Aerts, H.J.: Computational radiomics system to decode the radiographic phenotype. Cancer Res. **77**(21), e104–e107 (2017). October
9. Wilk, A., Borys, D., Fujarewicz, K., d'Amico, A., Suwiński, R., Świerniak, A.: Potential of radiomics features for predicting time to metastasis in nsclc. In: Nguyen, N.T., Tran, T.K., Tukayev, U., Hong, T.P., Trawiński, B., Szczerbicki, E. (eds.) Intelligent Information and Database Systems, pp. 64–76. Cham, Springer Nature Switzerland (2022)
10. Zhang, Y., Oikonomou, A., Wong, A., Haider, M.A., Khalvati, F.: Radiomics-based prognosis analysis for non-small cell lung cancer. Sci. Rep. **7**(1) (2017)
11. Zwanenburg, A., Valliè res, M., Abdalah, M.A., Aerts, H.J.W.L., Andrearczyk, V., Apte, A., Ashrafinia, S., Bakas, S., Beukinga, R.J., Boellaard, R., Bogowicz, M., Boldrini, L., Buvat, I., Cook, G.J.R., Davatzikos, C., Depeursinge, A., Desseroit, M.C., Dinapoli, N., Dinh, C.V., Echegaray, S., Naqa, I.E., Fedorov, A.Y., Gatta, R., Gillies, R.J., Goh, V., Götz, M., Guckenberger, M., Ha, S.M., Hatt, M., Isensee, F., Lambin, P., Leger, S., Leijenaar, R.T., Lenkowicz, J., Lippert, F., Losnegård, A., Maier-Hein, K.H., Morin, O., Müller, H., Napel, S., Nioche, C., Orlhac, F., Pati, S., Pfaehler, E.A., Rahmim, A., Rao, A.U., Scherer, J., Siddique, M.M., Sijtsema, N.M., Fernandez, J.S., Spezi, E., Steenbakkers, R.J., Tanadini-Lang, S., Thorwarth, D., Troost, E.G., Upadhaya, T., Valentini, V., van Dijk, L.V., van Griethuysen, J., van Velden, F.H., Whybra, P., Richter, C., Löck, S.: The image biomarker standardization initiative: Standardized quantitative radiomics for high-throughput image-based phenotyping. Radiology **295**(2), 328–338 (2020)

Assessing the Prognosis of Patients with Metastatic or Recurrent Non-small Cell Lung Cancer in the Era of Immunotherapy and Targeted Therapy

Seweryn Gałecki, Marzena Kysiak, Emilia Kozłowska, Agata Małgorzata Wilk, Rafał Suwiński, and Andrzej Świerniak

Abstract Locally advanced non-small cell lung cancer (NSCLC) is characterized by poor prognosis and is the leading cause of cancer-related deaths worldwide. The main problem in NSCLC is treatment resistance and a high potential for progression and metastasis, which results in low survival rates for patients. Immunotherapy (IO) alone or combined with chemotherapy and targeted therapies, while dedicated to selective subgroups of patients, replaced traditional cytotoxic therapies (radiotherapy or chemotherapy) in advanced stages of disease. Therefore, it's important to search for new features that may help predict the effectiveness of currently available therapies. The paper presents preliminary results of exploratory data and survival analyses (Kaplan-Meier survival curves, uni- and multivariate Cox proportional hazards models), indicating potential factors that may help stratify the group of patients to select the appropriate treatment.

Keywords Non-small cell lung cancer · Exploratory data analysis · Overall survival · Palliative treatment · Curative treatment

---

S. Gałecki (✉) · E. Kozłowska · A. M. Wilk · A. Świerniak
Department of Systems Biology and Engineering, Silesian University of Technology, Akademicka 16, 44-100 Gliwice, Poland
e-mail: seweryn.galecki@polsl.pl

M. Kysiak · R. Suwiński
II-nd Radiotherapy and Chemotherapy Clinic and Teaching Hospital, Maria Sklodowska-Curie National Research Institute of Oncology, Gliwice Branch, 44-102 Gliwice, Poland

A. M. Wilk
Department of Biostatistics and Bioinformatics, Maria Sklodowska-Curie National Research Institute of Oncology, Gliwice Branch, 44-102 Gliwice, Poland

P. Strumiłło et al. (eds.), *The Latest Developments and Challenges in Biomedical Engineering*, Lecture Notes in Networks and Systems 746,
https://doi.org/10.1007/978-3-031-38430-1_14

1 Introduction

Lung cancer (LC) is the leading cause of cancer-related death in Poland [1]. Due to the non-specific symptoms, patients are most often diagnosed at an advanced clinical stage. In 2020, 2.2 million cases of LC were diagnosed worldwide. Five-year survival rates are only 10–20%, and the presence of distant metastases [2] and treatment resistance are the most predictive factors of poor prognosis [3, 4]. Most patients at the time of diagnosis are inoperable cases due to the advanced clinical stage of the disease (III-IV) and the presence of comorbidities [5]. Lung cancers are divided into small cell lung cancer (SCLC) and non-small cell lung cancer (NSCLC) where the latter account for about 85% of all diagnoses [6].

Due to the high metastatic potential and frequent resistance to cytostatic drugs of LC, the treatment of patients is difficult and often depends mainly on the clinical stage assessed based on the TNM (Tumor, Node, Metastasis) classification and the performance status. Surgical treatment involving resection of the parenchyma is possible only in 25% of cases, therefore the mainstay of treatment in locally advanced disease is chemotherapy (CHT), radiotherapy (RT), or a combination of both. In lung cancer, CHT regimens are based on platinum compounds, most often in combination with vinorelbine or etoposide [5, 6]. In the past treatment of patients with stage IV disease was mainly palliative [7]. However, novel therapies such as immunotherapy or targeted therapies offer currently considerably prolonged survival and palliation of symptoms is no longer the main goal of therapy.

Molecular treatment Molecular therapies that target specific proteins are also used in LC treatment. However, this treatment concerns a small group of patients. The use of tyrosine kinase inhibitors (TKIs) is one of the methods of such treatment in patients with *EGFR* gene mutations, *ALK* gene rearrangements, or *ROS1* tyrosine kinase gene rearrangements [5, 8, 9]. Erlotinib, gefitinib, or afatinib are successfully used in the treatment of patients with *EGFR* (+) tumors, especially in non-smokers with the AC histological subtype [5]. This therapy is characterized by a more than 50% better response to treatment compared to chemotherapy alone [9]. Still, it is used in only about 13% of Caucasian patients due to the low frequency of *EGFR* mutations [5]. In patients with *ALK* (+) and *ROS1* abnormalities, crizotinib is used as a first- or second-line TKI and, in the case of resistance, ceritinib and alectinib, which are second-generation inhibitors with a higher tyrosine kinase affinity [9].

Immunotherapy In recent years, immunotherapy (IO) is a new, efficient method of treatment of patients with an advanced NSCLC, and its effect is based on the mobilization of the patient's immune system through the use of monoclonal antibodies that are inhibitors of the immune checkpoints. The effects of IO in patients with high expression of the PD-L1 (programmed death ligand 1) antigen are associated with a better prognosis compared to patients receiving conventional platinum-based chemotherapy. In addition, IO has fewer side effects. Currently, NSCLC therapy uses two programmed death receptor type 1 (PD-1) inhibitors registered in the European Union—nivolumab and pembrolizumab.

1.1 Goals and Hypotheses

High mortality in LC, caused by its high aggressiveness and often too-late diagnosis, makes it necessary to search for more efficient methods of treatment. The choice of individualized systemic, molecularly targeted treatment or the use of immunotherapy requires the detection of mutations in selected genes or the determination of their expression. However, many factors can modify the response to the chosen treatment method. It's increasingly recognized that common germline or somatic variants in the population and some clinico-demographic factors may influence the disease course and treatment response of various cancers, including NSCLC [10–12]. Therefore, it is worth trying to determine and investigate in more details which of the demographic and clinical factors at the individual level may influence the course of the disease. Hence, the study aims to verify the hypothesis that, demographic and clinical factors may help to better assess the prognosis of patients treated with novel treatment modalities.

2 Materials and Methods

2.1 Patients Cohort and Data Acquisition

The subject of the analysis is a relatively large cohort of advanced NSCLC patients (n = 342) that are at stages III or IV. The majority of the group present distant metastases at the diagnosis. Patients were treated with targeted molecular therapy (EGFRi, ALK/ROSi), immunotherapy (PD-L1i) or palliative chemotherapy. Most of the patients who received chemotherapy are historical controls treated before widespread implementation of immunotherapy or targeted therapy.

Clinical data were obtained from medical records and collected at Electronic Health Records (EHR) system. In addition to genetic records, the database includes extensive patient characteristics as well as overall (OS) and progression-free survival (PFS) time.

2.2 Statistical Analysis

The relationship between the clinico-demographic characteristics of patients and survival was analyzed using uni- and multivariate Cox proportional hazard models. To investigate whether the implementation of various therapeutic approaches affected survival time in patient subgroups, Kaplan-Meier curves were drawn and compared using a log-rank statistical test.

For all statistical analyses, the α of less than 0.05 was adopted as significant. The R programming environment was applied for all statistical analyses. The R package survival and survminer were applied for survival analysis of the data.

3 Results

3.1 Selection of Patients for Downstream Analysis

The cohort includes 342 patients with advanced non-small cell lung carcinoma. Following the data exploratory study, the 32 patients did not meet the criteria for inclusion in the statistical analysis. Thus, finally, 310 patients with complete demographic and clinical data were chosen. The reasons for exclusions are presented in Fig. 1.

The majority of excluded patients (n = 21) did not have complete medical documentation due to their treatment in another hospital. Also, four patients for whom information on overall survival was not available were not included in further analyses. From the study cohort, four cases without the TNM classification, two without the *EGFR* mutation data, and one with a histopathological subtype of large cell carcinoma were also excluded.

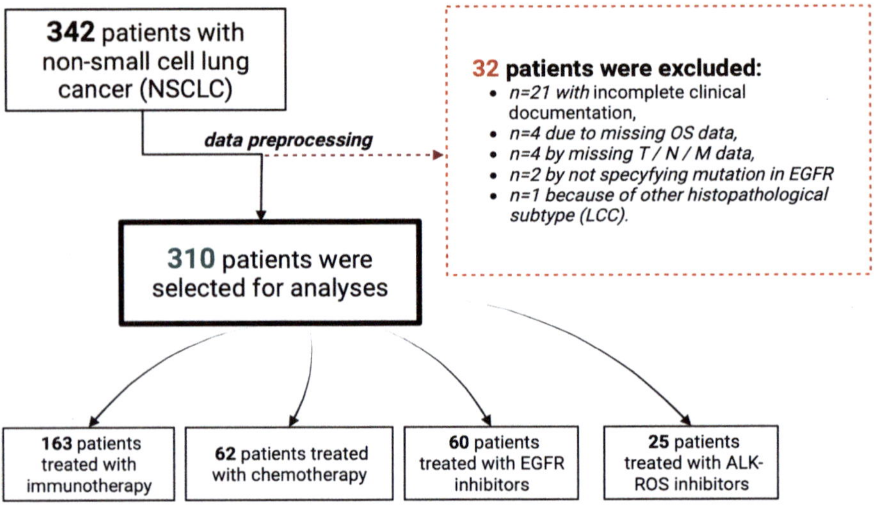

Fig. 1 Scheme of selection of NSCLC patients for statistical analysis, taking into account the division of the cohort depending on the therapeutic approach

3.2 Patient's Characteristics

The study group includes 310 patients with NSCLC, treated at Maria Sklodowska-Curie National Research Institute of Oncology (NRIO), Gliwice Branch. The mean age at diagnosis was 65 years (median 66, range 26–89). There were slightly more men in the study group (55%). All cases were in advanced clinical stages III or IV. In most cases, the tumor has spread to nearby tissues or organs, and numerous regional lymph nodes. Adenocarcinoma (AC) was diagnosed in 190 (61%) patients, squamous cell carcinoma (SCC) in 101 (33%) patients, and NSCLC not otherwise specified histopathological type (NOS) in 19 (6%) patients. Most patients underwent curative therapy using immunotherapy 163 (53%), *EGFR* inhibitors 60 (19%), or *ALK-ROS* inhibitors 25 (8%), while 62 (20%) patients received treatment in the form of platinum-based chemotherapy. The majority of the group (95%) were in good physical condition (1st or 2nd performance status according to the Zubrod scale). The detailed study group characteristics are shown in Table 1.

3.3 Kaplan-Meier Analysis

The Kaplan-Meier estimator computes the survival probability of patients over a specified period. Figure 2 shows a comparison of OS for four subgroups of NSCLC patients treated with all four types of treatment modalities. The x-axis represents the 5-year follow-up of the patients, while the y-axis shows the fraction of patients. The dashed lines show the median OS time. The risk table presents individual numbers of surviving patients in each subsequent year of observation.

A statistically significant association was observed between the OS and the various treatment type in the entire group ($p = 4.2 \times 10^{-4}$). The group receiving *EGFR* inhibitors characterize with a higher overall survival probability and had the longest survival time (median OS = 2.01 years), compared to patients treated with CHT (0.87 years), *ALK-ROS* inhibitors (1.02 years) or IO (1.22 years).

3.4 Log-Rank Statistical Test

A log-rank statistical test was performed, allowing the assessment of the impact of the studied factors on the endpoint by comparing the survival function of at least two independent groups. The principle of the log-rank test is based on the logarithm of the survival function, and its hypotheses assume that:

- H0: Compared survival curves do not differ from each other,
- Ha: There is a statistically significant difference between the compared survival curves.

Table 1 Characteristics of the study group

Factor		All patients	Treatment approach			
			IO	CHT	EGFRi	ALKi ROSi
		N (%)	N			
		310	163	62	60	25
Age at diagnosis	< 66 years	143 (46)	60	41	29	13
	≥ 66 years	167 (54)	103	21	31	12
Sex	Male	170 (55)	93	47	18	12
	Female	141 (45)	70	15	42	13
Histology NSCLC	AD	190 (61)	84	23	58	25
	SCC	101 (33)	72	29	–	–
	NOS	19 (6)	7	10	2	–
Tumor location	Right lung	164 (53)	94	29	26	15
	Left lung	146 (47)	69	33	34	10
Primary tumor (T)	T0	24 (8)	21	–	1	2
	T1	21 (7)	12	4	4	1
	T2	36 (11)	23	3	9	1
	T3	41 (13)	15	11	11	4
	T4	188 (61)	92	44	35	17
Regional lymph nodes (N)	N0	57 (18)	36	3	15	3
	N1	7 (2)	3	3	1	–
	N2	51 (17)	23	16	9	3
	N3	195 (63)	101	40	35	19
Distant metastasis (M)	M0	31 (10)	7	20	3	1
	M1a	118 (38)	68	22	20	8
	M1b	25 (8)	12	6	5	2
	M1c	136 (44)	76	14	32	14
Clinical stage	IIIa	7 (2)	1	5	1	–
	IIIb	12 (4)	3	8	1	–
	IIIc	12 (4)	3	7	1	1
	IVa	142 (46)	81	27	24	10
	IVb	137 (44)	75	15	33	14
Performance status	0	70 (22)	48	7	12	3
	1	226 (73)	114	47	44	21
	2	14 (5)	1	8	4	1

Figure 3 shows the cumulative results of the log-rank analysis, grouping the variables based on their p-values, starting with the highest (least significant). Statistically significant features are marked with green color. The applied treatments had the greatest impact on the diversification of the course of survival curves. Statistically significant results were also observed in the histology of the NSCLC patients, performance status, tumor location, gender, and T classification.

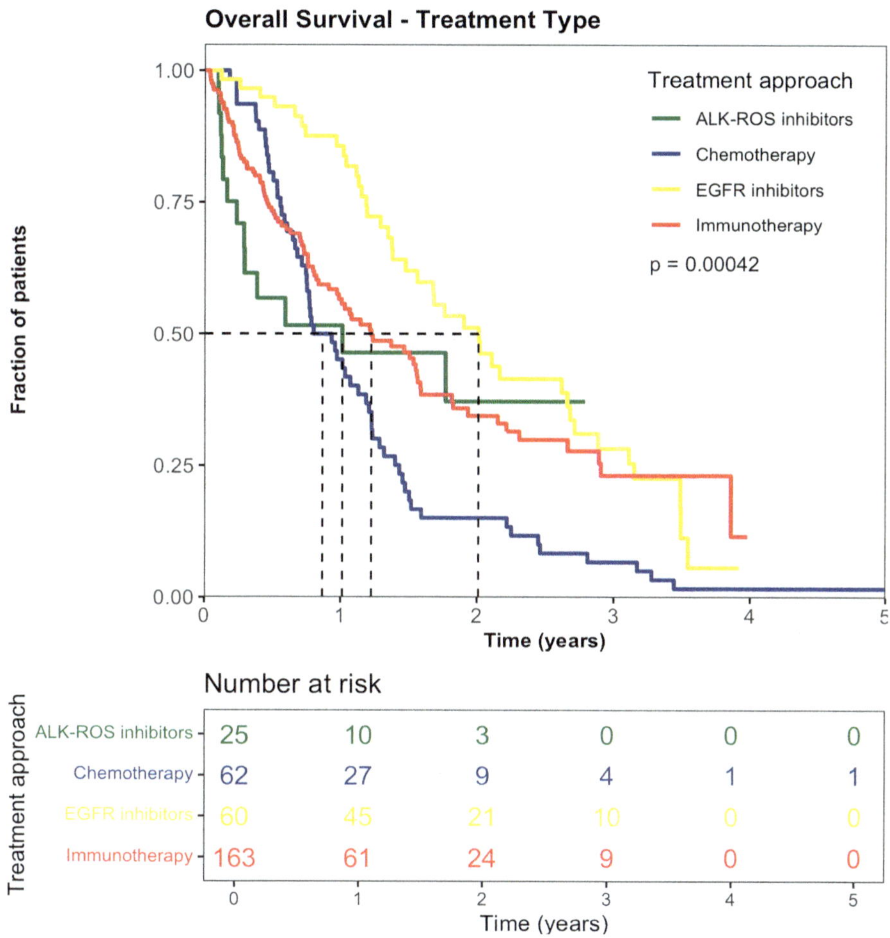

Fig. 2 The Kaplan-Meier curve for OS according to treatment type in the NSCLC patients group

3.5 *Univariate Cox Proportional Hazard Regression*

Univariate Cox proportional hazards analysis shows that for patients treated with IO, overall survival was significantly unfavorable for tumors located in the right lung (p = 0.023) and T4 stage (p = 0.018), while in the group receiving *EGFR* inhibitors, not otherwise specified histopathological subtype (NOS) was associated with a significantly higher risk of death (p = 0.011). In the remaining subgroups (receiving ALKi-ROSi and CHT), no statistically significant features were observed.

T4 tumor staging based on the TNM classification resulted in an almost two-fold increase in the risk of death (HR 1.75 95%CI 1.10-2.78) in the group receiving IO. However, an interesting observation is that patients with a tumor in the right lung

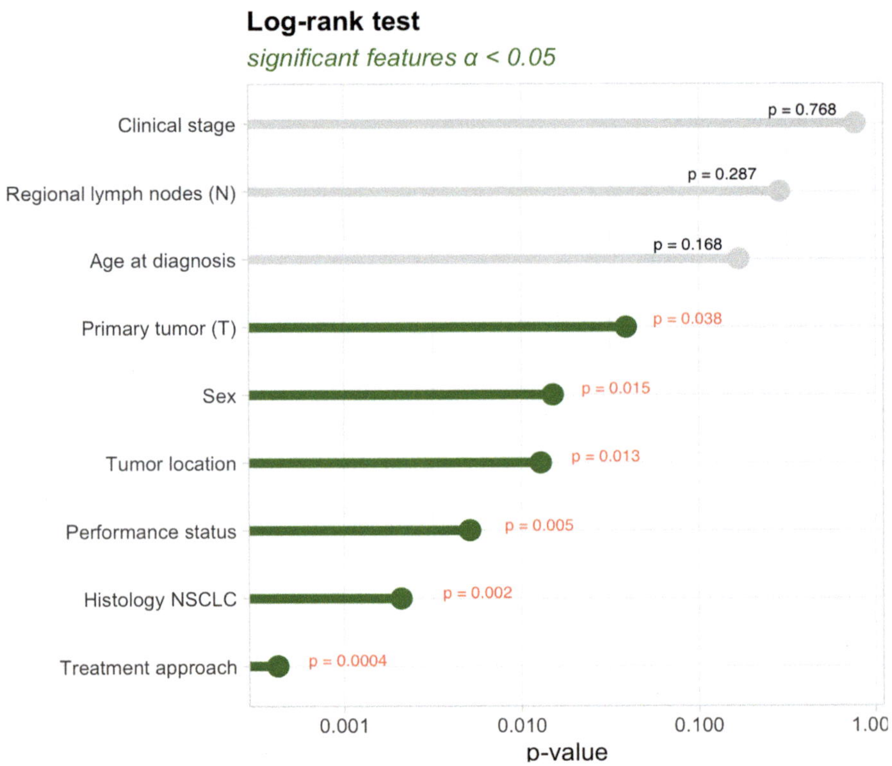

Fig. 3 Log-rank statistical test for the whole group of NSCLC patients

in the same subgroup had a 65% worse survival rate (HR 1.65 95%CI 1.07-2.54) compared to patients with a tumor in the other lung. A similar trend was observed in the group treated with *ALK-ROS* inhibitors, where a tumor localized in the right lung correlated with more than three times the risk of death (HR 3.06 95%CI 0.83-11.22). Patients with NOS in the group receiving *EGFR* inhibitors had an almost seven-fold higher risk of death, however, this subgroup was characterized by a low number of cases (n = 2) (Table 2).

3.6 *Multivariate Cox Proportional Hazard Regression*

Hazard ratios (HRs) with 95% confidence intervals (CIs) were estimated by Cox proportional hazards regression and their values are visualized in Fig. 4. HRs were adjusted for patient's age at diagnosis, gender, histology of NSCLC, T and N classification, performance status, clinical stage, and type of treatment.

Table 2 Univariate Cox proportional hazard regression in subgroups due to treatment approach (only factors with p less than 0.100 are shown)

Factor		Events/N	p	HR (95% CI)	p
Immunotherapy					
Age at diagnosis	< 66 years	35/60	0.088	1.0	
	≥ 66 years	55/103		0.68 (0.45–1.06)	0.094
Tumor location	Left lung	33/69	0.021	1.0	
	Right lung	57/94		1.65 (1.07–2.54)	0.023
Primary tumor (T)	T0–T2	27/56	0.052	1.0	
	T4	57/92		1.75 (1.10–2.78)	0.018
EGFR inhibitors					
Histology NSCLC	LUAD	37/58	0.003	1.0	
	NOS	2/2		6.96 (1.55–31.22)	0.011
ALK—ROS inhibitors					
Tumor location	Left lung	3/10	0.077	1.0	
	Right lung	10/15		3.06 (0.83–11.22)	0.093

In multivariate models adjusted for clinico-demographic factors, performance status had the strongest impact on patients' overall survival time. Patients in the 1st and 2nd stages were characterized by almost two and three times higher risk of death ($p = 0.008$ and $p = 0.003$ respectively), compared to those with stage 0.

The primary tumor located in the right lung was again associated with unfavorable OS ($p = 0.009$), shortening the survival of patients by 47% (HR 1.47 95%CI 1.10-2.0). Higher stages: T3 and T4 were also associated with shorter patient survival, ($p = 0.063$ and $p = 0.016$), reducing survival by 63 and 58%. The use of EGFR inhibitors showed a trend ($p = 0.063$) to a protective effect, reducing the risk of patient death by 37% (HR 0.63 95% CI 0.39-1.0) (Fig. 4).

4 Discussion and Future Work

Lung cancer is often diagnosed at an advanced stage leading to the question if a given patient with advanced lung cancer could be treated with a modern treatment modality or not. Of course, there are available genetic biomarkers such as the presence of the specific mutation (for example in *EGFR* gene) or gene expression level (such as expression of PD-L1). However, the presence of mutation only tells us about the higher probability that the patient will respond to a given treatment.

Here we performed exploratory data analysis of data from a cohort of advanced lung cancer patients. We evaluated using a log-rank test in which clinical covariates show predictive values. We showed that tumor histology and Zubrod performance status could help in overall survival prediction. Next, we performed univariate and

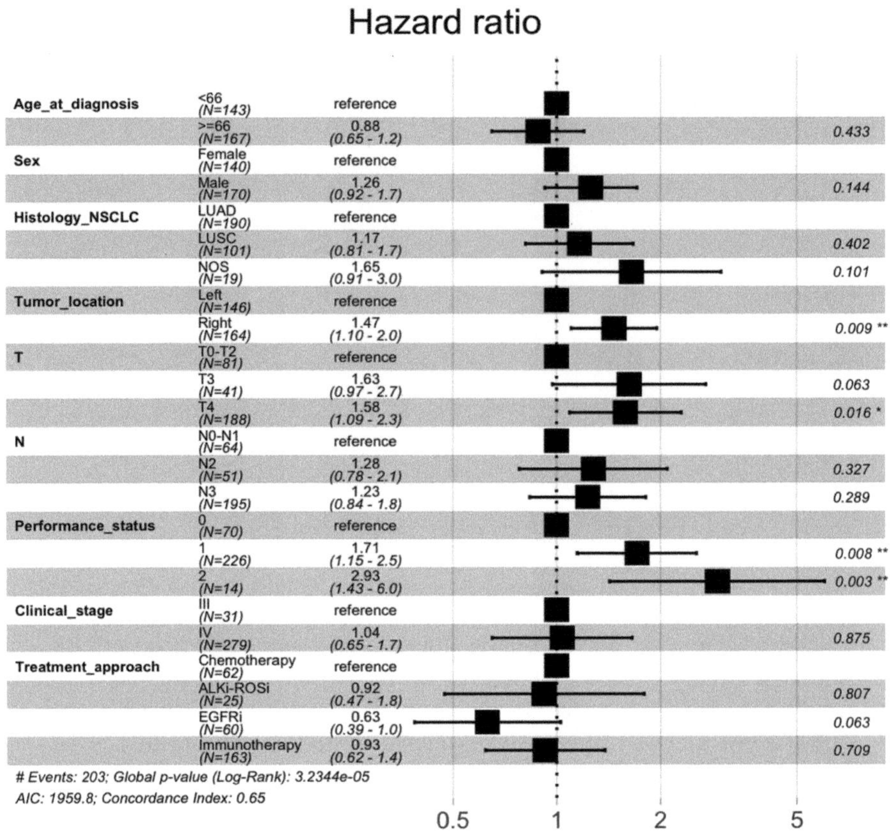

Fig. 4 Association between clinico-demographic characteristics and overall survival (OS) in a cohort of all NSCLC patients

multivariate Cox regression. Interestingly, we can predict the risk of death in patients treated with immunotherapy and EGFRi using two and one clinical covariates, respectively. The obtained results require validation in a larger group of patients and populations of different ethnic origins.

The study shows initial work of predicting the benefit of the administration of modern treatment modalities in comparison with platinum-based chemotherapy. Developed machine learning predictors could assist clinicians in deciding if patients with advanced lung cancer qualify for alternative treatment modalities or not. The developed predictor and the classificator will be next interpreted and evaluated by Marzena Kysiak.

Contributions

- Conceptualization: MK, EK, RS, AS, AMW, SG
- Data acquisition: MK, RS
- Data processing: MK, EK, SG
- Performance of statistical analysis: SG
- Preparation of the paper draft: SG

All authors have read and and agreed to the published version of the manuscript.

Acknowledgements This work was supported by the National Science Centre (NCN), Poland, grant no. 2020/37/B/ST6/01959 (AS, EK, SG, and AMW). The presented results and their interpretation will be part of Marzena Kysiak's doctoral dissertation.

References

1. Didkowska J., Wojciechowska U., Olasek P., Caetano dos Santos F., Michałek I.: Nowotwory złośliwe w Polsce w 2019 roku. ISSN 0867-8251, Warszawa (2021)
2. Kozłowska, E., Giglok, M., Dębosz-Suwińska, I., Suwiński, R., Świerniak, A.: Predicting metastasis-Free survival using clinical data in non-small cell lung. Recent Challenges in Intelligent Information and Database Systems: 14th Asian Conference, ACIIDS,: Ho Chi Minh City, Vietnam, November 28–30, 2022. Proceedings cancer, pp. 227–237 (2022)
3. Sung, H., Ferlay, J., Siegel, R.L., Laversanne, M., Soerjomataram, I., Jemal, A., Bray, F.: Global cancer statistics 2020: GLOBOCAN estimates of incidence and mortality worldwide for 36 cancers in 185 countries. CA Cancer J. Clin. **71**(3), 209–249 (2021)
4. Popper, H.H.: Progression and metastasis of lung cancer. Cancer Metastasis Rev. **35**(1), 75–91 (2016)
5. Kutkowska, J., Porębska, I., Rapak, A.: Non-small cell lung cancer - mutations, targeted and combination therapy. Postepy Higieny i Medycyny Doświadczalnej **71**, 431–445 (2017)
6. Krzakowski, M., Jassem, J.: Nowotwory płuca i opłucnej oraz śródpiersia. Onkologia w Praktyce Klinicznej - Edukacja **5**(1), 21–53 (2019)
7. Sierko E., Wojtukiewicz M.: Patofizjologiczne podstawy terapii ukierunkowanej na zahamowanie funkcji receptora czynnika wzrostu naskórka (EGFR). Onkologia w Praktyce Klinicznej, T. 6, nr 5, s. 217–227 (2010)
8. Chaitanya, T.K., Barsouk, A., Saginala, K., Sukumar, A.J., Barsouk, A.: Epidemiology of lung cancer. Contemporary Oncology/Współczesna Onkologia **25**(1), 45–52 (2021)
9. Zyśk, R., Krzakowski, M.: Personalised treatment of non-small-cell lung cancer patients—review of current evidence. Oncol. Clin. Pract. **14**, 23–34 (2018)
10. Butkiewicz, D., Krześniak, M., Gdowicz-Kłosok, A., Giglok, M., Marszałek-Zeńczak, M., Suwiński, R.: Polymorphisms in EGFR gene predict clinical outcome in unresectable non-small cell lung cancer treated with radiotherapy and platinum-based chemoradiotherapy. Int. J. Mol. Sci. **22**(11), 5605 (2021)
11. Shen, H.B., Li, J., Yao, Y.S., Yang, Z.H., Zhou, Y.J., Chen, W., Hu, T.J.: Impact of somatic mutations in non-small-cell lung cancer: a retrospective study of a chinese cohort. Cancer Manag. Res. **19**(12), 7427–7437 (2020). Aug

12. Kawaguchi, T., Takada, M., Kubo, A., Matsumura, A., Fukai, S., Tamura, A., Saito, R., Maruyama, Y., Kawahara, M., Ignatius, Ou.S., H.: Performance status and smoking status are independent favorable prognostic factors for survival in non-small cell lung cancer: a comprehensive analysis of 26,957 patients with NSCLC. J. Thorac. Oncol. **5**(5), 620–630 (2010)

Predicting the Risk of Metastatic Dissemination in Non-small Cell Lung Cancer Using Clinical and Genetic Data

Emilia Kozłowska, Agata Małgorzata Wilk, Dorota Butkiewicz, Małgorzata Krześniak, Agnieszka Gdowicz-Kłosok, Monika Giglok, Rafał Suwiński, and Andrzej Świerniak

Abstract The major cause of cancer patient death is the development of distant tumors (metastases) that leads to a disease that is hard to treat. Therefore, the emergence of metastasis is a key time-point in the management of cancer leading to a decision about more intensive treatment. One of the most metastatic cancer types is lung cancer, in which the most common subtype is non-small cell lung cancer (NSCLC). We selected a set of common genetic variants in 27 genes and identified them in DNA from the peripheral blood of 335 NSCLC patients. Our goal is to apply a predictive machine learning model to estimate the risk of metastasis based on clinical and SNPs data. We have found that Cox regression with LASSO regularization gives the best predictive accuracy with a concordance index equal to 0.63.

Keywords Single nucleotide polymorphisms · Non-small cell lung cancer · Metastasis-free survival · Risk of metastasis

---

E. Kozłowska (✉) · A. M. Wilk · A. Świerniak
Silesian University of Technology, Department of Systems Biology and Engineering, 44-102 Gliwice, Poland
e-mail: andrzej.swierniak@polsl.pl

A. M. Wilk
Department of Biostatistics and Bioinformatics, Maria Sklodowska-Curie National Research Institute of Oncology, Gliwice Branch, 44-102 Gliwice, Poland

D. Butkiewicz · M. Krześniak · A. Gdowicz-Kłosok
Center for Translational Research and Molecular Biology of Cancer, Maria Sklodowska-Curie National Research Institute of Oncology, Gliwice Branch, 44-102 Gliwice, Poland

M. Giglok · R. Suwiński
II-nd Radiotherapy and Chemotherapy Clinic and Teaching Hospital, Maria Sklodowska-Curie National Research Institute of Oncology, Gliwice Branch, 44-102 Gliwice, Poland

P. Strumiłło et al. (eds.), *The Latest Developments and Challenges in Biomedical Engineering*, Lecture Notes in Networks and Systems 746,
https://doi.org/10.1007/978-3-031-38430-1_15

1 Introduction

Lung cancer is the most common and deadliest cancer in the world [1]. There are two main types of lung cancer: small cell lung cancer (SCLC) and non-small cell lung cancer (NSCLC) where the latter accounts for about 85% of all lung cancer cases. Most lung cancers are diagnosed at an advanced stage leading to a poor prognosis and a few possible treatment options.

It is known that lung cancer patients die mostly due to distant metastases rather than primary tumors in the lung [REF]. Thus, the emergence of distant metastases is a turning point in lung cancer treatment. Patients with disseminated lung cancer have limited treatment options and often are treated with palliative intent.

Here, we collected retrospectively a cohort of about 400 NSCLC patients and their genetic data (31 single nucleotide polymorphisms, SNPs, and two gene deletions) to investigate whether selected inherited genetic factors affect the risk of metastasis [4]. We integrated clinical and genetic data to apply them to predict the distant metastasis risk.

We applied seven machine learning methods combined with automatic feature selection (regularization and boosting) together with three methods of data encoding (ordinal, one-hot, and target) for distant metastasis risk prediction. We have found that ordinal encoding with Cox LASSO regularization gives the best performance in the prediction of the risk of metastasis measured as a concordance index.

We show that a combination of clinical data and information about inherited gene polymorphisms could be applied in the prediction of the risk of distant metastasis emergence. The developed predictor, after validation with an independent cohort, could be applied by clinicians for treatment planning.

2 Materials and Methods

2.1 Patient Cohort

We collected a cohort of 436 advanced non-small cell lung patients from the National Research Institute of Oncology (NRIO), Gliwice Branch in Poland. Those patients were treated at the Institute with radiotherapy or radiotherapy combined with platinum-based chemotherapy (chemoradiation). The cohort included three subtypes of NSCLC: Lung Squamous Cell Carcinoma (LUSCC), Lung adenocarcinoma (LUAD), and not otherwise specified (NOS).

For all patients in the cohort, we computed metastatic-free survival (MFS) that was applied to predicting the risk of metastasis. MFS was obtained by measuring the elapsed time from the start of treatment to the detection of distant metastasis.

As we are here interested in predicting the risk of metastasis, we considered only those patients that did not have distant metastases at the diagnosis (M1 in the TNM classification). In addition, 19 patients had not full clinical and SNPs information.

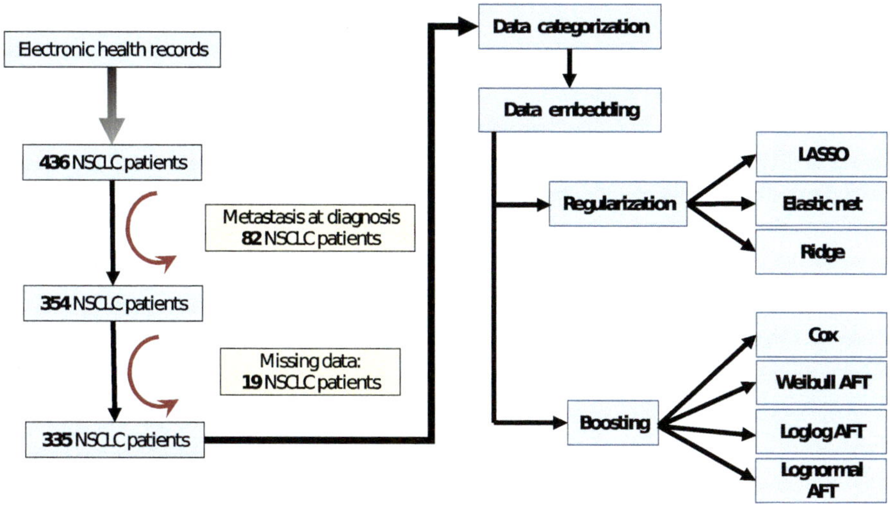

Fig. 1 The workflow of data acquisition and analysis

These patients were also excluded from downstream analysis. This leads to a cohort of 335 NSCLC patients that was applied for building the predictor (Fig. 1).

2.2 Genetic Data Acquisition

The analyzed common genetic variants included 31 SNPs and 2 gene deletions in 27 genes involved in, among others, DNA damage response and repair, angiogenesis, metastasis, and enzymatic antioxidant defense. The genes were selected based on a literature search [REF]. The selected genes were: *MMP-1*, *MMP-2*, *MMP-3*, *VEGF*, *VEGFR2*, *EGFR*, *XPA*, *XPD*, *APEX1*, *OGG1*, *XRCC1*, *RAD51*, *XRCC3*, *PARP1*, *DNA-PK*, *NBS1*, *SOD2*, *CAT*, *NQO1*, *GPx4*, *GSTM1*, *GSTT1*, *GSTP1*, *TP53*, *CCND1*, *FGFR4*, and *COX2*.

DNA was isolated from the frozen peripheral blood of all patients in the cohort collected before radiotherapy. DNA was isolated using a Genomic Maxi AX kit (A&A Biotechnology, Gdynia, Poland). The genotypes were determined using the polymerase chain reaction-restriction fragment length polymorphism (PCR-RFLP) and multiplex PCR methods. Genotyping was repeated in 50 randomly selected samples, and 100% concordance was found.

Each SNP was categorized into a more frequent reference homozygous genotype (marked as zero) or heterozygous and rarer homozygous genotypes combined (marked with one). The *GSTM1* and *GSTT1* gene deletions were grouped into two categories—gene absent (marked as zero) and at least one gene copy present (marked with one). In this form, the data were analyzed in downstream analysis.

2.3 Data Encoding Methods

Data encoding is a procedure often applied to categorical data such as tumor stage (that can be divided into three categories: I, II, and III). The goal of encoding is to convert categorical data into numerical ones while preserving the information that the data contain. This allows applying various machine learning models that often require numerical values instead of categorical ones.

Here, we tested three types of encoding: ordinal, one-hot, and target encoding which are explained below.

Ordinal encoding Ordinal encoding was performed in two steps. In the first one, the unique values of a given feature were sorted alphabetically. In the second step, for each value of the feature, a value from one to N is set depending on the location of sorted unique values. So, for example, all categorical values that belong to the first category (according to a sorted list of all categories) are set to one. The resulting data have the same number of columns (features) as the categorical data.

One-hot encoding This encoding codes each unique value of a given variable as a 0/1 depending if the value is present or not. This encoding leads to the number of columns (features) being equal to the number of unique values. Thus, if the number of categories is large, this encoding is not advisable.

Target encoding This encoding is very similar to the ordinal one. The difference is that instead of values from 1 to N, we replace the categorical values with a value of endpoint of interest inside the group. For example, we compute metastatic-free survival for the LUAD subtype and replacing the name of the subtype with the value of MFS for that subtype.

2.4 Survival Regression Methods

Survival regression is a method that is able to predict the risk of specific endpoints (such as the emergence of metastasis which is a focus of the article) based on values of selected features that could be numerical or categorical.

The most known survival regression method is Cox proportional hazard model which is a semi-parametric method that is widely applied in biomedical engineering. Cox proportional hazard model is semi-parametric as the model does not assume that hazard function follows a given probability distribution. Here, we also included parametric regression models: log-log, Weibull, and lognormal survival regression models. Below, we briefly describe all the models.

Cox survival regression The Cox survival regression model (known also as a proportional-hazard model) is the most commonly applied model for investigating the relationship between the survival time of patients and some predictor variables.

Let's denote by $X_i = (X_{i1}, \ldots, X_{ip})$ a list of covariates of patient i (a list of explanatory features). Then, the Cox model computes the so-called hazard:

$$h_i(t) = h_0(t) \cdot e(b_1 X_1 + b_2 X_2 + \cdots + b_p X_p) \tag{1}$$

where:

- t is the survival time,
- $(b_1, b_2, \ldots, b_p$ are model coefficients,
- h_0 is the baseline hazard ratio.

The hazard function written above gives us the potential of having an event at a time t, given survival up to that time.

There are many modifications of the Cox model and the most popular class are parametric hazard function models where the hazard functions follow some probability distribution. Here, we applied three of them: log-logistic [2], Weibull [10], and lognormal [8] model of hazard function that we describe below.

Log-logistic survival regression The hazard of the parametric log-logistic survival model equals:

$$h(t) = \frac{\lambda \cdot p \cdot t^{p-1}}{1 + \lambda \cdot t^p} \tag{2}$$

where $\lambda = e(b_0 + b_1 \cdot X_1 + b_1 \cdot X_1 \ldots b_p \cdot X_p)$. (intercept) and $p > 0$ (slope) are the model parameters to fit.

It leads to the following survival function:

$$S(t) = \frac{1}{1 + \lambda \cdot t^p} \tag{3}$$

Weibull survival regression

$$h(t) = \lambda \cdot p \cdot t^{p-1} \tag{4}$$

where $\lambda = e(b_0 + b_1 \cdot X_1 + b_1 \cdot X_1 \ldots b_p \cdot X_p)$

It leads to the following survival function:

$$S(t) = e(\lambda \cdot t^p) \tag{5}$$

Log-normal survival regression The survival function for the log-normal survival model is:

$$S(t) = 1 - \phi\left(\frac{ln(t) - \mu}{\sigma}\right) \tag{6}$$

where ϕ is cumulative density function, σ is a standard deviation, and μ is the mean.

The hazard function can be computed as:

$$h(t) = -\frac{dS(t)}{dt}/S(t). \tag{7}$$

2.5 Feature Selection Methods

We focus here on two methods that incorporate into model fitting a way to select the covariates that have the potential for predicting the given hazard function. Here, we applied two methods for automated feature selection regularization and boosting.

Regularization Regularization is a method that allows to reduce the model variance without a substantial increase in its bias.

This is a form of regression, that constrains/ regularizes or shrinks the coefficient estimates toward zero. This technique discourages learning a more complex model, so as to avoid the risk of overfitting.

Here, we applied the glmnet R package for Cox model regularization [9].

Boosting Boosting is an ensemble algorithm for reducing bias and variance in supervised learning. It is a family of machine learning algorithms that convert weak learners (the model with poor performance) to strong one through an iterative process.

Most boosting algorithms consist of iterative learning weak classifiers and adding them to a final strong classifier. When added, they are weighted so it is related to the weak learners' accuracy. After a weak learner is added, the data weights are readjusted, known as "re-weighting". Misclassified input data gain a higher weight and examples that are classified correctly lose weight. Thus, future weak learners focus more on the patients that previous weak learners misclassified.

Here, we consider component-wise gradient boosting, which is a method for optimizing model accuracy and for obtaining model estimates via gradient descent techniques [3, 7].

2.6 Performance Evaluation

The performance of the model was evaluated by calculation of the concordance index (also called c-statistics) [6]. It is an analogous metric to Area Under the Curve (AUC) that is commonly applied in classification problems. It can be interpreted as the probability of concordance between predicted and observed survival where the value 0.5 means that the predictor is random.

3 Results

3.1 Patients Characteristics

In the filtered cohort, there were 235 men and 100 women. Most of the patients were over 60 years old. The most common subtype of NSCLC was LUSCC. Patients were relatively at good performance status as the Zubrod score [5] (the score that

Table 1 Patient cohort characteristics

		N = 335
Sex	Male	235
	Female	100
Age	< 60	136
	≥ 60	199
Subtype	LUSCC	205
	LUAD	55
	NOS	75
Zubrod score	0	96
	1	219
	2	20
Stage	I–II	96
	III	219
Smoking	Yes	312
	No	23
Treatment	Chemoradiation	245
	Radiotherapy	90

measure the patient state) for most patients was one. Of the patients, 219 were at an advanced stage but did not have distant metastases at the diagnosis. As expected, over 90% of the patients smoked in the past or were currently smoking. All patients were treated with radiotherapy alone or a combination of radiotherapy and platinum-based chemotherapy. Table 1 summarizes the patient cohort.

3.2 *Data Categorization*

Before fitting the regression models, we performed data categorization that aim at the conversion of numerical values into some classes. As genetic data are categorical, to be more consistent, we categorized also clinical data. For example, age at the diagnosis was converted into two classes: $\{< 60; \geq 60\}$.

SNP data originally were encoded with one of three values: 0—common homozygote, 1—heterozygote, and 2—rare homozygote. We converted these values into a reference (coded as 0) and variant involving carriers of one or two polymorphic alleles (i.e. heterozygotes and rare homozygotes) (coded as 1).

As the first step in the computational analysis, we converted clinical data into numerical values as ordinal encoding (see Table 2).

Next, we integrated and visualized the data (see Fig. 2) using a heatmap. The heatmap shows the values of all clinical and genetic features for all patients in the cohort using ordinal encoding. Each row is a single patient and each column is a given feature. The data are clustered using the hierarchical clustering method. As

Table 2 Ordinal encoding of clinical data

Feature	From	To
Age	<60, ≥60	1, 2
Sex	Male, Female	1, 2
Hist	NOS, LUAD, LSCC	1, 2, 3
Stage	III, I/II	1, 2
Zubrod	0, 1, 2	1, 2, 3
CHT	Yes, No	1, 2
Intent	Curative, Palliative	1, 2

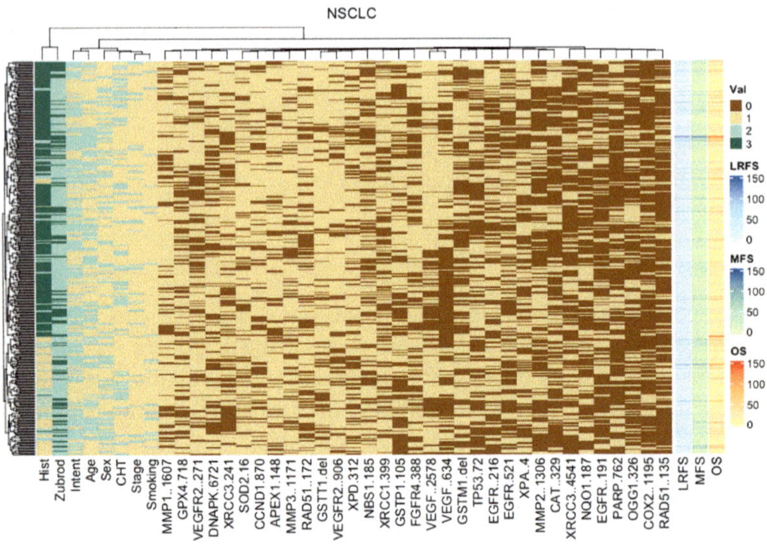

Fig. 2 The integration of clinical and SNPs data

we can see, the data do not contain visible clusters. It shows that using the original data we cannot say which patient has short and which patient has long MFS. Thus, it is important to select appropriate features for predicting the risk of metastasis and filter out unnecessary ones.

3.3 Performance of Several Encoding and Machine Learning Model Methods

We tested the performance of various machine learning models in the prediction of the risk of metastasis in NSCLC patients. We checked the effect of three types

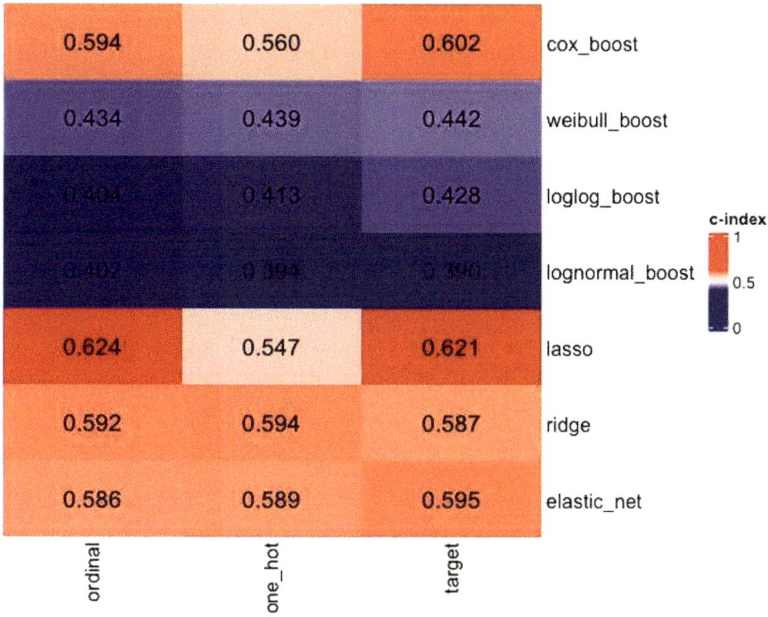

Fig. 3 Performance of survival regression models and encoding methods

of encoding (ordinal, one-hot, target) and two types of feature selection methods: boosting and regularization. We tested seven models with three types of encoding which give us a total of 21 tested models (Fig. 3).

The best performance was achieved with the LASSO Cox model where the data are in the form of ordinal encoding. LASSO regularization leads to a model where coefficients describing unnecessary features are set to zero. It means that unnecessary features are discarded in the model.

Interestingly, ordinal encoding gave the best performance, as looking at every model, ordinal encoding had the highest value of the concordance index.

3.4 Features Importance

Lastly, we checked the values of the selected model (Cox LASSO regression with ordinal encoding) coefficients. It gives us the importance of the features in the model where the highest value means that the feature is more important (Fig. 4).

The treatment intent shows the highest importance in the prediction of the risk of metastasis. In addition, two variants at two different genes were selected by the model: *XRCC1* Arg399Gln (rs25487) SNP and *GSTM1* gene deletion. The first gene (*XRCC1*) encodes a key DNA damage repair protein in the base excision repair (BER) mechanism and the second gene encodes glutathione S-transferase mu 1 participating

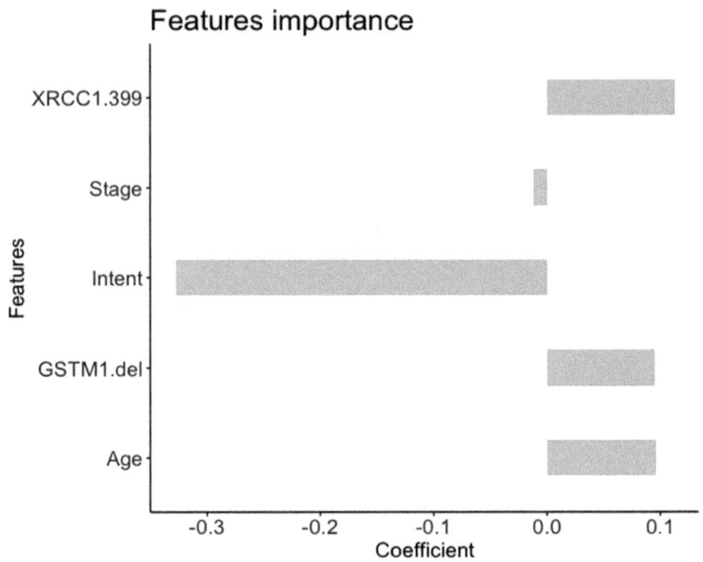

Fig. 4 The value of the coefficient selected by Cox LASSO regression with ordinal encoding features

in the detoxification of many compounds, including carcinogens and anticancer drugs (e.g., cisplatin). Interestingly patient's age was also selected by the Cox regression with LASSO regularization.

4 Discussion

Lung cancer is diagnosed mostly at the advanced stage due to non-specific symptoms and a lack of reliable biomarkers for lung cancer detection. One of the important questions during treatment planning is if the patient should be treated with curative or palliative intent. To facilitate answering this question it is crucial to predict if the patient will develop in a short time after diagnosis of distant metastases.

Here, we applied the combination of clinical and inherited genetic factors for the prediction of time to metastasis (measured as MFS) in non-small cell lung cancer patients. We applied machine learning and statistical models for the prediction of the risk of metastasis based on censored data. Together with the selected models, we applied three types of data encoding and two methods of feature selection. In total, 21 models were fitted.

The best performance was achieved with the Cox regression model with LASSO regression and the ordinal encoding (c-index equal to 0.624). We identified two genetic and three clinical features that have predictive power.

Presented herein results show the promise of non-invasive diagnostic methods in the prediction of the risk of metastasis. The developed model could potentially be applied as a pilot computational tool for better patient stratification and treatment personalization.

Acknowledgements This research was funded by the National Science Centre (NCN), Poland, grant number 2012/05/B/NZ5/01905 (to D.B.) and 2020/37/B/ST6/01959 (to A. S., E. K., A. M. W.).

References

1. Barta JA, Powell CA, Wisnivesky JP (2019) Global Epidemiology of Lung T. https://doi.org/10.5334/aogh.2419
2. Bennett, S.: Log-logistic regression models for survival data. Appl. Stat. **32**(2), 165 (1983). issn: 00359254. https://doi.org/10.2307/2347295
3. Bühlmann, P., et al.: Model-based Boosting 2.0. J. Mach. Learn. Res. 11(71), 2109–2113 (2010). issn: 1533-7928. http://jmlr.org/papers/v11/hothorn10a.html
4. Butkiewicz, D., et al.: Polymorphisms in egfr gene predict clinical outcome in unresectable non-small cell lung cancer treated with radiotherapy and platinum-based chemoradiotherapy. Int. J. Mol. Sci. **22**(11), 5605 (2021). issn: 14220067. https://doi.org/10.3390/IJMS22115605/S1, https://www.mdpi.com/1422-0067/22/11/5605/htm
5. ECOG Performance Status Scale—ECOG-ACRIN Cancer Research Group. https://ecog-acrin.org/resources/ecog-performance-status/. Visited 05 Nov 2023
6. Harrell, F.E., Lee, K.L., Mark, D.B.: Multivariable prognostic models: Issues in developing models, evaluating assumptions and adequacy, and measuring and reducing errors. Stat. Med. **15**(4), 361–387 (1996). issn: 02776715. https://doi.org/10.1002/(SICI)1097-0258(19960229)15:4<361::AID-SIM168> 3.0.CO;2-4, https://onlinelibrary.wiley.com/doi/epdf/10.1002/
7. Hothorn, T., Bühlmann, P.: Model-based boosting in high dimensions. Bioinformatics **22**(22), 2828–2829 (2006). issn: 1367-4803. https://doi.org/10.1093/BIOINFORMATICS/BTL462, https://academic.oup.com/bioinformatics/article/22/22/2828/197039
8. Royston, P.: The lognormal distribution as a model for survival time in cancer, with an emphasis on prognostic factors. Statistica Neerlandica **55**(1), 89–104 (2001). issn: 1467-9574. https://doi.org/10.1111/1467-9574.00158, https://onlinelibrary.wiley.com/doi/full/10.1111/14679574.00158
9. Simon, N., et al.: Regularization paths for cox's proportional hazards model via coordinate descent. J. Stat. Softw. **39**(5), 1–13 (2011). issn: 1548-7660. https://doi.org/10.18637/JSS.V039.I05, https://www.jstatsoft.org/index.php/jss/article/view/v039i05
10. Taketomi, N., et al.: Parametric distributions for survival and reliability analyses, a review and historical sketch. Mathematics **10**(20), 3907 (2022). issn: 22277390. https://doi.org/10.3390/MATH10203907/S1, https://www.mdpi.com/2227-7390/10/20/3907/htm

Metastasis Modelling Approaches—Comparison of Ideas

Artur Wyciślok and Jarosław Śmieja

Abstract Modelling of cancer growth with metastasis can be approached in different ways. The aim of this work is to compare two model types to answer a question of their equivalence and applicability. The first model is a compartmental one, described by ordinary differential equations, while the other is a distributed parameter model, given by partial differential equations. The comparison is based on the overall number of metastatic cells, changing in time. Various initial tumour sizes, growth rates and time horizons are taken into account.

Keywords Modeling of cancer growth · Metastasis

1 Introduction

Mathematical modelling of cancer growth has been extensively investigated for a long time, aiming at supporting clinical research (e.g. [10, 13]). One of the major difficulties in modeling and also one of factors of poor prognosis in many types of cancer is the process of creating distant tumours [9]. Incorporating therapeutic actions in the models of metastasis, should allow for prognoses of efficiency of different therapies [12]. Further benefits include, but are not limited to, identifying target groups for screening, determining interval of screening and risk-stratification [5, 6, 14].

Mathematical models of cancer growth can be divided according to approach used when creating equations. Main three of them are: ordinary differential equations (ODEs)—specifying growth of tumour mass, number of cancer cells in time; partial

A. Wyciślok (✉) · J. Śmieja
Department of Systems Biology and Engineering, Silesian University of Technology, Akademicka 16, 44-100 Gliwice, Poland
e-mail: artur.wycislok@polsl.pl

J. Śmieja
e-mail: jaroslaw.smieja@polsl.pl

P. Strumiłło et al. (eds.), *The Latest Developments and Challenges in Biomedical Engineering*, Lecture Notes in Networks and Systems 746,
https://doi.org/10.1007/978-3-031-38430-1_16

differential equations (PDEs)—accounting not only for time relations but also spatial dimension of tumour growth (or distributed parameters), and agent-based modeling.

Two first approaches—ODEs and PDEs—are in the scope of this work. A particular example of PDE model, defined in [7] (referred to later on as "Iwata model") and further analysed e.g. in [1–3, 8], is compared with single compartmental model (belonging to ODE approach) similar to one proposed e.g. in [11], but without inclusion of therapy effect.

Iwata model is concerned with prediction of number of distant metastatic tumours, as opposed with compartmental model describing number of metastatic cells. While it has been shown to be appropriate for prediction of metastatic sites, modifying it to incorporate therapy is difficult. Moreover, as shown in this paper, the tumour size obtained with that model is arguable. On the other hand, simple compartmental ODE models with therapy do not take into account multiple metastatic sites. Therefore, the aim of this work is to make a comparison of those two approaches in terms of overall number of metastatic cells to answer a question about their equivalence and applicability.

2 Mathematical Models

The choice of the tumour growth type in the ODE model was determined by assumptions of Iwata model (described in detail in Sect. 2.1). It was decided that Gompertzian growth of both tumours should be preserved for comparison. Also, colonisation rates for both models were taken according to Iwata model, so that only the differences in the structures of models would influence the comparison.

2.1 Iwata PDE Model

The model proposed in [7] was aimed at predicting number of metastatic tumours, to take into account that many sites may be colonised by metastatic cells [4]. Main assumptions underlying the model are that both primary and metastatic tumours grow according to the same growth rate $g(x)$ and emit single metastatic cells with rate $\beta(x)$. Those emitted cells are then colonising new sites.

Model operates with variable called colony size distribution—$\rho(x, t)$. Its changes in time are governed by a following partial differential equation:

$$\frac{\partial \rho(x,t)}{\partial t} + \frac{\partial g(x)\rho(x,t)}{\partial x} = 0 \tag{1}$$

The growth rate $g(x)$ is assumed to be a Gompertzian one:

$$g(x) = ax \ln\left(\frac{b}{x}\right) \tag{2}$$

where x denotes number of cells, a is growth rate parameter and b is the capacity of environment.
The initial and boundary condition for PDE in Eq. (1) is defined as follows [7]:

$$\rho(0, x) = 0 \tag{3}$$

$$g(1)\rho(1, t) = \int_{1}^{\infty} \beta(x)\rho(x, t)dx + \beta(x_p(t)) \tag{4}$$

Equation (4) indicates that a new metastatic site is created when a single cell emitted by other tumours successfully colonizes a distant site.

One of the main assumptions of the model is that colonization always leads to creation of new metastatic sites and emission of cells from both primary and secondary tumours. The emission rate $\beta(x)$, related to tumour vascularity is in the following form:

$$\beta(x) = mx^{\alpha} \tag{5}$$

The growth of primary tumour $x_p(t)$ is governed by a simple first order differential equation:

$$\frac{dx_p}{dt} = g\big(x_p(t)\big), x_p(0) = x_0 \tag{6}$$

Equation (6) is defined for any initial primary tumour size like in e.g. [1, 2].

2.2 A Simple Compartmental Model

The ODE-based model is using a simplified version of equation proposed in [12], to calculate growth of metastatic tumour:

$$\frac{dx_m}{dt} = ax \ln\left(\frac{b}{x}\right) + \beta(x_p(t)) \tag{7}$$

The number of metastatic cells is denoted as x_m. Both β and $x_p(t)$ are defined in the same way as in Sect. 2.1 (Eqs. (5) and (6)). It can me seen that first part of the r.h.s of Eq. (7) is $g(x)$ from Eq. (2).

3 Total Number of Metastatic Cells—Analytical Approach

Works concerning analysis of model described in Sect. 2.1, concentrate on number of created metastatic tumours and attempt to fit that number to clinical data. However, there exist no works focused at the size of metastatic tumours, thus facilitating comparison of different modeling approaches. To make this comparison possible, one should specify the number of tumour cells or its volume. Both models can be utilized for analysis in terms of either cell count or volume, assuming constant cell density of the tumour. Since Iwata model is originally described in terms of cell number, such approach was taken in the work.

Even though both models are continuous with solutions having values in real numbers, talking about cell number is a justified approximation given the analysis is made for cell counts of at least tens of thousands. As for solutions for small times, use of real values is a useful modelling tool, but additional caution must be taken to perform further analysis on that timescale. It is however not a case in this study, that is concerned with longer time scales.

In [7] it was specified that expression $\rho(x, t)dx$ defines number of metastatic tumours of size from x do $x + dx$. Given that x is defined in terms of number of cells, to obtain total number of metastatic cells following integration must be performed:

$$x_m(t) = \int_1^{\infty} x\rho(x, t)dx \tag{8}$$

Taking into account that growth of metastatic tumours is described by Gompertzian curve, we can limit the integral to finite range, defined by environment capacity parameter b:

$$x_m(t) = \int_1^{b} x\rho(x, t)dx \tag{9}$$

Further analysis of Eq. (9) in such form requires calculation of $\rho(x, t)$. Such solution for a particular initial condition was proposed in [7]. However additional features can be observed when looking not at total number of metastatic cells but its changes. Calculating time derivative of Eq. (9) we get:

$$\frac{dx_m}{dt} = \frac{d}{dt}\int_1^{b} x\rho(x, t)dx \tag{10}$$

Knowing that the only time-dependent element is colony size distribution we can rewrite the equation as follows:

$$\frac{dx_m}{dt} = \int_1^b x\frac{\partial \rho(x,t)}{\partial t}dx \tag{11}$$

We can now use the defining equation of the model (1), to substitute partial derivative over time with partial derivative over cell number.

$$\frac{\partial \rho(x,t)}{\partial t} = -\frac{\partial g(x)\rho(x,t)}{\partial x} \tag{12}$$

$$\frac{dx_m}{dt} = -\int_1^b x\frac{\partial g(x)\rho(x,t)}{\partial x}dx \tag{13}$$

It is worth noting that any additional term, associated e.g. with chemo or radiotherapy will show itself as another component added to $\frac{\partial g(x)\rho(x,t)}{\partial x}$ in the substitution.

Subsequently, integrating the resulting expression by parts we obtain:

$$\frac{dx_m}{dt} = -xg(x)\rho(x,t)\Big|_1^b + \int_1^b g(x)\rho(x,t)dx \tag{14}$$

$$\frac{dx_m}{dt} = -bg(b)\rho(b,t) + g(1)\rho(1,t) + \int_1^b g(x)\rho(x,t)dx \tag{15}$$

The growth function is defined as following Gompertz curve, therefore expression $g(b)$, with b being environment capacity is equal 0. Moreover expression $g(1)\rho(1,t)$ is the boundary condition of the problem, therefore we can substitute its form as specified in Eq. (4):

$$\frac{dx_m}{dt} = \int_1^b g(x)\rho(x,t)dx + \int_1^b \beta(x)\rho(x,t)dx + \beta(x_p) \tag{16}$$

The explicit form for variable x_p as a solution to Eq. (6) can be easily found, and for initial condition x_0 it is:

$$x_p(t) = x_0\left(\frac{b}{x_0}\right)^{(1-e^{-at})} \tag{17}$$

Considering the colonization rate defined in the model (Eq. (5)), the final form of equation for growth of total number metastatic cells that is considered in this work is:

$$\frac{dx_m}{dt} = \int_1^b (\beta(x) + g(x))\rho(x,t)dx + mx_0^{\alpha}\left(\frac{b}{x_0}\right)^{\alpha(1-e^{-at})} \tag{18}$$

Models given by (18) and Eq. (7) contain components that are similar in their interpretation. First is the expression relating growth of metastasis to size of primary tumour. If we adapt the same colonization rate for both models that part of both equations will be identical. Secondly, analysing the integral part of r.h.s of Eq. (18) we can see $g(x)\rho(x, t)$ being growth of existing metastatic tumours. Taking into account multiple metastatic tumours in Iwata model and single metastasis for compartmental one, that part is also very similar in both models describing cell growth. What differentiates expressions (18) and (7) the most, is cell migration from metastatic tumours present in Iwata model.

4 Simulation Results

In this section numerical results are presented comparing overall number of metastatic cells as predicted by models described in Sect. 2. Results are shown in different time scales and for different initial conditions.

For the simulations three time scales were considered, important from clinical point of view when talking about survival of patients, namely 1, 3 and 5 years. As for initial size of primary tumour four different values of $x_0 < b$ were selected: $x_{0_1} = 1$, $x_{0_2} = 10^3$, $x_{0_3} = 10^6$ and $x_{0_4} = 10^9$.

For actual numerical calculations procedure proposed in [1] was used.

4.1 *Impact of Different Initial Condition*

The impact that different initial tumour sizes have on number of metastatic cells for both analysed models was checked numerically. On Fig. 1 time courses of number of metastatic cells (x_m) and its growth ($\frac{dx_m}{dt}$) for compartmental ("Gompertz") model (defined in Sect. 2.2) are shown. Same characteristics for Iwata model (defined in Sect. 2.1) are shown in Fig. 2

As shown in Figs. 1 and 2, in the time horizon of 5 years, initial size of primary tumour influences only how fast the metastatic cancer grows, not the shape of the growth for both models. However, for Iwata model the differences after five years are large, contrary to compartmental model. Also interesting observation is that shapes of both x_m and $\frac{dx_m}{dt}$ for Iwata model are very similar, bearing resemblance to behaviour of exponential growth.

What was expected is seen for number of cells in Fig. 1. Following Gompertzian curve it approaches the environment capacity, resulting in significant decrease in growth rate. No similar behaviour is visible for Iwata model. That may indicate a potential problem with biological relevance of that model, as the tumour sizes are allowed to grow above what is considered a death condition.

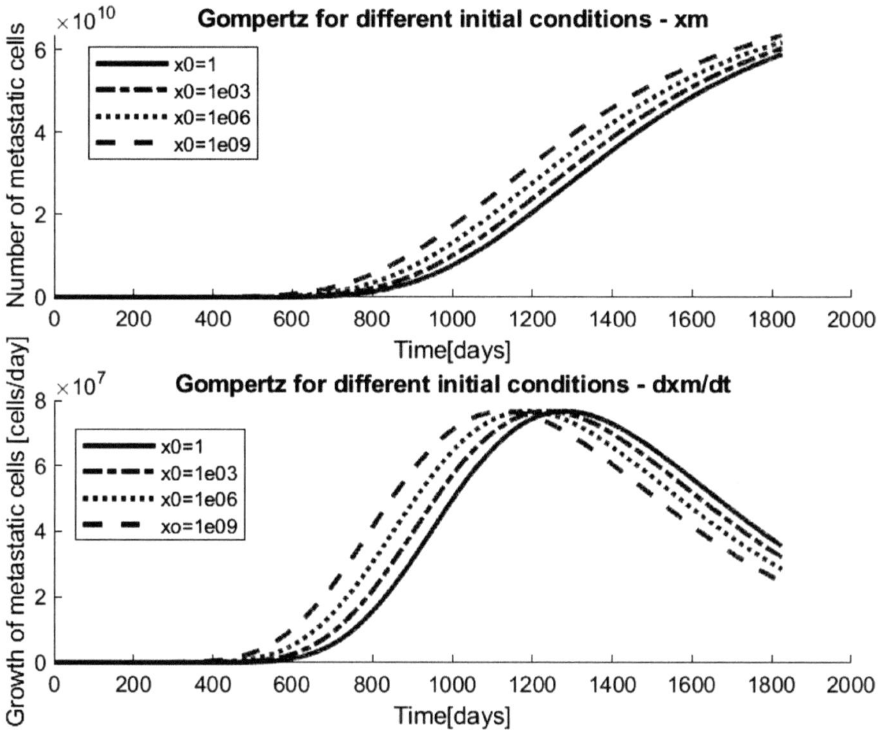

Fig. 1 Behaviour of Gompertz model for different initial conditions

4.1.1 Initial Condition—1 Cell

The first initial condition taken into account comes directly from [7]. Its interpretation is that metastatic processes begin as soon as the primary tumour is formed.

Looking at model responses after 1 year we can see that for compartmental model number of metastatic cells begins to grow much faster in the early stage than for Iwata model. Moreover looking at the time courses of the derivative we can see that throughout whole first year also the rate of growth is much higher in compartmental model.

However, when comparing those observations with what happens after 3 years, a significant increase in growth rate for Iwata model can be seen, resulting in total number of cells approaching that for compartmental model.

4.1.2 Initial Condition—10^3 Cells

The next initial size of primary tumour taken into consideration is 1000 cells.

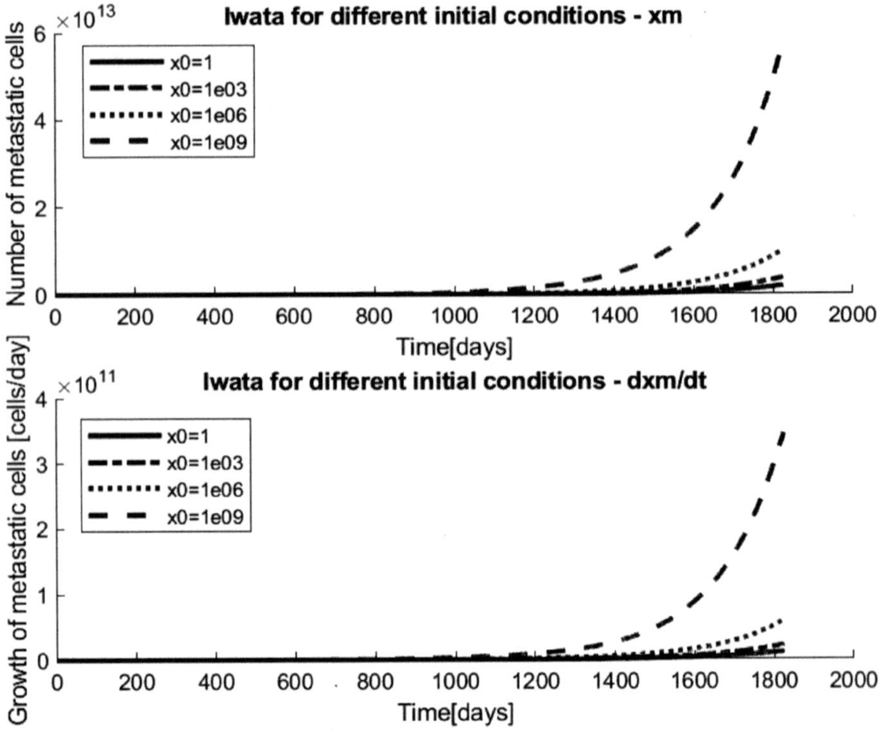

Fig. 2 Behaviour of Iwata model for different initial conditions

Just as in Sect. 4.1.1, we can observe that after 1 year (Fig. 4) number of cells and its growth for compartmental model are much larger than for Iwata model. What is interesting is behaviour after 3 years. When looking at number of cells we observe that both models provide similar values at the end of the time horizon. However, what makes the difference is the behaviour of the growth of number of cells. For compartmental model we can see on Fig. 4 that growth of number of cells is approaching its maximum value. At the same time growth rate for Iwata model is still increasing overtaking what is seen for compartmental model. That results in discrepancies between models' predictions after 5 years.

4.1.3 Initial Condition—10^6 Cells

For the third initial condition, 10^6 cells, changes seen between Sects. 4.1.1 and 4.1.2 are continued. Similarly, after one year the compartmental model yields larger number of cells and higher growth rates, however, the difference is visibly smaller than previously.

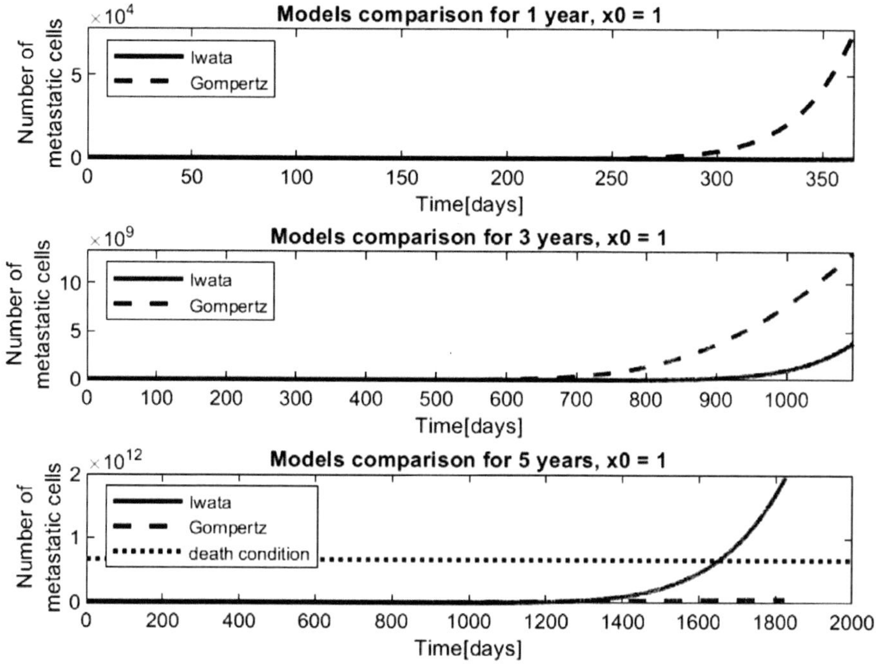

Fig. 3 Comparison of models for first initial condition

In horizon of 3 years, the moment when both models predict the same number of cells occurs almost 1 year faster than in Sect. 4.1.2. Also the growth rate for Iwata model becomes significantly larger than for compartmental one.

4.1.4 Initial Condition—10^9 Cells

When comparing results for the last initial condition—10^9 cells (Fig. 6) with what could be seen in Figs. 3, 4 and 5, first thing that can be noticed is that even for the shorter time horizon of 1 year, predictions for Iwata model are higher that those of compartmental one. That is contrary to what happened for smaller initial conditions, but at the same time it is in line with tendency seen for previous initial conditions—that the bigger initial primary tumour, the faster Iwata model overtakes compartmental one. As for comparison after 3 years the difference between both models is great both in terms of number of cells and in terms of growth rate.

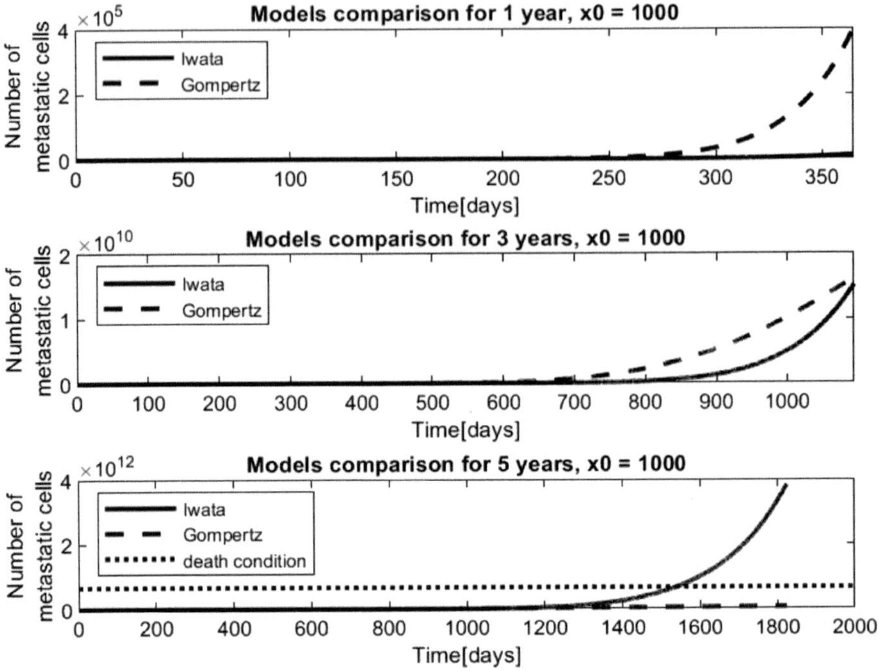

Fig. 4 Comparison of models for second initial condition

4.2 Changes of Parameters of Compartmental Model

In Sect. 4.1 it was shown that assuming the same values of parameters for both models their behaviour differs significantly. However, Iwata model assumes that even though single metastatic tumour is limited in its growth by environment capacity, overall number of possible metastatic tumours is not limited. Taking that into consideration for compartmental model, environment capacity for metastasis compartment may be adjusted to allow for growth of single metastatic tumour beyond maximum size defined in Iwata model. Moreover, for that model different metastatic sites grow independently. It may suggest that also growth rate for compartmental model can be adjusted to bring its behaviour closer to Iwata model. Therefore, parameters of metastasis growth for compartmental model (first expression in Eq. (7)) are changed for subsequent simulations.

In Fig. 7 we can see that increasing the environment capacity indeed results in number of metastatic cells comparable to those obtained in Iwata model. However, those numbers are reached much earlier.

On the other hand, in Fig. 8 it can be seen that by increasing the growth rate we also observe that compartmental model reaches its saturation much faster than without parameter change.

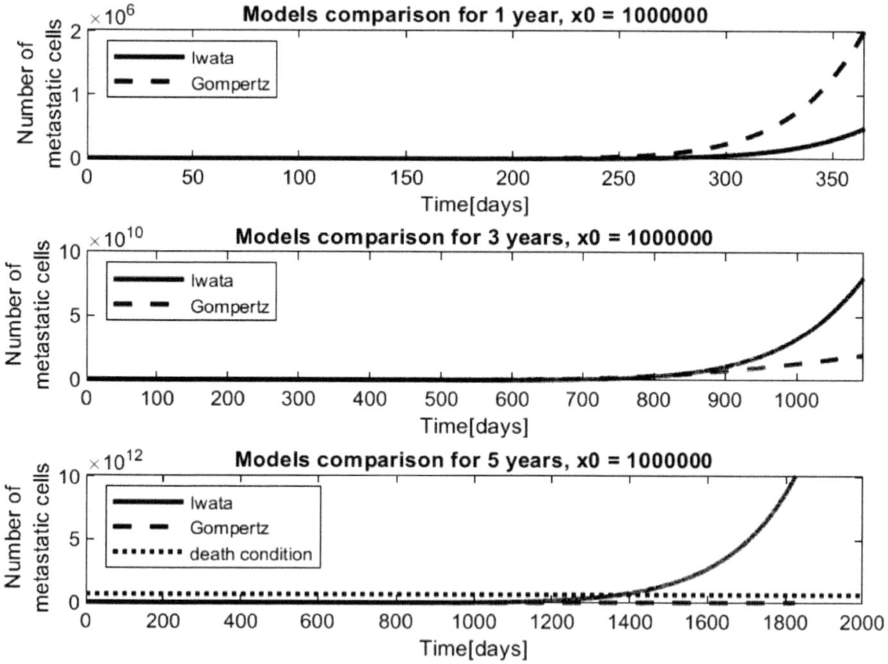

Fig. 5 Comparison of models for third initial condition

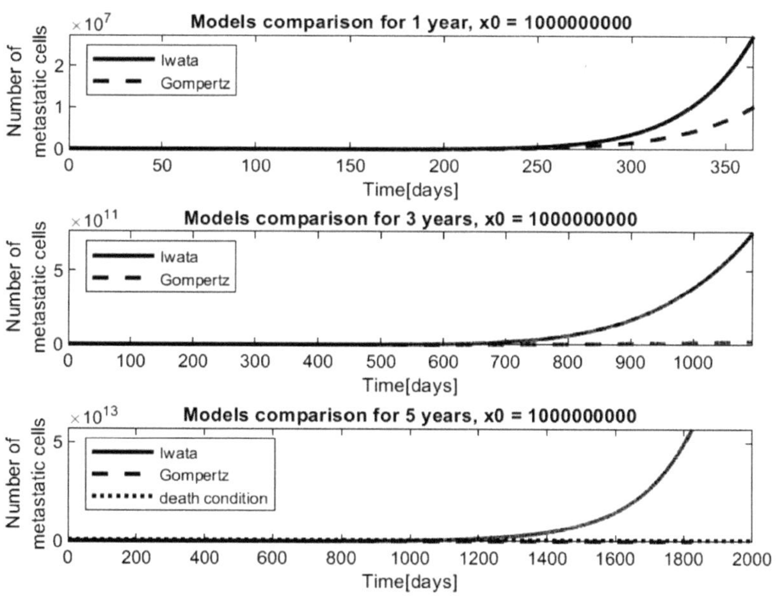

Fig. 6 Comparison of models for fourth initial condition

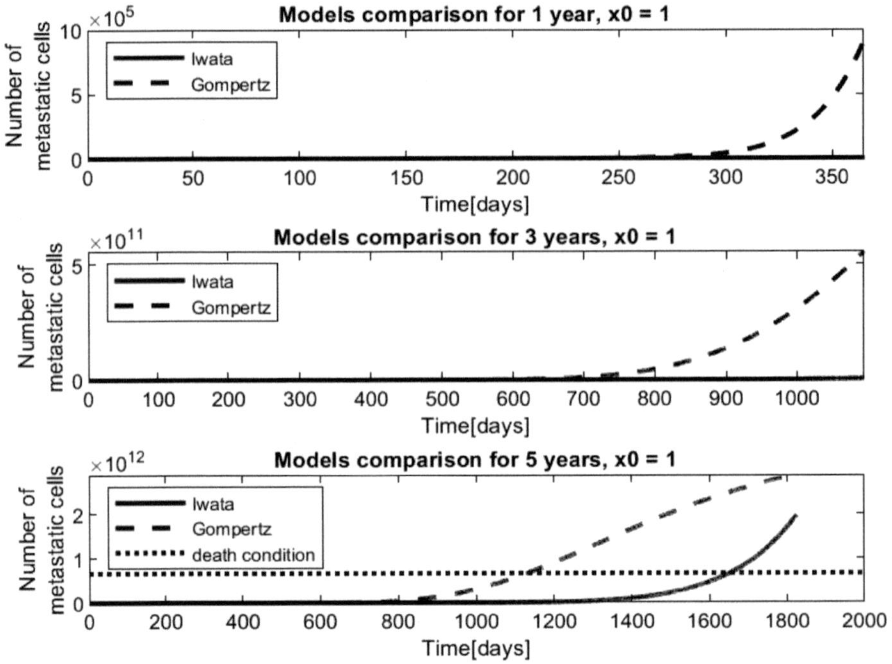

Fig. 7 Comparison of models for b 50 times higher than in Iwata model

Combining observations from Figs. 7 and 8, both parameters were adjusted so that behaviour of both models is similar. That similarity is required only up to a certain threshold—a number of metastatic cells leading to patient's death. Results of such adjustment, for two different initial conditions is shown in Figs. 9 and 10. It can be seen that results obtained by both models are comparable in given time horizon and for sizes below death condition. The difference is bigger for larger initial condition, however that can also be corrected with adjustments of parameters.

5 Conclusions

Though the Iwata model is widely referred to in modeling papers, it has not been compared explicitly to other models. To facilitate such comparison, in this work we derived a simple formula for calculation of total tumour size. As original Iwata model uses Gompertz growth model for both primary and metastatic tumours, compartmental (ODE-based) model making use of the same growth rule was chosen for comparison.

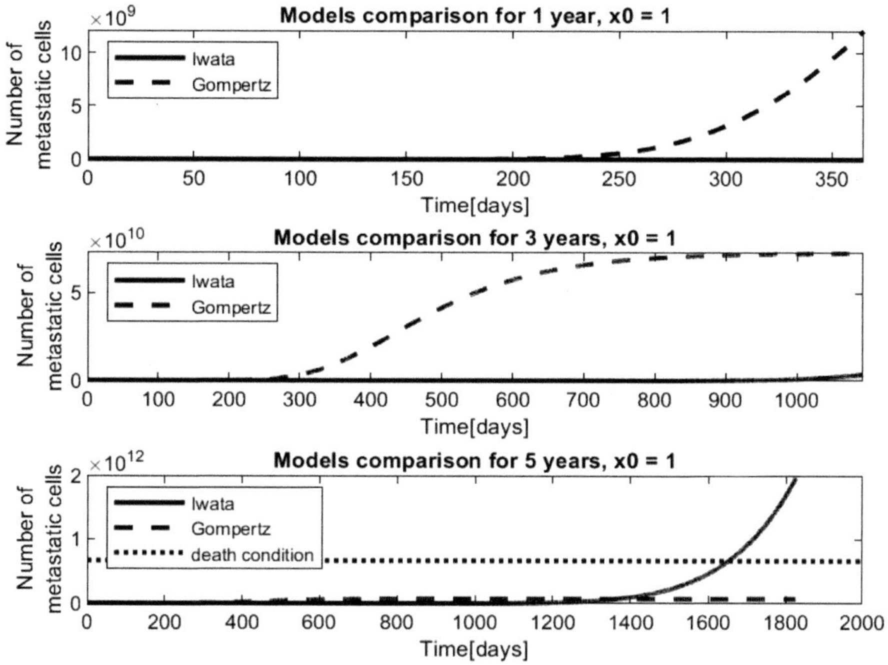

Fig. 8 Comparison of models for *a* 3 times higher than in Iwata model

Both models, for the same values of parameters, were simulated for different initial sizes of primary tumour. Results were analysed in three time horizons—1 year, 3 years and 5 years. Difference in sensitivity of models to initial primary tumour size shown in Sect. 4.1 was visible when models were compared for particular initial conditions in shorter horizons. For small initial conditions the increase of tumour size was significantly larger in the compartmental model. For medium ones both models produced similar results. Ultimately for the last initial condition the increase of tumour size was significantly smaller in the compartmental model.

Also, while growth rate for compartmental model exhibited standard Gompertz curve behaviour, as expected, growth rate for Iwata model resembled more of an exponential behaviour in spite of it being based also on Gompertzian growth (for a single metastatic tumour). It is a direct result of assuming an unlimited number of metastatic sites—though in each site there is a Gompertzian growth, there is no limiting factor for the total cancer growth that accumulates all metastatic sites. This might suggest that, while the Iwata model has a good predictive value, concerning prediction of metastatic sites, it might not be appropriate if the goal of modeling is to describe therapy effects and predict patient survival.

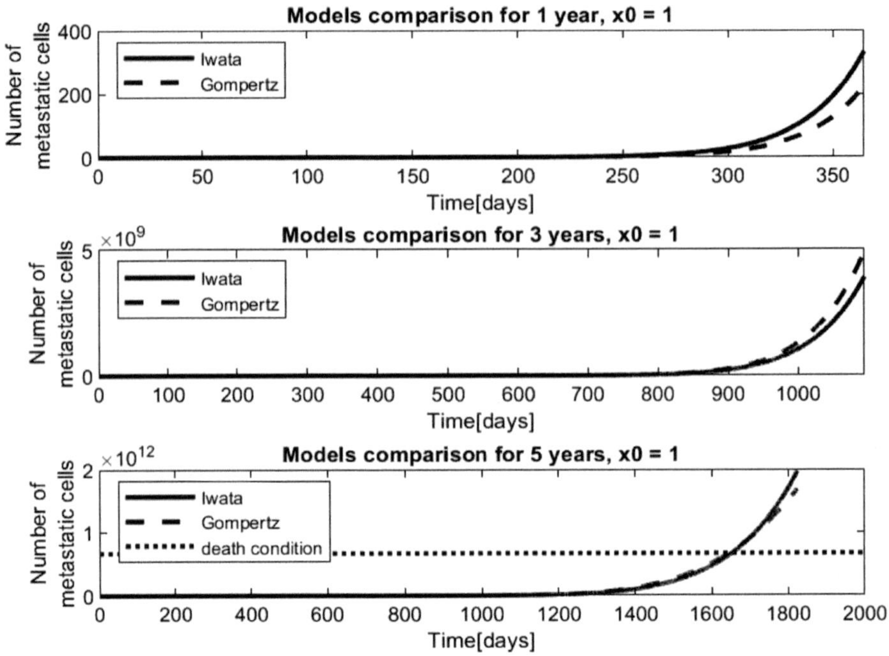

Fig. 9 Comparison of models for *b* 500 times higher than in Iwata model and *a* equal to 0.51 of growth rate in Iwata model—first initial condition

The differences between models can be significantly reduced when parameters of compartmental model are changed. Such approach is justified by unlimited number of possible metastatic sites in Iwata model and conclusions drawn from that (presented in Sect. 4.2). By appropriately selecting environment capacity and growth rate, predictions of compartmental model are comparable to those of Iwata model when analysed for number of metastatic cells lower than threshold associated with patient's death.

Discrepancies in results yielded for both models in different time scales and for different initial conditions are something that must be taken into account when choosing mathematical model for prediction of tumour growth.

It should be noted that incorporating anticancer therapy of any type is very difficult, if possible at all, in Iwata model, while it is relatively simple in the compartmental approach. The difficulties would include, among others, differentiating therapy effects among different metastatic sites.

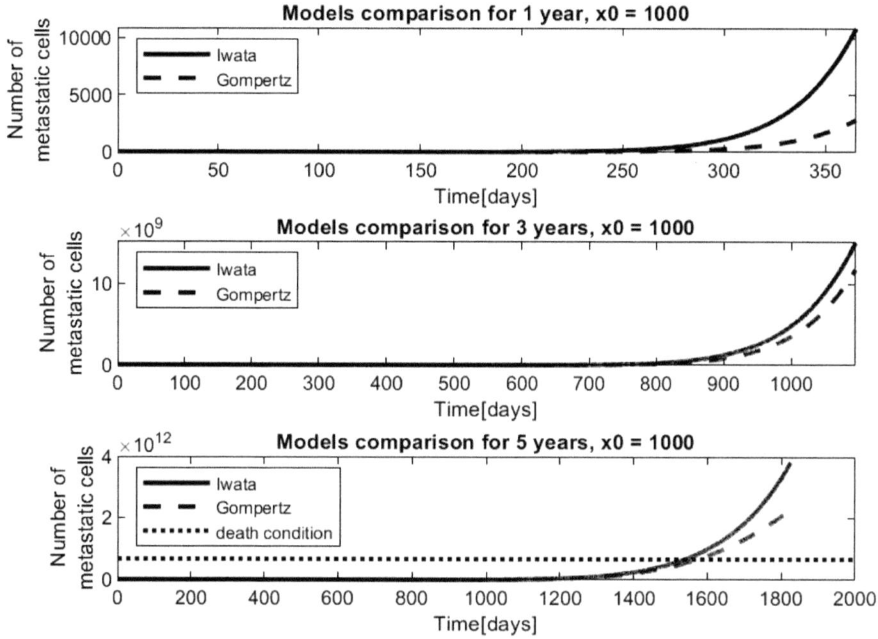

Fig. 10 Comparison of models for b 500 times higher than in Iwata model and a equal to 0.51 of growth rate in Iwata model—second initial condition

Acknowledgements This work was supported by the SUT internal grant for young researchers (AW) and the NCN grant DEC-2020/37/B/ST6/01959 (JS).

References

1. Barbolosi, D., Benabdallah, A., Hubert, F., Verga, F.: Mathematical and numerical analysis for a model of growing metastatic tumors. Math. Biosci. **218**(1), 1–14 (2009). https://doi.org/10.1016/j.mbs.2008.11.008
2. Benzekry, S.: Mathematical analysis of a two-dimensional population model of metastatic growth including angiogenesis. J. Evol. Equ. **11**(1), 187–213 (2010). https://doi.org/10.1007/s00028-010-0088-5
3. Benzekry, S.: Mathematical and numerical analysis of a model for anti-angiogenic therapy in metastatic cancers. ESAIM: Math. Model. Numer. Anal. **46**(2), 207–237 (2011). https://doi.org/10.1051/m2an/2011041.
4. Bilous, M., Serdjebi, C., Boyer, A., Tomasini, P., Pouypoudat, C., Barbolosi, D., Barlesi, F., Chomy, F., Benzekry, S.: Quantitative mathematical modeling of clinical brain metastasis dynamics in non-small cell lung cancer. Sci. Rep. **9**(1) (2019). https://doi.org/10.1038/s41598-019-49407-3.
5. Haaf, K., Aalst, C.M., Koning, H.J., Kaaks, R., Tammemägi, M.C.: Personalising lung cancer screening: an overview of risk-stratification opportunities and challenges. Int. J. Cancer **149**(2), 250–263 (2021). https://doi.org/10.1002/ijc.33578

6. ten Haaf, K., Jeon, J., Tammemägi, M.C., Han, S.S., Kong, C.Y., Plevritis, S.K., Feuer, E.J., de Koning, H.J., Steyerberg, E.W., Meza, R.: Risk prediction models for selection of lung cancer screening candidates: a retrospective validation study. PLOS Med. **14**(4), e1002,277 (2017). https://doi.org/10.1371/journal.pmed.1002277.
7. Iwata, K., Kawasaki, K., Shigesada, N.: A dynamical model for the growth and size distribution of multiple metastatic tumors. J. Theor. Biol. **203**(2), 177–186 (2000). https://doi.org/10.1006/jtbi.2000.1075
8. Liu, J., Wang, X.S.: Numerical optimal control of a size-structured PDE model for metastatic cancer treatment. Math. Biosci. **314**, 28–42 (2019). https://doi.org/10.1016/j.mbs.2019.06.001
9. Popper, H.H.: Progression and metastasis of lung cancer. Cancer Metastasis Rev. **35**(1), 75–91 (2016). https://doi.org/10.1007/s10555-016-9618-0
10. Schättler, H., Ledzewicz, U.: Optimal Control for Mathematical Models of Cancer Therapies. Springer, New York (2015). https://doi.org/10.1007/978-1-4939-2972-6, https://doi.org/10.1007/978-1-4939-2972-6
11. Smieja, J., Psiuk-Maksymowicz, K., Swierniak, A.: A framework for modeling and efficacy evaluation of treatment of cancer with metastasis. In: Biocybernetics and Biomedical Engineering—Current Trends and Challenges, pp. 88–97. Springer International Publishing (2021). https://doi.org/10.1007/978-3-030-83704-4_9.
12. Smieja, J., Swierniak, A., Kimmel, M.: A minimal model of cancer growth, metastasis and treatment. In: Recent Challenges in Intelligent Information and Database Systems, pp. 566–577. Springer Nature Singapore (2022). https://doi.org/10.1007/978-981-19-8234-7_44
13. Świerniak, A., Kimmel, M., Smieja, J., Puszynski, K., Psiuk-Maksymowicz, K.: System Engineering Approach to Planning Anticancer Therapies. Springer International Publishing (2016). https://doi.org/10.1007/978-3-319-28095-0
14. Yeo, Y., Shin, D.W., Han, K., Park, S.H., Jeon, K.H., Lee, J., Kim, J., Shin, A.: Individual 5-year lung cancer risk prediction model in korea using a nationwide representative database. Cancers **13**(14), 3496 (2021). https://doi.org/10.3390/cancers13143496

Model of Lung Cancer Progression and Metastasis—Need for a Delay

Krzysztof Psiuk-Maksymowicz

Abstract A deterministic model consisting of both an ordinary differential equation and delay differential equations was proposed to describe the problem of lung cancer growth and its metastatic progression. In order to represent the behaviour of a virtual population of affected individuals, some of the model parameters are randomly generated. This approach makes it possible to determine certain characteristics describing the population, such as the classification of the primary cancer, location of the metastasis or the estimation of survival curves.

Keywords Modeling of lung cancer growth · Metastasis · ODE · DDEs

1 Introduction

Despite the fact that lung cancer is the second most frequently detected type of cancer in US (after prostate cancer for males and breast cancer for females) and the number of patients is decreasing, it is expected that in 2022 the greatest number of deaths will be from cancers of the lung [1]. Nearly 80% of lung cancers are non-small cell lung cancer (NSCLC), and the remaining 20% are of small cell lung cancer (SCLC) type. Lung cancer spreads both through lymphatic and vascular system, and the metastases are responsible for many adverse effects of this disease. 60% of lung cancer cases are diagnosed at the metastatic stage. It most commonly spreads to the liver, the lymph nodes, the brain, the bones, and the adrenal glands.

Prognosis of the disease is based on the TNM cancer staging system [2]. The overall stage is determined after the cancer is assigned a letter and number, thus, T1-T4 stands for primary tumour's spreading, N0-N3 for dissemination to the lymph nodes (near and distant), and M0-M1 for absence or presence of distant metastases.

K. Psiuk-Maksymowicz (✉)
Department of Systems Biology and Engineering, Silesian University of Technology, Akademicka 16, 44-100 Gliwice, Poland
e-mail: krzysztof.psiuk-maksymowicz@polsl.pl

P. Strumiłło et al. (eds.), *The Latest Developments and Challenges in Biomedical Engineering*, Lecture Notes in Networks and Systems 746,
https://doi.org/10.1007/978-3-031-38430-1_17

The mathematical modelling of cancer growth itself has a history spanning several decades [3]. Dozens of models exist, taking into account different types of aspects of the disease itself, or different types of mathematical techniques. These include, for example, compartmental models [4], models with structure [5], models with delays [6], or models that take spatial dependencies into account [7]. The problem of modelling the metastasis process is not new, however, due to the nature of the process, it is not trivial. Among the works that deal with this problem, one may find e.g. [8, 9]. The work by Bilous et al. [10] touches explicitly the problem of lung cancer metastasis. However, these models do not provide information about the location of the metastases but their total number.

The aim of this study was to create a population model of lung cancer growth and metastasis, the model that would give not only an average value for the behaviour of a given population of individuals, but one that would give an insight into the behaviour of an individual patient and allow for a statistical description of considered virtual population. The introduction of delays in the model is intended to represent the situation when particular metastases start to develop in different regions of the body at different moments.

1.1 Clinical Data

We used clinical data from 659 patients with diagnosed lung cancer from the 2nd Radiotherapy and Chemotherapy Clinic, M. Sklodowska-Curie National Research Institute of Oncology, Gliwice Branch in Poland. The data contain age of the patients (mean and median equal to 63 years); sex (26% female, 74% male); location of primary tumour (42% left lung, 58% right lung); histopathology description; type of the cancer (86% NSCLC, 14% SCLC); TNM classification; type of therapy used; intend of treatment (palliative, curative); and complete blood count.

The TNM cancer staging system data show that most patients have diagnosed cancer in advanced stage (60% cases in T3 and T4 stages); most patients have regional and distant lymph node metastasis (76% cases in N2 and N3 stages), and only 14% of patients are in N0 stage. Metastasis description is more detailed than typical TNM classification, and it contains the exact location of the metastasis. 46% of the patients have metastasis, among those: 25% have multiple metastases, 21% brain, 18% second lung, and 17% bone metastasis.

For the purposes of this work data on the progression of the primary tumour and information about the location of metastases were used. Histogram of tumour classification according to the TNM system is present in Fig. 1. Histogram of metastasis location is presented in Fig. 2.

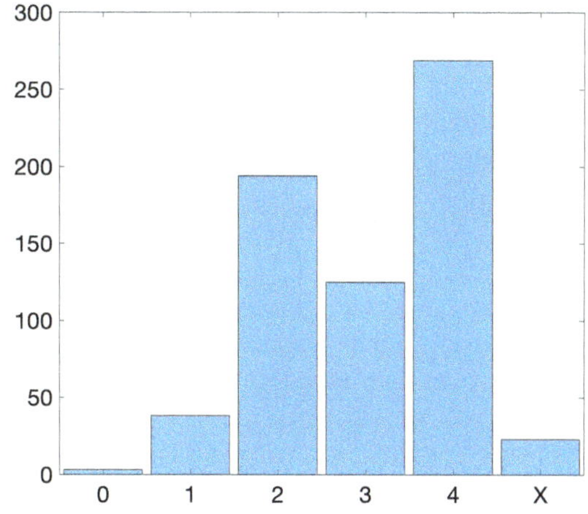

Fig. 1 Histogram of tumour classification

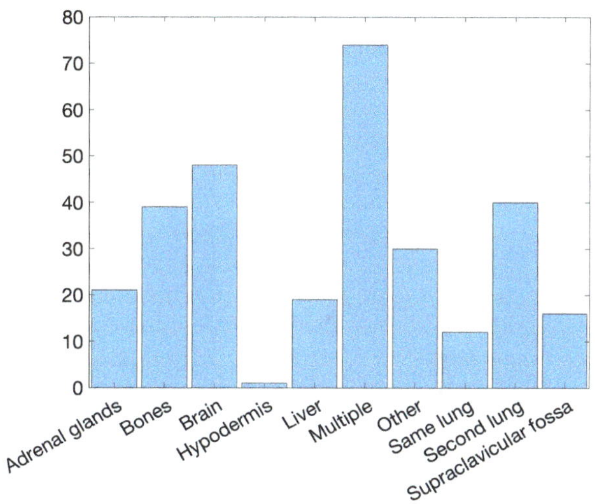

Fig. 2 Histogram of metastasis location

2 Model Formulation

The mathematical model is of compartmental type and consists of one ordinary differential equation (ODE) and 7 delay differential equations (DDEs). The first equation describes the growth of the primary tumour, the next 7 equations concern the progression of the metastasis in the organs where metastases are most frequently found. Lungs are treated as two separate compartments in order to be able to model metastasis arising in the lung from which the primary tumour originates and arising in the lung which was free of pathology. $N_j(t)$ represents the number of cells in different compartments, where $j \in \{P, L, B, V, K, M, G, S\}$ specifies their location, successively, primary tumour, first lung, brain, liver, kidney, muscle, gastrointestinal

(GI) tract, and second lung. The first lung is defined as the one in which the primary tumour was diagnosed.

Following Iwata et al. [8] we adopt Gompertzian growth rate for all equations and metastatic colonization rate mN_P^γ, where m is the colonization coefficient and γ is the fractal dimension of blood vessels infiltrating the tumour. The process of metastatic colonization of specific organ is assumed to be proportional to the amount of blood that flows through an organ and carries cancer cells, additionally it is also normalized. A very similar approach is used in modelling dynamics of drug molecules where a physiologically based pharmacokinetic (PBPK) models [11] are in use. However, we do not distinguish here cancer cells that are in the bloodstream, neither we create any additional compartment for this purpose. Instead, into the equations of the metastatic cells growth we introduce the delay associated with the transport of cancer cells to the tissue and their successful colonisation. The introduction of delays in the model was necessary, without it, it was very difficult to obtain a situation (data not shown) in which metastases appeared in the body at different, distant moments of time (e.g. one or two years). This was due to the initial structure of the model, where cells detaching from the primary tumour at the same time fed into all organ-related compartments. The full model is represented by the following set of equations:

$$\begin{cases} \frac{\mathrm{d}}{\mathrm{d}t} N_P = r_P N_P \ln \frac{K_P}{N_P} - mN_P^\gamma, \; N_P(0) = N_0 \neq 0, \\ \frac{\mathrm{d}}{\mathrm{d}t} N_i(t) = r_i N_i(t) \ln \frac{K_i}{N_i(t)} + \frac{w_i Q_i}{\sum_i w_i Q_i} mN_P^\gamma(t - T_i), \; t \geq T_i, \\ N_i(t) = 0, \; t < T_i, \; i \in \{L, B, V, K, M, G, S\}, \end{cases} \tag{1}$$

where r_j and K_j for $j \in \{P, L, B, V, K, M, G, S\}$ and T_i are positive constants, subsequent subscripts stands for particular compartments. For each growth term r_j represents growth rate, and K_j the carrying capacity. The flux of cancer cells entering metastatic cell compartments is proportional to blood flow rate Q_i in the respective tissues. The influence of the blood flow coefficient can be altered by the weighting factors w_i. They are introduced into the model to allow preferential colonisation of selected tissues by migrating cancer cells.

Model (1) is deterministic, therefore for a given set of initial conditions and parameters, its outcome is always the same and represents the average outcome for the entire population. In order to model the different progression of the lung cancer the metastasis in populations for many virtual patients selected parameters of the model are drawn from specific distributions. Thanks to this procedure, it is possible to observe differences at population level and compare the simulation results to the clinical data. Parameters for Model (1) are collected in Table 1. Parameters: r_P, r_i, and T_i are randomly generated, separately for each simulation, reflecting the individual character of each virtual patient. Parameters r_P, r_i are sampled from the normal distribution, and T_i from a uniform distribution.

Table 1 Model parameters

Symbol	Value	Name	Source
Tumour growth parameters			
$\bar{r}_P, \bar{r}_i$	$0.00286\ \text{day}^{-1}$	Mean growth rate	[8]
r_P	$N(\bar{r}_P, 0.1)$	Primary tumour growth rate	Assumed
r_i	$N(\bar{r}_i, 0.1)$	Metastasis growth rate	Assumed
K_P, K_i	7.3×10^{10} cells	Maximum size of tumour	[8]
m	$5.3 \times 10^{-8}\ (\text{cells day})^{-1}$	Colonization coefficient	[8]
γ	0.663	Fractal dimension of blood vessels	[8]
w_i	1	Weighting coefficients	Assumed
T_a	18 months	Time of metastasis appearance	[10]
T_b	30 months	Time of metastasis appearance	Assumed
T_i	$[T_a, T_b]$ months	Time of metastasis appearance	Assumed
Physiological parameters (L/h)			
Q_L	168.0	First lung blood flow rate	[11]
Q_B	42.0	Brain blood flow rate	[11]
Q_V	88.8	Liver blood flow rate	[11]
Q_K	74.4	Kidney blood flow rate	[11]
Q_M	45.0	Muscle blood flow rate	[11]
Q_G	66.0	GI tract blood flow rate	[11]
Q_S	168.0	Second lung blood flow rate	[11]

3 Results

All simulations were performed using Matlab environment ver. R2021b. Results of a single simulation of model (1) are shown in Fig. 3. It is possible to observe sigmoidal shape of all growth curves with a saturation at the level of parameter $K_P = K_i$. Due to the different delays, a phase shift in the growth functions of the respective metastases can be observed.

Defining certain volumetric thresholds for detection of metastasis or determination of patient death, we can convert them into thresholds expressed by number of cells. Similarly, we can convert the thresholds for belonging to particular T classes of the TNM system.

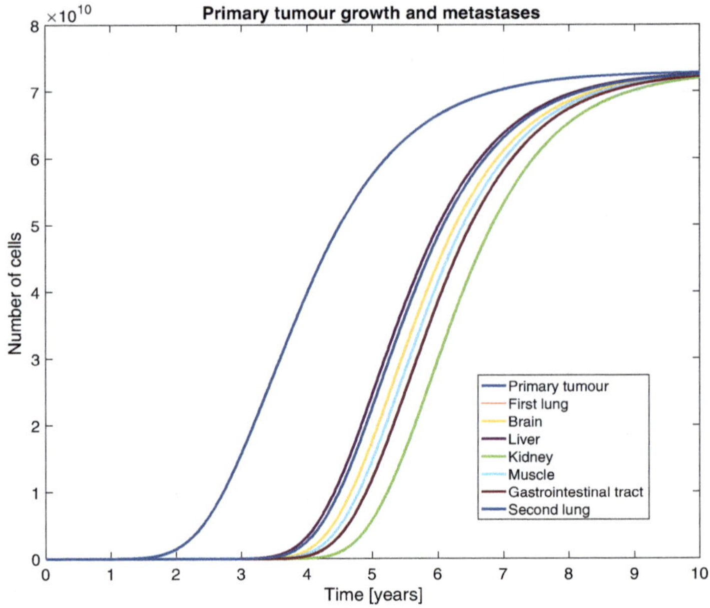

Fig. 3 Primary tumour and metastases growth for a single simulation

Results of 4 000 simulations, representing individual patients, of presented model are visible in the next two figures. Obviously, it is not possible to present 4,000 time courses of lung cancer growth, moreover, it would make no sense. Instead, it is possible to show statistics on the belonging of the lung cancer to the respective classes at the time of diagnosis. Figure 4 provides a histogram of the primary tumour belonging to the specific class. Such statistics are changing over time and, as the disease progresses, the last class becomes more and more numerous.

Another possible histogram presented in Fig. 5 shows occurrence of metastases in different organs two years after diagnosis.

Performed simulations also made it possible to construct the Kaplan-Meier survival curve (it is presented in Fig. 6) with a 99% confidence bounds. Estimated median survival of the patient is 5 years, 2 months and 14 d.

4 Discussion

The developed model is capable for simulating a virtual population of lung cancer patients. By changing the delay parameters, we can influence the frequency of metastases in an individual patient (c.f. Fig. 3). Similarly, by changing the parameters of the probability distributions, we can influence the population distributions of T-classification and the presence of metastases in respective organs (data not shown).

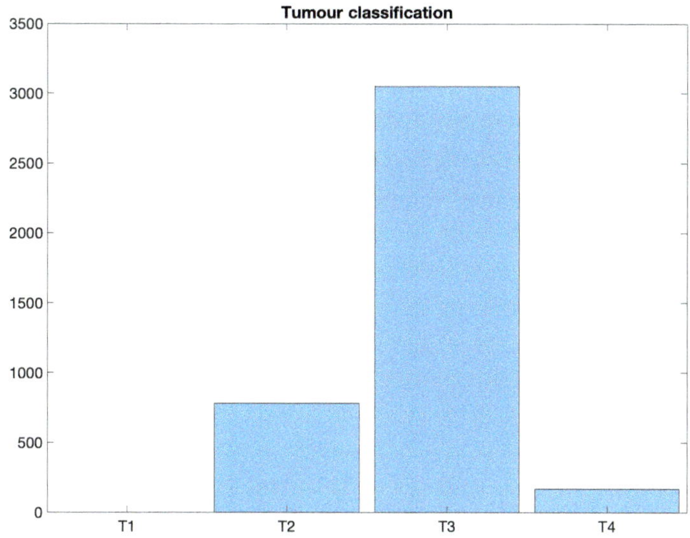

Fig. 4 Histogram of tumour classes resulting from the simulation

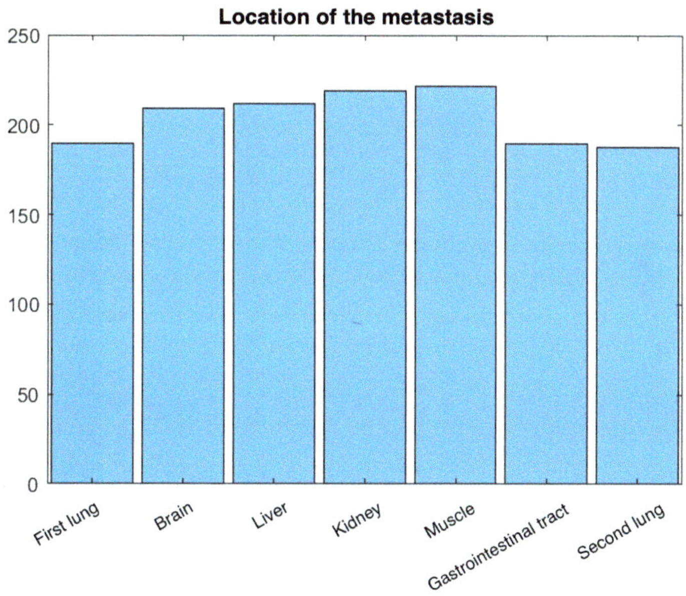

Fig. 5 Histogram of metastatic sites resulting from the simulations

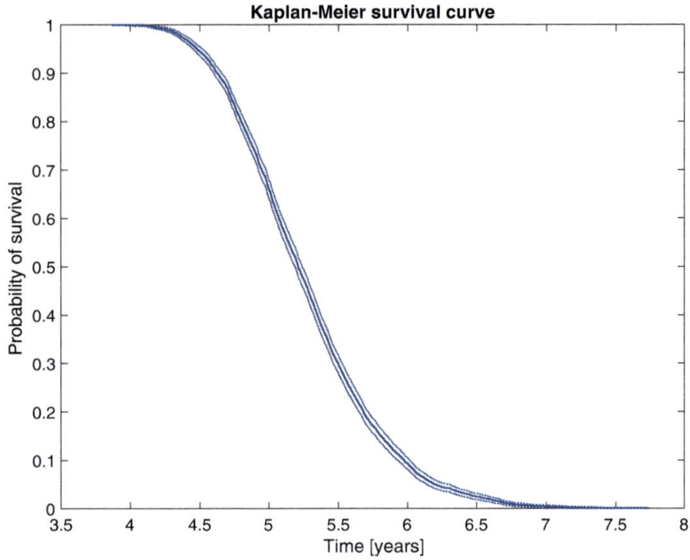

Fig. 6 Kaplan-Meier survival curve with 99% confidence bounds

Considering distribution of metastasis location our results are too uniformly distributed when compared to the clinical data. The formation of metastases in different organs has a different molecular basis, this dissimilarity could be reinforced by using different values for the weights w_i rather than fixed values as has been during simulations.

Relying on the statistic data obtained only at the time of diagnosis, we are forced to make a number of model assumptions, which can have a significant impact on overall results. For example we have to assume the time of the diagnosis of the patient from the development of the disease. This is a big problem because the doctor, at the time of diagnosis, does not know from when the tumour, whether primary or metastatic, has developed. Even if he had knowledge about previous medical imaging results of the patient, sufficiently small tumours would still not be detected. Nevertheless, a model of this kind can be useful to estimate the primary tumour growth and metastasis for an individual patient. Actually, having the input data in categorised form (e.g. T stage), the solution will not be one primary tumour growth curve but a whole family of them.

This work can be developed further in many directions. Some of the extensions may consider addition of the therapeutic actions; others additions of compartments related to the lymphatic system. Adding to the model the impact of the action of selected therapies (similarly to [12, 13]), e.g. platinum-based chemotherapy, will certainly bring the model closer to practical application.

Acknowledgements This work was supported by the NCN (National Science Centre, Poland) grant DEC-2020/37/B/ST6/01959.

References

1. Siegel, R.L., Miller, K.D., Fuchs, H.E., Jemal, A.: Cancer statistics, 2022. CA Cancer J. Clin. **72**(1), 7–33 (2022)
2. Detterbeck F.C. , Boffa D.J. , Kim A.W., Tanoue L.T.: The eighth edition lung cancer stage classification. Chest **151**(1), 193–203 (2017)
3. Araujo, R.P., McElwain, D.L.: A history of the study of solid tumour growth: the contribution of mathematical modelling. Bull. Math. Biol. **66**(5), 1039–1091 (2004)
4. Swierniak, A., Ledzewicz, U., Schattler, H.: Optimal control for a class of compartmental models in cancer chemotherapy. Int. J. Appl. Math. Comput. Sci. **13**(3), 357–368 (2003)
5. Swierniak, A., Kimmel, M., Smieja, J., Puszynski, K., Psiuk-Maksymowicz, K.: Structured models and their use in modeling anticancer therapies. In: System Engineering Approach to Planning Anticancer Therapies. Springer, Cham (2016)
6. Attia, E., Bodnar, M., Forys, U.: Angiogenesis model with Erlang distributed delays. Math. Biosci. Eng. **14**(1), 1–15 (2017)
7. Psiuk-Maksymowicz K., Multiphase modelling of desmoplastic tumour. J. Theor. Biol. Growth **329**, 52–63 (2013)
8. Iwata, K., Kawasaki, K., Shigesada, N.: A dynamical model for the growth and size distribution of multiple metastatic tumors. J. Theor. Biol. **203**, 177–186 (2000)
9. Franssen, L.C., Lorenzi, T., Burgess, A.E.F., Chaplain, M.A.J.: A mathematical framework for modelling the metastatic spread of cancer. Bull. Math. Biol. **81**, 1965–2010 (2019)
10. Bilous, M., Serdjebi, C., Boyer, A., Tomasini, P., Pouypoudat, C., Barbolosi, D., Barlesi, F., Chomy, F., Benzekry, S.: Quantitative mathematical modeling of clinical brain metastasis dynamics in non-small cell lung cancer. Sci. Rep. **9**(1), 13018 (2019)
11. Chen, Y., Zhao, K., Liu, F., Xie, Q., Zhong, Z., Miao, M., Liu, X., Liu, L.: Prediction of deoxypodophyllotoxin disposition in mouse, rat, monkey, and dog by physiologically based pharmacokinetic model and the extrapolation to human. Front. Pharmacol. **7**, 488 (2016)
12. Smieja, J., Psiuk-Maksymowicz, K., Swierniak, A.: A framework for modeling and efficacy evaluation of treatment of cancer with metastasis. In: Pijanowska, D.G., Zielinski, K., Liebert, A., Kacprzyk, J. (eds.) Biocybernetics and Biomedical Engineering - Current Trends and Challenges. Lecture Notes in Networks and Systems, vol. 293. Springer, Cham (2022)
13. Kozlowska, E., Suwinski, R., Giglok, M., Swierniak, A., Kimmel, M.: Mathematical model predicts response to chemotherapy in advanced non-resectable non-small cell lung cancer patients treated with platinum-based doublet. PLoS Comput. Biol. **16**(10), e1008234 (2020)

Classification of Recorded Electrooculographic Signals on Drive Activity for Assessing Four Kind of Driver Inattention by Bagged Trees Algorithm: A Pilot Study

Rafał Doniec, Szymon Sieciński, Natalia Piaseczna, Konrad Duraj, Joanna Chwał, Maciej Gawlikowski, and Ewaryst Tkacz

Abstract The act of engaging in secondary activities while driving can cause safety risks on public roads due to the driver's distracted attention. The objective of the research was to predict changes in driver concentration levels caused by secondary activities (eating, drinking, bending, and turning toward the rear seats) using the electrooculographic (EOG) signal. Four subjects, consisting of one male and three females between the ages of 23 and 57, performed distracting driving activities using a driving simulator. The EOG signals were recorded using JINS MEME Academic Pack smart glasses, and machine learning techniques (boosted trees, bagged trees, subspace discriminant, subspace KNN, RUSBoosted Trees) were used to classify the occurrence of secondary activities. The results show that the highest accuracy (87%) has been achieved for the bagged tree (ensemble classifier).

R. Doniec (✉) · S. Sieciński · N. Piaseczna · K. Duraj · J. Chwał · M. Gawlikowski · E. Tkacz
Faculty of Biomedical Engineering, Department of Biosensors and Processing of Biomedical Signals, Silesian University, Roosevelta 40, 41-800 Zabrze, Poland
e-mail: rafal.doniec@polsl.pl

S. Sieciński
e-mail: szymon.siecinski@polsl.pl; szymon.siecinski@uni-luebeck.de

N. Piaseczna
e-mail: natalia.piaseczna@polsl.pl

K. Duraj
e-mail: konrad.duraj@polsl.pl

J. Chwał
e-mail: joanna.chwal@polsl.pl

M. Gawlikowski
e-mail: maciej.gawlikowski@polsl.pl; etkacz@polsl.pl

S. Sieciński
University of Lübeck, Institute for Medical Informatics, Ratzeburger Allee 160, 23562 Lübeck, Germany

P. Strumiłło et al. (eds.), *The Latest Developments and Challenges in Biomedical Engineering*, Lecture Notes in Networks and Systems 746,
https://doi.org/10.1007/978-3-031-38430-1_18

Keywords Electrooculugraphy · Driving safety · Classification of EOG · Biomedical signal recognition · Automatic assessment

1 Introduction

Road safety remains one of the top priorities in society. The number of drivers distracted by smartphones or elderly drivers on the roads is increasing, for example, drivers aged 16–19 years and drivers over 85 years of age had per vehicle mile fatal crash rates of 4.7 and 7.6, respectively, compared to a rate of only 1.4 fatal crashes per 100 million miles traveled for those aged 45–49 years [1, 2].

Behavioral and physical restrictions at the age of retirement from professional activity significantly affect the level of road safety. To drive safely, you must be sufficiently aware of your surroundings, pay constant attention to the road and traffic, and be alert enough to react to unexpected circumstances. Age and inattention of the driver are still key factors that contribute to accidents and fatal accidents. Much attention has been paid to the dangers caused by the use of the phone while driving; here, the focus is on more common scenarios of secondary activities that are also causes of dangerous events on the roads [3–5].

Road safety is a crucial issue for seniors, who are at increased risk of accidents due to age-related declines in cognitive and physical abilities. According to the National Highway Traffic Safety Administration, in 2018, drivers 65 years and older accounted for 18% of all traffic fatalities in the United States. Machine learning has emerged as a promising approach to address road safety challenges for seniors using large-scale data to develop predictive models that can help identify drivers at risk and improve road safety [6].

According to the latest data from the National Highway Transportation and Safety Administration (NHTSA), distributed driving claimed the lives of more than 3,000 people in 2020 in America, accounting for 8.1% of all road accident deaths. Furthermore, approximately 324,000 people were injured due to driver inattention [7].

Basic driving activities refer to tasks that are directly related to maneuvering a vehicle, such as acceleration, braking, and steering. Human activity recognition (HAR) techniques can be used to automatically detect and classify basic driving activities using vehicle sensor data [8].

Secondary driver activities refer to actions that drivers perform while driving that are not directly related to driving the vehicle, such as using a cellphone, eating or drinking, adjusting the radio or air conditioning, and engaging in conversations with passengers. These activities can distract drivers and reduce their ability to operate the vehicle safely, increasing the risk of accidents [3, 5].

Observations show that 14% of the drivers were involved in some secondary activity, the most common activity being the use of a mobile phone (5%). Involvement in secondary work was significantly more common among drivers without passengers (15%), compared to 9% among drivers with passengers [9].

Human activity recognition (HAR) using machine learning is a challenging task due to several factors. HAR involves identifying and classifying human activities based on sensor data collected from wearable devices or other sources. Some of the challenges in HAR using machine learning include:

- Variability in human activities: Human activities can vary significantly in terms of duration, intensity, and complexity. HAR algorithms must be able to account for this variability to accurately classify activities [10, 11].
- Sensor noise and variability: The wearable sensors used in the HAR can produce noisy and unreliable data due to factors such as sensor placement, sensor drift, and environmental factors. This can make it difficult to accurately classify activities based on sensor data [12, 13].
- Data sparsity and class imbalance: HAR datasets are often sparse and imbalanced, and some activities occur less frequently than others. This can make it difficult to train machine learning models to accurately classify all activities [14].
- Subject variability: Different individuals may perform the same activity differently, making it difficult to generalize HAR models across different subjects [15].
- Context Awareness: Human activities are often performed in specific contexts, such as indoor or outdoor environments, and different contexts can affect the performance of HAR algorithms. HAR models must be able to take context into account to accurately classify activities [16–18].
- Real-time performance: HAR algorithms must often operate in real time, with low latency and high accuracy requirements. This can make it difficult to develop machine learning models that can classify activities quickly and accurately [19, 20].

Many machine learning techniques have been used to recognize human behavior. It is important to have appropriate labels for data for training and testing. Developing a reliable classifier to recognize activity patterns requires a large data set. First, a certain amount of data is needed to distinguish important characteristics for the neural network to distinguish important characteristics, and second, it is necessary to provide a wide range of classification of signal characteristics [21].

The purpose of the study was to automate the evaluation of machine learning classification results by developing the best cooperation of functions in the set to examine features corresponding to real driving activities based on EOG signals.

2 Material and Methods

2.1 Experiment Setup

The study was carried out in a controlled environment that consisted of a car simulator, JINS MEME ES_R smart glasses (JINS, Inc., Tokyo, Japan), and a laptop for the acquisition of signals. The study group consisted of four volunteers (one male and

Fig. 1 Driving simulator setup

three female) between the ages of 23 and 57. All of them had no health contraindications to drive a car, nor a driver's license to drive motor vehicles up to 3.5 tons. Four activities were considered in our study: eating, drinking, turning, and bending. We chose these activities because they can affect the attention of a driver and do not require the use of a car.

Each subject was asked to participate in second driver scenario which consisted of the following steps:

- Eating (1, class 0)—Taking a bite of any food;
- Drinking (2, class 1)—Taking a sip of water or any drink;
- Turning (3, class 2)—Turning back and reaching to a passenger's seat;
- Bending (4, class 3)—Bending and picking up a fallen object.

During the experiment, we sought to distinguish between secondary driving activities (activities not directly related to driving) in a controlled environment consisting of a car simulator. All participants had normal or corrected-to-normal vision. The setup of the experiment is shown in Fig. 1.

The outline of the experiment was approved by the Bioethics Committee of the Medical University of Silesia in Katowice (the resolution KNW/0022/KB1/79/18) and carried out following the rules of the Declaration of Helsinki of 1975, revised 2013.

2.2 Assessment of Results

The format of the data is as follows; first, the data acquisition parameters are presented, followed by the header describing the content of each column that contains the sample number, the date in the format: dd.mm.yyyy:hh:mm:ss, then the 3 channel accelerometer components:

- EOG_L—left eye raw EOG signal,
- EOG_R—right eye raw EOG signal,
- EOG_H—the difference between the left and right eye potential ($EOG\_L - EOG\_R$),
- EOG_V—negative arithmetical mean of the left and right eye potential—($EOG\_L + EOG\_R)/2$.

2.3 Preprocessing

Before feature extraction and classification, the signal was preprocessed by applying the median filter and bandpass filter.

The first step was to apply the median filter with a window width of 11 samples. The median filter was applied due to the efficiency in removing the spikes in the signal. The EOG signals were preprocessed using a digital Butterworth filter implemented using the scipy.signal module. A third-order low-pass filter with a cutoff frequency 20 Hz was used to remove high-frequency noise and artifacts. A band-pass filter with a passband from 0.1 30 Hz was used to remove DC offsets and electrode polarization. The filter coefficients were obtained using the "butter" function with a sampling frequency 100 Hz. The filtered signals were then evaluated to ensure that the filtering did not introduce any unwanted artifacts or distortion.

There are numerous potential reasons for this, such as electrode polarization or interference with background signals [22].

Using a technique called detrending by differencing a time series, this objective was accomplished. The new value at the current time step is determined as the difference between the current and previous observations.

Then, the signal data were normalized with the Z-score normalization and the first and last samples were discarded due to the artifacts. The mean and standard deviation of each signal were calculated, and the values of the samples are replaced with the newly determined values using the following formula:

$$x' = (x - \mu)/\sigma, \tag{1}$$

where μ is the mean of the signal, σ is the standard deviation, x is the current value of a sample and x is the new value, so that the new mean of all values is 0 and the standard deviation is 1.

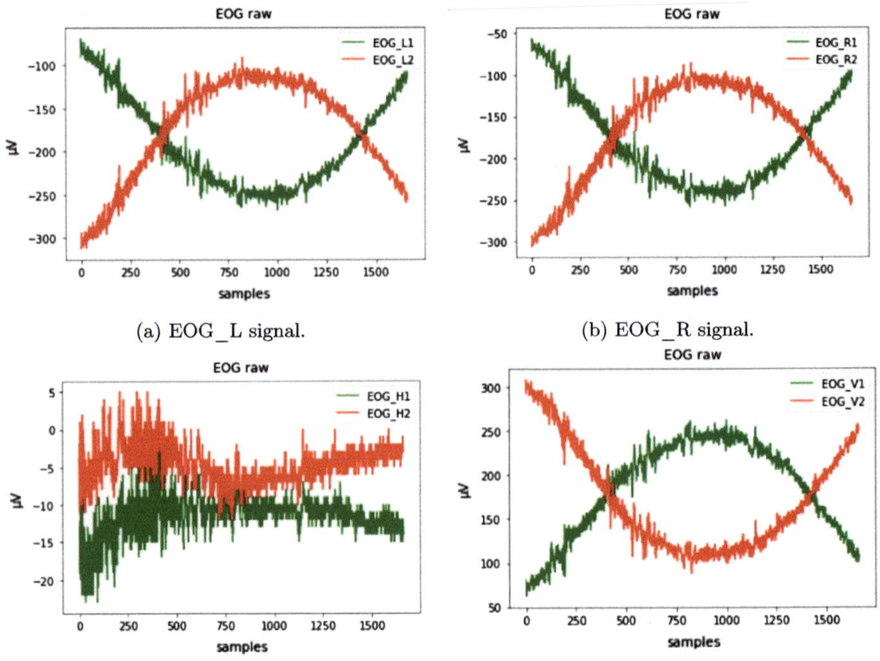

(a) EOG_L signal.

(b) EOG_R signal.

(c) EOG_H signal.

(d) EOG_V signal.

Fig. 2 Raw EOG signal of going straight at a roundabout

2.4 *Signal Processing*

The raw EOG signal presented in Fig. 2 contains different types of artifacts that must first be filtered out.

To reduce the noise of electricity lines, baseline wander and other potential types of noise in the EOG signal, we applied a second-order low-pass Butterworth filter with the pass band between x Hz and Y Hz. The filter was applied twice: once forward and once backward. This filter has twice the order of the initial filter and zero phase [23]. An example of the processed EOG signal is shown in Fig. 3.

3 Results

The results were obtained using four different classifiers: Boosted trees, RUSBoosted trees, bounded trees, subspace discriminant, and subspace KNN, based on 24 of 60 features, were chosen based on ANOVA results. The hyperparameters of the models are shown in Table 1 and their performance is shown in Table 2 and Figs. 4 and 5.

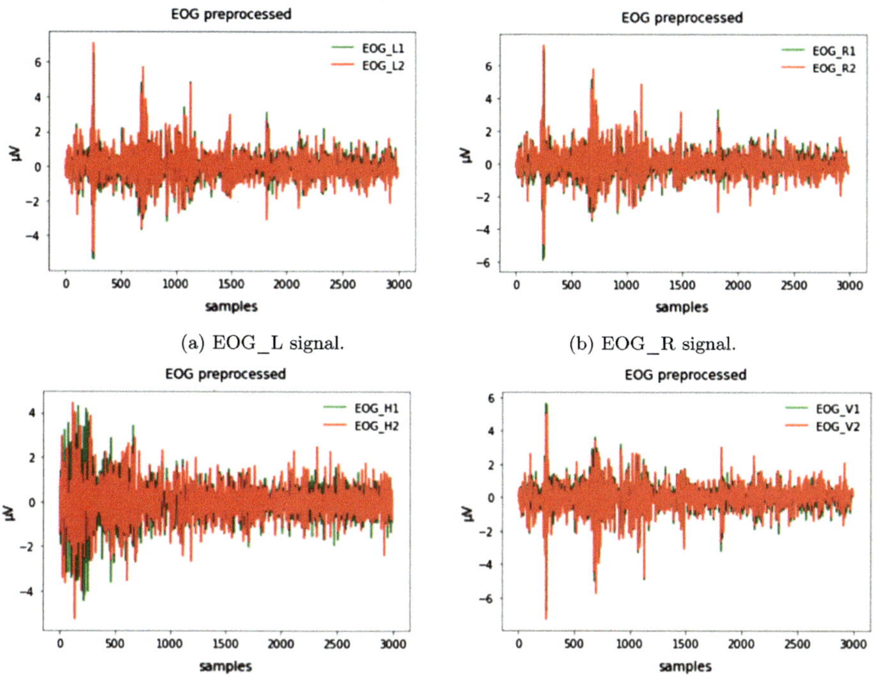

(a) EOG_L signal.

(b) EOG_R signal.

(c) EOG_H signal.

(d) EOG_V signal.

Fig. 3 Preprocessed EOG signal of going straight at a roundabout

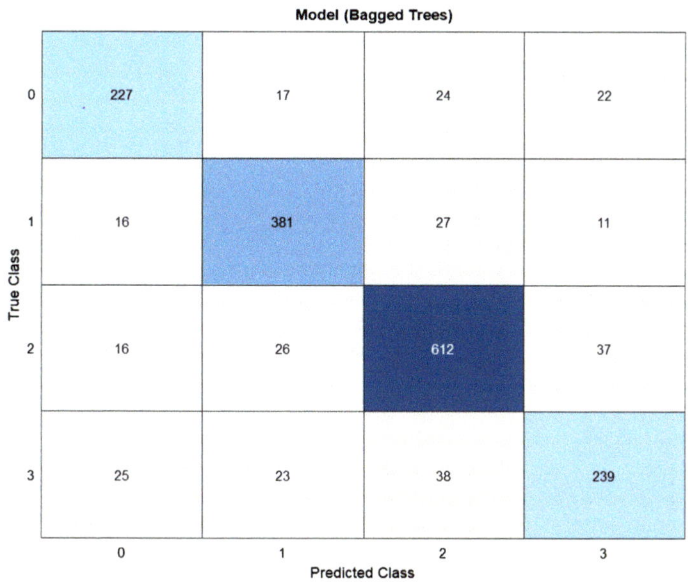

Fig. 4 The confusion matrix

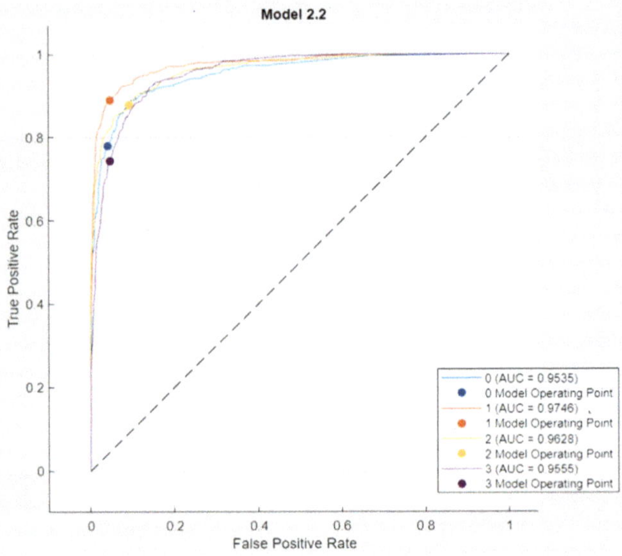

Fig. 5 ROC curves for four classes of activities

Table 1 Performance models hyperparameters for the validation and test data set

Preset	Ensemble method	Leaner method	Max. number of splits	Number of learners	Learning rate
Boosted trees	AdaBoost	Decision tree	20	30	0.1
Bagged trees	Bag	Decision tree	1740	30	–
Subspace discriminant	Subspace	Discriminant	-	30	–
Subspace KNN	Subspace	Nearest neighbors	–	30	–
RUSBoosted trees	RUSBoost	Decision tree	20	30	0.1

Table 2 Performance parameters for the different types of trees classifiers

Class	Accuracy	
	Validation	Test
Boosting	73.3%	73.1
Bagged	**83.8 %**	**87.0%**
Subspace discriminant	61.2 %	56.5%
Subspace KNN	64.4 %	55.4%
RUSBoosted	74.9 %	70.0%

Table 3 Performance parameters for the test dataset

Class	Accuracy	Recall	Specificity	Precision	F1 score
1	0.9174	0.8000	0.9509	0.7500	0.7741
2	0.9482	0.8979	0.9652	0.8979	0.8979
3	0.9223	0.8961	0.9397	0.9078	0.9019
4	0.9430	0.8378	0.9679	0.8611	0.8492
Overall	0.9327	0.8695	0.9559	0.8292	0.8558

To determine the performance of our classifier, we used several standard evaluation metrics derived from the confusion matrix of the classifier (shown in Fig. 3). This matrix shows the number of instances in the test dataset that were classified into a particular group (predicted label) compared to their actual class label (true label) (Fig. 5).

ROC curve (receiver operating characteristic curve) is a graph showing the performance of a classification model at all classification thresholds. This curve plots two parameters:

1. True Positive Rate,
2. False Positive Rate.

True Positive Rate (TPR) is a synonym for recall and is therefore defined as:

$$TPR = \frac{TP}{TP + FN}, \tag{2}$$

where TP is True Positives (the actual value is Positive and the predicted value is also Positive) and FN is False Negatives (the actual is Positive but the prediction is Negative, called Type 2 error) [24, 25].

False Positive Rate (FPR) is calculated as follows:

$$FPR = \frac{FP}{FP + TN}, \tag{3}$$

where FP is False Positives (the actual is negative but the prediction is positive, called the Type 1 error) and TN is True Negatives (the actual value is negative and the prediction is also negative) [25].

- Class 1, FPR = 0.0399724, TPR = 0.7793, AUC = 0.9535
- Class 2, FPR = 0.0459418, TPR = 0.8896, AUC = 0.9746
- Class 3, FPR = 0.0914286, TPR = 0.8784, AUC = 0.9628
- Class 4, FPR = 0.0459040, TPR = 0.7446, AUC = 0.9555.

Furthermore, we calculate the accuracy, recall, specificity, precision, and F1 score (see Tab. 3). Finally, we used the receiver operating characteristic (ROC) to evaluate the classifier precision (see Fig. 5).

Among the evaluated classification decision trees, the best-performing model was Bagged Trees (83.8% and 87.0% of accuracy for validation and test sets, respectively) and the worst performing model was the subsequence discriminant (61.2% and 56.5% of accuracy for validation and test sets, respectively).

4 Discussion and Conclusion

In this study, our objective was to determine whether it is possible to distinguish driving activities by tracking eyeball and head movements. Using simple light-weight preprocessing and machine learning methods, a series of secondary driving activities were recognized with high precision.

In the experiment, we provided 700 samples that were evenly distributed between classes. One participant provided 100 samples in each class, the remaining 25. In total, 2,100 samples were collected in this study. Signal flagging and acquisition, as well as real calculation data with timestamps and header indicating what is represented by the specified value and volunteer.

The model performed outstandingly with respect to binary classification and yielded very good results when grouping activities by type. The main objective of the study, recognizing all secondary behaviors, bending, drinking, eating and turning back, showed relatively good performance, precision, precision and the f1 score at the level of 87%, considering the high complexity of the imbalanced data.

The findings of our study demonstrate that detection of driver inattention is feasible using easy-to-wear commercial eyeglasses with an IMU affixed to the temples and EOG electrodes affixed to the bridge of the nose. All necessary components are readily available, portable and small in design, and the program is created entirely from open source parts, demonstrating that it is possible to avoid using expensive proprietary technologies [8, 26].

Although preliminary results on activity detection are promising from an application perspective, they still require confirmation in larger investigations. This study did not include participants over the age of retirement (in Poland, the age limit is 60 for women and 65 for men) and did not consider age as a variable. However, as a pilot study, its goal was to explore potential new routes for the next steps.

Future research should include examining a wider variety of subjects due to the user-dependent characteristics of the signal and the determination of cognitive and behavioral differences related to driving at the end of professional activity, that is often related to exclusion from social activities [27, 28]. Additionally, the model should also be trained and tested in a real world setting to improve robustness.

References

1. Wang, C.C., Carr, D.B.: Older driver safety: a report from the older drivers project. J. Am. Geriatr. Soc. **52**(1), 143–149 (2004)
2. Lyon, C., Mayhew, D., Granié, M.A., Robertson, R., Vanlaar, W., Woods-Fry, H., Thevenet, C., Furian, G., Soteropoulos, A.: Age and road safety performance: focusing on elderly and young drivers. IATSS Res. **44**(3), 212–219 (2020). October
3. Regan, M.A., Hallett, C., Gordon, C.P.: Driver distraction and driver inattention: definition, relationship and taxonomy. Accid. Anal. Prev. **43**(5), 1771–1781 (2011). Sep
4. Nel, F., Ngxande, M.: Driver activity recognition through deep learning. In: 2021 Southern African Universities Power Engineering Conference/Robotics and Mechatronics/Pattern Recognition Association of South Africa (SAUPEC/RobMech/PRASA), pp. 1–6 (2021)
5. Song, X., Cao, H.: Driver secondary driving task behavior recognition. In: Song, X., Cao, H. (eds.) Behavior Analysis and Modeling of Traffic Participants (Synthesis Lectures on Advances in Automotive Technology), pp. 71–89. Springer International Publishing (2022)
6. National Highway Traffic Safety Administration: Traffic safety facts 2018. https://crashstats.nhtsa.dot.gov/Api/Public/ViewPublication/812981. Accessed 23 March 2023
7. National Highway Traffic Safety Administration: Distracted Driving (2020) https://crashstats.nhtsa.dot.gov/Api/Public/ViewPublication/813309. Accessed 23 March 2023
8. Doniec, R., Konior, J., Sieciński, S., Piet, A., Irshad, M.T., Piaseczna, N., Hasan, M.A., Li, F., Nisar, M.A., Grzegorzek, M.: Sensor-based classification of primary and secondary car driver activities using convolutional neural networks. Sensors **23**(12), 5551. https://doi.org/10.3390/s23125551
9. Sagberg, F., Johansson, O.J., Sundfør, H.B.: Combining roadside interviews and on-road observation for assessing prevalence of driver inattention. Transp. Res. Part F: Traffic Psychol. Behav. **62**, 305–315 (2019). Apr
10. Ann, O.C., Theng, L.B.: Human activity recognition: a review. In: 2014 IEEE International Conference on Control System, Computing and Engineering (ICCSCE 2014), pp. 389–393 (2014)
11. Kulsoom, F., Narejo, S., Mehmood, Z., Chaudhry, H.N., Butt, A., Bashir, A.K.: A review of machine learning-based human activity recognition for diverse applications. Neural Comput. Appl. **34**(21), 18289–18324 (2022). Nov
12. Um, T.T., Pfister, F.M.J., Pichler, D., Endo, S., Lang, M., Hirche, S., Fietzek, U., Kulić, D.: Data augmentation of wearable sensor data for Parkinson's disease monitoring using convolutional neural networks. In: Proceedings of the 19th ACM International Conference on Multimodal Interaction. ICMI '17, Association for Computing Machinery, pp. 216–220 (2017)
13. Zhang, S., Li, Y., Zhang, S., Shahabi, F., Xia, S., Deng, Y., Alshurafa, N.: Deep learning in human activity recognition with wearable sensors: a review on advances. Sensors **22**(4), 1476 (2022). Jan
14. Hamad, R.A., Yang, L., Woo, W.L., Wei, B.: Joint learning of temporal models to handle imbalanced data for human activity recognition. Appl. Sci. **10**(15), 5293 (2020). Jan
15. Ferrari, A., Micucci, D., Mobilio, M., Napoletano, P.: Deep learning and model personalization in sensor-based human activity recognition. J. Reliab. Intell. Environ. **9**(1), 27–39 (2023). Mar
16. Cao, L., Wang, Y., Zhang, B., Jin, Q., Vasilakos, A.V.: GCHAR: an efficient group-based context-aware human activity recognition on smartphone. J. Parallel Distrib. Comput. **118**, 67–80 (2018). Aug
17. Khowaja, S.A., Yahya, B.N., Lee, S.L.: CAPHAR: context-aware personalized human activity recognition using associative learning in smart environments. Hum.-Centric Comput. Inf. Sci. **10**(1), 35 (2020). Aug
18. Omolaja, A., Otebolaku, A., Alfoudi, A.: Context-aware complex human activity recognition using hybrid deep learning models. Appl. Sci. **12**(18), 9305 (2022). Jan
19. Gao, C., Chen, Y., Jiang, X., Hu, L., Zhao, Z., Zhang, Y.: Bi-STAN: bilinear spatial-temporal attention network for wearable human activity recognition. Int. J. Mach. Learn. Cybern. (2023)

20. Ramasamy Ramamurthy, S., Roy, N.: Recent trends in machine learning for human activity recognition-a survey. WIREs Data Min. Knowl. Discov. **8**(4), e1254 (2018). Mar
21. Li, F., Shirahama, K., Nisar, M.A., Köping, L., Grzegorzek, M.: Comparison of feature learning methods for human activity recognition using wearable sensors. Sensors **18**(2), 679 (2018). Feb
22. Hao, W., Chen, Y., Xin, Y.: ECG baseline wander correction by mean-median filter and discrete wavelet transform. In: 2011 Annual International Conference of the IEEE Engineering in Medicine and Biology Society, IEEE (2011)
23. Gu, J.J., Meng, M.Q.H., Cook, A., Faulkner, M.G.: A study of natural eye movement detection and ocular implant movement control using processed EOG signals. In: Proceedings 2001 ICRA. IEEE International Conference on Robotics and Automation (Cat. No.01CH37164), vol. 2, pp. 1555–1560 (2001)
24. Visa, S., Ramsay, B., Ralescu, A., van der Knaap, E.: Confusion matrix-based feature selection. In: Visa, S., Inoue, A., Ralescu, A. (eds.) Proceedings of the Twenty-second Midwest Artificial Intelligence and Cognitive Science Conference, Cincinnati, Ohio, USA, Omni press, pp. 120–127 (2011)
25. Suresh, A.: What is a confusion matrix? (2011) https://medium.com/analytics-vidhya/what-is-a-confusion-matrix-d1c0f8feda5. Accessed 23 Feb 2023
26. Doniec, R., Piaseczna, N., Li, F., Duraj, K., Pour, H.H., Grzegorzek, M., Mocny-Pachońska, K., Tkacz, E.: Classification of roads and types of public roads using EOG smart glasses and an algorithm based on machine learning while driving a car. Electronics **11**(18), 2960 (2022). https://doi.org/10.3390/electronics11182960
27. Fakoya, O.A., McCorry, N.K., Donnelly, M.: Loneliness and social isolation interventions for older adults: a scoping review of reviews. BMC Public Health **20**(1) (2020)
28. Hansen, T., Kafková, M.P., Katz, R., Lowenstein, A., Naim, S., Pavlidis, G., Villar, F., Walsh, K., Aartsen, M.: Exclusion from social relations in later life: micro- and macro-level patterns and correlations in a european perspective. Int. J. Environ. Res. Public Health **18**(23), 12418 (2021). https://doi.org/10.3390/ijerph182312418

Monte-Carlo Modeling of Optical Sensors for Postoperative Free Flap Monitoring

Paulina Stadnik, Ignacy Rogoń, and Mariusz Kaczmarek

Abstract This work aims to develop a numerical tissue model and implement software to simulate photon propagation using the Monte Carlo method to determine design guidelines for a physical measurement system. C++ was used for the simulation program, and Python as a programming environment to create an interface that allows the user to customize individual simulation elements, allowing for increased accuracy and flexibility when simulating photon movement. This allows the user to customize the simulation to their specific requirements, ensuring the results are as accurate and reliable as possible. It also models the detector to determine if a given photon is in the desired location. The program simulates the propagation of light from a normal illumination medium with anisotropic scattering and records the escape of photons on the upper surface. The simulation also takes into account absorption and scattering coefficients for a given wavelength, and data regarding these parameters are read from a .csv file. The variance reduction technique is used to improve the efficiency of the simulation. The user interface allows users to define their own parameters, such as wavelength, anisotropy coefficient, refractive index, and layer thickness. In this paper, we simulate four photodiodes and different distances between the source and detector to determine the most suitable model for designing a physical sensor.

Keywords Monte-Carlo simulation · Light sensor · Photodiode · Optics · Photon · Light propagation · Skin tissue

Supported by Statutory Funds of Electronics, Telecommunications and Informatics Faculty, Gdansk University of Technology.

P. Stadnik · I. Rogoń · M. Kaczmarek (✉)
Biomedical Engineering Department, Gdańsk University of Technology, Narutowicza 11/12 Str., 80-233 Gdańsk, Poland
e-mail: markaczm@pg.edu.pl

I. Rogoń
e-mail: ignacy.rogon@pg.edu.pl

P. Strumiłło et al. (eds.), *The Latest Developments and Challenges in Biomedical Engineering*, Lecture Notes in Networks and Systems 746,
https://doi.org/10.1007/978-3-031-38430-1_19

1 Introduction

The International Agency for Research on Cancer reported that in 2020 breast cancer became the world's most commonly diagnosed cancer type. Over 2.26 million new cases of breast cancer and almost 685 000 deaths were reported. Breast cancer was the most common cause of cancer death in women and the fifth most common cause of cancer death overall [1]. Moreover, every year, 1.7 million people in Europe are diagnosed with cancer, almost half of whom die. Analyzing the above statistics, it can be seen that cancer is a huge problem not only in terms of health but also socially. It is estimated that in 2000 in European Union, around 25,000 women were diagnosed with breast cancer, which is about 25% of all cancers diagnosed in women [2]. In 2009 analysis [3] prepared by researchers at Maria Sklodowska-Curie Polish National Oncology Centre, incidence and mortality predictions for cancer in Poland up to the year 2025 were presented. The prognosis for the years 2010–2025 indicated an increasing trend of breast cancer incidence for all age groups. The forecast stated that in 2025, the number of new breast cancer cases is expected to be almost 50% higher than the one observed in 2006. The highest increase of new cancer cases is expected for women aged 50 years and more, with over half of the cases occurring between the 50th and 69th year of life. The researchers stated that the number of deaths due to breast cancer would increase, while the proportion of deaths for each age group would change. In 2006 about half of the deaths fell on women before the age of 70, 2025 one should expect a decrease in the number of deaths in this age group to about 30%, while almost 70% of deaths due to breast cancer would occur for the oldest women (after the age of 70). A mastectomy, supported by other therapies, is one of the basic methods of treating breast cancer. The consequence of mastectomy is a further treatment involving the reconstruction of the removed breast. This is done by various methods: implantation of silicone implants or by transplantation of own tissues. Tissue transplantation is used in surgical defect reconstructions when suturing existing tissues impossible. It is also used in the area of breast cancer diseases and is used to reconstruct the shape of the breast after the tumor is removed. One of the side effects of reconstructive surgery may be tissue ischemia. This means there is not enough blood flow, which can lead to tissue damage or even death. This phenomenon can have serious consequences for patients after breast cancer surgery, including an increased risk of postoperative complications and longer wound healing time. To prevent this, blood-supplying vessels should be monitored for a period of up to 72 h after the procedure. One of the most promising techniques in analyzing tissue blood flow is light-based methods [4]. The behavior of light to in tissues is an important topic in many fields, including medicine, where it can be used to diagnose diseases and monitor treatment processes. Light can behave in two main ways in tissues: scattering and absorption. The scattering coefficient measures the ability of particles to scatter photons, while the absorption coefficient measures the amount of photons absorbed at a given distance. These parameters are important in determining how light travels through tissues. Both mentioned processes are random, making the Monte Carlo method ideal for simulating them. In this paper, a Monte

Carlo method in simulations of photons behavior in modeled tissue is presented. A few existing sensors (photodiodes) were simulated to determine the most suitable for future system design.

2 Theoretical Basis

Most tissues in the human body are carriers with a high degree of scattering and variable absorbance both in the visible and non-visible range. In most cases, photons are scattered multiple times before being absorbed. Simulating the paths of photons is too complex to be modeled analytically. Often, instead of a manual and tedious process, Monte Carlo simulations are used to streamline this work [5, 6].

2.1 Skin Tissue Model

An anatomical tissue model is a virtual or physical representation of the tissue structure of the human or animal body, which is particularly useful in many fields, such as medicine, biology, and materials science. The model is made up of different layers that contain various elements, such as epithelial cells, connective tissue cells, collagen, and elastin, as well as information about the physical, chemical, and biological properties of the tissues. The skin consists of a layered epithelium containing differentiated keratinocytes (cells responsible for the production of keratin in the skin) that are cultured on a shrunken collagen matrix filled with dermal fibroblasts (connective tissue cells found in the dermis responsible for the production of collagen and elastin, the most important supporting proteins of the skin, determine its strength, density, flexibility, and elasticity) [7]. We assume that our model is nonuniform and consists of three layers—see Fig. 1a:

- skin—understood as two layers together: the epidermis and dermis, width = 3.5 mm,
- fat—laying unfed the skin layer, width 10 mm,
- muscle—deepest layer, width = 11.2 mm.

It should be emphasized that we model the structure of the skin and subcutaneous tissues for the thoracic region and the breast itself.

Tissue modeling One of the most important properties of tissues is their ability to absorb, scatter, or reflect light, which is important in many diagnostic and therapeutic techniques, such as laser treatment of eye diseases or photodynamic therapy of cancer, and in imaging by various technologies, such as computed tomography or magnetic resonance imaging. The tissue model can also be used in skin research to model and explain physical phenomena such as light absorption, reflection, and light scattering and to study the influence of external factors such as ultraviolet radiation, temperature, and humidity on skin properties. The Monte Carlo (MC) method is

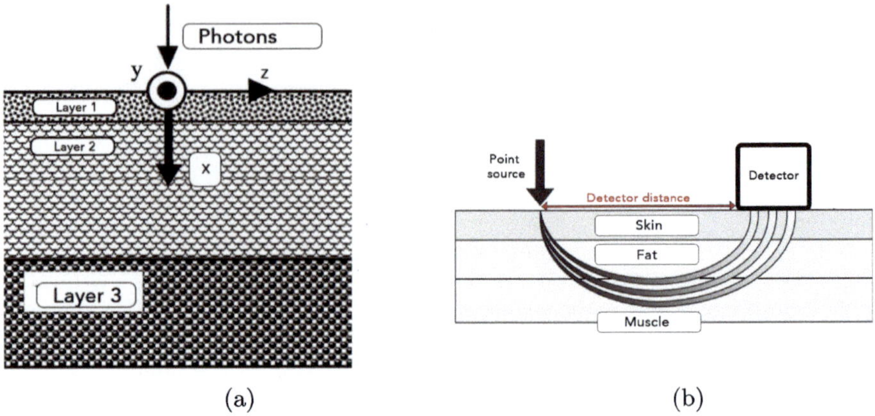

Fig. 1 **a** Numerical model of the skin tissue used in simulations. The y-axis points outward [9]; **b** Principle idea behind tracking the behavior of photons in the tissue. Please note that detector distance is measured to its closest edge

based on the random walk technique and is used to simulate the behavior of light in tissue. It is a valuable tool for medicine in diseases such as cancer. It is also useful for studying the effect of anisotropy on the propagation of light in the tissue and for determining the spatial distribution of optical parameters in the tissue.

There are methods for accurately mapping human organs. Modeled human tissues can be used in various stages of diseases if they faithfully reflect the shape, size, or, above all, the properties of the desired organ. Biomechanical modeling opens many areas of medical activity to new possibilities. Implementation of tissue modeling methods, however, often requires linear connections, and models do not always lead to satisfactory results. Non-linear models were introduced to overcome the limitations of linear solutions. The development of this method is facilitated by special frameworks dedicated to the creation of algorithms and simulators. Such tools allow researchers to focus on the essence of the problem [8].

Mathematical models of tissues make it possible to identify biological processes occurring in the human body and form the basis for simulations. Many methods of tissue modeling were found in the articles cited in the paper. In the available literature, methods such as the elastic-mass method and the finite element method were found. The former allows for complex real-time simulations. It describes the mechanical properties of tissue through a grid of points connected by weightless, flexible springs and optional shock absorbers. Each point has specific parameters, such as weight and position, and its trajectory is set according to Newton's law of motion. This method is used in surgical simulation. It is easy to implement and does not require a lot of computing power.

2.2 Light Behavior in Tissue

Scattering of light occurs to some degree in all media except vacuum and depends on the heterogeneity of the medium down to the atomic level. Charges in the matter are set in motion by passing electromagnetic waves. Moving charges emit secondary radiation in all directions, producing diffuse waves. The final radiation distribution is the sum of the incident and scattered waves.

The scattering coefficient is a measure of the particle's ability to scatter photons from a beam of light, while the absorption coefficient is a measure of how many photons have been absorbed. Each parameter is expressed as a number proportional to the number of photons scattered or absorbed over a distance.

If a beam of optical radiation propagates through a medium in which there is more than one scattering center, then optical radiation can be scattered not only singly but also many times. The scattering properties of such a medium depend not only on the properties of individual mediums but also on the spatial density of their occurrence and spatial order. Knowing the cross-sections of the scattering centers σ_s and their spatial density ρ_s, the scattering coefficient for this medium can be determined (1):

$$\mu_s = \sigma_s \cdot \rho_s \tag{1}$$

An alternative to light scattering is its conversion into heat, i.e., absorption. From the law of conservation of energy, when absorbing a photon, its energy depends on the frequency of the light (2):

$$\Delta E = h\nu = \frac{hc}{\lambda} = \frac{2\pi hc}{k}, \tag{2}$$

where h—Planck constant, c—speed of light in a vacuum, λ—length of the absorbed wavelength, ν—vibration frequency, k—wave number.

To simplify notation, the total interaction coefficient μ_t is sometimes used, which is the sum of the absorption coefficient μ_a and the dispersion coefficient μ_s. Consequently, the interaction coefficient denotes the probability of photon interaction per unit of infinitesimal path length.

By analogy to formula (1), the absorption coefficient for this medium can be determined (3):

$$\mu_a = \sigma_a \cdot \rho_a, \tag{3}$$

where σ_a are the cross-sections of the scattering centers, and their spatial density is ρ_a.

The extinction coefficient can be calculated in the same way (4):

$$\mu_e = \mu_s + \mu_a, \tag{4}$$

where μ_s and μ_a are the spatial densities of occurrence of extinction centers, respectively, the extinction coefficient is equal to the sum of the scattering and absorption coefficients.

In the matter of the behavior of light in tissues, the anisotropy coefficient is an important factor. This is the property of a material that allows it to change or take on different properties in different directions as opposed to isotropy. It can be defined as the difference, measured along different axes, in the physical or mechanical properties of a material (e.g., absorption).

The scattering anisotropy coefficient is defined as the mean value of the cosine of the scattering angle θ. Denoting this coefficient by g, we can determine it from the relationship (5):

$$g = \langle \cos\theta \rangle = \int p(\theta) \cos\theta \, d\omega\Omega, \tag{5}$$

where: $\langle ... \rangle$ stands for the average value, $p(s_2, s_1)$ is phase scattering function from angle s_1 to angle s_2 (6):

$$p(s_2, s_1) = \frac{\sigma_d(s_2, s_1)}{\sigma_d(s_1)} \tag{6}$$

The anisotropy factor can vary between -1 and 1 (where -1 means that all photons are scattered in the center of scattering in the direction opposite to the incident photon, and one means they are scattered in the direction of the incident photon). However, a value of $g=1$ means no scattered photons. Despite having the same direction, a scattered photon may differ from the incident photon in other parameters, such as polarization, phase, or frequency for inelastic scattering.

All tissues have specific layered structures. Each layer also has its specific texture, which results in different compositions, different concentrations of light absorbers, and thus different optical parameters characterizing them. It is difficult to accurately distinguish the baseline absorption coefficient values associated with meaningless epidermis and bloodless dermis. So the baseline absorption of both epidermis and dermis is often approximated to a singular value for certain wavelength [10].

The interaction of light with multi-layered and multi-component skin, such as breast skin, is a very complex process. The light beam is transformed into scattered light by microscopic non-uniformities at the boundary between air and the stratum corneum layer of the epidermis. Most of the reflected light is caused by backscattering in various skin layers (stratum corneum, epidermis, dermis, and microvasculature). The analysis of light propagation in such structures is more complicated than inhomogeneous objects.

Light source Delivery systems for light can function as an isotropic point source that emits spherically symmetrical light fields, a line source that emits cylindrically symmetrical light fields, or a plane source that emits light fields with a one-dimensional variation.

The source of photons in the simulation is a point source generating photons in all directions, located inside the scattering medium at point P with coordinates x, y, and z. Suppose we denote by s the unit vector describing the direction of the photon, in the case of a point source, for a given coordinate system Oxyz. In that case, it is sufficient to find two angles: θ and Ψ, which respectively are the polar angle of propagation direction, i.e., the angle between the z-axis and s, and the angle between the x-axis and the projection of the vector s onto the Oxy plane.

A point light source is a theoretical light source that emits light rays from a single point. In practice, such a spot can be represented by a very small light source, an LED, or a thinner optical fiber tip that can emit light in one direction.

The characteristic of a point light source is the emission of light evenly in all directions, which means that the light rays propagate in all directions with the same intensity. However, real point light sources emit light slightly unevenly in different directions in practice due to technical limitations. Due to detector placement regarding a point source, a banana-like characteristic (Fig. 1b) is to be expected. We should note, that multiple photons paths may occur, and can differ from the illustrated case. The surface tissue has been distinguished as the photons can bounce off it. Please note that detector distance is calculated from the point light source to the closest edge of a detector.

Photon movement simulation In the case of the Monte Carlo method, these are calculations performed on millions of photons, making speed necessary. That's why it was decided to use C++ language for this issue. The Monte Carlo method involves random sampling of variables from well-defined probability distributions. The accuracy of the result obtained by this method depends on the number of checks but, above all, on the quality of the pseudo-random number generator used.

Absorption, scattering, or reflection of a photon are physical phenomena that we want to simulate in the code. This program simulates light propagation from a normal irradiation medium with anisotropic scattering. It also records the escape of photons on the upper surface.

It was decided to use centimeters as the base unit of length throughout the simulation for consistency. The thickness of each layer and other distances from the user interface are given in mm. Absorption and scattering coefficients are given in cm^{-1}.

The program consists of individual blocks, as shown in Fig. 2. All of them are necessary to perform a correct simulation to analyze photon movement in tissues.

A simple variance reduction technique, implicit photon capture, is used to improve the performance of Monte Carlo simulations. This technique allows multiple photons to propagate equivalently as a packet along a specific path simultaneously. Each photon packet is initially assigned a weight. The photon is injected perpendicularly into the tissue at the starting point, corresponding to a collimated, arbitrarily narrow beam of photons.

The current position of the photon is given by Cartesian coordinates (x, y, z) as shown in Fig. 1a. The photon's current direction is given by the unit vector r, which can be equivalently described in terms of direction cosines(μ_{x_1}, μ_{y_1}, μ_{z_1}).

The photon position is initialized to (0, 0, 0), and the direction cosines are set to (0,

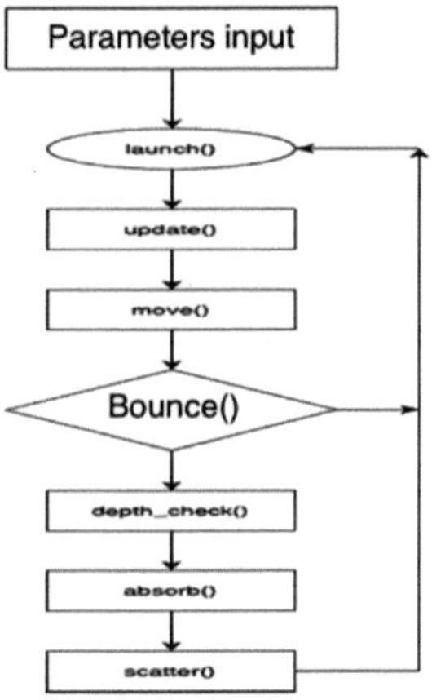

Fig. 2 Monte-Carlo simulation algorithm

0, 1). When a photon is fired, if there is a mismatched boundary on the surface of the tissue, there could be some specular reflection.

The move() function changes the photon's position in the tissue. The position of the photon packet is changed by a quantitative substitution, which can be represented as (7):

$$x \leftarrow x + d \cdot \mu_x;\ y \leftarrow y + d \cdot \mu_y;\ z \leftarrow z + d \cdot \mu_z, \tag{7}$$

where d is the components of the guiding vector.

The guiding vector equals –log((rand-f()+1,0)/($\mu_s + \mu_a$) is a function that returns non-negative double-precision floating-point values evenly distributed in the range from 0 to 1. To generate a random number, the $rnm()$ method is powered by a seed created based on the current system time so that the sequence of pseudo-random numbers differs each time the function is called. Then, a uniform_real_distribution class object is created with the name unif, which is initialized with the interval [0, 1]. This object is used to select a random number from a given range using the unif(rng) method.

Arrows indicate quantitative substitutions. Variables on the left have new values, and variables on the right have old values. The equals sign is used for this purpose in a real C program.

Once the photon takes a step, some mass attenuation of the photon due to absorption by the interaction site must be calculated. The local grid element will embed a fraction of the photon's current mass and weight.

After the photon packets have made their move and reduced their weight, the photon packet is ready for scattering. The probability distribution of the cosine of the deflection angle, $\cos\theta$, is described by the scattering function (8):

$$p(\cos\theta) = \frac{1 - g^2}{\left(1 + g^2 - 2g\cos\theta\right)^{3/2}}, \tag{8}$$

where the anisotropy, g, is defined as the average of the cosine of the scattering angle θ and is between -1 and 1.

The cosine of θ value is calculated using a function that returns pseudo-random numbers. With the anisotropy coefficient taking the value of 0, it is equal to the difference of double the pseudo-random number and one. For the rest of the value of the anisotropy coefficient, it is calculated based on the formula. To determine the new scattering direction, a new value of μ_2 is calculated using the coordinate values of the vector from the previous iteration. New direction values are computed in the scatter() function, and the following formulas are used (9), (10) and (11):

$$\mu_{x2} = \frac{\sin\theta}{\sqrt{1-\mu_{z1}^2}}\left(\mu_{x1}\mu_{x1}\cos\psi - \mu_{y1}\sin\psi\right) + \mu_{x1}\cos\theta \tag{9}$$

$$\mu_{y2} = \frac{\sin\theta}{\sqrt{1-\mu_{z1}^2}}\left(\mu_{y1}\mu_{z1}\cos\psi + \mu_{x1}\sin\psi\right) + \mu_{y1}\cos\theta \tag{10}$$

$$\mu_{z2} = -\sin\theta\cos\psi\sqrt{1-\mu_{z1}^2} + \mu_{z1}\cos\theta \tag{11}$$

where: $\mu_{x_1}, \mu_{y_1}, \mu_{z_1}$ are the previous coordinates of the versor, and $\mu_{x_2}, \mu_{y_2}, \mu_{z_2}$ calculated for later use in the next iteration.

If μ_{z_1} is either 1 or -1 in the previous iteration, the direction values can be described as:

$$\mu_{x_2} = \sin\theta\cos\phi, \tag{12}$$

$$\mu_{y_2} = A\sin\theta\sin\phi, \tag{13}$$

$$\mu_{z_2} = A\cos\theta, \tag{14}$$

where: A = 1 for $\mu_{z_1} = 1$, and A = -1 for $\mu_{z_1} = -1$.

In the simulation of photon motion, the coefficients necessary to carry out the calculations, apart from absorption or scattering, are the anisotropy mentioned above or the number of photons to be sent into the tissue. The settings function reads these and other data, such as the depth of skin layers, from the data.csv file and writes them to the file array. The downloaded data is used to calculate the rest of the parameters used in the simulation. For a tissue model to serve well as a research tool, it must have properties close to those of real human tissue. The proposed tissue parameters used in our simulation are summarized in the Table 1.

In each step of the loop, a single line is read in the main function and processed by the split function, which breaks it into individual elements separated by commas and writes them to the file array. After the data is read from the file, they are assigned to the appropriate variables. Then, the values of the absorption and scattering coefficients are read, and on their basis, the albedo is calculated depending on the data loaded in a given iteration.

Table 1 Absorption and scattering coefficient depending on wavelength

Wavelength	940 nm	850 nm	760 nm	565 nm
μ_a [cm^{-1}] of skin	0.1634	0.227	0.326	0.854
μ_a [cm^{-1}] of fat	1.351	0.324	0.466	1.296
μ_a [cm^{-1}] of muscle	0.432	0.3	0.384	1.991
μ_s[cm^{-1}] of skin	6.938	8.071	9.546	14.892
μ_s [cm^{-1}] of fat	0.0126	0.0001	0.0001	0.0001
μ_s [cm^{-1}] of muscle	1.351	0.3244	0.466	1.295

3 Implementation and Simulation

3.1 Sensor—Photodiodes and Their Parameters

Diode is an electronic component that allows current to flow in one direction. Photodiodes are semiconductor optoelectronic devices that change their electrical properties in response to light radiation. Photodiodes are used for detection, sensors, and control systems. The light emitted on the photodiode stimulates the current flow in a specific direction. This flow is measured and used for light-level determination or object detection. Infrared LEDs are widely used in industry, medicine, science, and consumer equipment such as remote controls or IR camera illuminators. Light-emitting diodes work mainly in the near-infrared range from 700 to 1500 nm nm. The index of relative spectral sensitivity to wavelength indicates how sensitive the diode is to the light of different wavelengths. The RSS value is calculated as the ratio of either detector's or sensor's maximum output to the brightness of the light source at a specific wavelength. RSS is essential for visual detectors and sensors as it helps determine how well a particular detector or sensor ig going to respond to visible light.

To illustrate the operation of various detectors, it was decided to perform simulations on numerous IR photodiodes. All four simulations are carried out for 1,000,000 photons, using the following parameters:

- The distance of the source from the detector—0.6 cm,
- Anisotropy coefficient—0.8, Refractive index—1.5,
- The thickness of the first layer of skin—3.5 mm,
- The thickness of the fat layer—10 mm,
- The thickness of the muscle layer—11.2 mm.

Absorption and scattering coefficients depending on the wavelength are given in Table 1.

BPW21R photodiode It is a round diode with a diameter of 5.9 mm [11]. For the simulation, we assume a diode with properties and dimensions (as a square, i.e.,

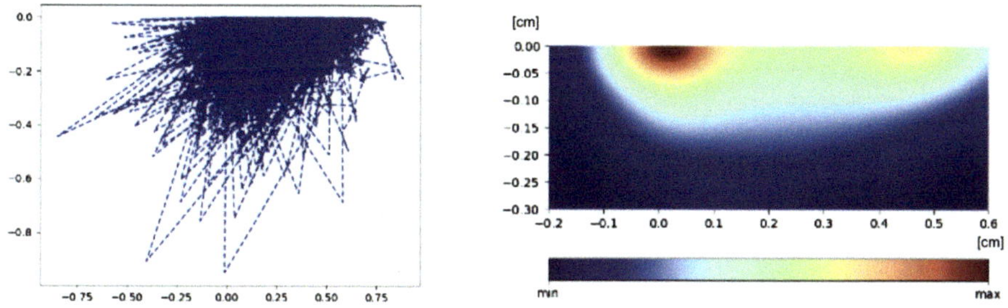

(a) (b)

Fig. 3 BPW21R photodiode; **a**—MC simulation; **b**—MC simulation heatmap

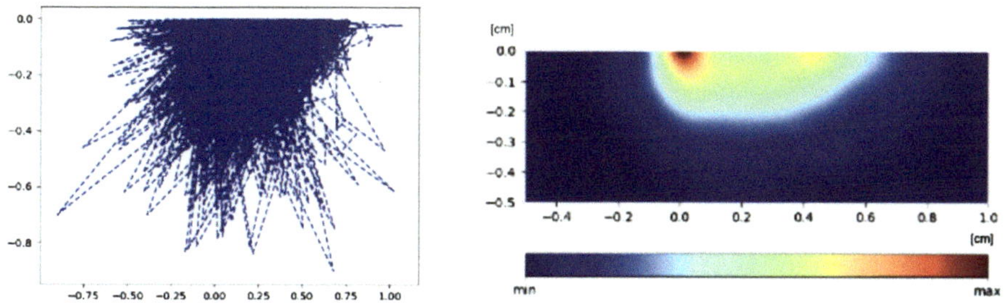

(a) (b)

Fig. 4 VBPW34S photodiode; **a**—MC simulation; **b**—MC simulation heatmap

2.8 × 2.8 mm, surface of 7.84 mm^2)) as the BPW21R diode. The wavelength that guarantees the highest sensitivity was used for the simulation. The simulation was performed for 565 nm. Traces of detected photons were plotted in Fig. 3a. Photons reached an average depth of 1.2 cm into the tissue. The intensity distribution of photon occurrences was presented as a heatmap in Fig. 3b.

VBPW34S photodiode The diode is rectangular and has a surface area of 7 mm^2 [12]. To simulate, the wavelength that provides the optimal sensitivity was chosen—940 nm. For the simulation, we assume a square with the same area but different dimensions of 2.65 × 2.65 mm. The photons detected by the detector reach up to 0.9 cm deep into the tissue, are shown in Fig. 4a, and clusters of light points are presented in Fig. 4b.

SFH 203 photodiode It is a through-hole diode with two contacts, dimensions 4.8 × 5.9 mm (surface 28.32 mm^2) [13]. A wavelength of 850 nm was chosen to achieve the diode's highest sensitivity. Photons detected that traveled more than a centimeter were shown in Fig. 5a. Photon concentration points are shown in Fig. 5b.

S10784 photodiode It is a through-hole diode with two contacts, dimensions 4.1 × 4.7 mm (surface 19.27 mm^2) [14]. To achieve the highest sensitivity of the diode, a wavelength of 760 nm was chosen. Photons detected by the detector reach

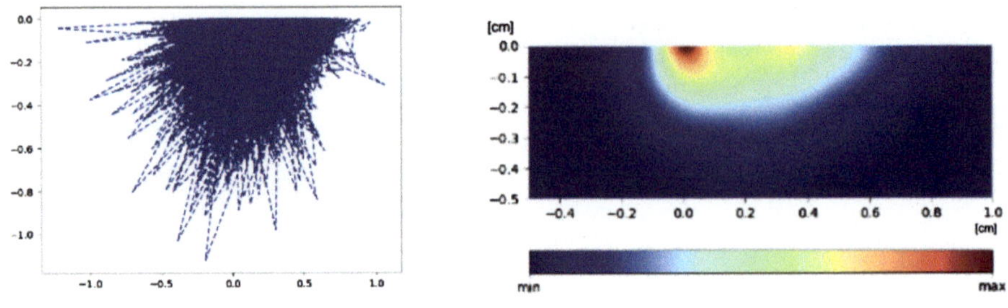

(a) (b)

Fig. 5 SFH203 photodiode; **a**—MC simulation; **b**—MC simulation heatmap

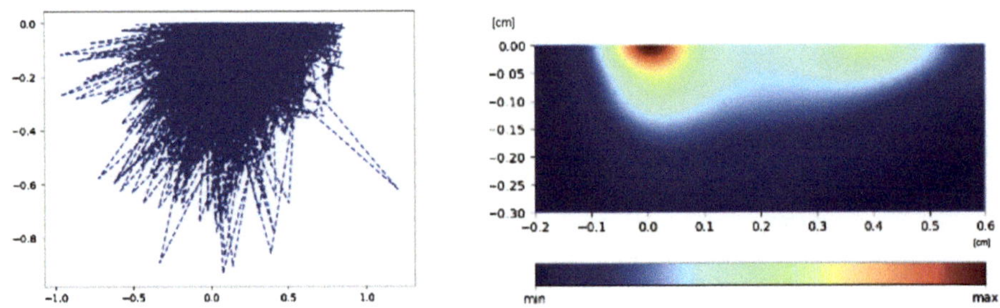

(a) (b)

Fig. 6 S10784 photodiode: **a**—MC simulation; **b**—MC simulation heatmap

about a centimeter deep into the tissue, shown in Fig. 6a. Clusters of light points can be seen in Fig. 6b.

For each photodiode, ten cycles of simulation were performed. A mean value and number of photons detected per 1 mm^2 (mean value of detected photons divided by photodiode active surface area) were later calculated. Results were presented in Table 2.

4 Results and Discussion

All four simulations performed on different types of IR photodiodes showed that the wavelength of the light affects the number of photons detected from the detector and the depth to which the radiation reaches. The lowest mean value of photons detected was observed for the BPW21R photodiode—an average of 298.8 (0.02988%) photons was detected. The SFH 203 photodiode was detected to be the highest value of photons averaging at around 2230.9 (0,22309%). As the simulated wavelength decreased, so did the value of the mean photon detected per 1 mm^2. The lower absorption coefficient of skin and fat layers caused it. As this SFH203 has proved

Table 2 Number of photons detected for different photodiodes during ten simulation cycles

Simulation no.	VBPW34S (940 nm)	SFH 203 (850 nm)	S10784 (760 nm)	BPW21R (565 nm)
1	762	2255	1077	295
2	784	2187	1053	307
3	794	2236	1083	298
4	786	2221	1066	302
5	776	2250	998	294
6	799	2195	1094	297
7	813	2228	1017	303
8	759	2294	1060	311
9	783	2213	1027	308
10	807	2230	1059	273
Mean	766.2	2230.9	1053.4	298.8
Mean photons value detected per 1 mm^2	109.46	71.71	54.67	38.112

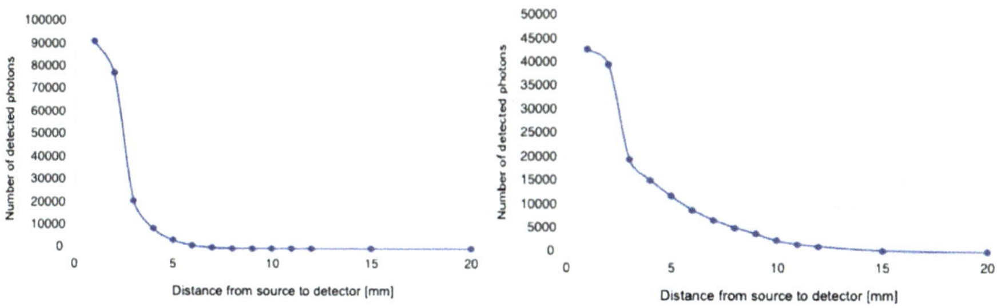

(a) Anisotropy = -0.8 (b) Anisotropy = 0.8

Fig. 7 Number of detected photons depending on the transmitter—detector distance for different anisotropy coefficient

to be the most effective in detecting photons under predetermined conditions, it was decided to perform simulations for different variants of the detector. The distance of the source from the detector was changed. The dimensions of the photodiode are 4.8×5.9 mm, and the wavelength of light used is 850 nm. In Fig. 7a, the simulation results for different distances of the photon source from the detector were presented.

Also, simulations for an anisotropy coefficient equal to 0.8 (instead of -0.8) were performed. Results are shown in Fig. 7b. Even though for small anisotropy values number of detected photons was greater as the source-detector distance grew, the number of detected photons drastically decreased. Higher values of anisotropy allowed for simulations of many further regions of tissue. A low anisotropy coefficient is associated with a high degree of non-uniform distribution of photons in space.

Due to this parameter and its low value, photons spread evenly to both sides of the source, not just one, as it would be in the case of a higher value of this parameter. During further experiments, the influence of the anisotropy factor on the detection ratio should be taken into account.

5 Conclusions

The aim of the work was to develop a tissue model and implement software for simulating the motion of photons using the Monte Carlo method. The simulation model takes into account factors such as the absorption, scattering, and reflection of photons by skin tissue, as well as the diffusion of photons through the skin. This allows researchers to study the effects of different wavelengths of light on blood flow and oxygenation in the skin.

We conclude that the Monte-Carlo simulation is a powerful tool for understanding the complex physics of photon interaction with skin tissue and can help researchers develop new strategies for improving blood circulation and tissue oxygenation. Depending on the parameters selected by the user, the program reads the appropriate absorption and dispersion coefficients and simulates photon behavior in user-defined tissue. As a final result, information about the distribution of light intensity in the tissue and the number of photons absorbed in individual tissue locations is provided. This information can then be used to create visualize behavior of light in tissue or to study various optical properties of the tissues. A series of simulations for four models of photodiodes were performed. The following design guidelines were proposed:

- Since the ratio of detected to emitted photons is relatively small, a very sensitive measurement system should be designed. In sensor design, photodiodes with larger surface areas should be more effective.
- Photodiodes with their peak Relative Spectral Sensitivity within higher wavelengths exhibit higher photon detection ratio per 1 mm^2 ratio. Thus light sources and photodiodes working in NIR spectra should be considered
- Since the distance between the source and the detector highly affects the number of photons detected, it should be easily adjustable.

References

1. IARC Homepage. https://www.iarc.who.int/cancer-type/breast-cancer/. Accessed 21 Mar 2023
2. Mould R.: Cancer statistics overview with special reference to prostate, colon and rectum, lung, breast and cervix uteri. J. Oncol. **58**(3) (2008)
3. Didkowska, J., Wojciechowska, U., Zatoński, W.: Prediction of cancer incidence and mortality in Poland up to the year 2025, Nationwide cancer database (2009)
4. Repež, A., Oroszy, D., Arnež, ZM.: Continuous postoperative monitoring of cutaneous free flaps using near infrared spectroscopy. J. Plast. Reconstr. Aesthetic Surg. **61**, 71–77

5. Rogoń, I., Tojza, P., Wittbrodt, E., Łuszczek, M., Jankau, J.: Monte-Carlo modeling of optical sensor for post-operative free flap monitoring. In: Information Technology in Biomedical Engineering, pp. 141–153 (2021)
6. Hossain, S., Kim, K.-D.: Non-invasive in vivo estimation of HbA1c using monte carlo photon propagation simulation: application of tissue-segmented 3D MRI stacks of the fingertip and wrist for wearable systems. Sensors **23**, 540 (2023)
7. Carlson, M., Alt-Holland, A., Egles, C., Garlick, J.: Three-dimensional tissue models of normal and diseased skin 1,2,3, (2008)
8. Micheels, J., Alsbjorn, B., Sorensen, B.: Laser doppler flowmetry. A new non-invasive measurement of microcirculation in intensive care? Resuscitation (1984)
9. Metropolis, N.: The beginning of the Monte Carlo method, Los Alamos science issue (1987)
10. Jacques S.: Skin optics. In: Oregon Medical Laser Center News (1998)
11. Vishay, Silicon Photodiode BPW21R, 81519 datasheet, Rev. 1.7 (2011)
12. Vishay, Silicon PIN Photodiode VBPW34S, VBPW34SR, 81128 datasheet (2011)
13. Osram, SFH203 Radial T1 3/4 Silicon PIN Photodiode, Rev. 1.3 (2018)
14. Hamamatsu, Si PIN photodiodes S10783, S10784, KPIN1079E02 datasheet (2012)

3D-Breast System for Determining the Volume of Tissue Needed for Breast Reconstruction

Gabriela Małyszko, Julia Czałpińska, Andżelika Janicka, Katarzyna Ostrowska, and Mariusz Kaczmarek

Abstract This article presents methods for surface reconstruction and volume determination based on the point cloud created by 3D imaging. Such a system would be used to accurately estimate breast volume in patients classified for breast reconstruction surgery at plastic surgery centers. To develop such a system, various methods of determining volume, based on images from the Intel D435i camera, were tested. In addition, an application and a measuring station tailored to clinical needs were developed. Finally, we stated that 3D imaging systems can effectively determine breast volume for surgical procedures.

Keywords Breast reconstruction · 3D modelling · Depth camera

1 Introduction

Breast reconstruction surgery is the most often performed procedure after mastectomy in patients with advanced breast cancer. In recent years, techniques that allow breast reconstruction have developed significantly [1]. An increasingly common method for this type of surgery is a technique using the patient's own tissues which allows for avoiding complications associated with breast implants such as inflammatory reactions and capsular contracture. The procedure involves transferring a dermal-muscular flap called TRAM from the abdomen and forming the patient's breast from it. As a result, the reconstructed breast has a shape and texture similar to the natural breast. In the TRAM procedure (Transverse rectus abdominis muscle), the lower abdomen's skin, muscle and fat are used to recreate the breast. Tissue from the abdomen used in TRAM flap reconstruction can be taken in two different ways:

Supported by Statutory Funds of Electronics, Telecommunications and Informatics Faculty, Gdansk University of Technology.

G. Małyszko · J. Czałpińska · A. Janicka · K. Ostrowska · M. Kaczmarek (✉)
Biomedical Engineering Department, Gdańsk University of Technology, Narutowicza 11/12 Str., 80-233 Gdańsk, Poland
e-mail: mariusz.kaczmarek@pg.edu.pl

P. Strumiłło et al. (eds.), *The Latest Developments and Challenges in Biomedical Engineering*, Lecture Notes in Networks and Systems 746,
https://doi.org/10.1007/978-3-031-38430-1_20

using a tunneling, “pedicle,” TRAM flap or using a free, “microsurgical,” TRAM flap. In the pedicle (“tunneling”) approach to this procedure, this tissue separates from its original location (not fully), flips up and tunnels under the abdomen. It passes up and out through the mastectomy site (or scar, depending on when reconstruction was done). The tissue is then modelled to look as similar as possible to the other breast. The lower abdomen is then stitched together. In the case of a double mastectomy, tissue from the lower abdomen can be used to create both breasts. The “Free flap” (microsurgery) TRAM usually uses the same tissue as the “tunnel” procedure described above. The main difference with free TRAM reconstruction is that the tissue is not attached but is completely removed from the body. After removal from the abdomen, the tissue is transferred to the mastectomy site. The arteries and veins that supply the flap tissue must also be moved to do this. The flap is connected to the thoracic circulation as the tissue is brought to the site of the mastectomy. The patient remains under close observation for several weeks after such a procedure. In addition, the amount of own tissue harvested, that is, an accurate assessment of the volume of the breast, will determine the recovery period of such a patient [1, 2]. To date, healthcare professionals have used various available measurement techniques to determine breast volume, such as medical imaging (e.g., Magnetic Resonance Imaging (MRI), Computer Tomography (CT), mammography), as well as anthropometric measurements and casts. Despite the widespread use of the above methods, they are subject to high errors and uncertainties in determining breast tissue volume [3]. Therefore, alternative methods are constantly being sought. 3D imaging is gaining popularity around the world. Three-dimensional imaging of the breast surface has the potential to aid in treatment planning and more effectively predict outcomes. The undoubted advantages of this method are computational accuracy, versatility, short execution time, ease of use and non-invasiveness to the patient. It can be useful in planning surgery more efficiently and ensuring less asymmetry in the reconstructed breast. Unfortunately, this method is not widely used clinically due to the high cost and spatial requirements of 3D scanning equipment [4]. In this study, we aim to investigate whether 3D imaging with the portable Intel D435i camera is a reliable tool for estimating breast volume in clinical practice.

2 Review of the Literature

There are many publications that indicate that 3D medical imaging applications may have a high potential for clinical use. Geometric 3D models of the breast are capable of assisting in assessment, prediction and treatment planning, but unfortunately, these methods are expensive and require extensive infrastructure. One of the 3D methods that have been used to reconstruct the breast surface is a template-free nonrigid reconstruction pipeline running only on a standard PC, taking a noisy RGBD input video from a MS Kinect-style camera. The method can reconstruct clinical-quality surface models despite varying degrees of postural sway during data capture [5]. The authors of the article [6] investigated 3D breast surface reconstruction from RGBD video provided by MS Kinect depth-sensing cameras. They demonstrated that MS Kinect cameras are capable of 3D scanning of the breast surface by streaming depth and colour data of moderate quality at high frame rates in a lightweight acquisition setup. The limitation of this method was time, which was about 2 h for a complete reconstruction. From the surgeons' perspective, it is desirable that the generated results be available immediately [6].

Another approach to solving the problem of 3D object reconstruction was to use a system called 3D Builder [7]. The methodology in this platform is based on Zhang's algorithm. In this method, acquiring a sequence of two images was necessary. The first was acquired by keeping the camera intact, and in the second case the object was placed on a simple rotating device. The model of the object under study was created using Generalized Voxel Coloring (GVC). Based on voxels that replicate surface points in the scene, the algorithm models it based on consistent colours in the input images. GVC allows input cameras to be placed anywhere on the stage [8]. The reconstruction process started with a three-dimensional voxel bounding box that contained the object to be reconstructed. For each voxel of this frame, corresponding pixels were built. A voxel was then cut from the reconstructed surface when it protruded beyond one of the silhouette images. The method from this article [7] was limited depending on the accuracy of the camera calibration results and in segmenting the object in the images used. In addition, the aforementioned GVC method is computationally demanding and therefore slow, which is a disadvantage for real-time applications.

3D modeling techniques are finding their way into virtual reality and holographic displays. 3D modeling can enable intuitive viewing of anatomical images, leading to better diagnoses, education, surgical planning and outcomes. The creation of 3D models enables oral and maxillofacial surgeons to successfully achieve the goal of surgery by uwtting surgical templates, determining the dimensions of required implant plates, and shortening surgery time. In addition, medical students also receive modelled and 3D-printed organ simulants by 3D printers, which form the basis for training future doctors [9].

3 Methods and Materials

The purpose of this project was to prepare a system and an application which will help doctors estimate breast volume. Therefore, it should be uncomplicated to use and intuitive. The main idea is to use a depth camera as a sensor and a PC computer with an appropriate application to calculate the required volume of the reconstructed breast. The initial study was conducted at home, so measurements were taken on four volunteers who were part of the research team. Ethical approvals were therefore not sought.

3.1 Camera

The research presented in this paper was used 3D visualization technology. Cameras with a 3D imaging function were used to take a point cloud of the object under study. The functionality of two Intel cameras, the D435i and the D455, was tested [10]. The D455 camera uses stereoscopic depth technology and works well at distances from 0.6m to 6m. The depth output resolution of this camera is 1280×720 with a maximum frame rate of 90 fps. The device uses the Intel RealSense Vision D4 video processor. The D455 extends the distance between the depth sensors to 95mm, improving the depth error to less than 2 percent at 4 mm. The camera also integrates an Inertial Measurement Unit (IMU) to allow applications to refine depth awareness in any situation where the camera moves. The second camera, the D435i, also operates using the RealSense Vision Processor D4, as well as a stereo vision module with two depth sensors, an RGB sensor and an infrared projector. The camera also has an IMU module for motion and pan detection at 6 °C of freedom. The device operates at resolutions up to 1920×1080 px with active depth detection in stereo. The depth detection range for this camera is from 0.1m to 10m. Despite the slight differences in the parameters of the two cameras, the D435i was chosen for further post-testing because its field of view is smaller than that of the D455. This is very important when one wants to cover only a woman's breast with the camera's eye, without visualizing surrounding elements. In addition, the D435i has a longer working range, which increases the possibility of setting up a measuring station.

3.2 Preparation of the Measuring Station and Data Collection

For the initial trials and testing of the volume determination system, a suitable measuring station had to be prepared. A suitable phantom was created to mimic a woman's breast. The camera's performance was tested, as well as its effectiveness in generating a point cloud under different environmental conditions (among other lighting). In addition, the effect of reflective materials, including, for example, film, was also examined. Preliminary tests have shown that lighting conditions are the basis for correct measurements. Shadows or unevenly incident sunlight are capable of disrupting the point cloud. In addition, taking into account the clinical use of the application and the conditions of the medical facility, it was decided to perform measurements under artificial lighting, additionally assisted by a circline-led lamp (Fig. 1), which evenly illuminates the test area.

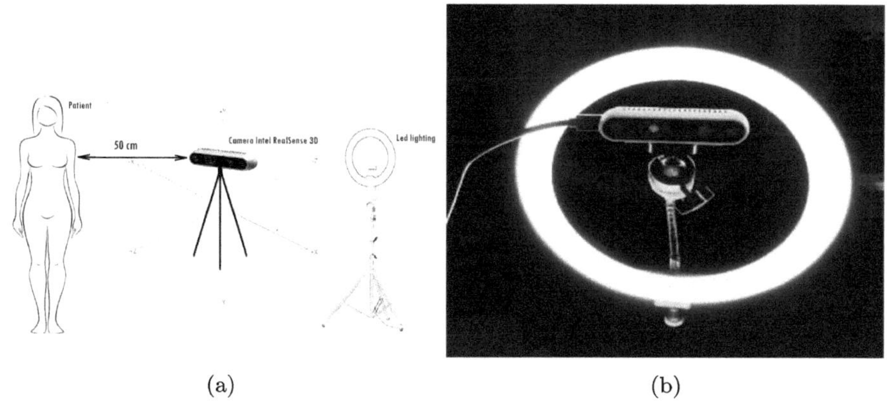

Fig. 1 Measuring set-up—idea and photo; **a** Positioning of the patient for measurement; **b** Camera and lighting setup

3.3 *Acquisition of a 3D Image of the Surface*

All 3D images were acquired with the Intel RealSense Depth Camera D435i which has a 3D imaging function. With the addition of the IMU, the RealSense D435i enables to tackle a new set of challenges related to tracking movement and depth. The built-in IMU provides an additional data point for dense 3D model reconstructions and can be used in applications such as gaming, pointing devices and image stabilization.

3.4 *Surface Reconstruction Methods*

The work examined three known point cloud surface reconstruction algorithms and analyzed their ability against breast reconstruction.

The first algorithm used was Poisson Surface Reconstruction [11]. The approach to surface reconstruction in this case is based on the observation that the normal field of a solid boundary can be interpreted as the gradient of an index function. Given such a relationship, given a set of points, a grid on their surface can be created by transforming the oriented points into a continuous vector field, finding a scalar function whose gradients best match the vector field, and by extracting the corresponding iso-surface.

The second algorithm was Ball-Pivoting Algorithm (BPA). This algorithm computes a triangle mesh interpolating a given point cloud [12]. Typically, the points are surface samples acquired with multiple range scans of an object. The principle of the BPA is very simple: Three points form a triangle if a ball of a user-specified radius p touches them without containing any other point. Starting with a seed triangle, the ball pivots around an edge (i.e., it revolves around the edge while keeping in contact with the edge's endpoints) until it touches another point, forming another triangle. The process continues until all reachable edges have been tried, and then starts from another seed triangle until all points have been considered. The process can then be repeated with a ball of a larger radius to handle uneven sampling densities.

The third algorithm was Alpha Shape. This algorithm is a method derived from Delaunay triangulation (DT) methods [13, 14]. The main difference between DT and the alpha shape algorithm is an additional constraint in the latter - the radius of each circle passing through the three vertices of the triangles must be smaller than a predetermined value of alpha (α). The value of alpha ranges from zero to infinity and generates different shapes, called alpha shapes, depending on the value of alpha. For $\alpha = 0$, the alpha shape is the point itself. When α is chosen large enough (infinity), the alpha shape is a convex hull, which is the result of the DT algorithm. The value of alpha can be determined experimentally to achieve a balance between size and shape/surface contours [15].

Ultimately, none of the described reconstruction algorithms was used in the presented algorithm. However, they allowed us to delve into the problem of surface reconstruction and allowed us to draw conclusions that will help in further work.

4 Proposed Algorithm

The goal of the proposed algorithm is to achieve the most accurate volume of tissue needed for breast reconstruction. Therefore, the most important step is fast and precise surface reconstruction. The proposed pipeline is presented in Fig. 2. After experimentation and camera positioning, the most optimal patient positioning is shown in Fig. 1a. In addition, it is recommended to use lighting in the form of a circline-led lamp (Fig. 1b) and position the patient at a certain distance from the camera, which is 50cm.

A. *3D image taking*

In the first step, 3D image is taken (Fig. 3). The infrared camera in the RealSense camera allows us to see the distance between objects and to separate them from the background layer. This makes it possible to distinguish a convex breast from a chest.

B. *Point cloud generation*

A point cloud is a multi-million set of points that is a geometric representation of the scanned object. Point clouds are usually so dense that they appear to form a continuous surface. Only after considerable magnification does it become apparent that it is a collection of individual points. However, the points of the point cloud are not interpreted as unrelated objects, but as a coherent structure. Many programs, thanks to the ability to determine the normal, give the ability to search for planes, cylinders, spheres and even very complex solid objects in the cloud. Using either a camera built into the scanner or an external one, the point cloud can be given the

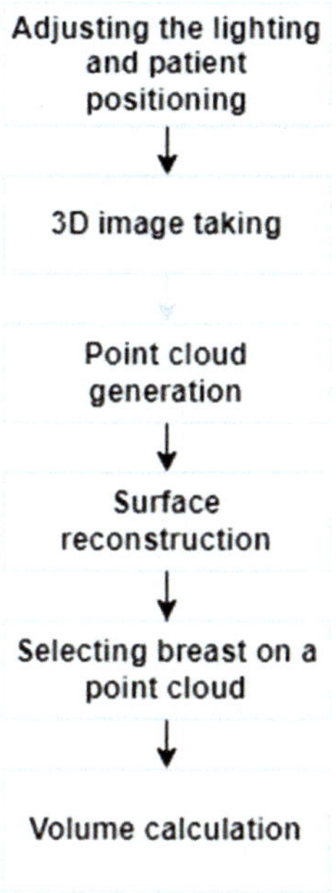

Fig. 2 Measurement procedure diagram

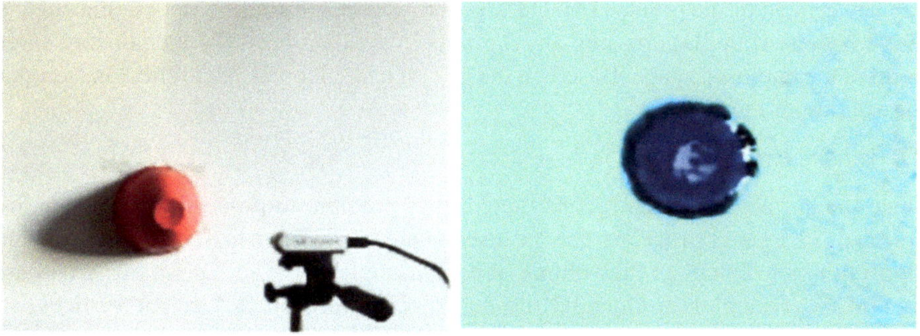

(a) (b)

Fig. 3 Example of phantom experiments; **a** A real photo of the phantom used in the experiments; **b** a 3D picture from D435i Intel Depth Camera

RGB values of the pixels from the camera's sensor. This gives the cloud a realistic feel and makes it ideal for visualization [16]. The obtained point cloud is shown in Fig. 4.

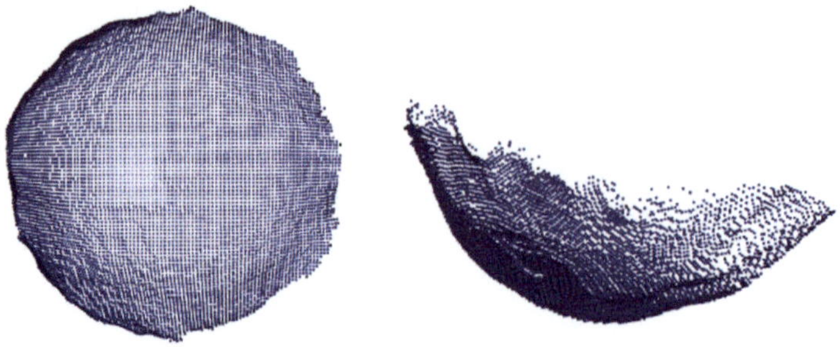

Fig. 4 Generated point cloud from example 3b

C. *Surface reconstruction*

It is not possible to compute the volume from a Point Cloud (PCD), a mesh is needed. Surface reconstruction algorithms must be used to move from the point cloud to the mesh. A projection of the PCD onto the XY plane was made by removing the Z value. A triangularization was constructed to reconstruct the surface in 2D space. Triangulation of 2D points was obtained using the Delaunay algorithm. The triangles obtained by Delaunay were then preserved, however, 3D vertices were used instead of 2D. The Delaunay triangulation (DT) method has been shown to be efficient and accurate for dense and noise-free data [14]. It generates a mesh of triangles which can cover most of the points [14]. DT algorithms check for the "empty circle" test: a circle (sphere if in 3D) crossing any 3 points in the data must not include any other point in the enclosed space. Based on the grid of triangles was later relied on to count the volume of the object.

D. *Volume calculation*

Once the surface has been reconstructed, it is possible to proceed to calculate the volume. Various approaches can be used. One of them, which was used in this algorithm, is to calculate the volume of each triangle of the surface to the XY plane (height 0). Triangles in Open3D meshes specify vertex indexes, not vertices, so translate the triangles so that their values are actually 3D points, not indexes to a list of vertices.

Next, a function to calculate the volume under each triangle is needed. Using an article by Bourbaki mathematician [16], if we label A1 and A2 as the areas of triangles P2P3P4 and P1P2P4 projected onto a fixed z-plane (Fig. 5), it follows that the volume under the designated triangle can be determined from the formula:

$$V = A1(z3 - z2)/3 - A2(z2 - z1)/3 + (A1 + A2)z2 \tag{1}$$

where:

$$A1 = [(x4y2 - x2y4) + (x3y4 - x4y3) + (x2y3 - x3y2)]/2 \tag{2}$$

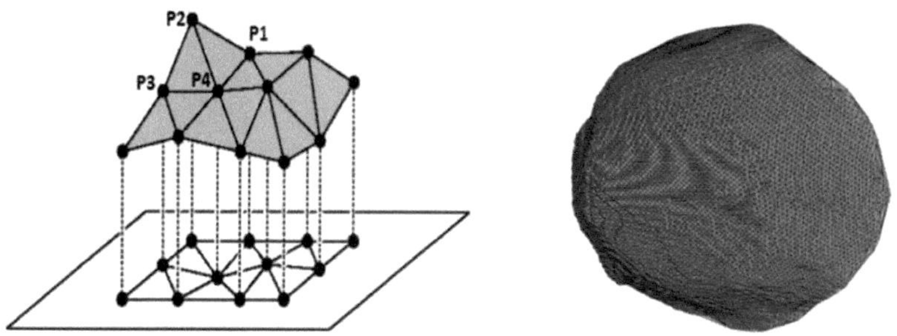

(a) (b)

Fig. 5 Delaunay Triangulation; **a** Visualization of the principles [18]; **b** Surface reconstruction

$$A2 = [(x1y2 - x2y1) + (x4y1 - x1y4) + (x2y4 - x4y2)]/2 \tag{3}$$

and the values of x4 and y4 are given by

$$x4 = x1 + ((z2 - z1)/(z3 - z1))(x3 - x1) \tag{4}$$

$$y4 = y1 + ((z2 - z1)/(z3 - z1))(y3 - y1) \tag{5}$$

Substituting into the equation for the volume gives the result

$$V = (z1 + z2 + z3)(x1y2 - x2y1 + x2y3 - x3y2 + x3y1 - x1y3)/6 \tag{6}$$

The last step is to determine the volume of all triangles.

E. *Selection of the voxel size*

The key parameter that determined the value of the volume was the selection of the appropriate size of the voxels. It describes the size of the polyhedrons that are used to complete the surface. The smaller the voxel size, the more polyhedrons are used for reconstruction and the more accurate the algorithm's response in terms of the returned volume. Figure 5 shows the surface reconstructions from example 3b. This is a finite element mesh reconstruction that is so dense that it is hard to see the individual elements, which are tetrahedrons with edge lengths of 0.001mm, in the visualization. Changing this parameter caused the result to change from the actual volume value by up to 4%. The effect of changing the voxel size on the speed of the program was also investigated. At the tested values, which gave satisfactory results, there were no noticeable changes in the execution time of the algorithm.

4.1 Application

The application was created using functionality from the available Open3D Visualizer solution. The application allows users to take pictures based on the live stream from the camera (Fig. 6). After taking the photo, the user has the option to crop part of the image to reduce the background of the photo. In such a prepared image, a background cropping algorithm is applied to leave only the breast. Finally, the volume of the object is calculated. The user can access the created images and view them as 3D visualizations. In addition, the user can access the automatically generated report.

Both the algorithms required to determine the volume of the breast and the GUI of the application were implemented in the Python programming environment. The program mainly uses the Open3D library. The real-time use of the software is enabled by the OpenCV library. Our system is compatible with Windows software and can be used on computers with Intel Core processors.

5 Results

The tests were carried out on a processor Intel Core i7-1255U 1.70 GHz with an embedded graphics card Intel Iris Xe Graphics. There were 3 consecutive trials of the experiment on the same objects with known volume and on the same people. On this basis, we know the actual volume and the three measured values, which made it possible to calculate the average volume and the standard deviation value. The trials were repeated under the same conditions, meaning without changing the camera settings and lighting. Volume calculations are available in real time, which means that the user obtained the result during the patient's consultation time. This time is

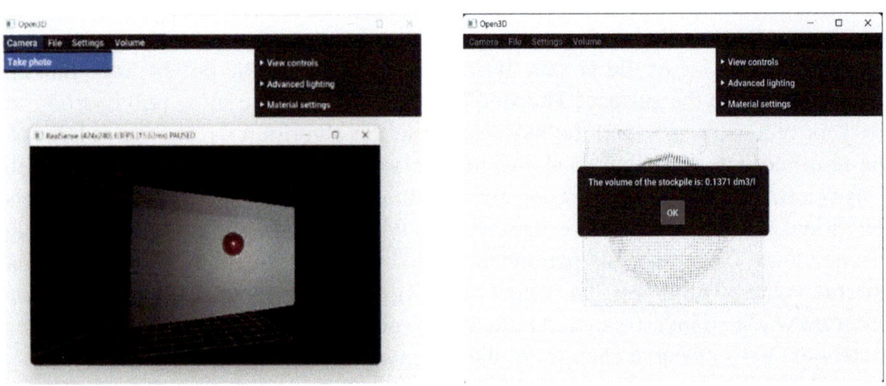

(a) (b)

Fig. 6 View of the application; **a** GUI of the app; **b** Estimated volume

Table 1 Phantom volume measurements in liters [l]

	Real [l]	Est1[l]	Est2[l]	Est3[l]	Mean[l]	Error[%]	STD
Obj1	0.75	0.725	0.747	0.714	*0.729*	*2.8*	*0.0137*
Obj2	1.8	1.669	1.785	1.693	*1.716*	*4.66*	*0.05*
Obj3	3	3.004	2.894	3.053	*2.984*	*0.53*	*0.066*

Table 2 Human breast volume measurements in liters [l]

	Real[l]	Est1[l]	Est2[l]	Est3[l]	Mean[l]	STD
Person1	0.340	0.346	0.351	0.347	*0.348*	*0.006*
Person2	0.272	0.267	0.276	0.282	*0.275*	*0.002*
Person3	0.204	0.206	0.202	0.199	*0.202*	*0.001*
Person4	0.124	0.118	0,121	0.121	*0.120*	*0.003*

determined by a few minutes. It takes more time to properly position the patient in relation to the camera than to get the results.

Results on phantoms The tests were carried out in three series (Est1–Est3) on three objects of different but known volumes fixed to a flat surface. The objects used in the experiment were kitchen bowls of a specific shape and known volume. The tests were repeated under the same conditions, without changing the configuration, distance or lighting conditions. The results obtained are summarized in Table 1.

Human test results Tests were conducted on several women with different breast sizes and shapes. These were only tests conducted under home conditions; advanced clinical tests could not be conducted. The tests were carried out on the previously described test stand in three measurement series (Table 2). Obtained reconstructed breast of one patient is shown in Fig. 7.

The results from Table 1 show that the largest value of the standard deviation is 0.066, with an actual volume of 3l. On the other hand, the smallest error appears at the smallest actual volume for object 1 and is 0.75 l. In addition, Error, that is, the difference between the actual volume and the average oscillates within 0.5–5%. For Table 2, the standard deviation oscillates between values of 0.002–0.009. In summary, these are acceptable errors by surgeons. In addition, each of the measurements deviated little from the mean value, a testament to the accuracy of the apparatus. These are certainly better results than the doctors' estimates. Improving the errors can be done by expanding the module with additional cameras.

The genesis of the errors can be taken from the insufficient number of cameras, to cover the breast grid from several planes. Besides, the Delaunay (DT) algorithm used itself does not control the size of the triangles. It creates larger triangles in sparsely populated regions where points are far apart, and smaller triangles in densely popu-

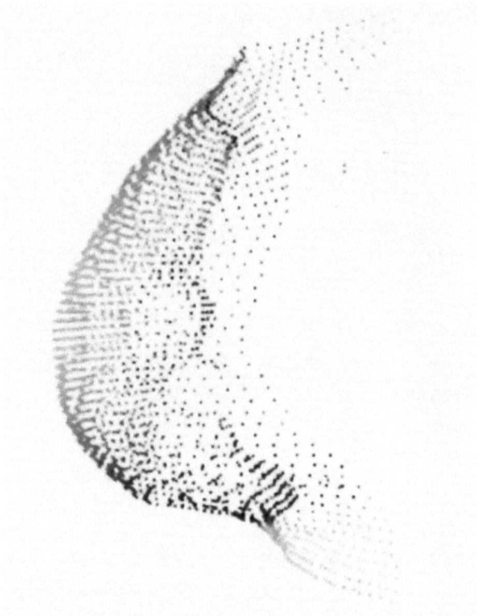

Fig. 7 Reconstructed breast from point cloud

lated regions where points are very close together. From there, all unwanted points are taken into account for surface reconstruction. This can cause the reconstructed surface to resemble a larger or smaller convex shell with a different measured area than the actual object [16].

6 Conclusion

In this paper different methods for surface reconstruction have been compared and a system for determining the volume of tissue needed for breast reconstruction has been implemented. The reconstruction method has been chosen based on various object experiments. Extensive research on methods used in similar applications has been conducted and also a literature review of 3D imaging and volume determination algorithms.

The method presented in this paper can be used in real time system. It has been proved that determining the volume of tissue needed for breast reconstruction can be done based on a depth camera. During the work, many trials were carried out, but in order to present the results, a few of them were selected to represent a representative sample. Based on these trials and the results presented, we are able to assess that the presented system gives reliable results. Further research should focus on optimizing the method's parameters for volume determination to achieve greater accuracy. Measurement can be improved by automatically selecting feature points in the image and thus no manual cropping of the photo will be necessary. Another improvement

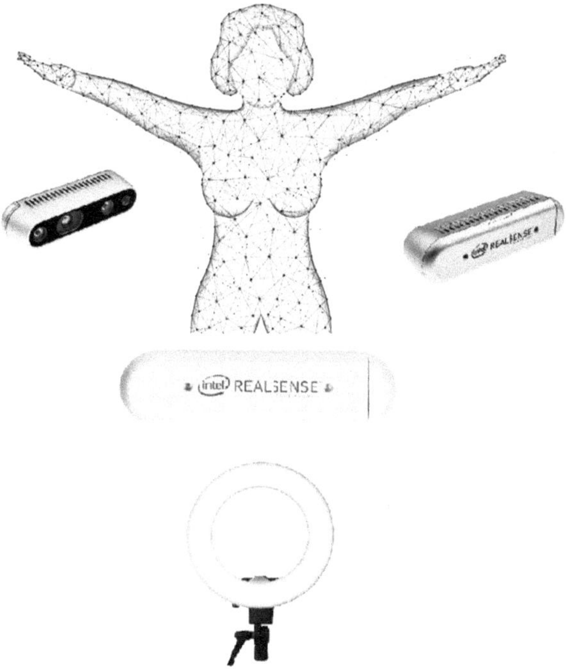

Fig. 8 Prototype of improved system with three depth cameras

could be the addition of a database of implants available on the market. This would make it possible to select the right implant for a patient at the same time.

The prototype system for determining the volume of tissue needed for breast reconstruction uses only one depth camera, while one or two additional cameras could be added in further work. This would make it possible to take several images, each from a different perspective, and then combine them together (Fig. 8). Taking images of the patient from two or three perspectives should improve the quality of the resulting point cloud due to the lack of holes which are created while using only one camera. As a result, surface reconstruction would be more accurate, and consequently so would the estimated breast volume. In conclusion, we recommend the designed system, but it is still in the preliminary stage of testing. One of its main advantages is the inexpensive cost of building such a system. The cost of the camera is not exorbitant and computers with the required parameters are widely used. We plan further testing and expansion of the system so that it can be used in breast reconstruction clinics.

Acknowledgements Research supported by Statutory Funds of Electronics, Telecommunications and Informatics Faculty, Gdansk University of Technology. We would also like to thank doctors from the Medical University of Gdansk, Plastic Surgery Clinic for their substantive support and consultations.

References

1. Teymouri, H., Stergioula, S., Eder, M., Kovacs, L., Biemer, E., Papadopulos, N.: Breast reconstruction with autologous tissue following mastectomy. Hippokratia **10**(4), 153–162 (2006)
2. Sachs, D., Winters, R.: Unipedicled Tram Breast Reconstruction, StatPearls. StatPearls Publishing (2022). https://www.ncbi.nlm.nih.gov/books/NBK436009/
3. Jeon, F.H.K., Griffin, M., Almadori, A., Varghese, J., Bogan, S., Younis, I., Mosahebi, A., Butler, P.: Measuring differential volume using the subtraction tool for three-dimensional breast volumetry: a proof of concept study. Surg. Innov. **27**(6), 659–668 (2020). https://doi.org/10.1177/1553350620945563
4. Gaisberger, C., Steininger, P., Mitterlechner, B., Huber, S., Weichenberger, H., Sedlmayer, F., Deutschmann, H.: Three-dimensional surface scanning for accurate patient positioning and monitoring during breast cancer radiotherapy. Strahlenther Onkol. **189**(10), 887–893 (2013). https://doi.org/10.1007/s00066-013-0358-6
5. Lacher, R.M., Vasconcelos, F., Williams, N.R., Rindermann, G., Hipwell, J., Hawkes, D., Stoyanov, D.: Nonrigid reconstruction of 3D breast surfaces with a low-cost RGBD camera for surgical planning and aesthetic evaluation. Med. Image Anal. **53**, 11–25 (2019). ISSN 1361-8415, https://doi.org/10.1016/j.media.2019.01.003
6. Lacher, R.M.: 3D breast surface reconstructions from consumer-grade RGB-D cameras. Doctoral thesis (Ph.D), UCL (University College London) (2019)
7. Azevedo, T., Tavares, J., Vaz, M.: 3D object reconstruction from uncalibrated images using an off-the-shelf camera, advances in computational vision and medical image processing. In: Computational Methods in Applied Sciences, vol. 13. Springer (2009). https://doi.org/10.1007/978-1-4020-9086-8_7
8. Culbertson, W., Malzbender, T., Slabaugh, G.: Generalized voxel coloring. In: Triggs, B., Zisserman, A., Szeliski, R. (eds.) Vision Algorithms: Theory and Practice, pp. 100–115. Springer (2001). ISBN 3-540-67973-1. https://doi.org/10.1007/3-540-44480-7_7
9. Mathew, N., Gandhi, S., Singh, I., Solanki, M., Bedi, N.S.: 3D models revolutionizing surgical outcomes in oral and maxillofacial surgery: experience at our center. J. Maxillofac. Oral Surg. **19**(2), 208–216 (2020). https://doi.org/10.1007/s12663-019-01275-0
10. Intel® RealSenseTM Camera 400 Series Product Family Dataseet, Mar. 2022
11. Kazhdan, M., Hoppe, H.: Screened Poisson surface reconstruction. ACM Trans. Graph. **32**(3), 13 (2013). https://doi.org/10.1145/2487228.2487237. Article 29
12. Maiti, A., Chakravarty, D.: Performance analysis of different surface reconstruction algorithms for 3D reconstruction of outdoor objects from their digital images (2016). Article 2016 Jun. 30. https://doi.org/10.1186/s40064-016-2425-9
13. Bernardini, F., Bajaj, C.L., Chen, J., Schikore, D.: A triangulation-based object reconstruction method. In: Proceedings of the Thirteenth Annual Symposium on Computational Geometry, pp. 481–484 (1997). https://doi.org/10.1145/262839.263098
14. Teichmann, M., Capps, M.: Surface reconstruction with anisotropic density-scaled alpha shapes. In: Proceeding IEEE Visualization, pp. 67–72 (1998). https://doi.org/10.1109/VISUAL.1998.745286
15. Edelsbrunner, K., Seidel, R.: On the shape of a set of points in the plane. IEEE Trans. Inf. Theory **29**(4), 551–559 (1983)
16. Brown, K.: Volume Under a Triangle. mathspages.com (25 Jan. 2023) (2018)
17. Mohammad, R. Maddah, C.G.L.: Application of the Alpha Shape Method to Visualize and Analyze Surgical Motion (2017). https://doi.org/10.4236/ss.2017.811052
18. UCI Homepage. https://www.ics.uci.edu/~goodrich/teach/geom/notes/DT.pdf. Accessed 23 Mar. 2023

Preeclampsia Risk Prediction Using Machine Learning Methods Trained on Synthetic Data

Magdalena Mazur-Milecka, Natalia Kowalczyk, Kinga Jaguszewska, Dorota Zamkowska, Dariusz Wójcik, Krzysztof Preis, Henriette Skov, Stefan Wagner, Puk Sandager, Milena Sobotka, and Jacek Rumiński

Abstract This paper describes a research study that investigates the use of machine learning algorithms on synthetic data to classify the risk of developing preeclampsia by pregnant women. Synthetic datasets were generated based on parameter distributions from three real patient studies. Four models were compared: XGBoost, Support Vector Machine (SVM), Random Forest, and Explainable Boosting Machines (EBM). The study found that the XGBoost and EBM consistently outperform the other models. An analysis of patient subsets based on their pregnancy history was also conducted, revealing that the group of patients in their first pregnancy achieved the highest prediction accuracy. Additionally, the study explored the efficacy of risk prediction based on various parameters and found that the results vary depending on the models used and the degree of class balance in the database. Finally, an additional test was performed on the dataset annotated by physicians.

Keywords Preeclamspia risk · Machine learning · Pregnancy · Women health · Digital twin

---

M. Mazur-Milecka (✉) · N. Kowalczyk · M. Sobotka · J. Rumiński
Gdansk University of Technology, Gdansk, Poland
e-mail: magdalena.milecka@pg.edu.pl

K. Jaguszewska · D. Zamkowska · D. Wójcik · K. Preis
Gameta Clinic, Rzgów, Poland

K. Jaguszewska
Medical University of Gdansk, Gdansk, Poland

H. Skov · P. Sandager
Aarhus University Hospital, Aarhus, Denmark

S. Wagner
Aarhus University, Aarhus, Denmark

P. Strumiłło et al. (eds.), *The Latest Developments and Challenges in Biomedical Engineering*, Lecture Notes in Networks and Systems 746,
https://doi.org/10.1007/978-3-031-38430-1_21

1 Introduction

The state of pregnancy is accompanied by various physiological and pathological alterations. Among these changes, one of the most dangerous is preeclampsia syndrome (PE) which is recorded in 2–8% of all pregnancies [1–4]. This disease is a cause of more than 50,000 maternal deaths and results in over 500,000 fetal deaths worldwide [5]. Hypertensive disease during pregnancy is associated with an increased risk of developing long-term maternal cardiovascular diseases, including ischemic heart disease and hypertension [6]. The cost of preeclampsia was found to be $2.18 billion to the U.S. healthcare system for the first 12 months after delivery [7]. Recently, in some countries, four times as many women are dying from preeclampsia than in 2012–14 [8]. Once preeclampsia is diagnosed, it cannot be cured. Therefore, it is very important to monitor women at risk effectively. While the only curative treatment for PE during pregnancy is delivery of the baby, studies from recent years have shown that treatment with aspirin from the first trimester can reduce the rate of severe PE [9, 10].

Many studies have confirmed that administration of low-dose aspirin at 16 week of gestation or earlier during high risk pregnancies has the potential to reduce the incidence of PE [11–13]. The main challenge of modern obstetrics is the early identification of women at high risk of developing PE and taking the necessary measures to reduce the incidence of this disease [14, 15]. The traditional approach to identifying women at high risk involves following the guidelines of organizations such as the National Institute for Health and Care Excellence (NICE) or the American College of Obstetricians and Gynecologists (ACOG) [16, 17]. According to these guidelines, risk factors for PE include hypertensive disease in a previous pregnancy, chronic hypertension, chronic renal disease, diabetes mellitus, nulliparity, advanced maternal age, autoimmune disease, obesity, prolonged interpregnancy interval, and family history of PE. Bartsch et al. in [18] presented a systematic review of studies analyzing the relationship between clinical risk factors occurring before the 16th week of gestation and the risk of PE. The significant factors turned out to be (in order of importance): prior history of PE and chronic hypertension, nulliparity, maternal age >35 years, chronic kidney disease, conception by assisted reproductive technology, prepregnancy body mass index (BMI) of >30 kg/m^2 and preexisting diabetes mellitus.

1.1 FMF Calculator

Considering the various parameters by which the risk of preeclampsia can be assessed, The Fetal Medicine Foundation (FMF) has made a competing risk model and made the calculator available online [19]. It can be used to evaluate this risk in early pregnancy (i.e., from 11+0 to 14+1 weeks). Risks can be derived from maternal history and any combinations of biomarkers. Maternal characteristics contain typical information like height, weight, race, smoking during pregnancy, and a family history of preeclampsia, as well as medical history (e.g., chronic hypertension, diabetes mellitus, autoimmune diseases) and obstetric history. The calculator also consists of the type of pregnancy (singleton, monochorionic or dichorionic twins). Biophysical measurements and biochemical markers include mean arterial pressure (MAP), uterine artery pulsatility index (UTPI), serum placental growth factor (PLGF), and pregnancy-associated plasma protein-A (PAPP-A). The obtained risk of developing PE is determined based on a maternal characteristic only or in various combinations depending on the given biomarkers (maternal characteristic + MAP, maternal characteristics + MAP + UTPI, maternal characteristics + UTPI, etc.).

1.2 Contribution

This preliminary study is a part of the long-term international project (WODIA) focused on efficient, early, and cost-effective prediction of preeclampsia risk for pregnant women. The main contributions of this study are:

- New synthetic datasets were generated based on estimated distributions of parameters from three different observational studies on real patients.
- New models were trained and validated using the knowledge-based FMF expert system, demonstrating high accuracy above 96%.
- The machine learning models were validated on ground truth data showing the highest F1 score equal to 86%.
- The Explainable Boosting Machine (EBM) model was adapted and trained to demonstrate the individual contributions of each feature that could provide explainable AI solutions for preeclampsia risk prediction and future digital twins simulations.
- A final hypothesis was formulated that the machine learning models (contrary to simple expert systems) could continuously learn from multi-center data to improve the explainability of rare diseases.

2 Related Work

Several studies, such as [20–22], utilize statistical analysis as means of predicting preeclampsia. However, due to the availability of larger datasets and the increasing prevalence of machine learning techniques, such approaches are also being employed for preeclampsia prediction. In [23], a predictive model for preeclampsia has been proposed. The dataset utilized in this study contained more than 1,600 cases and consisted of 25 different features, including gestational age, medical risk factors, congenital anomalies, and the mother's marital status. Fourteen different machine learning models were evaluated to identify the best-performing model. Random Forest and AdaBoost Classification Trees were the most accurate, with accuracy and ROC above 82%. The five most significant features affecting the results were the duration of a completed pregnancy in weeks, poverty, water retention/edema during pregnancy, toxemia, and completed years of schooling (education).

In [24], a study on predicting the risk of late-onset PE using machine learning models was conducted. The study compared several state-of-the-art algorithms, including logistic regression (LR), decision tree model (DT), naive Bayes classification (NBC), and stochastic gradient boosting method (SGB). The dataset used in the study consisted of clinical and medical history data from over 11,000 pregnant women, which included features such as age, blood pressure (BP), height, weight, gestational age, maternal and family medical history of hypertension, diabetes, and previous preeclampsia, and a series of biochemical laboratory data. The input data for the models were selected through pattern recognition and cluster analysis. The most influential variables were systolic blood pressure, serum blood urea nitrogen, creatinine levels, platelet counts, serum potassium level, white blood cell count, serum calcium level, and urinary protein. The SGB algorithm achieved the highest accuracy of above 97% among the tested models.

A solution based on deep neural networks for early prediction of preeclampsia was proposed in [25]. The model was validated using three large databases of over 30,000, 84,000, and 360,000 cases, respectively. The Chi-square test was employed to identify the most significant features in each database. Obesity and pre-existing diabetes mellitus were the most influential features in any complete dataset. The proposed model was tested on the full datasets and smaller ethnic groups extracted from them. The cost-sensitive deep neural network (CSDNN) yielded satisfactory results, especially for minority populations.

A model for predicting early preeclampsia using machine learning was proposed in [26]. The study used data from 5,245 women, among whom only 561 were diagnosed with PE. The data was divided into two sets, one for predicting PE and another for predicting early-onset PE. The proposed solution is based on the elastic net algorithm, which uses only the most informative features from all variables. For the preeclampsia prediction model, the most important features were hypertension, history of PE, insulin, and mean systolic BP. In contrast, for the early on-set PE model, the most important features were hypertension, history of PE, prednisone taking, and mean systolic BP. The resulting models achieved an area under the curve of 0.79 for preeclampsia classification and 0.89 for early-onset preeclampsia.

3 Methods

3.1 Data Generation

The data used in this study was synthetically generated due to the scarcity of sufficiently large public datasets. A synthetic data generator based on three different estimations of parameter distributions was created. All estimations were based on observations of real patient data.

For the first dataset, hereinafter referred to as *DS_10k*, the generator employed the population characteristics presented in [27] to produce 10,000 cases of 5,000 individuals without pre-eclampsia and 5,000 who developed it. Each case was characterized by 20 features, including weight, height, presence of diseases (such as type I or II diabetes, hypertension, and Systemic Lupus Erythematosus), previous pregnancy history, as well as mean arterial pressure (MAP), and uterine artery pulsatility index (UTPI). Parameters not described in the population characteristics, such as pregnancy type or information about the last pregnancy (e.g., interpregnancy interval or gestation weeks and days at delivery), were randomly generated with equal probability of occurrence.

Similarly, for the second dataset (called *DS_Brazil*), the characteristics of the Brazilian population presented in [28] were used. It was also decided to randomly generate parameters with the same probability of occurrence for such data as race, method of conception, type of pregnancy, and information about the last previous pregnancy. Based on these characteristics, 498 cases of people with PE and 703 without PE were generated.

Distribution characteristics described in the literature were separated for patients with and without preeclampsia. The origin of the sample characteristics, i.e., information about whether the sample data was drawn from healthy patients' distributions or patients at risk distribution, has been stored in the database and referred to as *ground truth* throughout the paper.

The last dataset (*DS_Denmark_test*) was generated based on distribution characteristics from a preeclampsia screening project in Denmark [29]. A total of 1,000 cases were generated based on the characteristic provided for a cohort with four risk markers—complete information regarding risk markers included in the FMF calculator.

Subsequently, using a calculator created by the Fetal Medicine Foundation [19], described in Sect. 1.1, four risk values were obtained for every simulated instance generated from the three artificial datasets. The risk values were based solely on maternal characteristics and their combination with parameters including MAP and UTPI.

Using the obtained risk values, an additional division into classes was introduced in relation to the obtained values. A division has been introduced that assigns the risk to one of two classes: low (with a risk value of $1 \geq 150$ or higher) and increased (with a risk value below $1 < 150$). The threshold value of 150 was chosen based on the proposed risk cut-off value [30], which covers 94% of the detection rate for early PE in the white population. By introducing a division into classes, a more accurate classification can be achieved, providing information about the level of risk as well.

The final training dataset was created from a combination of *DS_10k* and *DS_Brazil*, and consisted of 11,201 cases and 19 parameters (age, height, weight and racial origin of the mother, information about smoking, twins in the family, previous preeclampsia, previous pregnancies, conception method, chronic hypertension, existing diabetes type 1 or 2, as well as gestational age, fetal rump lengths,

systemic lupus erythematosus, antiphospholipid syndrome, MAP and UTPI). The amount of data was correspondingly smaller for the group of women in the first pregnancy (9805) and for the group with previous pregnancies (1396). The last group of patients had additional parameters showing the time interval since the last pregnancy, the gestation weeks at delivery of the previous pregnancy, and whether there was preeclampsia in the previous pregnancy.
Model testing was performed using three different test sets:

- *DS_10k_test*—test subset as 20% (2,241) of randomly selected examples from the training set; this dataset also had information about the origin of the sample (*ground truth*);
- *DS_Denmark_test*—1000 cases;
- *DS_med_test*—a set of selected 20 cases with different parameters' values. The cases were independently annotated by two physicians with over 10 years of experience in the field of gynecology, without using the calculator.

3.2 Data Preprocessing

Preprocessing of the generated data consisted in coding the data into integer values, converting the interpregnancy interval into the number of days, and rejecting data relating to specific dates of measurements. The women's age was calculated based on their date of birth.

Due to the different possible patient data configurations, missing values appear in the dataset. They occur for features that cannot be determined for each patient (e.g., crown-rump length for the second fetus in a twin pregnancy or data on the previous pregnancy), and it is impossible to complete them with known techniques. To prevent data loss, the dataset was additionally divided into three subsets, women in their first pregnancy, women in a subsequent pregnancy, and all women excluding data on previous births.

3.3 Models

Four supervised learning classifiers were used: XGBoost, Support Vector Machine (SVM), Random Forest, and Explainable Boosting Machines (EBM).

XGBoost [31] is an efficient gradient boosting framework that utilizes decision trees as base learners. The method sequentially builds an ensemble of decision trees, where each subsequent tree is trained to correct the errors of the previous tree, achieving high predictive accuracy. SVM—a binary classification algorithm—aims to find the optimal hyperplane for separating the data into classes by mapping the data into a high-dimensional feature space. SVM is particularly effective in problems with a clear margin of separation between the classes. Random Forest constructs a multitude of decision trees. The algorithm has several advantages, including its ability to handle high-dimensional data and robustness to missing values and outliers. EBM [32] is an algorithm that creates accurate and interpretable models, making it useful in domains where the interpretability of the model is important, such as healthcare or finance. EBM has been shown to perform well in several benchmark datasets.

3.4 Training

For the purposes of this research, we have performed the following comparative analyses:

1. comparison of the prediction quality for four different classifier models (*XGBoost, SVM, RandomForest, EBM*);
2. comparison between three groups of women: in the first pregnancy (*first_pregn*), with the previous pregnancies (*prev_pregn*), and all together (*all*);
3. comparison of prediction quality based on four groups of parameters (that align with the risk calculated by the FMF calculator): maternal characteristic only (*risk_maternal*), maternal characteristic + MAP data (*risk_MAP*), maternal characteristics + UTPI (*risk_UTPI*) and maternal characteristics + MAP + UTPI data (*risk_MAP_UTPI*);

The models' performance evaluation parameter was accuracy, described by Eq. 1. It measures the proportion of correctly classified instances: positive (TP) and negative (TN) among all instances in the validation set (TP + TN + false positives (FP) + false negatives (FN)).

$$Acc = \frac{TP + TN}{TP + TN + FP + FN} \tag{1}$$

However, when the classes are imbalanced, accuracy can be misleading, as the model may predict the majority class most of the time and still achieve high accuracy. Our database exhibited unbalanced groupings in certain instances. The F_1 score (Eq. 2) is a better metric in such cases. F_1 score balances precision and recall equally, making it suitable for imbalanced classes.

$$F_1\,score = \frac{2TP}{2TP + FP + FN} \tag{2}$$

Furthermore, we have trained models for *ground truth* data, without the use of prediction results from the FMF calculator. Additionally, extensive testing of the trained models was performed on the *DS_Denmark_test* and *DS_med_test* datasets.

4 Results

Multiple training was performed for different sets of parameters. Table 1 presents an example comparison of the classifiers' performance for selected training configurations. Accuracy values are presented for balanced distributions of classes for a given subset of parameters. The F_1 score values are presented otherwise. Results are presented for four models: XGBoost, SVM, Random Forest, and EBM. Based on the results, it can be observed that the EBM model consistently performs better than the other models in all cases. Its classification accuracy varies between 0.96 and 0.98 and is only slightly better than the second-best model—XGBoost, which achieves an accuracy of 0.95–0.96. The results for Random Forest and SVM are the worst, around 0.9. For imbalanced data, only EBM achieves an F_1 score above 0.9. The lowest score, 0.83, belongs to the Random Forest.

In the context of subsequent pregnancies, a subset of patients' data was found to possess crucial information pertaining to their previous gestational experiences, including the date of their most recent delivery and the incidence of preeclampsia during prior pregnancies. Given the significance of these distinct data points, a discrete patient cohort was created (*prev_pregn*) to isolate and analyze this subset of patients. The remaining patients (in their first pregnancy) constituted the second subset (*first_pregn*). The third group encompassed *all* patients, but in the event of a subsequent pregnancy, data from previous pregnancies were omitted, as certain algorithms are not adjusted to accommodate such a large amount of missing data.

Figure 1 shows the F_1 score for classifying the risk of preeclampsia based on maternal characteristics of the patient (Fig. 1a) and maternal characteristics + MAP + UTPI (Fig. 1b). Due to the imbalance observed in the group of patients with earlier pregnancies, the F_1 score was selected and used as the metric of choice. The figure illustrates that the most accurate classification is achieved for the patient group undergoing their first pregnancy, with the best-performing models displaying F_1 scores ranging between 0.95 and 0.98. The second most successful classification was observed for the *all* patients subset with F_1 scores ranging between 0.95 and

Table 1 Exemplary test accuracy and F_1 score for all models

	XGBoost	SVM	RandomForest	EBM
Accuracy				
risk_MAP_UTPI first_pregn	0.95	0.91	0.92	**0.97**
risk_UTPI first_pregn	0.96	0.9	0.93	**0.98**
risk_UTPI all	0.95	0.89	0.92	**0.96**
F_1 score				
risk_MAP prev_pregn	0.89	0.87	0.83	**0.91**

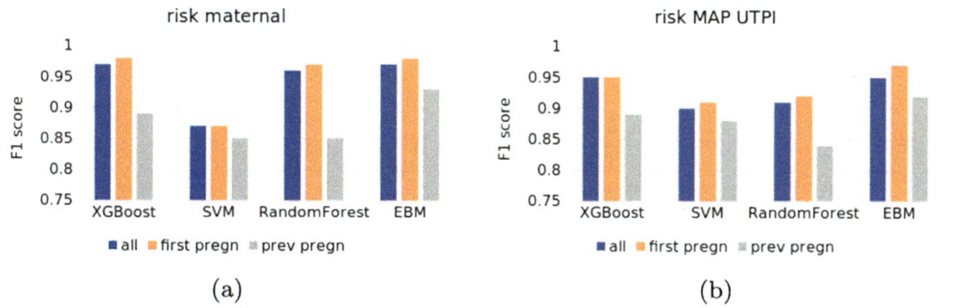

Fig. 1 $F_1 score$ result for calculating risk only from maternal characteristics (**a**) and from all parameters (**b**). Patients groups are color-coded: gray—patients with a previous pregnancy, orange—patients in their first pregnancy, blue—all patients

0.97. Conversely, the group of patients with prior pregnancies displayed the least favorable outcome (only EBM above 0.9); however, it should be noted that this may be attributable to this group's comparatively smaller sample size (1396).

The FMF Calculator can calculate risk based on various parameters, starting with the maternal characteristic alone to the incorporation of MAP and UTPI biomarkers. In Fig. 2, a comparison of the efficacy of risk prediction across all four models is presented based on four distinct groups of parameters. The most homogeneous results are achieved by EBM, with high levels of accuracy (roughly 0.98, Fig. 2a) and F_1 scores ranging between 0.91 and 0.93 (Fig. 2b). Risks based solely on maternal characteristics (blue in Fig. 2) and with an additional MAP measurement (orange) were found to have the top two best accuracies in almost all cases. On the other hand, the highest F_1 scores were achieved by predictions that considered all parameters (yellow) and maternal characteristics with UTPI (gray). The described models were also used to classify the group from which the generated patient data sample comes. For this purpose, as ground truth, we used the information stored in the database without using the results of the FMF calculator. Table 2 shows the F_1 score results for the test set. The prediction outcomes for all models were comparatively poorer for the *first pregnancy* and *all* patients (F_1 between 0.68 and 0.77). However, the group with previous pregnancies attained a prediction outcome similar to the risk calculated by the FMF calculator.

Table 2 Test F_1 score for GT group

	XGBoost	SVM	RandomForest	EBM
All	0.75	0.72	0.76	**0.77**
First pregnancy	0.71	0.68	0.72	**0.73**
With previous pregn.	**0.86**	0.85	0.85	0.83

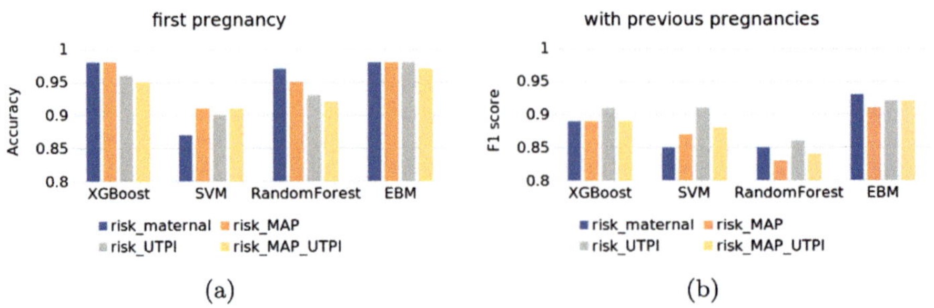

Fig. 2 Classification results based on: only maternal characteristics (blue), maternal characteristics + MAP (orange), maternal characteristics + UTPI (gray), and maternal characteristics + MAP + UTPI (yellow). **a** Accuracy for the first pregnancy group, **b** F_1 score for a group with previous pregnancies

Table 3 Test F_1 score for *DS_Denmark_test*

	XGBoost	SVM	RandomForest	EBM
risk_MAP_UTPI				
All	0.72	0.55	0.65	0.69
First pregnancy	0.91	0.74	0.79	**0.94**
risk_MAP				
All	0.56	0.46	0.57	0.66
First pregnancy	0.90	0.79	0.98	**0.99**
risk_UTPI				
All	0.79	0.61	0.66	0.78
First pregnancy	0.97	0.82	0.9	**0.98**
risk_maternal				
All	0.63	0.31	0.64	0.63
First pregnancy	**0.97**	0.39	0.94	0.96

Additional tests described in more detail in Sects. 3.1 and 3.4 were performed for the *DS_Denmark_test* and *DS_med_test* datasets. Table 3 shows the results that were obtained specifically for the *DS_Denmark_test* set. Due to a large amount of missing or incompatible data regarding previous pregnancies, in this case, classifier tests were performed only for women in their first pregnancy, and all patients with missing data on previous pregnancies were excluded. A significant discrepancy in results is apparent between the two groups: the classification for women in their first pregnancy achieves a F_1 score above 0.9 for the best models, while classification for all patients drops even below 0.6. The SVM model achieves the poorest test results.

Table 4, in turn, presents the tests for a small dataset annotated by physicians (*DS_med_test*). The highest agreement was achieved for risk assessment based solely on maternal characteristics (above 0.9). As the number of parameters considered in

Table 4 Test accuracy for physicians annotations (*DS_med_test*)

	XGBoost	SVM	RandomForest	EBM
risk_MAP_UTP				
All	0.6	**0.8**	0.6	**0.8**
first pregnancy	0.75	0.75	0.58	0.75
with previous pregn.	0.62	0.62	0.62	0.62
risk_MAP				
All, first pregnancy and with previous pregn.	0.75	0.75	0.75	0.75
risk_UTPI				
All	0.75	0.75	0.75	0.75
First pregnancy	**0.83**	**0.83**	**0.83**	**0.83**
With previous pregn.	0.62	0.62	0.62	0.62
risk_maternal				
All	0.95	0.85	0.95	0.85
First pregnancy	0.92	0.92	0.92	0.92
With previous pregn.	**1**	**1**	**1**	**1**

risk prediction increased, accuracy decreased. However, it should be noted that the database size is small, consisting of only 40 cases, which significantly lowers its credibility compared to the datasets described earlier.

5 Discussion

The presented results are preliminary studies on the early prediction of preeclampsia in pregnant women, which could enable earlier initiation of necessary treatment and reduce instances of unnecessary medication administration to healthy women.

The very limited preeclampsia datasets were described in the literature. All of them are private datasets. In order to overcome the lack of training data the authors generated synthetic data following the distributions of existing studies using specific parameters' distribution from each study. All variables were generated independently, which could have a limiting effect on the validity of the produced algorithms. Furthermore, the simulated data sets are mainly based on North European and South American populations, and cannot be generalized beyond these geographical areas.

Based on the synthetic data, many analyses were performed using four machine learning classification methods. The EBM and XGBoost models produced more accurate results.

Maric et al. [26] showed that for their private data, the best machine learning models achieved an accuracy of 89% for preeclampsia classification. In the work [24] over 97% accuracy was achieved for the stochastic gradient boosting model. The results of the experiments presented in our study are accuracy over 98% for the top (EBM) model. However, these results cannot be directly compared due to different data categories. The high performance of the EBM can be attributed to its extreme ability to capture complex non-linear relationships between features and target variables more effectively. XGBoost and EBM can also identify and use important features, including interactions between features, which is critical for accurate predictions. They are also less sensitive to noisy data and can better handle missing data and outliers. In addition, EBM performs the best among all models on imbalanced datasets, as shown in Table 1.

A comparison of the classification for different groups of patients showed that a significantly different group was patients with repeated pregnancies. This group, in most cases, achieved F_1 below 0.9, while the results for the other groups of *all* patients and patients in the first pregnancy always clearly exceeded this threshold. However, this may be due to the small number of data for patients with repeated pregnancies.

The high-risk classification results based only on maternal characteristics are related to the methodology employed in the analysis. The prediction of preeclampsia risk based on medical history is unreliable [30], and caution is advised for physicians making clinical decisions on this basis. One of the objectives of this study was to identify patterns in the generated data using the FMF calculator able to predict risk only on maternal characteristics. The models developed in this work can be interpreted as "surrogate" models of the FMF calculator. However, the machine learning models could be further trained (e.g., tuning, continual learning, federated learning, etc.), which offers possible higher generalization properties in comparison to rule-based systems. This is especially important for rear medical cases. For one clinic, only a limited number of such rare cases are usually observed. Therefore, machine learning models that could be collaboratively trained are very important.

The observed high accuracy of risk prediction based on maternal characteristics indicates that the trained models effectively captured the relationships among the parameters considered in the first (and relatively unreliable) risk calculation by the FMF calculator. High accuracy and F_1 score (at the level of about 0.9) in all risks classification (*risk_maternal*, *risk_MAP*, *risk_UTPI* and *risk_MAP_UTPI*) indicate that the models correctly found the relationships between the input parameters used in the calculator. Slightly lower results of the classification of the sample origin group (Table 2) indicate that there are dependencies between the values of the input parameters for healthy patients and patients with preeclampsia. In this case, the prediction of the distribution from which the sample was generated, not risk classes, is performed.

Models trained on many examples (different real studies) can potentially offer higher generalization properties and can be useful in developing effective risk assessment models for preeclampsia. In the future, the Digital Twin agent could be prepared, so each woman can introduce her parameters and simulate how the change of each of the parameters can influence her status or related risk. Future studies should take into account also other parameters such as serum placental growth factor or pregnancy-associated plasma protein-A [19], in order to investigate their individual influence.

6 Conclusion

In this work, new synthetic datasets were generated based on estimated distributions of parameters from three different observational studies on real patients and the FMF calculator. The best models trained on these datasets demonstrate very high accuracy of preeclampsia risk prediction. The best results were obtained for the EBM method which offers high explainability properties, important in medical diagnostics. Future machine learning models should be continuously and collaboratively trained to improve the explainable risk prediction of rare diseases.

Acknowledgements This work has been partially supported by IFD, NCBR, and UEFISCDI in the framework of the ERA PerMed, project WODIA—Personalized Medicine Screening and Monitoring Programme for Pregnant Women Suffering from Preeclampsia and Gestational Hypertension.

References

1. Duley, L.: The global impact of pre-eclampsia and eclampsia. In: Seminars in Perinatology, vol. 33, no. 3, pp. 130–137. WB Saunders (2009)
2. Abalos, E., Cuesta, C., Grosso, A.L., Chou, D., Say, L.: Global and regional estimates of preeclampsia and eclampsia: a systematic review. Eur. J. Obstet. Gynecol. Reprod. Biol. **170**(1), 1–7 (2013)
3. Harmon, Q.E., Huang, L., Umbach, D.M., Klungsøyr, K., Engel, S.M., Magnus, P., Skjærven, R., Zhang, J., Wilcox, A.J.: Risk of fetal death with preeclampsia. Obstet. Gynecol. **125**(3), 628–635 (2015)
4. Groot, M., Hofmeyr, C.J., Thangaratinam, S., Magee, L.A., de Groot, C.J.M., Hofmeyr, G.J.: Pre-eclampsia. Lancet **387**, 999–1011 (2016)
5. MouliSaha, M., Mukhyopadhyay, M., Bhattacharyya, M.: A study on maternal and fetal cell free DNA (cffDNA) for predicting the adverse pregnancy outcomes (2022)
6. Poon, L.C., Magee, L.A., Verlohren, S., Nicolaides, K.H., Shennan, A., von Dadelszen, P., et al.: A literature review and best practice advice for second and third trimester risk stratification, monitoring, and management of pre-eclampsia. Int. J. Gynecol. Obstet. **154**(S1), 3–31 (2021)
7. Stevens, W., Shih, T., Incerti, D., Ton, T.G., Lee, H.C., Peneva, D., Macones, G.A., Sibai, B.M., Jena, A.B.: Short-term costs of preeclampsia to the United States health care system. Am. J. Obstet. Gynecol. **217**(3), 237–248 (2017)
8. Knight, M., Bunch, K., Patel, R., Shakespeare, J., Kotnis, R., Kenyon, S., Kurinczuk, J.J.: On behalf of MBRRACE-UK. Saving lives, improving mothers' care core report-lessons learned

to inform maternity care from the UK and Ireland confidential enquiries into maternal deaths and morbidity 2018–20 (2022)

9. Rolnik, D.L., Wright, D., Poon, L.C., et al.: Aspirin versus placebo in pregnancies at high risk for preterm preeclampsia. N. Engl. J. Med. **377**(7), 613–622 (2017)
10. Van Doorn, R., Mukhtarova, N., Flyke, I.P., Lasarev, M., Kim, K., Hennekens, C.H., et al.: Dose of aspirin to prevent preterm preeclampsia in women with moderate or high-risk factors: a systematic review and meta-analysis. PLOS One **16**(3) (2021)
11. Bujold, E., Roberge, S., Lacasse, Y., et al.: Prevention of preeclampsia and intrauterine growth restriction with aspirin started in early pregnancy: a meta-analysis. Obstet. Gynecol. **116**(2), 402–414 (2010)
12. Roberge, S., Villa, P., Nicolaides, K., et al.: Early administration of low-dose aspirin for the prevention of preterm and term preeclampsia: a systematic review and meta-analysis. Fetal Diagn Ther **31**(3), 141–146 (2012)
13. Bujold, E., Roberge, S., Nicolaides, K.H.: Low-dose aspirin for prevention of adverse outcomes related to abnormal placentation. Prenat. Diagn. **34**(7), 642–648 (2014)
14. Norwitz, E., Bellussi, F.: Early pregnancy prediction of preeclampsia. UpToDate (2018)
15. Han, L., Holland, O.J., Da Silva Costa, F., Perkins, A.V.: Potential biomarkers for late-onset and term preeclampsia: a scoping review. Front. Physiol. **14**, 338 (2023)
16. LeFevre, M.L., US Preventive Services Task Force: Low-dose aspirin use for the prevention of morbidity and mortality from preeclampsia: U.S. Preventive Services Task Force recommendation statement. Ann. Intern. Med. **161**(11), 819–826 (2014)
17. ACOG Committee Opinion No: 743: low-dose aspirin use during pregnancy. Obstet Gynecol **132**(1), e44–e52 (2018)
18. Bartsch, E., Medcalf, K.E., Park, A.L.: Clinical risk factors for preeclampsia determined in early pregnancy: systematic review and meta-analysis of large cohort studies. BMJ **353**, i1753 (2016)
19. The Fetal Medicine Foundation, Risk assessment, Risk for preeclampsia. https://fetalmedicine.org/research/assess/preeclampsia/First
20. Poon, L.C., Nicolaides, K.H.: Early prediction of preeclampsia. Obstet Gynecol Int. **2014**, 297397 (2014)
21. O'Gorman, N., Wright, D., Syngelaki, A., Akolekar, R., Wright, A., Poon, L.C., Nicolaides, K.H.: Competing risks model in screening for preeclampsia by maternal factors and biomarkers at 11–13 weeks gestation. Am. J. Obstet. Gynecol. **214**(1), 103-e1 (2016)
22. Suksai, M., Geater, A., Phumsiripaiboon, P., Suntharasaj, T.: A new risk score model to predict preeclampsia using maternal factors and mean arterial pressure in early pregnancy. J. Obstet. Gynaecol. **42**(3), 437–442 (2022)
23. Martínez-Velasco, A., Martinez-Villaseñor, L., Miralles-Pechuán, L.: Machine learning approach for pre-eclampsia risk factors association. In: Proceedings of the 4th EAI International Conference on Smart Objects and Technologies for Social Good, pp. 232–237 (2018)
24. Jhee, J.H., Lee, S., Park, Y., Lee, S.E., Kim, Y.A., Kang, S.W., Kwon, J.Y., Park, J.T.: Prediction model development of late-onset preeclampsia using machine learning-based methods. PLoS One **14**(8), e0221202 (2019)
25. Bennett, R., Mulla, Z.D., Parikh, P., Hauspurg, A., Razzaghi, T.: An imbalance-aware deep neural network for early prediction of preeclampsia. PLoS One **17**(4), e0266042 (2022)
26. Marić, I., Tsur, A., Aghaeepour, N., Montanari, A., Stevenson, D.K., Shaw, G.M., Winn, V.D.: Early prediction of preeclampsia via machine learning. Am. J. Obstet. Gynecol. MFM **2**(2), 100100 (2020)
27. Cuenca, D., Rolle, V., de Paco Matallana, K., Valiño, N., Revello, R., Adiego, B., Mendoza, M., Santacruz, B., del Mar Gil, M.: Risk factors for preeclampsia: results from a cohort of over 5000 pregnancies in Spain. Matern.-Fetal Med. **3**(02), 100–106 (2021)
28. de Freitas Leite, J., Lobo, G.A.R., Nowak, P.M., Antunes, I.R., Júnior, E.A., da Silva Pares, D.B.: Prediction of preeclampsia in the first trimester of pregnancy using maternal characteristics, mean arterial pressure, and uterine artery Doppler data in a Brazilian population. Obstet. Gynecol. Sci. **62**(6), 391–396 (2019)

29. Riishede, I., Rode, L., Sperling, L., Overgaard, M., Ravn, J.D., Sandager, P., Skov, H., Wagner, S.R., Nørgaard, P., Clausen, T.D., Juel Jensen, C.A., Pihl, K., Jørgensen, F.S., Munk, J.K., Zingenberg, H.J., Pedersen, N.G., Andersen, M.R., Wright, A., Wright, D., Tabor, A., Ekelund, C.K.: Pre-eclampsia screening in Denmark (PRESIDE): national validation study. Ultrasound Obstet. Gynecol. (2023)
30. Tan, M.Y., Syngelaki, A., Poon, L.C., Rolnik, D.L., O'Gorman, N., Delgado, J.L., Akolekar, R., Konstantinidou, L., Tsavdaridou, M., Galeva, S., Ajdacka, U., Molina, F.S., Persico, N., Jani, J.C., Plasencia, W., Greco, E., Papaioannou, G., Wright, A., Wright, D., Nicolaides, K.H.: Screening for pre-eclampsia by maternal factors and biomarkers at 11–13 weeks' gestation. Ultrasound Obstet. Gynecol. **52**(2), 186–195 (2018)
31. Tianqi, C., Guestrin, C.: XGBoost: a scalable tree boosting system. In: Proceedings of the 22nd ACM SIGKDD International Conference on Knowledge Discovery and Data Mining, San Francisco, CA, USA, August 13-17, pp. 785–794. ACM (2016)
32. Lou, Y., Caruana, R., Gehrke, J., Hooker, G.: Accurate intelligible models with pairwise interactions. In: Proceedings of the 19th ACM SIGKDD, pp. 623–631 (2013)

Computational Approach for Verification of Aortic Wall Tear Size on CT Contrast Distribution in Patients with Type B Aortic Dissection—The Preliminary Study

Andrzej Polanczyk, Aleksandra Piechota-Polanczyk, Ludomir Stefańczyk, Julia Balcer, and Michal Strzelecki

Abstract The purpose of this research was to prepare a preliminary mathematical approach for the identification and analysis of the gap distribution using changes in brightness on CT and spatial analysis of the aorta. We used AngioCT data from three male patients, 45 to 65 years of age, with acute type IIIb aortic dissection that originated near the left subclavian artery and involved the renal arteries. The analysis was performed on CT slides from the descending aorta to the diaphragm. Each time Feret D_F diameter, width of gap in dissection and average brightness B_{AV} were analyzed. Finally, to describe the influence of the gap distribution in the area of aortic wall dissection, a mathematical function was calculated that combined the difference in brightness value and the gap width for each CT scan. The results showed that there is a positive correlation between the width of the gap (tear) within the dissected aorta and the brightness coefficient. The gathered results also stress that the most accurate data are when the gaps in the dissected wall are equally distributed throughout the dissection and are not concentrated in a few areas within the dissection. In summary, our data show that there may be a correlation between the width of the gap and the brightness coefficient calculated from CT scans that may also be used for the prognosis of hemodynamic changes within the dissection. In addition, we also highlight the need for the appropriate number of patients incorporated for

A. Polanczyk (✉)
The Main School of Fire Service, Warsaw, Poland
e-mail: andrzej.polanczyk@gmail.com

A. Piechota-Polanczyk
Department of Medical Biotechnology, Jagiellonian University, Krakow, Poland

L. Stefańczyk
Department of Radiology and Diagnostic Imaging, Medical University of Lodz, Lodz, Poland

J. Balcer · M. Strzelecki
Institute of Electronics, Lodz University of Technology, Lodz, Poland

P. Strumiłło et al. (eds.), *The Latest Developments and Challenges in Biomedical Engineering*, Lecture Notes in Networks and Systems 746,
https://doi.org/10.1007/978-3-031-38430-1_22

computational analysis of spatial configuration of gaps in the dissected aorta, as too few cases can lead to misleading results.

Keywords Gap analysis · Aortic dissection · Brightness analysis · Gap distribution

1 Introduction

The dissection of the aorta is a pathology that causes a disruption of the intimal layer of the aortic wall [3]. Aortic dissection with tear appearance in the inner lining of the aorta is one of the most serious cardiovascular diseases associated with a considerable death rate. Depending on the spatial distribution of the dissection, type A is present in the ascending aorta, which is usually treated with surgical interventions, and type B is present in the descending aorta and it is characterized by progressive dilatation of the lumen, and aortic rupture [18]. Studies speculate that blood flow intensity is plays a fundamental role in the development and progression of dissection of the aortic wall [17].

Currently, different numerical models have been proposed to understand the work of the cardiovascular system and the development of cardiovascular diseases [8, 11]. Among the tools of artificial intelligence describing the mechanical properties of veins and artificial vessels, the technique of computational fluid dynamics (CFD) can be mentioned. [7, 13, 16]. Numerical tools provide information, crucial for the assessment of blood flow distribution for patients affected by vascular diseases, e.g., aortic dissection [2]; however, they require the import of medical images [9]. Different medical imaging techniques, for example, computational tomography (CT) or magnetic resonance imaging (MRI), allow one to visualize the spatial configuration of cardiovascular system dysfunction, for example, presentation of aortic dissection [17]. The images obtained with these methods should be analyzed using appropriate algorithms to obtain a model of the vessel network [15]. The quantitative assessment of such algorithm's accuracy can be verified using tools that generate synthetic images, where true geometry of vessels is known (like a framework for computer modeling of MR angiography [6]). However, none of the medical imaging techniques permit one to predict blood hemodynamics, for example, after surgical procedure [10]. Therefore, the processing of medical imaging data may reflect patient-specific in vivo analysis conditions, e.g. a reliable spatial configuration of the human cardiovascular system [1]. Thus, it seems reasonable to use computational methods for the diagnosis and/or treatment of patients with aortic dissection, as mathematical approaches may mimic similar hemodynamic parameters, as in patients, within 3D models and carry important information on hemodynamic changes within the true and false lumen [7]. The aim of the study was to use the quantitative analysis of computed tomography image data to analyze how the distribution of fissures in the wall of the dissected aorta affects the hemodynamic properties of blood in the dissection area. The paper has four sections: Section II describes medical data, the numerical model, and its verification; Section III—the results and the discussion; Section IV—conclusions.

2 Materials and Methods

The data was from three male CTA patients aged 45 to 65 (GE Light-Speed 64 VCT; GE Healthcare, Fairfield, CT, USA) who were treated at Barlicki Hospital No. 2 in Łódź, Poland in 2016. All participants provided written informed consent to participate in this study. The study protocol was approved by the local ethics committee of the Lodz Medical University (approval no.: RNN/126/07/KE). Inclusion criteria for the study included an acute type IIIb aortic dissection that started proximally to the left subclavian artery and involved the renal arteries. Each patent had an entry tear at the proximal of the left subclavian artery (LSA) (zone number 4 according to Fillinger et al. [4]), and the end of the dissection at the right iliac artery (zone number 9 according to Fillinger et al. [4]).

Figure 1 presents an example of DICOMs for Patient 1 for true and false ducts separately as well as for the gap connecting true and false ducts.

Digital Imaging and Communications in Medicine (DICOM) data (512 × 512 × 270 voxels, in-plane resolution of 0.78 × 0.78 mm, slice thickness 0.625 mm) for recognition of wall dissection areas was used [12, 14]. Feret D_F (Eq. 1) was calculated to describe the spatial configuration of the aorta in each CT scan. [5]. Moreover, for each CT slice, the gap width was calculated.

$$D_F = (D_v + D_h)/2 \quad (1)$$

where:

D_F—Feret diameter, [mm]

D_v—the highest distance between two points on the perimeter of the analyzed aorta, calculated in the vertical direction, [mm]

D_h—the highest distance between two points on the perimeter of the analyzed aorta, calculated in the horizontal direction, [mm],

The mean value of BAV brightness (Eq. 2) of the area that represented the blood inside the aorta was calculated using the Osirix software (Pixmeo SARL, Bernex, Switzerland). Also, the distance from the descending aorta to the diaphragm was analyzed.

$$B_{AV} = \sum B_i / \sum APixel \quad (2)$$

where:

B_{AV}—average brightness value, [pixel]

B_i—brightness value for i Pixel, [pixel]

$\sum APixel$—number of Pixels in the analyzed area, [−]

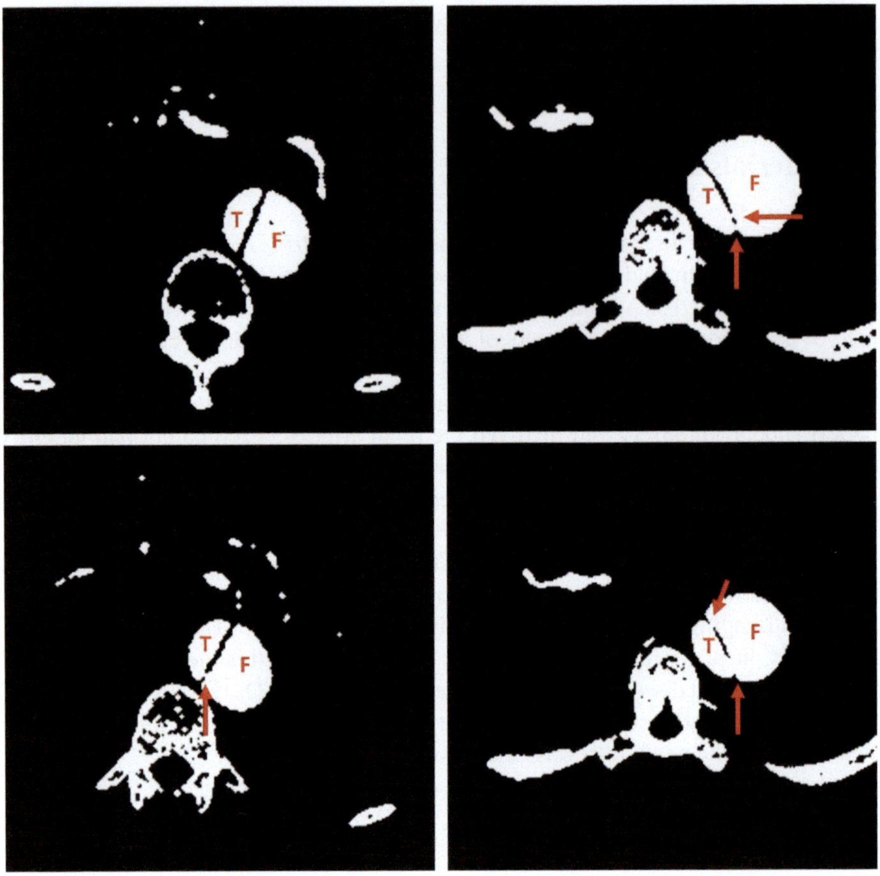

Fig. 1 An example of DICOMs presenting true (T) and false (F) ducts separately, as well as gap-connected true and false ducts (arrows)

The GraphPad Prism 8 was used to calculate statistical analysis. Data were presented as relative values or mean ± SD. Moreover, the Spearman correlation was used to check the compatibility between the Feret diameter of the common duct and Gap width. Linear regression was calculated.

3 Results and Discussion

To describe the spatial configuration of the analyzed domains, Feret diameter and Gap width together with brightness were considered. The distribution of the width of the gaps along the CT scans is presented in Fig. 2. In Patient 1 gaps were distributed along the entire length of the dissected wall (Fig. 2a). Furthermore, there were few areas where there was a lack of communication between false and true channels. For Patient 2 (Fig. 2b), fewer gaps were observed between the true and false channels. While Patient 3 (Fig. 2c) was characterized by the smallest number of connections between the false and true channels. This indicates that the gaps within the true and false channels in aneurysmal patients vary greatly between individuals, and further analysis must be performed separately for each patient.

Figure 3 presents the correlation between the patient-specific gap distribution and the brightness coefficient in the common duct of the aortic aneurysm. It may be noticed that the distribution of the brightness coefficient data in the common duct of aneurysmal patients varies significantly between patients (Fig. 3). Therefore, the brightness coefficient as a function of the gap width should not be present as average values.

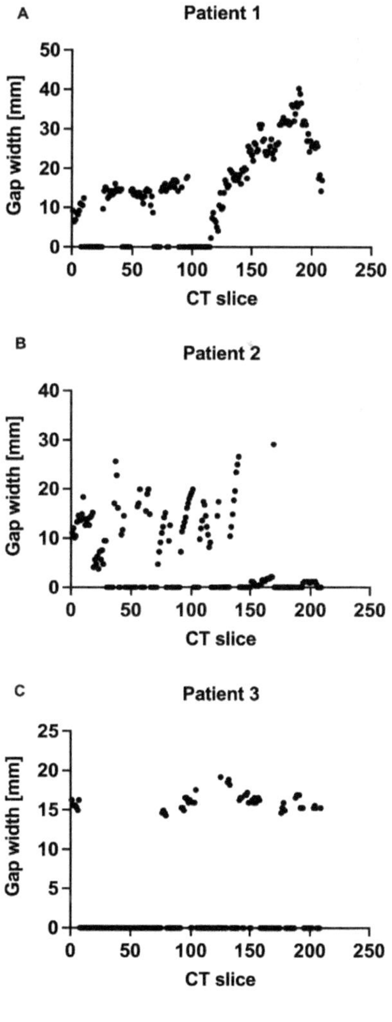

Fig. 2 Gap width distribution for patients analyzed. **a** Patient 1, **b** Patient 2, **c** Patient 3

Fig. 3 The patient-specific gap distribution strongly influences the brightness coefficient in the common duct of the aortic aneurysm. **a** Distribution of the brightness coefficient data in each of the patients

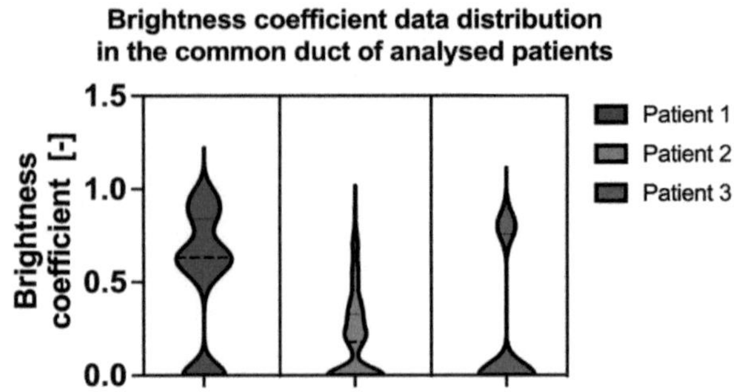

Furthermore, we observed a strong positive linear regression for Patient 1 (Fig. 4a). It was observed that when the Gap width increased, the brightness coefficient values were increasing. The highest value of the gap width was equal to 40.27 mm, while the average Feret diameter for the entire analyzed domain was equal to 45.76 mm. According to the data presented, areas appeared where both channels (true and false) were fully communicated.

For Patient 2, a weak positive linear regression was observed (Fig. 4b). It was observed that for not all points presented, increasing the value of Gap width was related to an increase in brightness coefficient. The area of brightness coefficient (0.1–0.5) where increase of the gap width was associated with a decrease in the value of the brightness coefficient. The highest value of the gap width was equal to 29.11 mm, while the average Feret diameter for the entire domain was equal to 31.64 mm.

Similarly, as in Patient 1, a strong positive linear regression was observed for Patient 3 (Fig. 4c). It was observed that together with Gap width increase, the brightness coefficient value was increasing. The highest value of the gap width was equal to 19.17 mm, while the average Feret diameter for the entire domain was equal to 32.51 mm. According to the data presented, places appeared where both channels (true and false) were fully communicated (Fig. 5).

Furthermore, Spearman's rank correlation coefficients were determined between Feret diameter and Gap width (Fig. 4a–c). There was a positive correlation between the average Feret diameter of the common duct and the width of the gap for Patient 1 (rho = 0.4152) Patient 2 (rho = 0.8231) and Patient 3 (rho = 0.9817).

3.1 Limitation to the Study

Even though in our study we propose a new method of describing the hemodynamics of blood at the aortic wall where fractures are present within the aortic wall, we see some limitations of our conclusions. First, we performed our calculations only for three patients with type IIIb acute aortic dissection. Second, the accuracy of the simulations is influenced by the resolution of the analyzed medical data; therefore,

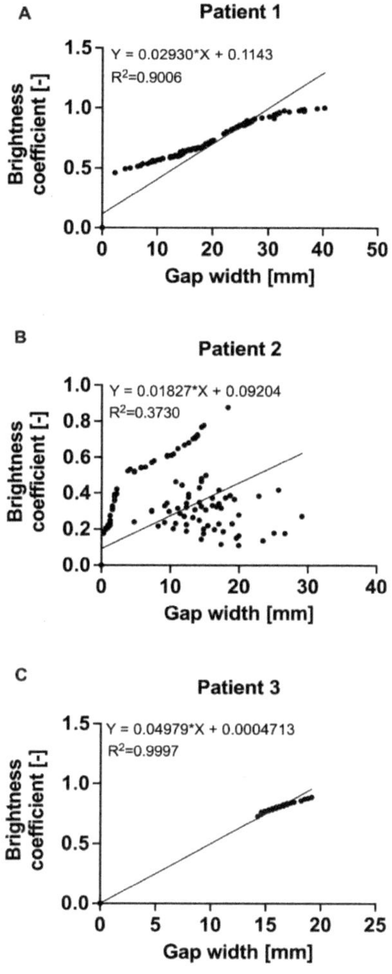

Fig. 4 Results of the linear regression for the brightness coefficient in the common duct and the width of the vascular gap for each patient **a** Patient 1; **b** patient 2; **c** patient 3

in the next steps, the proposed solution should be confirmed in a larger number of cases and the different resolution of medical images. In addition, our studies are limited to patients with one type of dissection, which may not be the case in patients with another type of aortic wall injury.

4 Conclusions

Taken together, these results indicate that the data gathered for Patient 1 and Patient 2 are more accurate for the analysis of brightness changes and gap width and for further conclusions. The data for Patient 3 should be re-analyzed before being used

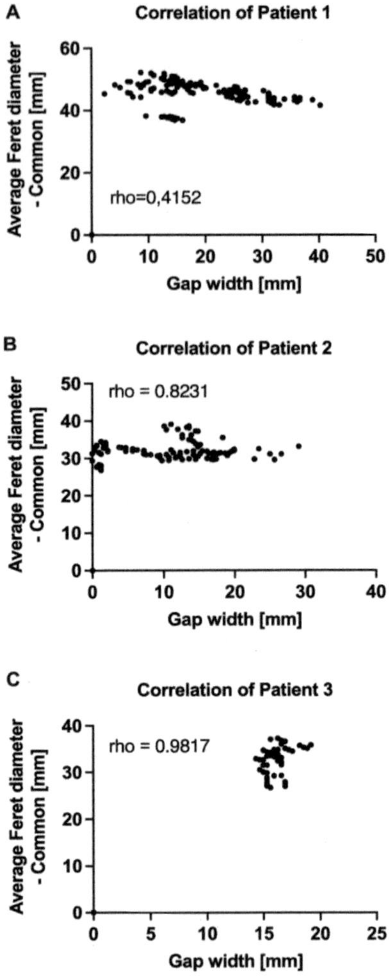

Fig. 5 Scatterplot graphic representation of Spearman correlation factors rho (–) for patients for the mean Feret diameter as a function of the width of the gap: **a**—Patient 1 (rho = 0.4152), **b**—Patient 2 (rho = 0.8231), **c**—Patient 3 (rho = 0.9817)

for the analysis of correlations between brightness coefficient and gap width and further prognosis of its usefulness for radiologists and cardiologists. This study also highlights the need for a sufficient number of patients for computational analysis of the spatial configuration of gaps (tears) in the dissected wall, as too few cases can lead to misleading results.

References

1. Ben Ahmed, S., Dillon-Murphy, D., Figueroa, C.A.: Computational study of anatomical risk factors in idealized models of type B aortic dissection. Eur. J. Vasc. Endovasc. Surg. **52**, 736–745 (2016). https://doi.org/10.1016/j.ejvs.2016.07.025

2. Bonfanti, M., Balabani, S., Alimohammadi, M., Agu, O., Homer-Vanniasinkam, S., Díaz-Zuccarini, V.: A simplified method to account for wall motion in patient-specific blood flow simulations of aortic dissection: Comparison with fluid-structure interaction. Med. Eng. Phys. **58**, 72–79 (2018). https://doi.org/10.1016/j.medengphy.2018.04.014
3. Fatma, K., Carine, G.-C., Marine, G., Philippe, P., Valérie, D.: Numerical modeling of residual type B aortic dissection: Longitudinal analysis of favorable and unfavorable evolution. Med. Biol. Eng. Comput. **60**, 769–783 (2022). https://doi.org/10.1007/s11517-021-02480-1
4. Fillinger, M.F., Greenberg, R.K., McKinsey, J.F., Chaikof, E.L.: Reporting standards for thoracic endovascular aortic repair (TEVAR). J. Vasc. Surg. **52**, 1022-1033.e5 (2010). https://doi.org/10.1016/j.jvs.2010.07.008
5. Harris, C., Alcock, A., Trefan, L., Nuttall, D., Evans, S.T., Maguire, S., Kemp, A.M.: Optimising the measurement of bruises in children across conventional and cross polarized images using segmentation analysis techniques in Image J, Photoshop and circle diameter measurements. J. Forensic Leg. Med. **54**, 114–120 (2018). https://doi.org/10.1016/j.jflm.2017.12.020
6. Klepaczko, A., Szczypiński, P., Dwojakowski, G., Strzelecki, M., Materka, A.: Computer simulation of magnetic resonance angiography imaging: Model description and validation. PLoS ONE **9**(4), e93689 (2014). https://doi.org/10.1371/journal.pone.0093689
7. Li, D., Zheng, T., Liu, Z., Li, Y., Yuan, D., Fan, Y.: Influence of distal re-entry tears on false lumen thrombosis after thoracic endovascular aortic repair in type B aortic dissection patients: A computational fluid dynamics simulation. Cardiovasc. Eng. Technol. **12**, 426–437 (2021). https://doi.org/10.1007/s13239-021-00532-z
8. Li, X., Qiao, H., Shi, Y., Xue, J., Bai, T., Liu, Y., Sun, L.: Role of proximal and distal tear size ratio in hemodynamic change of acute type A aortic dissection. J. Thorac. Dis. **12**, 3200–3210 (2020). https://doi.org/10.21037/jtd-20-1920
9. Menichini, C., Cheng, Z., Gibbs, R.G.J., Xu, X.Y.: A computational model for false lumen thrombosis in type B aortic dissection following thoracic endovascular repair. J. Biomech. **66**, 36–43 (2018). https://doi.org/10.1016/j.jbiomech.2017.10.029
10. Osswald, A., Karmonik, C., Anderson, J.R., Rengier, F., Karck, M., Engelke, J., Kallenbach, K., Kotelis, D., Partovi, S., Böckler, D., Ruhparwar, A.: Elevated wall shear stress in aortic type B dissection may relate to retrograde aortic type A dissection: A computational fluid dynamics pilot study. Eur. J. Vasc. Endovasc. Surg. **54**, 324–330 (2017). https://doi.org/10.1016/j.ejvs.2017.06.012
11. Polanczyk, A., Piechota-Polanczyk, A., Huk, I., Neumayer, C., Balcer, J., Strzelecki, M.: Computational fluid dynamic technique for assessment of how changing character of blood flow and different value of HCT influence blood hemodynamic in dissected aorta. Diagnostics **11**, 1866 (2021). https://doi.org/10.3390/diagnostics11101866
12. Polanczyk, A., Piechota-Polańczyk, A., Stefanczyk, L., Strzelecki, M.: Shape and enhancement analysis as a useful tool for the presentation of blood hemodynamic properties in the area of aortic dissection. J. Clin. Med. **9**, 1330 (2020). https://doi.org/10.3390/jcm9051330
13. Polanczyk, A., Podgorski, M., Polanczyk, M., Piechota-Polanczyk, A., Neumayer, C., Stefanczyk, L.: A novel patient-specific human cardiovascular system phantom (HCSP) for reconstructions of pulsatile blood hemodynamic inside abdominal aortic aneurysm. IEEE Access **6**, 61896–61903 (2018). https://doi.org/10.1109/ACCESS.2018.2876377
14. Polanczyk, A., Podgorski, M., Wozniak, T., Stefanczyk, L., Strzelecki, M.: Computational fluid dynamics as an engineering tool for the reconstruction of hemodynamics after carotid artery stenosis operation: A case study. Medicina (Mex.) **54**, 42 (2018). https://doi.org/10.3390/medicina54030042
15. Polanczyk, A., Wozniak, T., Strzelecki, M., Szubert, W., Stefanczyk, L.: Evaluating an algorithm for 3D reconstruction of blood vessels for further simulations of hemodnamic in human artery branches. Signal Processing—Algorithms, Architectures, Arrangements, and Applications Conference Proceedings, SPA, 2016, pp. 103–107, 7763595 (2016)
16. Takeda, R., Sato, F., Yokoyama, H., Sasaki, K., Oshima, N., Kuroda, A., Takashima, H., Li, C., Honda, S., Kamiya, H.: Investigations into the potential of using open source CFD to analyze the differences in hemodynamic parameters for aortic dissections (healthy versus stanford type A and B). Ann. Vasc. Surg. **79**, 310–323 (2022). https://doi.org/10.1016/j.avsg.2021.08.007

17. Yu, S.C.H., Liu, W., Wong, R.H.L., Underwood, M., Wang, D.: The potential of computational fluid dynamics simulation on serial monitoring of hemodynamic change in type B aortic dissection. Cardiovasc. Intervent. Radiol. **39**, 1090–1098 (2016). https://doi.org/10.1007/s00270-016-1352-z
18. Zhang, X., Chen, D., Wu, M., Dong, H., Wan, Z., Jia, H., Liang, S., Shao, J., Zheng, J., Xu, S., Xiong, J., Guo, W.: Functional evaluation of embedded modular single-branched stent graft: Application to type B aortic dissection with aberrant right subclavian artery. Front. Cardiovasc. Med. **9**, 869505 (2022). https://doi.org/10.3389/fcvm.2022.869505

Signal Processing

Using Frequency Correction of Stethoscope Recordings to Improve Classification of Respiratory Sounds

Adam Biniakowski, Krzysztof Szarzyński, and Tomasz Grzywalski

Abstract Recent advancements in artificial intelligence have brought some spectacular innovations, including systems for automatic analysis of respiratory sounds. These solutions, however, usually have a very limited set of supported signal sources and it raises question if these can be adapted to work with signals that they have not been trained on. This work explores the possibility to automatically discover the characteristics of unknown source of body signals and convert them to match that of a known source. Next, it answers a question of how well the AI can perform given signals processed in this way. Our proposed method is used to adapt the ICBHI 2017 Respiratory Sounds Challenge recordings to match the expected input signal characteristic of commercially available AI solution for lung sounds analysis. The method achieves an ICBHI score of 60.99% which is an improvement of 2.7% points over the current challenge leader. The presented results show that existing AI solutions can solve congenial tasks even when fed with signals from unknown sources and even achieve superior performance to dedicated solutions. In this case the advantage of using larger and more precisely described training data during AI development outweighs potential imperfections of the audio data transformation leading to higher accuracy. In fact, the true performance of the proposed solution might be even higher because some ICBHI recordings were found to be mislabeled, which could also explain limited success of models developed based on this data.

Keywords Stethoscope · Pathological breath sounds · Audio characteristics conversion · Filterbanks

---

A. Biniakowski · K. Szarzyński (✉) · T. Grzywalski
StethoMe sp. z o.o., ul. Winogrady 18a, 61-663 Poznań, Poland
e-mail: szarzynski@stethome.com
URL: https://www.stethome.com/

P. Strumiłło et al. (eds.), *The Latest Developments and Challenges in Biomedical Engineering*, Lecture Notes in Networks and Systems 746,
https://doi.org/10.1007/978-3-031-38430-1_23

1 Introduction

1.1 Respiratory Diseases, Auscultation, Adventitious Breath Sounds

Respiratory diseases are one of the leading causes of death and disability in the world. Nearly 200 million people, or 3% of the world's population, have COPD and 3.2 million die of it each year [20], making it the third-leading cause of death worldwide. Asthma affects more than 262 million people worldwide [22] and is the most common chronic disease of childhood worldwide. Pneumonia kills more than 2.4 million people annually [18] and is a leading cause of death among children younger than 5 years outside the neonatal period and adults older than 65 years. More than 6 million people develop tuberculosis and 1.4 million die of it each year, making it the most common lethal infectious disease next to the COVID-19 pandemic [3].

The most widespread and cost-effective medical examination that allows for diagnosis and monitoring of respiratory diseases is auscultation. Lung auscultation is usually performed with a stethoscope which helps to isolate and amplify sounds generated in the respiratory tract [12]. The interpretation of lung sounds is considered one of the key skills of any general practitioner, pediatrician, pulmonologist or allergologist. This requires a lot of training and years of practice, and despite its commonness and unquestionable importance in healthcare, is not adequately grasped by all practitioners [1, 9, 19].

The most important aspect of respiratory system auscultation is the evaluation of adventitious respiratory system sounds. According to the generally accepted standards [17, 26], additional abnormal breath sounds can be divided into two main categories: continuous and transient. The continuous breath sounds can be further subdivided into wheezes and rhonchi and transient sounds (also called crackles) are often differentiated into fine and coarse.

1.2 Recording of Respiratory Sounds

Electronic devices used for recording auscultation sounds have different characteristics that depend on their design. Most important element of their construction is the transducer that converts vibration of air or skin surface into electric signal. Transducers like piezoelectric can be placed directly on the skin, but others, like microphones, require to be coupled with air cavity and membrane which makes their construction similar to traditional stethoscope. Signal from electric form is converted into digital samples, then it's preprocessed, transmitted and stored as an audio file. Each construction element changes properties of the final signal such as signal-to-noise ratio, frequency response, dynamics and distortions. Respiratory sound level differences between devices may exceed 20 dB [28].

However, assuming each device provides recordings without clipping and excessive noise, not compressed and stored with sufficient sampling frequency and bit depth, their signals can be corrected with appropriate filtering to get close to any other device's characteristics.

1.3 Automated Detection of Adventitious Respiratory Sounds in Recordings

Historically many attempts have been made to automatically detect adventitious respiratory sounds. Early approaches tackled the problem using simple time-frequency analysis [24]. As an example of a more advanced approach, Lin and Lin [15] used frequency cepstral coefficients (MFCCs) in combination with a gaussian mixture model (GMM) to detect the presence of wheezes. With the advent of deep convolutional neural networks (CNN), researchers' attention shifted towards utilization of this successful model. Examples of such attempts include [2, 14, 16]. However, the progress in the field was limited by the lack of a large-scale database of lung sounds which was required for the deep models to work to their full potential.

To address this issue, in 2017 a large-scale dataset of stethoscope recordings [25] was released to the public as part of the 3rd International Conference in Biomedical and Health Informatics (ICBHI). Aim of the challenge was to develop a solution for automatic detection of breathing cycles with wheezes and/or crackles. The dataset consists of a total of 5.5 h of recordings containing 6898 respiratory cycles, of which 1864 contain crackles, 886 contain wheezes, and 506 contain both crackles and wheezes. In total, the dataset consists of 920 audio samples from 126 subjects annotated by respiratory experts. The recordings were collected using heterogeneous equipment and their duration ranges from 10s to 90s. According to the challenge rules, 539 (59%) of audio samples can be used for the development of the classifier (training subset) while the remaining 381 (41%) are restricted for evaluation (test subset).

The availability of the challenge data and the existence of a platform for creative competitiveness sparked a new interest of the research community in the topic of breath sounds classification which resulted in many algorithmical advancements in the field. A publicly available web service [4] holds a record of the challenge participants and their official challenge scores.

As of 7th of March, 2023 leading the poll is the submission made by Nguyed and Pernkopf who published their results in [23]. Their proposed solution was trained on two datasets: ICBHI challenge dataset (training data) and an additional dataset developed by the authors that included a total of 6196 respiratory cycles. Their solution is based on a ResNet50 CNN model [11], pre-trained on a large-scale dataset of natural images (ImageNet) [5]. This model was fine-tuned for the target task using co-tuning for transfer learning and stochastic normalization. The method achieved a challenge score of 58.29% (an average of 37.24% sensitivity and 79.34% specificity).

In contrast the current second place holders, Moummad and Farrugia [21] used only the ICBHI challenge data for training. Their baseline model, a relatively small CNN (4 convolutional blocks, 4.3 million trainable parameters), was pre-trained on AudioSet [7]. In this case high accuracy was achieved by using supervised contrastive learning which is said to work well in case of unbalanced and noisy data. This method achieved an ICBHI score of 57.55% based on 39.15% sensitivity and 75.95% specificity. Other notable participants of the challenge include Xu et al. [29] who achieved a score of 56.76% using a combination of residual block with channel-spatial attention, Gairola et al. [6], who achieved a score of 56.20% using device specific fine-tuning, concatenation-based augmentation, blank region clipping, and smart padding, and Wand et al. [27] who achieved a score of 55.30% using a domain transfer-based data augmentation method.

1.4 Commercial Solutions for Adventitious Breath Sounds Detection

Amongst the most notable examples of commercially available solutions for breath sounds detection is the Polish company *StethoMe* which in 2019 released to the market two CE-certified medical devices: a home-use stethoscope (*StethoMe Smart Stethoscope*) and *StethoMe AI*, a cloud-based service for automatic analysis of stethoscope recordings. The *StethoMe AI* is capable of detecting four basic types of adventitious breath sounds (wheezes, rhonchi, fine and coarse crackles) including their precise location within each breathing cycle. In its current version, *StethoMe AI* uses a deep convolutional-recurrent neural network to detect physiological and pathological breath sounds. The model processes a time-frequency representation of the input signal through a series of eight 2D convolutional layers and three bidirectional recurrent layers and provides sound labels predictions every 10ms. In total the model consists of 7.4 million trainable parameters. The solution has been verified in a clinical setting during both internal [8] and independent external [13] investigations and was recently shown to have practical application for controlling asthma [10].

Despite an overall similar indication for use, the *StethoMe AI* shows many differences in comparison to the solutions developed as part of the ICBHI challenge. Most importantly, the *StethoMe AI* was trained on a different dataset of stethoscope recordings that doesn't contain any recordings from the ICBHI dataset. The *StethoMe AI* training data included over 10.000 stethoscope recordings (compared to 539 recordings available in the IBCHI dataset) registered by only two devices—*Littmann 3200* (approx. 5.000 recordings) and the company's own *StethoMe Smart Stethoscope* (approx. 5.000 recordings)—resulting in a much smaller heterogeneity. Moreover the *StethoMe AI* training data was described using much more fine-grained labels and the description process included an additional verification step performed by a trained acoustician [10].

2 Aim

In this work we present a method for converting audio signals registered by one body sound registering device to match the characteristics of a different device. The purpose of this conversion is to be able to use a commercially available AI-based medical device to detect adventitious respiratory sounds in recording registered with unknown devices (i.e. devices which were not used during training of the AI-based system).

3 Materials and Methods

3.1 Approach

Our method was tested on the ICBHI dataset by first taking the 539 training samples to discover the acoustical characteristics of each of the four devices represented in the dataset. Next, using the same approach we calculated the characteristic of the device whose recordings were used during training of the commercially available AI solution and designed a transfer function that converts the ICBHI dataset recordings to sound as if they were recorded with the smart stethoscope device.

In the next step the 391 recordings from the ICBHI test set were converted accordingly and processed with *StethoMe AI* version 3.3.0.1 (the publicly available solution not modified in any way for the purpose of this task). Next, the AI analysis results, provided in a form of labeled time intervals, were generalized and translated into full breathing cycle labels to match the ICBHI dataset labeling format. Finally the sensitivity, specificity and challenge score were calculated and compared to the current challenge leaders to evaluate the feasibility of our proposed method.

3.2 Devices

ICBHI dataset The ICBHI dataset contains recordings from six chest locations which are roughly consistent with *StethoMe AI* training recordings' locations, with additional point located near the trachea. Tracheal sounds contain more higher frequency content and are unknown from *StethoMe AIs* perspective.

Four devices were used to create the ICBHI dataset. Two of them are electronic stethoscopes that record signals in digital form—*Meditron* and *Littmann 3200*. There are few issues that may affect our final results. *Meditron* recordings contain hum noise and discontinuities, they are also clipped in some cases, but fortunately they are only a small part of entire the dataset. Recordings from *Littmann 3200* are very unevenly split with only 5 out of 60 recordings available for training. Finally, *Littmann 3200*

Table 1 Characteristics of devices

Name	Meditron	Littmann 3200	Littmann C2SE	AKG C417L	StethoMe
Manufacturer	Welch Allyn	3M	3M	AKG	StethoMe
Full name	Meditron Master Elite Plus Stethoscope Model 5079-400	Littmann 3200	Littmann Classic II SE	C 417 PP	StethoMe Smart Stethoscope
Description	Electronic stethoscope	Electronic stethoscope with piezoelectric transducer	Traditional stethoscope modified by installing a microphone in the main tube	Lavaliere microphone with electret condenser transducer coupled with custom capsules made from Teflon	Electronic stethoscope
Bit rate [bits]	16	16	16 or 24	16 or 24	16
Sampling rate [Hz]	4000 or 10000	4000	44100	44100	12000
Recordings type	Single channel	Single channel	Single and multi channel	Multi channel	Single channel
Other remarks	Hum noise, clipping, discontinuities	Applied dynamic compression, filtering	–	High level, clipping	–

sounds are processed (filtered) and they are not directly compatible with *StethoMe AI*.

Littmann Classic II SE (*Littmann C2SE*) is a traditional stethoscope that was modified by putting a microphone in its main tube. Recordings from this device were technically the most correct without clipping and other noises. Last device with the highest number of recordings is *AKG C417L* which is a custom device made from lavalier microphones and capsules made of teflon. These recordings are often heavily distorted due to very high levels of low frequency components. Signals from the last two devices are neither dynamically compressed nor filtered [25]. Device characteristics are summarized in Table 1.

ICBHI dataset authors made mistakes in labeling some of the recordings and they provided a file with correct mapping between files and devices. This was the case for 91 recordings labeled as *Meditron*. According to the provided file 37 recordings should have been labeled as *AKG C417L* and 54 as *Littmann C2SE*. However, our preliminary analysis suggested that all 91 recordings should be labeled as *Littmann C2SE*. The aforementioned 37 recordings, unlike most *AKG C417L* recordings, did not have the zero padding at the beginning and had noticeable lower levels, much closer to *Littmann C2SE*. In all subsequent analysis we used labels that were consistent with these findings.

StethoMe AI *StethoMe AI* was trained on signals without optional processing such as lungs/heart filters and dynamic compression (with exception for *Littmann 3200* where compression couldn't be turned off). As a reference for level and spectrum corrections 1500 *StethoMe Smart Stethoscope* recordings were used. Examples were randomly chosen from the *StethoMe's* database which contains recordings acquired by *StethoMe* system users.

3.3 Signal Processing

Recordings were grouped by device and analyzed using two methods—*level normalization* and *frequency spectrum correction*. Before any analysis all recordings were resampled 12000 Hz used in *StethoMe* recordings. Only recordings from the training set of ICBHI dataset were used to compute reference level values.

Level normalization For each of the recording's signal y of length N samples, root mean square (RMS) was computed using (1) where y_i represents the i-th sample of the signal.

$$y_{rms} = \frac{1}{N}\sum_{i=1}^{N}\sqrt{y_i^2} \tag{1}$$

Signal levels presented in this article are converted to decibel scale calculated using (2).

$$L = 20 \times \log_{10}(y_{rms}) \tag{2}$$

Median RMS value for each device was used as representative for all recordings in the device's dataset. *StethoMe* recording level was computed in similar manner and used as point of reference. Each recording level correction was applied using (3).

$$L_{Correction} = L_{Stethome} - L_{Device} \tag{3}$$

Frequency spectrum correction For signal correction a bank of filters was used. Total of 25 filters (1 low-pass, 23 band-pass, 1 high-pass) were designed to cover the whole spectrum range of signals used by *StethoMe AI*. Frequency cutoffs f_{cut} for filters were computed using (4).

$$f_{cut}(n) = 2^{\frac{n}{3}} \times 1000, \quad n \in \{-19, -18, \ldots, 3\} \tag{4}$$

First and last filters (low-pass and high-pass respectively) were 8-th order Butterworth filters. Band-pass filters were a combination of low-pass and high-pass of the same type. All operations used zero-phase filtering that preserves the original phase

Table 2 Composition of the test set of the ICBHI Challenge with number of recordings and annotated cycles for each device

Device	No. recordings	Annotated cycles
All Devices	381	2756
AKG C417L	285	1836
Littmann C2SE	32	347
Littmann 3200	55	461
Meditron	9	112

and allows the signal to be reconstructed without phase issues by adding components from each filter bank. Median RMS value of each of 25 outputs for all recordings in the device's dataset was used as a reference level. Similarly to level normalization, levels were compared to *StethoMe* and differences were used for further correction.

3.4 Ground Truth

The challenge rules stated that the prediction results will be evaluated on the test set consisting of 381 recordings. Composition of the test set is described in Table 2.

For each recording the organizers provided annotations of beginning and end of respiratory cycles. Each cycle was labeled with two values: *presence of crackles* and *presence of wheezes*. The goal of the challenge was to create a system that, for given recording and location of the respiratory cycle (its beginning and end), determines whether any or both phenomena are present. This can be treated as a four class classification problem with following classes:

wheezes—cycle contains only wheezes,
crackles—cycle contains only crackles,
both—cycle contains both wheezes and crackles,
none—cycle does not contain wheezes nor crackles.

During analysis we noticed that labels were incomplete and inconsistent, even within single recordings. Examples of these, including comments from a researcher with auscultation analysis experience, can be found in Figs. 1, 2 and 3. Labeling of data is not subject of this research, but these issues might have affected the overall score significantly and might have hidden the real performance achieved by our proposed method as well as other researchers' methods. One possible reason for this situation might be lack of internal verification of labels. Another issue we found is that, in the case of two patients, recordings were split between training and testing sets which is incorrect due to the correlation of these recordings.

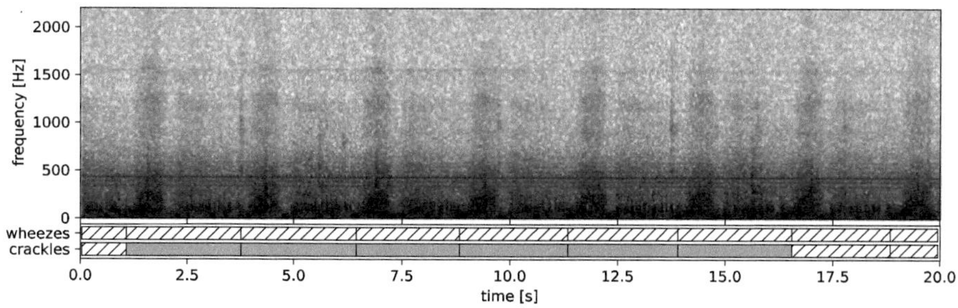

(a) anterior right

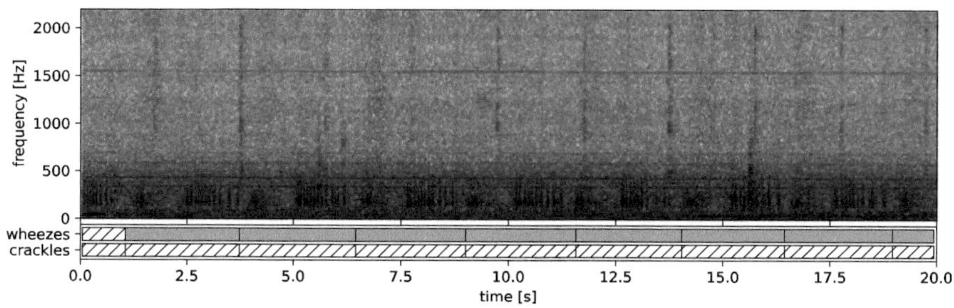

(b) lateral left

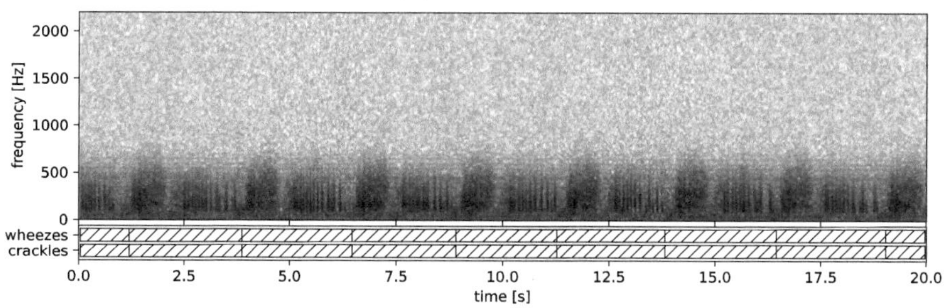

(c) posterior right

Fig. 1 Gold standard labels provided for recordings of patient 146 using AKG C417L device. All recordings are from the same multichannel set. All recordings contain crackles, especially on expirations, represented by vertical lines in range 0–500 Hz. In anterior right recording (**a**) most of the cycles are labeled as crackles, but the last one was omitted despite similarity. Recordings from lateral left (**b**) and posterior right (**c**) are labeled using different rules—as wheezes or as none

Lastly, for *Littmann 3200*, only 5 recordings were used for training and these were very different from the 55 test set recorded with this device. In the test set every recording (with one exception) contained coughs and the patient's voice. From the *StethoMe AI* standpoint cough sounds are not subject to analysis and they are usually treated as noise, thus results from this device need to be treated with caution.

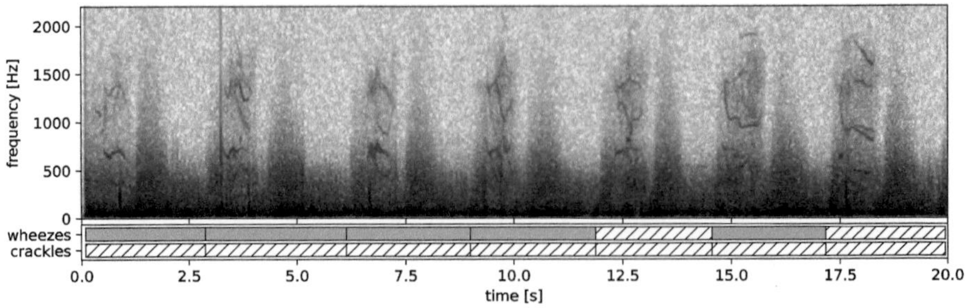

Fig. 2 Gold standard labels provided for recordings of patient 130 using AKG C417L. Wheezes are present in all cycles. They can be discriminated as horizontal lines, mainly in the range of 500–1500 Hz. Last and third from last cycles are marked differently as none (without wheezes) without any clear reason

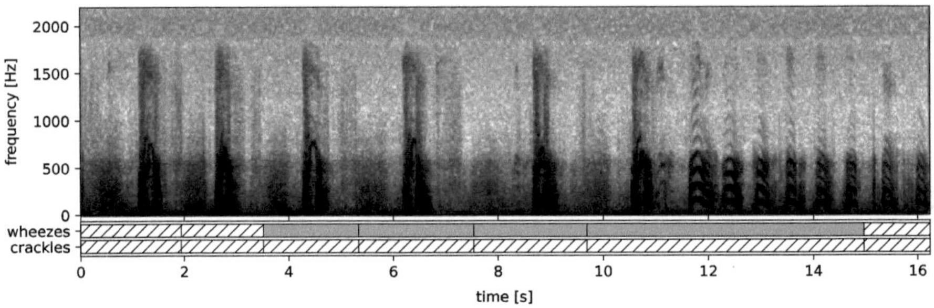

Fig. 3 Gold standard labels provided for recordings of patient 139 using Littmann 3200. First six cycles contain coughing. They are marked as none (first two cycles) or wheezes (third and later). It is unclear why two first cycles and later are labeled differently despite being very similar

3.5 Evaluation

The scoring system calculated *Se* sensitivity, *Sp* specificity, and *Sc* final score. Equations (5), (6), (7) are used to calculate the results where P_c is the number of true positives for class c, and T_c is the total number of examples of class c.

$$Se = \frac{P_{crackles} + P_{wheezes} + P_{both}}{T_{crackles} + T_{wheezes} + T_{both}} \cdot 100\% \tag{5}$$

$$Sp = \frac{P_{none}}{T_{none}} \cdot 100\% \tag{6}$$

$$Sc = \frac{Se + Sp}{2} \tag{7}$$

StethoMe AI provides more detailed labeling than one provided and evaluated by challenge organizers. First, the results are more granular and detected phenomena

Table 3 Summary of recordings and devices used in the preprocessing with number of recordings, cycles available in training set and respective median signal level, interquartile range and correction value in decibels

Device	Number of recordings	Cycles	Median level [dB]	IQR [dB]	Correction [dB]
StethoMe	1500	–	–45.9	8.0	–
AKG C417L	361	2572	–13.7	9.0	–32.2
Littmann C2SE	146	1066	–22.8	11.4	–23.0
Littmann 3200	5	41	–23.2	7.3	–22.6
Meditron	27	463	–9.6	1.7	–36.2

are localized with a 10 ms resolution (i.e. the detected pathological sound can be present in just part of the respiratory cycle). Second, the provided classes of labels are more detailed. *StethoMe AI* detects two types of continuous pathological sound types—*wheezes* and *rhonchi*, and two types of transient pathological phenomena: *fine and coarse crackles*. However, for the purpose of the challenge we used a more general labeling that the system provides—*general wheezes* and *general crackles*. For provided locations of respiratory cycles we analyzed *StethoMe AI*'s output and classified the cycle as containing *wheezes* (*crackles* respectively) if at least one 10 ms fragment of the cycle was classified as containing *general wheezes* (*general crackles*).

We tested *StethoMe AI* on the original ICBHI challenge recordings, recordings with levels normalized, and recordings with their frequency corrected to match characteristics of the *StethoMe Smart Stethoscope*. Recordings were processed using a Python script and saved as WAV files. Then they were processed by a local instance of *StethoMe AI*, the obtained results were processed and analyzed using a Python script.

4 Results

Preprocessing results Differences in levels between devices were significant. All devices used in the challenge had higher output levels than expected by *StethoMe AI*. This was convenient since normalization correction values were always negative and it did not require additional handling of potential clipping of signals. Signal levels for each device with applied correction levels are shown in Table 3.

Frequency corrections applied for *AKG C417L* and *Littmann C2SE* recordings were the most accurate. There are only minor differences between them and *StethoMe* as can be seen in Fig. 4. In the case of *Meditron* there are some differences above 1000 Hz due to the presence of noise in recordings, but recordings from *Littmann 3200* are far from the specified reference. This highlights issues with unbalanced training and testing datasets.

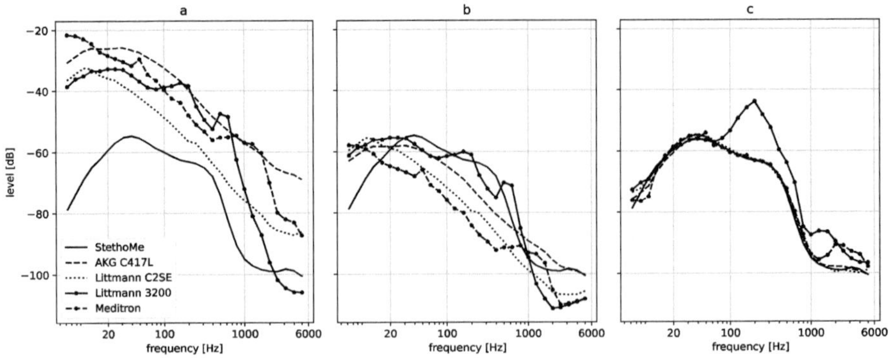

Fig. 4 Comparison of median spectra of recordings from devices after preprocessing: after resampling (**a**), after resampling and level correction (**b**), after resampling and frequency spectrum correction (**c**)

Table 4 Detailed scores for *StethoMe AI* on the ICBHI challenge evaluation dataset for all preprocessing variants, grouped by device type. All values are given in percentages

Device	No preprocessing			Level normalization			Frequency correction		
	Se	*Sp*	*Sc*	*Se*	*Sp*	*Sc*	*Se*	*Sp*	*Sc*
All devices	**8.24**	**86.26**	**47.25**	**29.65**	**85.50**	**57.57**	**45.28**	**76.69**	**60.99**
AKG C417L	1.12	97.87	49.49	30.47	92.34	61.40	45.31	80.11	62.71
Littmann C2SE	22.54	88.04	55.29	18.31	90.94	54.63	29.58	87.32	58.45
Littmann 3200	45.59	51.08	48.33	44.85	59.69	52.27	47.79	59.38	53.59
Meditron	12.16	86.84	49.50	2.70	97.37	50.04	55.41	63.16	59.28

Challenge results According to challenge rules, when using the original recordings with no preprocessing *StethoMe AI* achieved score *Sc* of 47.52%. When the recordings levels were normalized to match the *StethoMe Smart Stethoscope* the score increased to 57.57%. Even more, when the frequency spectrum was corrected the achieved *Sc* score was 60.99%. Detailed results with values for *Se*, *Sp*, and *Sc* for all preprocessing variants with distinction between devices in the evaluation dataset are presented in Table 4.

5 Conclusions

In this work we asked the following question: is it possible to detect adventitious respiratory sounds in recordings from unknown sources using an existing AI solution with very specific input signal requirements. Towards this end we have developed an audio signal conversion method and tested it on the ICBHI Respiratory Sound Database which was processed with *StethoMe AI*, a commercially available AI solution for lung and heart sounds analysis.

The obtained results clearly suggest a positive answer to the question asked. Using the proposed approach we achieved a new state-of-the-art ICBHI challenge score of 60.99%, which is an improvement of almost three percentage points over the best solution proposed thus far. It can be observed that the biggest gain in performance, as compared to the raw (unprocessed) ICBHI recordings, is due to increased sensitivity when using frequency spectrum correction.

6 Discussion

The fact that an independently-built AI system is able to detect adventitious respiratory sounds with higher accuracy than dedicated, highly sophisticated solutions clearly implies that the key to achieving good performance is the size and quality of the training data. As already pointed out, *StethoMe AI* was trained on a dataset that is almost 20 times larger than the ICBHI training set, which was also described using a much more extensive labeling protocol. Additionally the ICBHI dataset suffers from at least two major problems: (1) some recordings are mislabeled in terms of the originating device, and (2) some cycles are mislabeled in terms of presence of adventitious lung sounds, which was confirmed by an expert in the field of identification of such sounds. All these factors contribute to the fact that the ICBHI dataset in itself is insufficient for training of a highly accurate pathological breath sounds detection system. Moreover, because the same issues are found in the ICBHI test set, achieving a very high challenge score, even with a highly accurate system such as *StethoMe AI*, is not possible.

Luckily, presented results suggest that developing a dedicated system for each body sounds registering device is not necessary. Using a relatively simple audio characteristics conversion method, signals coming from external devices can become suitable for processing by an existing, certified cloud-based AI solution.

However, it should be noted that such application of *StethoMe AI* is considered out scope of the device's intended use until sufficient clinical evidence is presented regarding safety and accuracy and the input device becomes officially supported by the AI system. In order to confirm full compatibility of the existing AI system with new signal sources a more comprehensive testing procedure needs to be developed and performed, in particular one that tests the whole scope of analysis performed by the AI. Developing such procedures and tools will be one of the main focuses of the authors of this work in the near future.

References

1. Arts, L., et al.: The diagnostic accuracy of lung auscultation in adult patients with acute pulmonary pathologies: a meta-analysis. Sci. Rep. **10**(1) (2020). https://doi.org/10.1038/s41598-020-64405-6
2. Aykanat, M., Özkan Kılıç, Kurt, B., Saryal, S.: Classification of lung sounds using convolutional neural networks. EURASIP J. Image Video Process. **2017**(1) (2017). https://doi.org/10.1186/s13640-017-0213-2
3. Bagcchi, S.: WHO's global tuberculosis report 2022. Lancet Microbe **4**(1), e20 (2023). https://doi.org/10.1016/s2666-5247(22)00359-7
4. with Code, P.: Audio classification on ICBHI respiratory sound database. https://paperswithcode.com/sota/audio-classification-on-icbhi-respiratory. Accessed 07 Mar. 2023
5. Deng, J., et al.: ImageNet: a large-scale hierarchical image database. In: 2009 IEEE Conference on Computer Vision and Pattern Recognition. IEEE (2009). https://doi.org/10.1109/cvpr.2009.5206848
6. Gairola, S., et al.: RespireNet: a deep neural network for accurately detecting abnormal lung sounds in limited data setting. In: 2021 43rd Annual International Conference of the IEEE Engineering in Medicine & Biology Society (EMBC). IEEE (2021). https://doi.org/10.1109/embc46164.2021.9630091
7. Gemmeke, J.F., et al.: Audio set: an ontology and human-labeled dataset for audio events. In: 2017 IEEE International Conference on Acoustics, Speech and Signal Processing (ICASSP). IEEE (2017). https://doi.org/10.1109/icassp.2017.7952261
8. Grzywalski, T., Piecuch, M., Szajek, M., Bręborowicz, A., Hafke-Dys, H., Kociński, J., Pastusiak, A., Belluzzo, R.: Practical implementation of artificial intelligence algorithms in pulmonary auscultation examination. Eur. J. Pediatr. **178**(6), 883–890 (2019) https://doi.org/10.1007/s00431-019-03363-2
9. Hafke-Dys, H., Bręborowicz, A., Kleka, P., Kociński, J., Biniakowski, A.: The accuracy of lung auscultation in the practice of physicians and medical students. PLOS One **14**(8), e0220,606 (2019). https://doi.org/10.1371/journal.pone.0220606
10. Hafke-Dys, H., Kuźnar-Kamińska, B., Grzywalski, T., Maciaszek, A., Szarzyński, K., Kociński, J.: Artificial intelligence approach to the monitoring of respiratory sounds in asthmatic patients. Front. Physiol. **12** (2021). https://doi.org/10.3389/fphys.2021.745635
11. He, K., et al.: Deep residual learning for image recognition. In: 2016 IEEE Conference on Computer Vision and Pattern Recognition (CVPR). IEEE (2016). https://doi.org/10.1109/cvpr.2016.90
12. Henrik, P., Svend, N.: Stethoscope-over 200 years. J. Pulmonol. Respir. Res. **3**(1), 001–008 (2019) https://doi.org/10.29328/journal.jprr.1001010
13. Kevat, A., Kalirajah, A., Roseby, R.: Artificial intelligence accuracy in detecting pathological breath sounds in children using digital stethoscopes. Respir. Res. **21**(1) (2020). https://doi.org/10.1186/s12931-020-01523-9
14. Kim, Y., et al.: Respiratory sound classification for crackles, wheezes, and rhonchi in the clinical field using deep learning. Sci. Rep. **11**(1) (2021). https://doi.org/10.1038/s41598-021-96724-7
15. Lin, B.S., Lin, B.S.: Automatic wheezing detection using speech recognition technique. J. Med. Biol. Eng. **36**(4), 545–554 (2016) https://doi.org/10.1007/s40846-016-0161-9
16. Lin, B.S., Wu, H.D., Chen, S.J.: Automatic wheezing detection based on signal processing of spectrogram and back-propagation neural network. J. Healthc. Eng. **6**(4), 649–672 (2015). https://doi.org/10.1260/2040-2295.6.4.649
17. Loudon, R., Murphy, R.L., Jr.: Lung sounds. Am. Rev. Respir. Dis. **130**(4), 663–673 (1984)
18. Lozano, R., et al.: Global and regional mortality from 235 causes of death for 20 age groups in 1990 and 2010: a systematic analysis for the global burden of disease study 2010. Lancet **380**(9859), 2095–2128 (2012). https://doi.org/10.1016/s0140-6736(12)61728-0

19. Mangione, S., Nieman, L.: Pulmonary auscultatory skills during training in internal medicine and family practice. Am. J. Respir. Crit. Care Med. **159**(4), 1119–1124 (1999). https://doi.org/10.1164/ajrccm.159.4.9806083
20. Meghji, J., et al.: Improving lung health in low-income and middle-income countries: from challenges to solutions. Lancet **397**(10277), 928–940 (2021). https://doi.org/10.1016/s0140-6736(21)00458-x
21. Moummad, I., Farrugia, N.: Supervised contrastive learning for respiratory sound classification (2022). https://doi.org/10.48550/ARXIV.2210.16192
22. Network, G.A.: Global asthma network: the global asthma report 2022. http://globalasthmareport.org/foreword/summaries.php. Accessed 07 Mar. 2023
23. Nguyen, T., Pernkopf, F.: Lung sound classification using co-tuning and stochastic normalization. IEEE Trans. Biomed. Eng. **69**(9), 2872–2882 (2022). https://doi.org/10.1109/tbme.2022.3156293
24. Palaniappan, R., et al.: Computer-based respiratory sound analysis: a systematic review. IETE Tech. Rev. **30**(3), 248 (2013). https://doi.org/10.4103/0256-4602.113524
25. Rocha, B.M., et al.: An open access database for the evaluation of respiratory sound classification algorithms. Physiol. Meas. **40**(3), 035,001 (2019). https://doi.org/10.1088/1361-6579/ab03ea
26. Sarkar, M., Madabhavi, I., Niranjan, N., Dogra, M.: Auscultation of the respiratory system. Ann. Thorac. Med. **10**(3), 158 (2015). https://doi.org/10.4103/1817-1737.160831
27. Wang, Z., Wang, Z.: A domain transfer based data augmentation method for automated respiratory classification. In: ICASSP 2022-2022 IEEE International Conference on Acoustics, Speech and Signal Processing (ICASSP). IEEE (2022). https://doi.org/10.1109/icassp43922.2022.9746941
28. Weiss, D., et al.: An in vitro acoustic analysis and comparison of popular stethoscopes. Med. Devices Evid. Res. **12**, 41–52 (2019). https://doi.org/10.2147/mder.s186076
29. Xu, L., et al.: ARSC-net: adventitious respiratory sound classification network using parallel paths with channel-spatial attention. In: 2021 IEEE International Conference on Bioinformatics and Biomedicine (BIBM). IEEE (2021). https://doi.org/10.1109/bibm52615.2021.9669787

Bioimpedance Spectroscopy—Niche Applications in Medicine: Systematic Review

Ilona Karpiel, Mirella Urzeniczok, and Ewelina Sobotnicka

Abstract Bioelectrical impedance analysis (BIA) consists in measuring the total resultant electrical resistance of the body, which is a derivative of the resistance and reactance, by using a set of surface electrodes, connected to a computer analyzer and using a current of a given frequency and intensity. Bioimpedance spectroscopy (BIS) is an extension of BIA method that involves measuring the impedance of the test object over a range of frequencies. The aim of this systematic review was to evaluate prevalence of bioelectrical spectroscopy in medicine, especially (a) for measuring blood glucose level (b) classification of pulmonary nodules (c) detection and estimation of lymphedema and breast cancer-related lymphedema (BCRL). We focus on finding a new area of activity, where we can use BIS and BIA, which is a critical aspect in the context of rapidly developing medicines. A systematic research on electronic databases (PubMed, Cochrane) from inception to January 2018 was performed according using PICO strategy. The quality of the publication (clinical trials) was evaluated by using Jadad checklist. Through the initial literature search and after removing duplicates and excluding papers by screening titles and abstracts, 140 potentially relevant studies were examined. Fourteen studies met the inclusion criteria. The review confirmed that BIA and BIS are used as a tool to assist in the diagnosis of patients to a very limited extent. In addition, it is worth mentioning the lack of clinical reviews based on the application of machine learning or deep learning to narrow pathologies.

Keywords Bioimpedance spectroscopy · BIS · Bioelectrical impedance · BIA · Systematic review

---

I. Karpiel (✉) · M. Urzeniczok · E. Sobotnicka
Krakow Institute of Technology, The Centre for Biomedical Engineering, Zakopianska Str. 73, 30-418 Krakow, Poland
e-mail: Ilona.karpiel@kit.lukasiewicz.gov.pl

M. Urzeniczok
Faculty of Biomedical Engineering, Department of Biosensors and Biomedical Signals Processing, Silesian University of Technology, Zabrze, Poland

P. Strumiłło et al. (eds.), *The Latest Developments and Challenges in Biomedical Engineering*, Lecture Notes in Networks and Systems 746,
https://doi.org/10.1007/978-3-031-38430-1_24

Abbreviations

BIA	Bioelectrical impedance analysis
TEB	Thoracic electrical Bioimpedance
PhA	Bioelectrical impedance analysis-derived phase angle
BCRL	Breast cancer-related lymphedema
T2DM	Type 2 diabetes mellitus

1 Introduction

Bioelectrical impedance analysis (BIA) is a reliable, non-invasive, safe and effective way to study body composition in healthy people and those suffering from diabetes, hypertension, obesity and other diseases. It can be used by people of both sexes, at any age, practically regardless of the state of health. The results are easy to obtain, highly reproducible and the equipment needed to perform it is portable and relatively inexpensive.

BIA of a particular part of an object is estimated by measuring the voltage signal developed across that body part by injecting a constant current signal to the object [1]. It is being used for studying the frequency response of the electrical impedance of biological tissues, which depends on the tissue composition and frequency of the applied alternating current signal. It varies changes due to the variation in tissue structure, composition, and health status. Therefore, the complex electrical impedance analysis of biological tissues is found to be an efficient tool for noninvasive investigations of physiological or pathological status [1]. The bioimpedance measurements could be made by either the two-electrode or four-electrode method. In the two-electrode method the same electrodes are used for the injecting the constant current signal and for the voltage measurements of the returned signal. In the four-electrode method a low and constant amplitude ($\leq$1 mA) alternating current (usually 50 kHz) is injected through two electrodes called current electrodes and the voltages are measured on two different electrodes called voltage electrodes [1]. The main problem with the two-electrode method is related to the voltage drop of the measurement data due to the contact impedance, so the four-electrode method is recommended for bioimpedance measurements [1]. Bioimpedance spectroscopy (BIS) is an extension of BIA method that involves measuring the impedance of the test object over a range of frequencies.

Impedance methods, in biomedical research have been used for many years, and the electrical properties of tissues have been known for almost a century and a half—Herman wrote about them as early as 1871 [2]. They are used in many areas of research related to natural tissues for both in vitro and in vivo measurements [3]–[6]. Bioimpedance measurements are divided according to the frequencies used: low-frequency bioimpedance, including alpha ($\sim$10 Hz to 10 kHz) and beta ($\sim$10 kHz to 10 MHz) ranges; high-frequency bioimpedance, including delta

(~100 MHz to 10 GHz) and gamma (>10 GHz) ranges [7]. As the path of applied current through cells is dependent on applied frequency, the alfa and beta frequencies, when current travels around cells in the intracellular space, are the most widely used in bioimpedance studies of biological tissues, provide in-depth information about the tissue and its structure. At high frequencies, current affects the space inside the cells entirely, what is used in pathology diagnostics, including cancer detection (as the cancerous tissue shows a less impedance than normal tissue). High-frequency bioimpedance measurements can also yield information about the electrical properties in internal cellular components, as bioimpedance in this range is not influenced by the plasma membrane or polar properties of water.

Due to the ease of execution, bioimpedance measurements are widely used to test, analyze and evaluate body fat and body fat free mass. BIA and BIS also have medical applications in the detection and estimation of lymphedema and breast cancer-related lymphedema (BCRL) [8, 9], cardiovascular evaluation and other hemodynamic parameters [10]–[14], detection and classification of pulmonary nodules [15, 16] or blood glucose measurements [17]–[21]. An interesting study combining bioelectrical impedance analysis-derived phase angle (PhA) with the possibility of early detection of sarcopenia has appeared in 2021 [22]. The data from the selected by authors papers proved that PhA is decreased in sarcopenic subjects and the prevalence of sarcopenia is higher in subjects with low PhA. We can also find the first works combining bioimpedance measurements and artificial intelligence to predict the occurrence of disease exacerbations [23, 24]. BIS also has applications in detecting volume overload, which is common among hemodialysis (HD) patients and is a key risk factor for all-cause and cardiovascular mortality [24].

Facing this background, we aimed to evaluate whether and to which extent bioimpedance spectroscopy is used: a) especially for measuring glucose, b) classification of pulmonary nodules, c) detection and estimation of lymphedema and breast cancer-related lymphedema (BCRL). We focus on finding a new area of activity, where we can use BIS and BIA, which is a critical aspect in the context of rapidly developing medicines.

2 Materials and Methods

A systematic review of the literature was conducted to obtain reliable information on the possibility use bioimpedance method especially bioimpedance spectroscopy for measuring glucose, classification of pulmonary nodules and detection and estimation of lymphedema and breast cancer-related lymphedema. The systematic review presented in this paper was carried out in accordance with the guidelines of the Cochrane Collaboration and Health Technology Assessment (HTA) and consisted of several closely related stages. The first stage consisted in defining a research problem, which was determined based on the PICO scheme (population (P), intervention (I), comparison (C), outcome (O)). P—the population in which the intervention will be used; I—proposed intervention; C—proposed comparator; O—health outcomes against which clinical effectiveness will be assessed. The study focused on the population of patients with lung diseases, breast cancer and diabetes (P). The intervention was performed with bioimpedance spectroscopy (I), while the comparator was conventional measurement methods used in the diagnosis of a given disease (C). The end point was the improvement in diagnostic effects (O).

2.1 Search Strategy

In the PubMed and Cochrane database, after applying the query:

((("pulmonary nodules"[All Fields] AND ("Multiple Pulmonary Nodules"[MeSH Terms] OR "pulmonary nodule multiple"[All Fields] OR "pulmonary nodules multiple"[All Fields])) OR ("breast cancer"[All Fields] AND ("Breast Neoplasms"[MeSH Terms] OR "Breast Neoplasm"[All Fields] OR "Breast Tumors"[All Fields])) OR (("diabete"[All Fields] OR "Diabetes Mellitus"[MeSH Terms] OR ("diabetes"[All Fields] AND "mellitus"[All Fields]) OR "Diabetes Mellitus"[All Fields] OR "diabetes"[All Fields] OR "Diabetes Insipidus"[MeSH Terms] OR ("diabetes"[All Fields] AND "insipidus"[All Fields]) OR "Diabetes Insipidus"[All Fields] OR "diabetic"[All Fields] OR "diabetics"[All Fields] OR "diabets"[All Fields]) AND ("Diabetes Mellitus"[MeSH Terms] OR "Diabetes Insipidus"[All Fields] OR "Glucose Intolerance"[All Fields]))) AND ("bioimpedance spectroscopy"[All Fields] OR ("impedance spectroscopy"[All Fields] AND ("Dielectric Spectroscopy"[MeSH Terms] OR "Impedance Spectroscopies"[All Fields] OR "spectroscopy impedance"[All Fields])) OR ("Bioelectrical Impedance"[All Fields] AND ("Electric Impedance"[MeSH Terms] OR "Bioelectrical Impedance"[All Fields] OR "impedance bioelectrical"[All Fields])) OR ("blood glucose measurement"[All Fields] OR "body composition analysis"[All Fields]))) AND "machine learning" [All Fields] OR "deep learning" [All Fields] AND ((ffrft[Filter]) AND (2018/1/9:2023/1/9[pdat]) AND (english[Filter])), from 09 January 2018 to 09 January 2023, a total of 129 (PubMed) and 11 (Cochrane) publications were obtained.

The number of the total searched results (140) was reduced by increasing the sensitivity and specificity of the search. The criteria responsible for increasing sensitivity and specificity stood for, from the very beginning, searching for multi-word phrases in quotation marks, searching for systematic reviews to obtain publications with a high degree of credibility, and narrowing the set of publications to the last 5 years. The above search strategy resulted in fourteen publications.

2.2 Inclusion and Exclusion Criteria

The inclusion and exclusion criteria related to the characteristics of the literature sought and indicate the type of intervention used and the population of people covered by the study. For the analysis of publications related to the bioimpedance methods of diabetic, pulmonary and with breast cancer patients, the following were indicated as inclusion criteria, for example: Pulmonary nodules, Breast cancer, Neoplasm, Diabetes, Diabetic, Bioimpedance spectroscopy, BIS, body impedance, bioelectrical impedance, BIA, Blood glucose measurement, Body composition analysis, BMI. In contrast, topics related to invasive and biochemical measurement were exclusion criteria.

2.3 Data Collection

A search of the PubMed and Cochrane medical database was conducted based on queries formed from appropriately selected keywords means diseases and measurement method connected by Boolean Logic Operators: AND, OR. The number of the searched results was reduced by increasing the sensitivity and specificity of the search. The criteria responsible for increasing sensitivity and specificity were search for multi-word phrases in quotation marks, search for articles with full texts to gain an insight into the article's relevance, search for publications only in English and Polish, as these are the languages understood by the authors compiling this review.

2.4 Data Selection

Two analysts carried out the selection of the identified scientific reports and the extraction of the results; the final outcome was determined by consensus. The review was conducted in two stages. In the first stage, literature was rejected if its relevance to endorse the claim was negatively assessed after reviewing the title of the paper. Next, publications were rejected that proved to be unhelpful after reviewing the abstract. Publications relating to conventional measured methods and not related to bioimpedance were rejected too. In the next step, literature was rejected if its usefulness was negatively assessed after reviewing the full text and for which full content was not available. After applying a comprehensive search strategy with inclusion and exclusion criteria, fourteen publications (closely related to the topic at hand) were obtained and used as a reliable source of information to endorse the claim that bioimpedance possibility use for measuring glucose, classification of pulmonary nodules and detection and estimation of lymphedema and breast cancer-related lymphedema. Figure 1 presents the steps in the search process for systematic review.

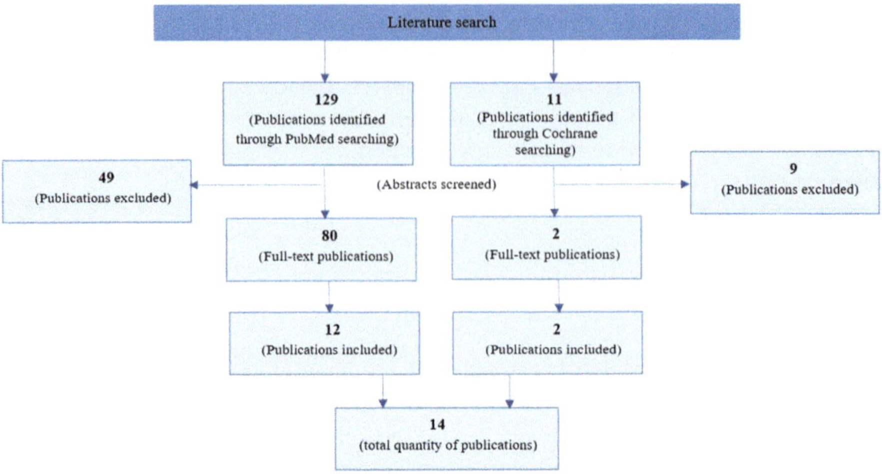

Fig. 1 Flowchart on the search and selection of articles to be included in the systematic review

3 Results

A search of the PubMed database yielded a total of twelve publications [25]–[36] and in Cochrane yielded a total of two publications [37, 38], which were subjected to a selection process in the next step. Presented strategy resulted in fourteen publications (Fig. 1).

The selection of the identified scientific reports resulted in fourteen publications. Detailed results are shown in the Table 2. For 3 clinical study [33, 37, 38], a credibility assessment and, consequently, a clinical trials quality assessment (according to the Jadad scale) was carried out. The Jadad scale is a scale for measuring the quality of randomized clinical trials [39]. The clinical trials can receive values from 0 (low quality) to 5 (highest quality). A score of three points or more indicates high quality. In ten publication no credibility analysis was performed according to the Jadad scale because it is not a clinical trial, but a research paper, review and article (the Jadad scale is not applicable in this case). For all clinical trial using Jadad scale, we obtained 3-point score. The result of the analysis indicates for high quality. The Jadad score for clinical trials was presented in Table 1.

The trials are randomized but not blinded, so studies lost one point each. Clinical study [40] and [37] proved that thanks bioimpedance spectroscopy BIS detected more lymphedema than standard tape measure (TM) method. In study obtained higher sensitivity of BIS than TM. In [32] study compare the risk of subclinical breast cancer–related lymphedema (sBCRL) using bioimpedance spectroscopy BIS or TM by the extent of axillary surgery and regional nodal irradiation (RNI). The obtained results also indicate higher sensitivity of BIS than TM. Whereas in paper [38] authors described the analysis of body composition using BIA techniques. Obtained relevant

Table 1 Publication evaluation according to the Jadad scale

Question	Ridner et al. [37]		Ridner et al. [38]		Boyages et al. [33]	
	YES/NO answer	Scoring	YES/NO answer	Scoring	YES/NO answer	Scoring
Was the study described as randomised?	YES	1	YES	1	YES	1
Was the study described as double-blind?	NO	0	NO	0	NO	0
Is patient loss from the study and follow-up reported?	YES	1	YES	1	YES	1
Should add 1 point for the given description of randomization and the correct method?	YES	1	YES	1	YES	1
Should you add 1 point for the given description of blinding and the correct method?	NO	0	NO	0	NO	0
Should 1 point be deducted for the wrong method of randomization?	NO	0	NO	0	NO	0
Should you deduct 1 point for the wrong blinding method?	NO	0	NO	0	NO	0
Sum	3 × YES 4 × NO	**3**	3 × YES 4 × NO	**3**	3 × YES 4 × NO	**3**

findings suggest that BIA can be considered appropriate for body composition analysis in type 2 diabetes mellitus (T2DM) complicated by sarcopenia/muscle loss. Papers [25]–[27] and [29] aimed to establish a subclinical standard that can screen high-risk patients for upper limb lymphedema early using BIA. This method is highly sensitive, can be used as a basis to establish standardized criteria, and can be used to measure extracellular space. Bioelectrical impedance technology is a common technique used for the early detection of BCRL. The conducted research gave satisfactory results consisting of the detection of a greater number of patients with lymphedema, which was statistically proven. Additional in works [33] and [35] authors proof that

Table 2 Main results of the studies included in the systematic review

Authors (Year)	Diseases	Describe
Ridner et al. [37]	Breast cancer-related lymphedema (BCRL)	The study proved that BIS detected more lymphedema than standard tape measure (TM)
Ridner et al. [38]	Breast cancer-related lymphedema (BCRL)	BIS was recommended by the authors as a possible measurement method to evaluate subclinical lymphedema
Sprignadello et al. [25]	Diabetes	BIA can be used to analyze body composition in diabetic and sarcopenia patients
Zhuang et al. [26]	Breast cancer-related lymphedema (BCRL)	Bioelectrical impedance analysis in early screening of subclinical upper limb lymphedema: a case–control study
Liu et al. [27]	Breast cancer-related lymphedema (BCRL)	Studies have confirmed the effectiveness and safety of using BIA and BIS to detect lymphoedema
Jung et al. [28]	Breast cancer	Evidence was obtained for the effectiveness of the method for the detection of lymphedema (used the Inbody 3.0 system)
Albalat et al. [29]	Diabetes	Sensor for non-invasive glucose measurement based on impedance difference is presented
Lim et al. [30]	Breast cancer-related lymphedema (BCRL)	Use of SFBIA ratio can serve as an alternative method to monitor and/or diagnose BCRL
Jun et al. [32]	Diabetes	PhA bioimpedance values provide a non-invasive tool for monitoring diabetes progression
Boyages et al. [33]	Breast cancer	The lower triggering rates with BIS and its better discrimination of the risk of sBCRL by receipt and type of regional nodal irradiation compared with tape measure. Authors propose the use of posttreatment surveillance to detect sBCRL and to initiate early intervention
Matias et al. [34]	Breast cancer	Study aimed to investigate whether phase angle (PhA) derived from the electrical properties of the body tissues is a predictor of muscular strength in breast cancer survivors (BCS). Described that PhA is a potential indicator of disease-related functionality in BCS
Jurca-Simina et al. [35]	Diabetes	Investigating whether BIA for measuring body percent fat in a healthy medical students' population could be a reliable method to improve risk estimation for developing T2DM
Bundred et al. [36]	Breast cancer	BIS alone is not appropriate for lymphoedema screening or diagnosis

BIS is the criterion to diagnose lymphoedema. In paper [28] authors described non-invasive monitoring sensors to measure blood glucose continuously. This sensor uses bioimpedance to measure glucose. This is important information that work is underway on the use of bioimpedance spectroscopy for glucose measurements. And finally, bioelectrical impedance is useful in screening for diabetes mellitus and monitoring disease progression [31]. The obtained results prove the effectiveness and safety of the BIA and BIS methods.

4 Discussion

A systematic review related to bioimpedance spectroscopy, bioimpedance spectroscopy analysis and conditions such as diabetes, lung nodules and breast cancer was conducted to gain general knowledge about the use of this method. With the development of medicine, there is a growing need for new areas of research, especially to find methods that are cheaper than those available on the market and, most importantly, non-invasive.

Equally important is evaluate the suitability of the BIA and BIS in the assessment of blood glucose, considering patients with diabetes, for the diagnosis of selected diseases, based on only a clinical review. One of the main applications that we find widely described in the literature is the use of BIS for body composition analysis [41], therefore, we have decided to exclude these applications and focus on less common applications in medicine.

Paper [40] presents the non-invasive technology that can measure glucose concentrations from passively expressed human body sweat by using sweat sensor platform and electrochemical impedance spectroscopy. The supervised machine learning and the decision tree regression algorithm used samples analytical reference values from an electrochemical sweat sensor. The authors argue that the results were able to correctly capture the trend in glucose progression. The described technology offers hope for real-time glucose reporting from passively expressed human eccrine sweat. It can therefore be concluded that the presented research area requires further development.

The first, recently discovered, but appears to be promising application, is identification of pulmonary nodules. In the paper [15] the authors designed and built a bioimpedance probe with spherical electrodes, which examined the tissues from 38 patients, with a frequency spectrum in the range of 50 kHz to 5 MHz. The authors used the classification with support vector machine (SVM), linear discriminant analysis (LDA), and K-nearest neighbors algorithm (KNN) to create the intelligent system which can distinguish between healthy and cancerous lung tissue. Conducted research offers the potential for the future development of the intelligent system for localizing invisible pulmonary nodules using the bioimpedance spectrum of lung tissue, which will be safe for patients. This method could be potentially beneficial in the localization of invisible and even nonpalpable in-depth pulmonary nodules in thoracic surgeries.

Over the past years the non-invasive glucose level monitoring by skin impedance has been studied. The results of these studies have also been commercialized, for example the device in the form of wristwatch called Pendra. Pendra was produced by Pendragon Medical ltd, Zurich, Switzerland [42]. The device was approved by the CE (Conformite Europeenne) in May 2003, as the standard glucometer for detecting blood glucose level. Later studies, however, showed the bad correlation in comparison to standard (invasive) glucometers. In addition, glucose readings varied from person to person, the device required additional calibration due to differences in skin and tissue thickness, the measurement tape had to be changed every 24 h, and the device had to be reattached at the same place each time [42]. Due to all above difficulties production of Pendra was stopped in 2005. Quoted example shows the complexity of the problem of non-invasive blood glucose measurements. As the key challenge of the blood glucose estimation systems is not only the measurement technique, but also the proper signal processing, concentrating on minimalizing the interferences from patients, such as their physiological characteristics, as well as individual specify of glucose levels. Many papers take up the topic of bioimpedance signal processing, for example the paper [17] highlighted various methods to improve specificity and minimize interferences e.g., reducing the effect from sensor's attachment to the skin surface, optimizing the multiparameter predicting models and using better calibration methods. When it comes to personal characteristics of blood glucose levels, the use of artificial intelligence can be a significant help.

Measurement of glucose concentration is the next step that affects the development in medicine. It is worth noting that there are still too few clinical trials, and in combination with AI we have zero results, which shows that this research area may be another undiscovered part of medicine that is worth focusing on.

Overall, evidence of the literature reasonably supports our thesis. Monitoring of volume overloads in patients with chronic heart failure using BIS is now well known and commonly used. However, combining these measurements with a heart failure (HF) hospitalization occurrence prediction algorithm, developed based on artificial intelligence, seems to be a completely new perspective on the problem, which the authors of the article [23] took up. They created the disposable, remote monitoring system patch consisted of the few sensors, which collects continuous ECG waveform, skin impedance and other informations on heart rate, activity and posture, to create a predictive algorithm, which is able to reflect whether there is a significant difference in the observed remote monitoring data, compared with the data expected based on the training period personalized baseline. In the results the platform was able to detect precursors of hospitalization for HF exacerbation with 76% to 88% sensitivity and 85% specificity, and the median time between initial alert and readmission was about 6 days [23].

The review confirmed that BIA and BIS are useful tools to assist in the diagnosis of patients with BCRL, diabetes or breast cancer. It is worth mentioning the lack of clinical reviews based on the use of machine learning or deep learning for the detection of selected pathologies using BIA or BIS. Chosen fields of areas need further development, therefore implementation of bioimpedance into clinical practice as another, more accessible method, appears to be forward-looking.

Conflict of Interest None declared.

References

1. Bera, T.K.: Bioelectrical impedance methods for noninvasive health monitoring: a review. J. Med. Eng. **2014**, 1–28 (2014). https://doi.org/10.1155/2014/381251
2. Hermann, L.: 'Ueber eine Wirkung galvanischer Ströme auf Muskeln und Nerven. Pflüg Arch. Für Gesammte Physiol. Menschen Thiere **5**(1), 223–275 (1872). https://doi.org/10.1007/BF01675805
3. Heinroth, K.M., et al.: Impedance cardiography: a useful and reliable tool in optimization of cardiac resynchronization devices. Europace **9**(9), 744–750 (2007). https://doi.org/10.1093/europace/eum086
4. Schloerb, P.R., Forster, J., Delcore, R., Kindscher, J.D.: Bioelectrical impedance in the clinical evaluation of liver disease. Am. J. Clin. Nutr. **64**(3), 510S-514S (1996). https://doi.org/10.1093/ajcn/64.3.510S
5. Zou, Y., Guo, Z.: A review of electrical impedance techniques for breast cancer detection. Med. Eng. Phys. **25**(2), 79–90 (2003). https://doi.org/10.1016/S1350-4533(02)00194-7
6. Impedance cardiography: A valuable method of evaluating haemodynamic parameters | Sodolski | Cardiol J. https://journals.viamedica.pl/cardiology_journal/article/view/21717 (accessed Jan. 05, 2023)
7. Abasi, S., Aggas, J.R., Garayar-Leyva, G.G., Walther, B.K., Guiseppi-Elie, A.: Bioelectrical impedance spectroscopy for monitoring mammalian cells and tissues under different frequency domains: a review. ACS Meas. Sci. Au **2**(6), 495–516 (2022). https://doi.org/10.1021/acsmeasuresciau.2c00033
8. Sutherland, A., et al.: Is bioimpedance spectroscopy a useful tool for objectively assessing lymphovenous bypass surgical outcomes in breast cancer-related lymphedema? Breast Cancer Res. Treat. **186**(1), 1–6 (2021). https://doi.org/10.1007/s10549-020-06059-6
9. Ward, L.C., Koelmeyer, L.A., Moloney, E.: Staging breast cancer-related lymphedema with bioimpedance spectroscopy. Lymphat. Res. Biol. **20**(4), 398–408 (2022). https://doi.org/10.1089/lrb.2021.0013
10. Spiess, B.D., Patel, M.A., Soltow, L.O., Wright, I.H.: Comparison of bioimpedance versus thermodilution cardiac output during cardiac surgery: evaluation of a second-generation bioimpedance device. J. Cardiothorac. Vasc. Anesth. **15**(5), 567–573 (2001). https://doi.org/10.1053/jcan.2001.26533
11. Rajput, R.S., Das, S., Chauhan, S., Bisoi, A.K., Vasdev, S.: Comparison of cardiac output measurement by noninvasive method with electrical cardiometry and invasive method with thermodilution technique in patients undergoing coronary artery bypass grafting. World J. Cardiovasc. Surg. **04**(07), 123–130 (2014). https://doi.org/10.4236/wjcs.2014.47019
12. Drazner, M.H., et al.: Comparison of impedance cardiography with invasive hemodynamic measurements in patients with heart failure secondary to ischemic or nonischemic cardiomyopathy. Am. J. Cardiol. **89**(8), 993–995 (2002). https://doi.org/10.1016/S0002-9149(02)02257-9
13. Siedlecka, J., Siedlecki, P., Bortkiewicz, A.: Impedance cardiography – Old method, new opportunities. Part I. Clinical applications. Int J Occup Med Environ Health (2015), doi: https://doi.org/10.13075/ijomeh.1896.00451
14. Karpiel, I., Richter-Laskowska, M., Feige, D., Gacek, A., Sobotnicki, A.: An effective method of detecting characteristic points of impedance cardiogram verified in the clinical pilot study. Sensors **22**(24), 9872 (2022). https://doi.org/10.3390/s22249872
15. Baghbani, R., Shadmehr, M.B., Ashoorirad, M., Molaeezadeh, S.F., Moradi, M.H.: Bioimpedance spectroscopy measurement and classification of lung tissue to identify pulmonary nodules. IEEE Trans. Instrum. Meas. **70**, 1–7 (2021). https://doi.org/10.1109/TIM.2021.3105241

16. Yang, D., et al.: Electrical impedance analysis for lung cancer: a prospective, multicenter, blind validation study. Front. Oncol. **12**, 900110 (2022). https://doi.org/10.3389/fonc.2022.900110
17. Huang, J., Zhang, Y., Wu, J.: Review of non-invasive continuous glucose monitoring based on impedance spectroscopy. Sens. Actuators Phys. **311**, 112103 (2020). https://doi.org/10.1016/j.sna.2020.112103
18. Geng, Z., Tang, F., Ding, Y., Li, S., Wang, X.: Noninvasive continuous glucose monitoring using a multisensor-based glucometer and time series analysis. Sci. Rep. 7(1) Art. no. 1 (2017), doi: https://doi.org/10.1038/s41598-017-13018-7
19. Pedro, B.G., Marcôndes, D.W.C., Bertemes-Filho, P.: Analytical model for blood glucose detection using electrical impedance spectroscopy. Sensors **20**(23), E6928 (2020). https://doi.org/10.3390/s20236928
20. Głuchowska, Ż: Model czujnika poziomu glukozy we krwi opartego na metodzie spektroskopii impedancyjnej. Elektron.—Konstr. Technol. Zastos. **1**(1), 33–35 (2017). https://doi.org/10.15199/13.2017.1.6
21. Osiecka, X., Pałko, T.: Microwave and impedance spectroscopy as a useful tool for testing dielectric properties of glucose solutions. In: World Congress on Medical Physics and Biomedical Engineering 2018, L. Lhotska, L. Sukupova, I. Lacković, and G. S. Ibbott, Eds., in IFMBE Proceedings, vol. 68/2. Singapore: Springer Singapore, pp. 911–915 (2019). doi: https://doi.org/10.1007/978-981-10-9038-7_168
22. Di Vincenzo, O., Marra, M., Di Gregorio, A., Pasanisi, F., Scalfi, L.: Bioelectrical impedance analysis (BIA) -derived phase angle in sarcopenia: A systematic review. Clin. Nutr. **40**(5), 3052–3061 (2021). https://doi.org/10.1016/j.clnu.2020.10.048
23. Stehlik, J., et al.: Continuous wearable monitoring analytics predict heart failure hospitalization. Circ. Heart Fail. **13**(3), e006513 (2020). https://doi.org/10.1161/CIRCHEARTFAILURE.119.006513
24. Sandys, V., Sexton, D., O'Seaghdha, C.: Artificial intelligence and digital health for volume maintenance in hemodialysis patients. Hemodial. Int. Int. Symp. Home Hemodial. **26**(4), 480–495 (2022). https://doi.org/10.1111/hdi.13033
25. Sbrignadello, S., Göbl, C., Tura, A.: Bioelectrical impedance analysis for the assessment of body composition in sarcopenia and type 2 diabetes. Nutrients **14**(9), 1864 (2022). https://doi.org/10.3390/nu14091864
26. Zhuang, L., et al.: Bioelectrical impedance analysis for early screening of upper limb subclinical lymphedema: a case–control study. PLoS ONE **17**(9), e0274570 (2022). https://doi.org/10.1371/journal.pone.0274570
27. Liu, S., et al.: Determination of bioelectrical impedance thresholds for early detection of breast cancer-related lymphedema. Int. J. Med. Sci. **18**(13), 2990–2996 (2021). https://doi.org/10.7150/ijms.53812
28. Jung, M., Jeon, J.Y., Yun, G.J., Yang, S., Kwon, S., Seo, Y.J.: Reference values of bioelectrical impedance analysis for detecting breast cancer-related lymphedema. Medicine (Baltimore) **97**(44), e12945 (2018). https://doi.org/10.1097/MD.0000000000012945
29. Albalat, A. L., Begona Sanz Alaman, M., Dejoz Diez, M. C., Martinez-Millana, A., Salcedo, V. T.: Non-invasive blood glucose sensor: a feasibility study. In: 2019 41st Annual International Conference of the IEEE Engineering in Medicine and Biology Society (EMBC), Berlin, Germany: IEEE, Jul. 2019, pp. 1179–1182. doi: https://doi.org/10.1109/EMBC.2019.8857261
30. Lim, S.M., Han, Y., Kim, S.I., Park, H.S.: Utilization of bioelectrical impedance analysis for detection of lymphedema in breast Cancer survivors: a prospective cross sectional study. BMC Cancer **19**(1), 669 (2019). https://doi.org/10.1186/s12885-019-5840-9
31. Genua, I., et al.: Accuracy of flash glucose monitoring in patients with Diabetes mellitus on hemodialysis and its relationship with hydration status. J. Diabetes Sci. Technol. **15**(6), 1308–1312 (2021). https://doi.org/10.1177/1932296820975057
32. Jun, M., Ku, B., Kim, J., Kim, K.H., Kim, J.U.: Mediation effect of the duration of diabetes mellitus on the decrease in bioimpedance phase angles in ethnically Korean people: A multicenter clinical study. J. Diabetes Investig. **12**(5), 790–802 (2021). https://doi.org/10.1111/jdi.13399

33. Boyages, J., Vicini, F.A., Shah, C., Koelmeyer, L.A., Nelms, J.A., Ridner, S.H.: The risk of subclinical breast cancer-related lymphedema by the extent of axillary surgery and regional node irradiation: a randomized controlled trial. Int. J. Radiat. Oncol. **109**(4), 987–997 (2021). https://doi.org/10.1016/j.ijrobp.2020.10.024
34. Matias, C.N., et al.: Phase angle as a marker of muscular strength in breast cancer survivors. Int. J. Environ. Res. Public. Health **17**(12), 4452 (2020). https://doi.org/10.3390/ijerph17124452
35. Jurca-Simina, I.-E., et al.: 'What if body fat percentage association with FINDRISC score leads to a better prediction of type 2 diabetes mellitus?'.
36. The Investigators of BEA/PLACE studies *et al.*: Increases in arm volume predict lymphoedema and quality of life deficits after axillary surgery: a prospective cohort study. Br. J. Cancer 123(1):17–25 (2020). doi: https://doi.org/10.1038/s41416-020-0844-4
37. Ridner, S.H., et al.: A randomized trial evaluating bioimpedance spectroscopy versus tape measurement for the prevention of lymphedema following treatment for breast cancer: interim analysis. Ann. Surg. Oncol. **26**(10), 3250–3259 (2019). https://doi.org/10.1245/s10434-019-07344-5
38. Ridner, S.H., et al.: A prospective study of L-Dex values in breast cancer patients pretreatment and through 12 months postoperatively. Lymphat. Res. Biol. **16**(5), 435–441 (2018). https://doi.org/10.1089/lrb.2017.0070
39. Li, J., et al.: The quality of reports of randomized clinical trials on traditional Chinese medicine treatments: a systematic review of articles indexed in the China National Knowledge Infrastructure database from 2005 to 2012. BMC Complement. Altern. Med. **14**(1), 362 (2014). https://doi.org/10.1186/1472-6882-14-362
40. Sankhala, D., et al.: A machine learning-based on-demand sweat glucose reporting platform. Sci. Rep. **12**(1), 2442 (2022). https://doi.org/10.1038/s41598-022-06434-x
41. Campa, F., Toselli, S., Mazzilli, M., Gobbo, L.A., Coratella, G.: Assessment of Body Composition in Athletes: A Narrative Review of Available Methods with Special Reference to Quantitative and Qualitative Bioimpedance Analysis. Nutrients **13**(5), 1620 (2021). https://doi.org/10.3390/nu13051620
42. Wentholt, I.M.E., Hoekstra, J.B.L., Zwart, A., DeVries, J.H.: Pendra goes Dutch: lessons for the CE mark in Europe. Diabetologia **48**(6), 1055–1058 (2005). https://doi.org/10.1007/s00125-005-1754-y

Evaluation of Neurological Disorders in Isokinetic Dynamometry and Surface Electromyography Activity of Biceps and Triceps Muscles

Anna Roksela, Anna Poświata, Jarosław Śmieja, Dominika Kozak, Katarzyna Bienias, Jakub Ślaga, and Michał Mikulski

Abstract Diagnosing neurological disorders and tracking rehabilitation progress requires objective, quantitative measures. Isokinetic exercises are commonly employed to evaluate muscle strength and activity. However, repetitive, quantitative assessment of the tests performed during these exercises can be difficult, especially for patients with neurological disorders. Usually, either dynamometric or sEMG measurements are used. This study is concerned with the evaluation of the applicability and reliability of these two measurement techniques, based on analysis of elbow flexion and extension supported by the Luna EMG rehabilitation robot, aimed at the evaluation of the activity of the biceps and triceps muscles. Dynamometric and surface electromyography measurements are compared with respect to their repeatability between sessions and days, and their ability to differentiate impaired limbs from healthy ones.

Keywords Rehabilitation robotics · Neurological disorders · Electromyography · Dynamometry · Isokinetic tests

A. Roksela (✉)
Faculty of Automatic Control, Electronics, and Computer Science, Silesian University of Technology, 44-100 Gliwice, Poland
e-mail: aroksela@polsl.pl

A. Roksela · A. Poświata · D. Kozak · K. Bienias · M. Mikulski
EGZOTech Sp. z o.o., 44-100 Gliwice, Poland

J. Śmieja
Faculty of Automatic Control, Electronics, and Computer Science, Silesian University of Technology, 44-100 Gliwice, Poland

J. Ślaga
Department of International Learning Centre, Wellness Progressive Group Ltd., 63/66 Hatton Garden, Fifth Floor, Suite 23, London EC1N 8L, UK

P. Strumiłło et al. (eds.), *The Latest Developments and Challenges in Biomedical Engineering*, Lecture Notes in Networks and Systems 746,
https://doi.org/10.1007/978-3-031-38430-1_25

1 Introduction

Stroke is ranked as the second leading cause of death worldwide, however, the high morbidity also results in 50% of survivors suffering from chronic disability [1]. One of the most important areas affected by stroke is motor skills. Precise, sensitive muscle strength and activity testing methods are needed to evaluate muscle functioning in patients with neurological disorders. Biomedical engineering provides valid, reliable, and sensitive evaluation tools that, when used alongside clinical measures, can support clinical decision-making and supply richer data on patient health state and progress [2]. There is a clear need for guidelines for clinicians and researchers regarding technology-based assessment and application of clinical measures and procedures [3].

An isokinetic dynamometer is one of the devices that can be used for the quantitative measurement of muscle strength. It facilitates the assessment of a dynamic muscle contraction, in which the velocity, the resistance, and the joint position are tightly controlled [4]. While it is a reliable source of measurement data, it cannot provide that when muscle activity is not strong enough. Surface electromyography (EMG) can overcome this drawback, enabling a non-invasive monitoring of muscle condition even for patients with no detectable activity by traditional clinical assessments [5].

Isokinetic evaluation methods have achieved good reliability and sensitivity for lower limb (knee and ankle) assessment in a patient with orthopaedic disorder [9]. On the other hand, the use of isokinetic tests has been hardly discussed with regard to patients with neurological disorders. As it is a common method for health and sports subjects, such analysis can be a valuable reference for rehabilitation practitioners.

Rehabilitation robots provide various tools to analyse muscle conditions. They are efficiently used for rehabilitation [6] and assessment [7, 8] by clinicians. Therefore, one such device, Luna EMG, has been used in this pilot study. It is aimed to evaluate the impairments in stroke patients and healthy subjects to provide clues on how the disorder affects the results of the isokinetic dynamometry and surface electromyography tests. Additionally, we aimed to investigate the usability and reliability of the rehabilitation robots for the assessment of neurological disorders.

2 Material and Methods

The pilot study was run in a group of ten subjects with a neurological disorder resulting from stroke (Group 1) and a group of ten healthy subjects (Group 2). The characteristics of both groups are presented in Table 1. Patients in the study group had a stroke from 0.5 to 18 years before the study. Due to the small size of the group they were not additionally stratified with respect to how long ago the stroke took place. Exercise tests were organized into three sessions, two of which were performed on the same day (S1 and S2) and the third—several days later (S3). The subjects were

Table 1 Subjects' characteristic

	Group 1	Group 2
Number of subjects	10	10
Age (years)	55,50 ± 16,42	52,40 ± 10,62
Sex (male, female)	5 M 5 F	6 M 4 F
Handedness	9 right 1 left	9 right 1 left
Impaired upper limb	7 right 3 left	–

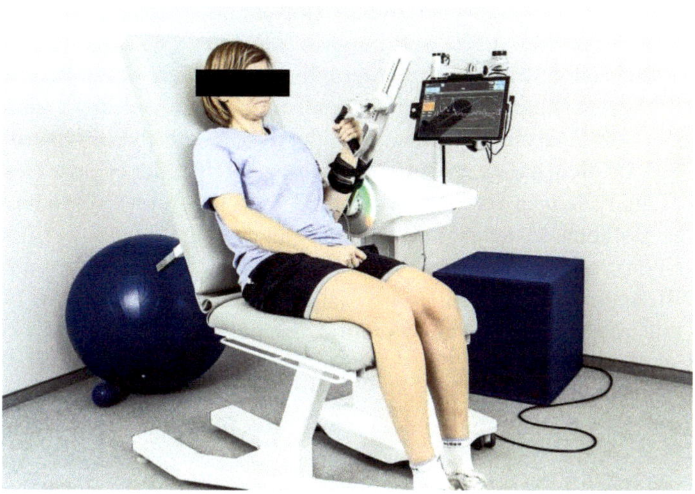

Fig. 1 Subject during the isokinetic test in Luna EMG

involved in no concurrent therapy or physical exercise and their functional conditions were stable. All subjects received comprehensive information about the study goals and procedures and gave their informed consent to participation. All testing took place at "AMED" Rehabilitation Clinic in Katowice, Poland.

The study's measurements were performed using a rehabilitation robot Luna EMG (EGZOTech Sp. z o.o., Gliwice, Poland) equipped with a dynamometer and sEMG data acquisition unit. Subjects were seated, with the upper limb fixed to the robot (Fig. 1). The robot was positioned either at the left or right side, according to the limb being tested. The range of motion (ROM) was set from 30° to 150° (ROM 120°) in the elbow joint. The bioelectrical activity of the triceps brachii (CH1) and biceps brachii (CH2) was recorded according to SENIAM guidelines [10] with surface electrodes (Sorimex, Toruń, Poland).

The test consisted of three one-minute continuous passive motion exercises with speeds 10°/s, 30°/s, and 50°/s for warm-up of the subject, one-minute rest interval, and the isokinetic test of five repetitions of maximum flexions and extensions in the elbow joint at 50°/s, and one-minute interval for rest. Luna EMG isokinetic test relies on a constant measurement of the position (angle) of the elbow joint and torque produced by the patient and then modifies the resistance produced by the dynamometer in order to control the velocity.

An isokinetic test in concentric mode was performed on the biceps brachii and triceps brachii of both arms. Mean torque (Torque) and peak torque (Peak T) in Newton-meters and the position of the upper limb in degree for isokinetic elbow flexion and extension were analysed with a sampling frequency 500 Hz. The minus sign in torque informs about the counterclockwise direction of the applied force. The EMG signal was registered at a sampling 500 Hz with 24-bit accuracy using Luna EMG during the whole test. A bandstop and a bandpass filters were used (with cutoff frequency 48–52 Hz and passing frequency 28–138 Hz, respectively). Root mean values (RMS) of electromyography signals were computed with a time window of 100 [ms]. Mean RMS EMG signals in [μV] for isokinetic elbow flexion and extension were analysed. All data were analysed with Python-based software. An example of a patient's data chart for a neurological disorder is presented in Fig. 2.

Statistical analysis was performed using STATISTICA 13.3 software (StatSoft, Poland). The T student test was used to determine the significance of differences in the measurement variables between the study groups and intragroup for neurological patients. Differences were considered statistically significant if the level of the test was lower than the assumed level of significance ($p < 0.05$). Pearson's correlation coefficient was calculated between the results of research sessions (S1 vs. S2, S1 vs. S3, S2 vs. S3) to check their repeatability and reliability. The following scale was used to assess the correlation coefficient: $r = 0$ means that variables are not correlated, $0 < r < 0.1$ means slender correlation, $0.1 \leq r < 0.3$ weak correlation, $0.3 \leq r < 0.5$ average correlation, $0.5 \leq r < 0.7$ high correlation, $0.7 \leq r < 0.9$ very high correlation, $0.9 \leq r < 1$ nearly full correlation.

3 Results

Table 2 shows the mean, standard deviation (STD), and coefficient of variability (CV) of the performed isokinetic tests of flexion (flex) and extension (ext) of elbow joint for the group with healthy subjects, and Table 3 contains data obtained for the group of subjects with neurological disorders for the impaired and non-impaired (non-im) upper limb.

The healthy group reached higher results in all measured variables, but a significant difference was not found in the EMG of triceps brachii (EMG CH1) results in extension and the EMG of biceps brachii (EMG CH2) results in flexion for the non-impaired (healthy) arm of neurological patients in comparison with left and right arms of the healthy group. Additionally, there wasn't a significant difference in the EMG of biceps brachii (EMG CH2) results in extension and torque in flexion in comparison between the non-impaired arm and the left arm (Tables 4 and 5). All results of the impaired limb had significant differences with the healthy group (Tables 6 and 7). Moreover, the impaired arm outcomes were significantly lower than

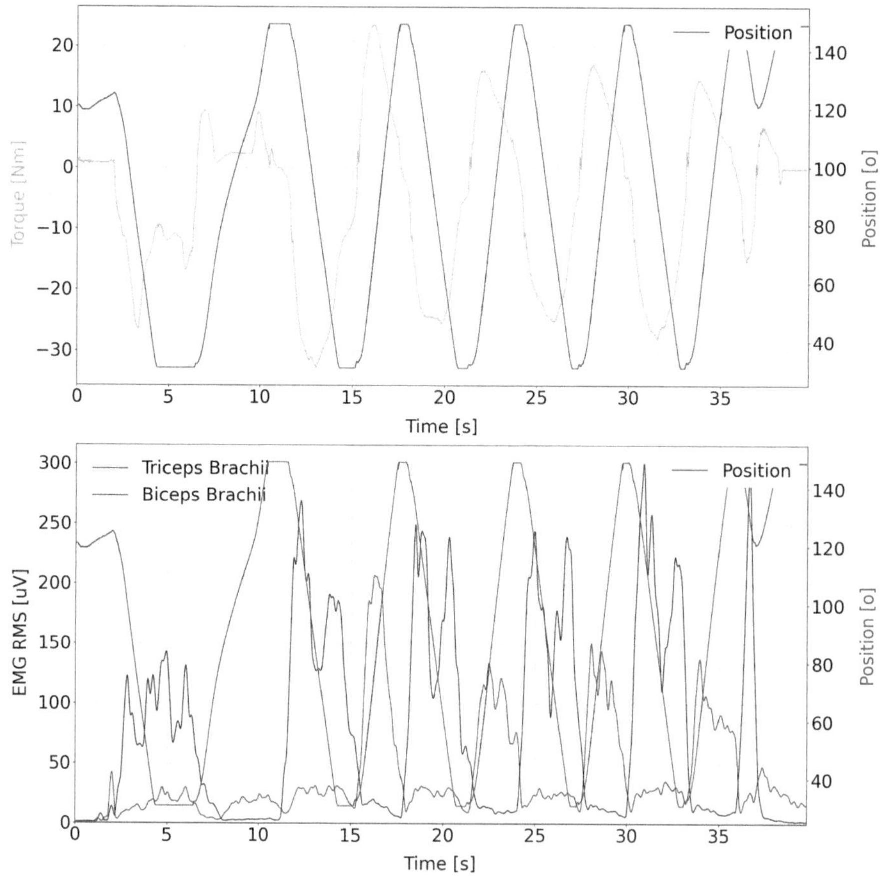

Fig. 2 Sample measurement results of an isokinetic test for a patient with neurological disorder: Torque and Position (upper panel), EMG of triceps brachii and biceps brachii and Position (lower panel)

the non-impaired arm ones besides mean EMG of the triceps brachii during elbow extension and peak torque for both flexion and extension (Table 8).

All calculated parameters were reliable for the healthy group except for the peak torque of the left limb during flexion between the first and the third research trials (Table 9). The highest correlation was reached by dynamometry measurements—mean torque (Torque) and peak torque (Peak T). In the group with the neurological disorder the mean RMS EMG for the triceps brachii (EMG CH1) impaired arm between the first research trial and the third one on another day, and the second research and the third one did not reach correlated results. Impaired limbs did not achieve similar results between the second and the third research trial in mean torque during flexion and peak torque during flexion and extension (Table 10).

Table 2 Results for the healthy group during each research session

			EMG CH1		EMG CH2		TORQUE		PEAK T	
			μ V		μ V		Nm		Nm	
			Right	Left	Right	Left	Right	Left	Right	Left
S 1	Flex	MEAN	63.21	94.60	201.96	187.48	–16.80	13.02	–28.55	25.80
		STD	36.70	63.70	100.00	109.23	7.60	5.94	12.94	11.97
		CV	0.43	0.67	0.50	0.58	0.45	0.46	0.45	0.46
	Ext	MEAN	164.24	169.47	64.98	53.92	11.03	–13.17	23.44	–23.59
		STD	119.14	104.91	44.48	31.67	4.13	6.21	8.60	10.62
		CV	0.73	0.62	0.68	0.59	0.37	0.47	0.37	0.45
S 2	Flex	MEAN	78.08	205.14	260.19	195.84	–21.59	15.18	–34.67	27.73
		STD	36.70	152.93	120.73	73.57	9.23	6.68	15.26	12.08
		CV	0.47	0.75	0.46	0.38	0.43	0.44	0.44	0.44
	Ext	MEAN	213.21	78.08	87.55	54.76	13.42	–15.31	28.10	–26.82
		STD	135.97	36.70	63.05	25.41	4.94	6.27	10.44	9.76
		CV	0.64	0.47	0.72	0.46	0.37	0.41	0.37	0.36
S 3	Flex	MEAN	55.78	61.25	198.90	201.18	–18.98	14.88	–29.04	25.63
		STD	24.12	21.08	100.00	128.61	10.49	7.31	14.49	11.08
		CV	0.43	0.34	0.50	0.64	0.55	0.49	0.49	0.43
	Ext	MEAN	209.03	191.50	62.33	58.25	14.24	–13.91	27.06	–24.93
		STD	172.98	129.48	24.31	43.48	7.95	7.84	14.14	10.82
		CV	0.83	0.68	0.39	0.75	0.56	0.56	0.52	0.43

4 Discussion

This study aimed to evaluate neurological impairments in order to provide clues on how disorders affect the results of tests with Luna EMG and how clinicians could measure patient progress during the rehabilitation process.

The brain's control of muscular activity can be disrupted when a stroke occurs. Depending on the location of the stroke and its severity, this can often lead to changes in muscle tone and function. Loss of muscle strength applies after a stroke, since muscle weakness or paralysis may cause you to be unable to move certain muscles for extended periods of time.

There were significant differences in the results of impaired arm and non-impaired ones except for the mean EMG of triceps brachii during extension movement and peak torque. The impairment of the limb influenced the intersession correlation of the test results between days, especially electromyography parameters. The comparison of the non-impaired arm of the neurological group and the left or right arm of the healthy group did not show a significant difference in all measured parameters. The used measurement device, alignment of the dynamometer axis to the axis of the joint, the position of the tested subject, range of movement, and angular velocity are all

Table 3 Results for the group with neurological disorders during each research session

			EMG CH1		EMG CH2		TORQUE		PEAK T	
			μ V		μ V		Nm		Nm	
			Impaired	Non-im	Impaired	Non-imp	Impaired	Non-im	Impaired	Non-im
S 1	Flex	MEAN	35.17	47.74	102.80	188.37	–8.33	11.12	–15.25	20.81
		STD	21.73	28.68	68.27	169.46	5.42	5.41	9.44	9.39
		CV	0.62	0.60	0.66	0.90	0.65	0.49	0.62	0.45
	Ext	MEAN	120.93	152.69	34.25	49.01	6.35	–9.19	15.47	–18.82
		STD	129.46	110.47	20.55	34.47	5.24	5.06	10.29	7.94
		CV	1.07	0.72	0.60	0.70	0.82	0.55	0.67	0.42
S 2	Flex	MEAN	30.69	44.30	99.15	195.40	–9.15	11.74	–14.69	21.15
		STD	17.96	26.85	85.15	189.08	7.52	7.29	14.15	9.77
		CV	0.59	0.61	0.89	0.97	0.82	0.62	0.96	0.46
	Ext	MEAN	128.51	151.62	27.84	44.08	6.89	–9.86	15.76	–18.88
		STD	172.99	116.17	18.28	39.98	6.71	5.28	9.45	9.45
		CV	1.35	0.77	0.66	0.91	0.97	0.54	0.60	0.50
S 3	Flex	MEAN	29.80	44.57	88.72	149.87	–7.01	10.11	–12.75	16.54
		STD	12.33	29.77	56.70	139.07	3.64	5.46	6.34	8.81
		CV	0.41	0.67	0.64	0.93	0.56	0.54	0.50	0.53
	Ext	MEAN	90.27	141.76	29.91	47.97	6.24	–8.58	12.69	–16.74
		STD	36.41	90.68	16.40	34.99	3.52	4.85	7.02	7.86
		CV	0.40	0.64	0.55	0.73	0.56	0.57	0.55	0.47

factors that have an influence on muscle strength [13]. Therefore a comparison of different studies' results is difficult.

Previous studies on isokinetic dynamometry showed that they might not be able to detect clinically significant changes in neuromuscular disorders muscle power and strength, and that results from isokinetic dynamometry should be viewed cautiously [11]. On the other hand, Ekstrand et al. [12] in their study claimed that upper limb muscle strength can be reliably measured in people with chronic stroke. Based on that, this study has added the measurement of electromyography to achieve more precise results.

It has been reported that monitoring with EMG may help clinicians quantify muscle activity, track recovery, and inform rehabilitation [5]. It should be noted that our study has some limitations. First of all, only individuals with mild to moderate paresis in the upper extremity after stroke were included. A further limitation is that it included only 10 stroke survivors. Moreover, Harbo et al. reported in [13] that muscle strength varied significantly across muscle groups depending on age, height, and body mass on muscle strength which varied significantly across muscle groups. These factors have not been considered in this study. Therefore, the results cannot be generalized to the entire stroke population. To ascertain the reliability and normative

Table 4 T student test results of the non-impaired limb of neurological patients G1 versus left limb of healthy group G2

		MEAN		t	df	p	N		STD		VARIANCE	
		G1	G2				G1	G2	G1	G2	F ratio	p
EMG CH1	Flex	45.54	86.89	−3.26	55	**0.00191**	30	27	28.99	62.43	4.637	0.0001
	Ext	148.69	188.39	−1.23	55	0.222082	30	27	108.27	134.13	1.535	0.2635
EMG CH2	Flex	177.88	194.13	−0.43	55	0.672219	30	27	171.21	105.62	2.628	0.0150
	Ext	47.02	55.35	−0.88	55	0.382693	30	27	37.25	33.86	1.210	0.6261
Torque	Flex	10.99	14.30	−1.92	55	0.060695	30	27	6.26	6.79	1.178	0.6656
	Ext	−9.21	−14.15	3.08	55	**0.00323**	30	27	5.18	6.88	1.765	0.1391
Peak T	Flex	19.50	26.47	−2.49	55	**0.01569**	30	27	9.73	11.37	1.366	0.4137
	Ext	−18.15	−25.13	2.82	55	**0.00669**	30	27	8.65	10.05	1.350	0.4321

Table 5 T student test results of the non-impaired limb of neurological patients G1 versus right limb of healthy group G2

		MEAN		t	df	p	N		STD		VARIANCE	
		G1	G2				G1	G2	G1	G2	F ratio	p
EMG CH1	Flex	45.54	66.79	−2.621	55	**0.01130**	30	27	28.99	32.24	1.237	0.5764
	Ext	148.69	193.99	−1.342	55	0.18509	30	27	108.27	145.49	1.806	0.1242
EMG CH2	Flex	177.88	222.74	−1.169	55	0.24765	30	27	171.21	107.71	2.527	0.0194
	Ext	47.02	72.65	−2.186	55	**0.03309**	30	27	37.25	50.85	1.863	0.1057
Torque	Flex	10.99	19.14	−3.878	55	**0.00028**	30	27	6.26	9.43	2.273	0.0337
	Ext	9.21	12.75	−2.408	55	**0.01944**	30	27	5.18	5.90	1.299	0.4931
Peak T	Flex	19.50	30.94	−3.615	55	**0.00065**	30	27	9.73	13.98	2.065	0.0602
	Ext	18.15	26.11	−3.106	55	**0.00300**	30	27	8.65	10.67	1.521	0.2731

Table 6 T student test results of the impaired limb of neurological group G1 versus right limb of healthy group G2

		MEAN		t	df	p	N		STD		VARIANCE	
		G1	G2				G1	G2	G1	G2	F ratio	p
EMG CH1	Flex	31.85	66.79	–4.97	53	**0.00001**	28	27	18.25	32.24	3.120	0.00450
	Ext	113.78	193.99	–2.14	53	**0.03668**	28	27	131.89	145.49	1.217	0.61491
EMG CH2	Flex	97.17	222.74	–5.11	54	**0.00000**	29	27	74.22	107.71	2.106	0.05643
	Ext	30.69	72.65	–4.14	54	**0.00012**	29	27	19.08	50.85	7.100	0.00000
Torque	Flex	–8.20	–19.14	5.22	54	**0.00000**	29	27	5.98	9.43	2.492	0.01994
	Ext	6.50	12.75	–4.11	54	**0.00014**	29	27	5.47	5.90	1.164	0.69227
Peak T	Flex	14.71	26.11	–4.26	54	**0.00008**	29	27	9.35	10.67	1.302	0.49387
	Ext	–14.28	–30.94	5.00	54	**0.00001**	29	27	10.83	13.98	1.665	0.18901

Table 7 T student test results of the impaired limb of neurological patients G1 versus left limb of healthy group G2

		MEAN		t	df	p	N		STD		VARIANCE	
		G1	G2				G1	G2	G1	G2	F ratio	p
EMG CH1	Flex	31.85	86.89	–4.473	53	**0.00004**	28	27	18.25	62.43	11.702	0.00000
	Ext	113.78	188.39	–2.080	53	**0.04237**	28	27	131.89	134.13	1.034	0.92956
EMG CH2	Flex	97.17	194.13	–3.997	54	**0.00020**	29	27	74.22	105.62	2.029	0.07045
	Ext	30.69	55.35	–3.388	54	**0.00132**	29	27	19.08	33.86	3.148	0.00374
Torque	Flex	8.20	14.30	–3.574	54	**0.00075**	29	27	5.98	6.79	1.292	0.50715
	Ext	6.50	14.15	–4.621	54	**0.00002**	29	27	5.47	6.88	1.582	0.23662
Peak T	Flex	14.28	26.47	–4.107	54	**0.00014**	29	27	10.83	11.37	1.102	0.79881
	Ext	14.71	25.13	–4.019	54	**0.00018**	29	27	9.35	10.05	1.155	0.70682

Table 8 T student test results of impaired limb (G1) versus non-impaired limb (G2) of group with neurological disorder

		MEAN		t	df	p	N		STD		VARIANCE	
		G1	G2				G1	G2	G1	G2	F ratio	p
EMG CH1	Flex	31.85	45.54	–2.134	56	**0.03720**	28	30	18.25	28.99	2.523	0.0179
	Ext	113.78	148.69	–1.105	56	0.27390	28	30	131.89	108.27	1.484	0.2995
EMG CH2	Flex	97.17	177.88	–2.335	57	**0.02311**	29	30	74.22	171.21	5.321	0.0000
	Ext	30.69	47.02	–2.108	57	**0.03944**	29	30	19.08	37.25	3.810	0.0007
Torque	Flex	–8.20	10.99	–12.040	57	**0.00000**	29	30	5.98	6.26	1.097	0.8092
	Ext	6.50	–9.21	11.328	57	**0.00000**	29	30	5.47	5.18	1.116	0.7707
Peak T	Flex	14.71	19.50	–1.927	57	0.05893	29	30	9.35	9.73	1.082	0.8362
	Ext	–14.28	–18.15	1.515	57	0.13523	29	30	10.83	8.65	1.568	0.2348

Table 9 Intersession correlations for healthy group

		Pearson correlation R							
		EMG CH1		EMG CH2		TORQUE		PEAK T	
		Right	Left	Right	Left	Right	Left	Right	Left
S1 versus S2	Flex	**0.90**	**0.86**	**0.63**	**0.74**	**0.86**	**0.92**	**0.95**	**0.83**
	Ext	**0.79**	**0.66**	**0.94**	**0.34**	**0.81**	**0.95**	**0.82**	**0.96**
S1 versus S3	Flex	**0.87**	**0.65**	**0.94**	**0.87**	**0.95**	**0.63**	**0.95**	0.28
	Ext	**0.86**	**0.94**	**0.80**	**0.83**	**0.90**	**0.87**	**0.79**	**0.92**
S2 versus S3	Flex	**0.78**	**0.60**	**0.88**	**0.90**	**0.95**	**0.86**	**0.98**	**0.59**
	Ext	**0.90**	**0.73**	**0.60**	**0.69**	**0.92**	**0.80**	**0.47**	**0.87**

Table 10 Intersession correlations for the group with neurological disorders

		Pearson correlation R							
		EMG CH1		EMG CH2		TORQUE		PEAK T	
		Impaired	Non-im	Impaired	Non-im	Impaired	Non-im	Impaired	Non-im
S1 versus S2	Flex	**0.96**	**0.92**	**0.89**	**0.99**	**0.92**	**0.90**	**0.89**	**0.90**
	Ext	**0.99**	**0.88**	**0.87**	**0.97**	**0.97**	**0.66**	**0.97**	**0.92**
S1 versus S3	Flex	–0.14	**0.90**	**0.62**	**0.89**	**0.33**	**0.65**	**0.37**	**0.60**
	Ext	0.15	**0.86**	**0.50**	**0.94**	**0.52**	**0.94**	**0.37**	**0.78**
S2 versus S3	Flex	0.12	**0.85**	**0.55**	**0.93**	0.14	**0.48**	0.23	**0.36**
	Ext	0.13	**0.69**	**0.40**	**0.92**	**0.45**	**0.57**	0.26	**0.62**

results of such evaluation in a group of patients with a neurological disorder, more studies are needed, particularly involving a larger subjects group and also patients with weaker muscle strength after stroke.

5 Conclusion

Our study has highlighted the promise and challenges of using the isokinetic dynamometry and surface electromyography test to evaluate muscle strength and activity. Isokinetic testing with a rehabilitation robot, facilitating both types of measurements is recommended to provide outcome measures for the progress evaluation of neurological disorders, valuable for clinicians.

Acknowledgements The research presented in this paper is co-financed by the European Union from the European Regional Development Fund, Smart Growth Operational Programme, grant no. POIR.01.01.01-00-1859/20 "Development of an innovative robot for automatic neurorehabilitation and occupational therapy of the hand using electromyography." (AR, AP, DK, KB, MM) and SUT internal grant 02/040/BK_23/1035 (JSm).

References

1. Donkor, E.S.: Stroke in the 21st Century: A Snapshot of the Burden, Epidemiology, and Quality of Life. Hindawi, Stroke Research and Treatment (2018). https://doi.org/10.1155/2018/3238165
2. Cifrek, M., Medved, V., Tonkovic, S., Ostojic, S.: Surface EMG based muscle fatigue evaluation in biomechanics. Clin. Biomech. **24**, 327–340 (2009). https://doi.org/10.1016/j.clinbiomech.2009.01.010
3. Schwid, S., Thornton, C., Pandya, S., Manzur, K., Sanjak M., P.M., McDermott, M., Goodman, A.: Quantitative assessment of motor fatigue and strength in MS. Neurology, 743 (1999)
4. Hislop, H.J., Perrine, J.J.: The isokinetic concept of exercise. Phys. Ther. **47**(2), 114–7 (1967). PMID: 6045281
5. Steele, K.M., Papazian, C., Feldner, H.A.: Muscle activity after stroke: perspectives on deploying surface electromyography in acute care. Front. Neurol. **11** (2020). https://doi.org/10.3389/fneur.2020.576757
6. Lewandowska-Sroka, P., Stabrawa, R., Kozak, D., Poświata, A., Łysoń-Uklańska, B., Bienias, K., Roksela, A., Kliś, M., Mikulski, M.: The influence of EMG-triggered robotic movement on walking, muscle force and spasticity after an ischemic stroke. Medicina **57**, 227 (2021). https://doi.org/10.3390/medicina57030227
7. Oleksy, Ł, Królikowska, A., Mika, A., Reichert, P., Kentel, M., Kentel, M., Poświata, A., Roksela, A., Kozak, D., Bienias, K., et al.: A reliability of active and passive knee joint position sense assessment using the Luna EMG rehabilitation robot. Int. J. Environ. Res. Public Health **19**, 15885 (2022). https://doi.org/10.3390/ijerph192315885
8. Stańczyk, K., Poświata, A., Roksela, A., Mikulski, M.: Assessment of muscle fatigue, strength and muscle activation during exercises with the usage of robot Luna EMG, among patients with multiple sclerosis. In: Information Technology in Biomedicine. Springer International Publishing, Cham, Switzerland (2019). https://doi.org/10.1007/978-3-030-23762-2_11
9. Tiffreau, V., Ledoux, I., Eymard, B., Thévenon, A., Hogrel, J.Y.: Isokinetic muscle testing for weak patients suffering from neuromuscular disorders: a reliability study. Neuromuscul. Disord.: NMD **17**(7), 524–531 (2007). https://doi.org/10.1016/j.nmd.2007.03.014
10. Hermens, H.J., Freriks, B., Disselhorst-Klug, C., Rau, G.: Development of recommendations for SEMG sensors and sensor placement procedures. J. Electromyogr. Kinesiol. **10**, 361–374 (2000). https://doi.org/10.1016/S1050-6411(00)00027-4
11. Woude, D., Ruyten, T., Bartels, B.: Reliability of muscle strength and muscle power assessments using isokinetic dynamometry in neuromuscular diseases: a systematic review. Phys. Ther. **102**(10), pzac099 (2022). https://doi.org/10.1093/ptj/pzac099
12. Ekstrand, E., Lexell, J., Brogårdh, C.: Isometric and isokinetic muscle strength in the upper extremity can be reliably measured in persons with chronic stroke. J. Rehabil. Med. **47**(8), 706–713 (2015). https://doi.org/10.2340/16501977-1990
13. Harbo, T., Brincks, J., Andersen, H.: Maximal isokinetic and isometric muscle strength of major muscle groups related to age, body mass, height, and sex in 178 healthy subjects. Eur. J. Appl. Physiol. **112**(1), 267–275 (2012). https://doi.org/10.1007/s00421-011-1975-3

EMG Mapping Technique for Pinch Meter Robot Extension

Marcel Smolinski, Michal Mikulski, and Jaroslaw Śmieja

Abstract Rehabilitation of patients with impaired grip functions involves standard exercise that should be repeated in a predefined sequence. They are aimed to gradually improve patients' ability to grip and operate objects used in daily life. Rehabilitation robots, employing surface EMG (sEMG) as a diagnostic and feedback signal may be used to support these exercises and trace the rehabilitation progress. However, to properly employ exercise regimen, information about what force should be used to hold and operate various objects is needed. This work is focused on finding the correlation between the value of sEMG signal and the force required to grip a particular object. Calibration plots obtained that way will reduce the need for additional equipment, such as specialized gloves, and at the same time provide information needed for better exercise protocols.

Keywords EMG · Mapping · Force · Pinch meter · Robotics

1 Introduction

SEMG (Surface Electromyography) is a non-invasive measurement technique that facilitates analysis of skeletal muscle activity by recording electrical signals generated by muscle fibers during contractions. The method involves placing electrodes on the skin surface, over the muscles being studied. SEMG allows for the assessment of muscle activation timing, intensity, and coordination.

SEMG has many applications in various fields such as sports science, rehabilitation, ergonomics, and neurophysiology (see, e.g. the review in [1]). It is used to study, e.g., muscle function, movement, and fatigue. SEMG can provide valuable informa-

M. Smolinski · M. Mikulski
EGZOTech Sp. z o.o.,44-100 Gliwice, Poland

M. Smolinski · J. Śmieja (✉)
Faculty of Automatic Control, Electronics and Computer Science, Silesian University of Technology, Akademicka 16, 44-100 Gliwice, Poland
e-mail: jaroslaw.smieja@polsl.pl

P. Strumiłło et al. (eds.), *The Latest Developments and Challenges in Biomedical Engineering*, Lecture Notes in Networks and Systems 746,
https://doi.org/10.1007/978-3-031-38430-1_26

tion about muscle activity during different activities, such as walking, running, or weightlifting.

The results of SEMG measurements are used in a variety of ways, including understanding the mechanisms of muscle contractions, monitoring the progression of muscle disease treatment, assessing the impact of therapy and rehabilitation on muscle function, and analyzing movement and ergonomic work [2–4]. The relative ease of signal acquisition makes SEMG an ideal candidate for support of rehabilitation devices.

In recent years rehabilitation robots gained a lot of interest. With that came the possibility of functional training, that is, training focused more on improving patients' skills in daily activities [5, 6], than on performing simple workouts to, for example, increase muscle mass. An example of such action could be using an upper limb to pick and hold an object. However, various objects manipulated in daily activities, require application of different forces or torques needed to lift, twist, push or pull them. Quantitative information about these forces and torques is needed to train motor skills needed for specific functions. The focus of this paper is on using sEMG to acquire that information that could subsequently be used by a rehabilitation robot during exercises aimed to improve patient motor skills [7]. It would be an alternative to exoskeleton devices [8].

The main idea is to create a dataset of values characteristic for specific actions that would be used as setpoints during training. These setpoints would determine the force or torque generated in a handle or other element during an exercise. By reorienting a training to be task-specific a patient would become more motivated and reliable information about patient's progress would become available both for the patient and physicians, trainers and caretakers, providing quantitative feedback that is often missing and whose lack may lead to decreased patient's motivation.

In this work we concentrate on exercises consisting in using a key-type handle, aiming at improving gripping and turning motor skills. This type of a handle may represent various objects, such as a lock key, TV remote, a credit card, a fork, a medium-sized cup, a smartphone, a small or large plate. Each of these objects requires the so-called scissor or pincer grip but with a different force or torque to be applied during operation. The final grip is bounded by approximately 15 mm. It should be noted that to support functional training aimed at using other objects, as e.g., a screwdriver or medicine vials, whose gripping diameter reaches approximately 30–40 mm, a separate set of measurements would be required. Otherwise, skin tensions might lead to misinterpretation of sEMG signals, as electrodes could read signals from muscle regions other than expected.

Successful rehabilitation would mean, in that case, that a patient is capable of independently performing daily activities involving these objects.

The object measurement method proposed by the author involves using the results to implement into the memory of an upper limb rehabilitation robot with CPM and CAM functionality. The combined solution will make it possible to help patients struggling with problems such as: – flexor and extensor tenolysis, – aponeurectomies for Dupuytren disease, – metacarpophangeal arthrolysis, – open reduction and internal fixation of intra-articular, diaphyseal, metaphyseal and epiphyseal fractures of the phalanges, – prosthetic replacement of the MCP, PIP and DIP joints, – rheumatoidal/neurological or after-burn stiffness, – capsulotomies, arthrolysis and tenolysis for post-traumatic stiffness of MCP, PIP and SIP joints, – synovectomies, – distal radius fractures (stable), – open Carpal Tunnel procedures, – ganglionectomy, – following arthrotomy and drainage of acute septic arthritis providing infection is controlled.

2 Training Exercises Setup

In this section, the calibration measurement setup is described. It is needed to assign specific grip force values to signal levels acquired by sEMG.

The calibration performed within this study is performed for the upper limb with the arm positioned in parallel to the body and the forearm bent at the elbow at the right angle (Fig. 1). That way, the forearm is parallel to the ground. The measurement is taken from the *i*nterosseus dorsalis muscle of the arm, using a single EMG channel. Surface electrodes, measuring 40×55 mm were used, attached with the conductive gel as depicted in Fig. 2. For repeatable fixation, the Mezos SIT rehabilitation chair was used together with a fixed arm support and a fixed measurement beam mount.

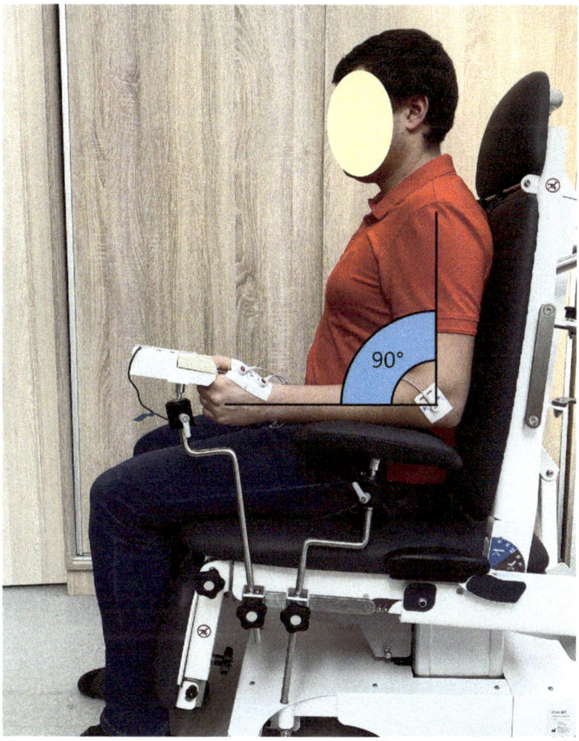

Fig. 1 User position during calibration and testing

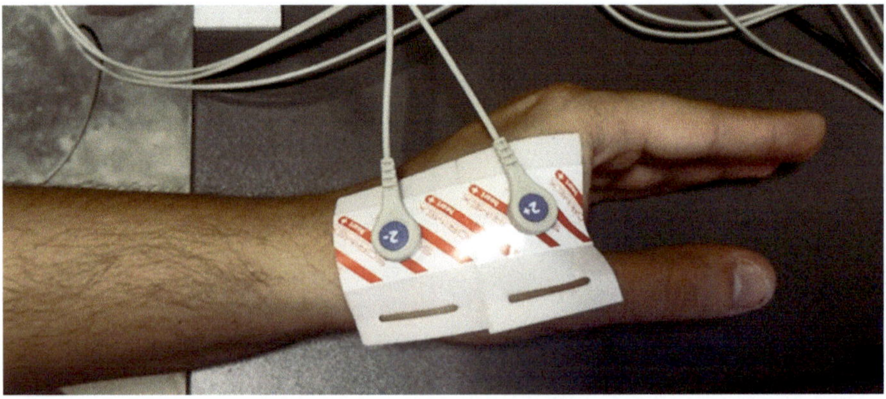

Fig. 2 Placement of surface electrodes

The electrodes were attached throughout the calibration and tests with a dedicated test bench (Fig. 3) and were not disassembled until a specific test was not finished.

Localization of the electrodes was determined in a separate experiment, in which value of sEMG RMS was measured for different electrode's placements and the localization yielding maximum value of signal was chosen. Also such placement was the same as in [9]. The selected measuring surface was also chosen because of the better freedom of operation of the hand. When squeezing the hand on the pinch meter, electrode's points located e.g. inside of the hand would cause discomfort and measurements could be affected by an intentionally unnatural squeeze.

One of the critical issues in the setup of tests described in this paper is providing repeatable results. The first problem is associated with an accurate placement of

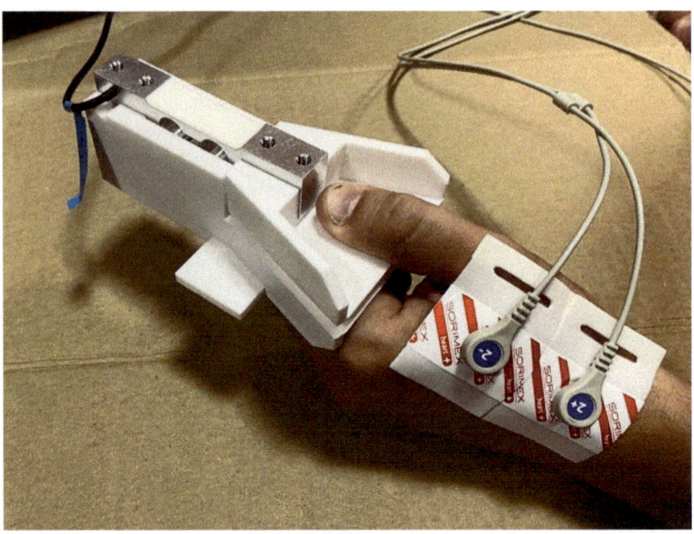

Fig. 3 Tensiometer bench during calibration process

the electrodes, which should be the same across multiple measurements taken for various objects being manipulated. Even a few millimeter difference in electrodes localization would lead to loss of correlation between the EMG signal value and the gripping force. Physiological constraints of muscle activity may influence test repeatability as well, with muscle fatigue being the major factor. Even assuming a very accurate electrode placement, muscle fatigue after a whole day of activity (or muscle mass gain) may lead to quantitatively different strength of EMG signal, associated with a specific action. Therefore, calibration should precede each series of measurements taken with a specific test bench.

3 Calibration Experiments

A tensometer beam allowing to measure the force of a scissor grip was used together with calibrated weights of 100 g, 2 × 200 g, 1000 g and 2000 g for all load combinations (Fig. 4). The beam is used within its elasticity range of deformation.

The tests were performed according to the following procedure (each stage lasted 10 s):

1. START—muscle relaxation.
2. Scissor grip of 1N.
3. Muscle relaxation.
4. Scissor grip of 2N.
5. Muscle relaxation.
6. Scissor grip of 5N.

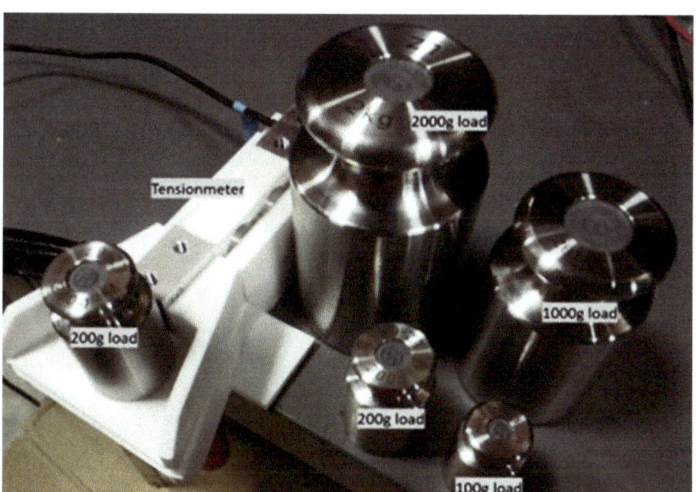

Fig. 4 A tensometer with scissors grip clamp and calibrated loads

7. Muscle relaxation.
8. Scissor grip of 10N.
9. Muscle relaxation.
10. Scissor grip of 20N.
11. Muscle relaxation.
12. Scissor grip of 30N.
13. END—Muscle relaxation.

SEMG signal was acquired with the medical device Stella BIO, which works within the EMG signal range of ±6000 μV, with accuracy of 0.1 μV and baseline noise <0.5 μV RMS (Root Mean Square). Duration of each step of the procedure was timed by the Stella Bio device, which provided voice commands to start and stop.

Data was collected for a healthy single individual. The results are shown in Fig. 5.

Taking into account highly variable signal both at the beginning and the end of each effort, RMS was calculated for the time period [0.5; 9.5] [s] only. The values are presented in Table 1. It should be noted that the relation between the load and sEMG RMS is not linear for the load range that was used in the experiment, though it may be considered to be linear for smaller loads (Fig. 6).

Having checked the strength of sEMG signal for predefined load values, typical everyday objects were chosen for calibration. Corresponding sEMG RMS values are presented in Table 2. The corresponding load was determined basing on the calibration plot (Fig. 6), approximated by linear sections linking subsequent points. Exercise setup is presented in Figs. 7 and 8. Measurements obtained for each of the objects shown in Fig. 7 are presented in Fig. 9.

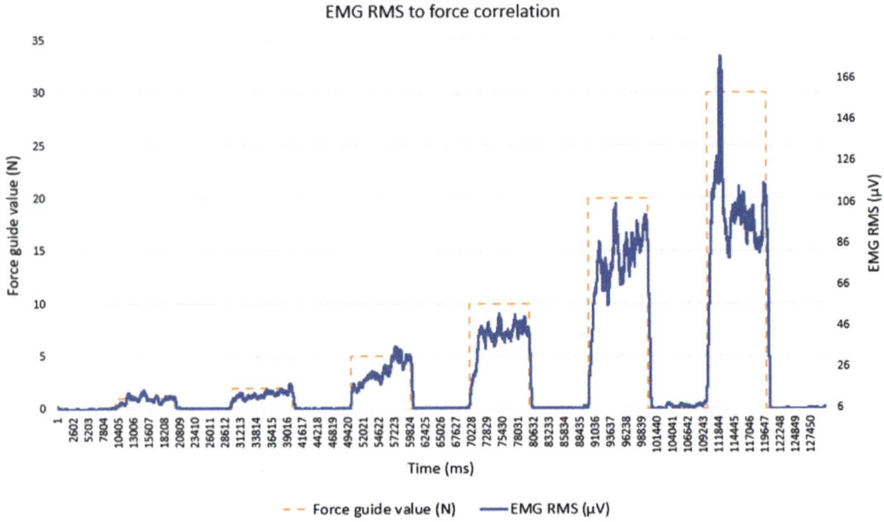

Fig. 5 Calibration test results: load set point [N] and sEMG ([μV)] (red and blue lines, respectively)

Table 1 EMG RMS to estimated force values

Force (N)	EMG RMS (μV)
1	10,9
2	14,3
5	28,5
10	44,5
20	82,7
30	96,0

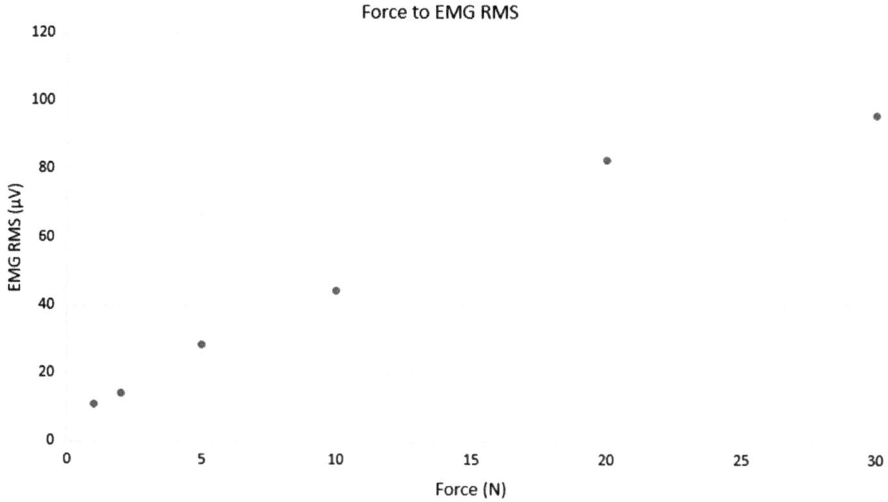

Fig. 6 Load and EMG RMS correlation

Table 2 sEMG RMS values for sample everyday objects

No.	Object	Object thickness [mm]	Object mass [g]	sEMG RMS [μV]	Estimated force [N]
1	A key door	2.3	13	12.57	149
2	A TV remote	12.5	82	29.38	528
3	A fork	2.0	29	17.11	260
4	A credit card	12.0	7	11.08	105
5	A smartphone	9.5	197	16.43	245
6	A small plate (205 mm in diameter)	4.0	333	20.66	335
7	A large plate (267 mm in diameter)	5.1	744	54.03	1248
8	A 250 ml cup	14.9	429	46.78	1058

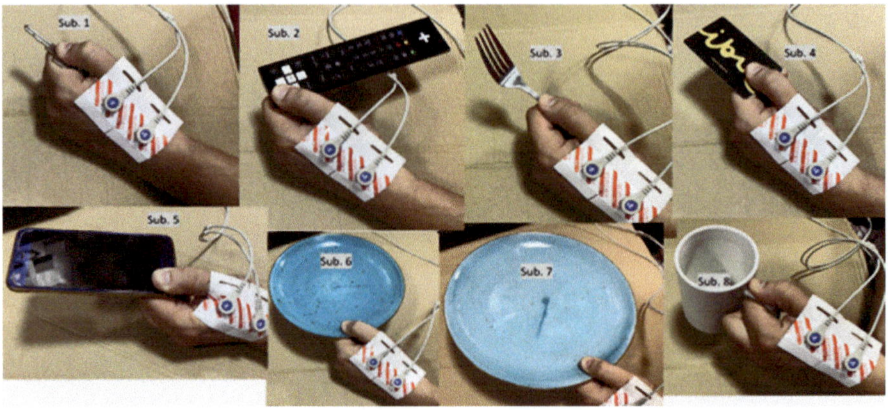

Fig. 7 Test objects and the way of holding them

Fig. 8 User holding smartphone object during test procedure

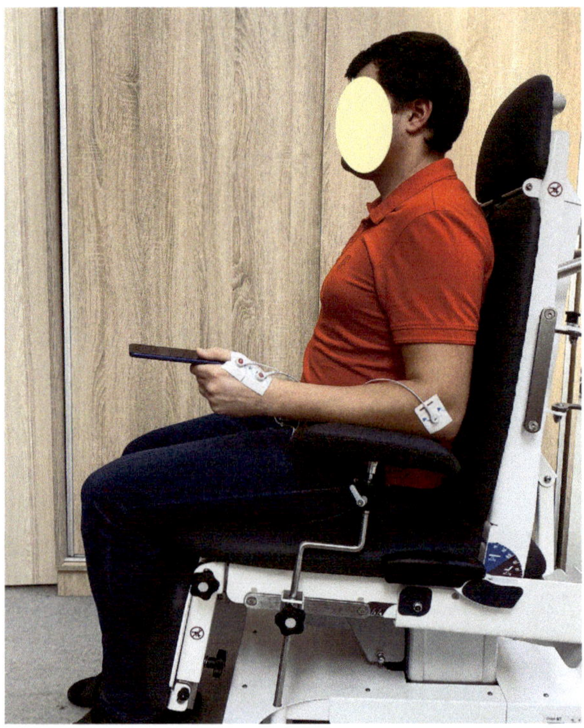

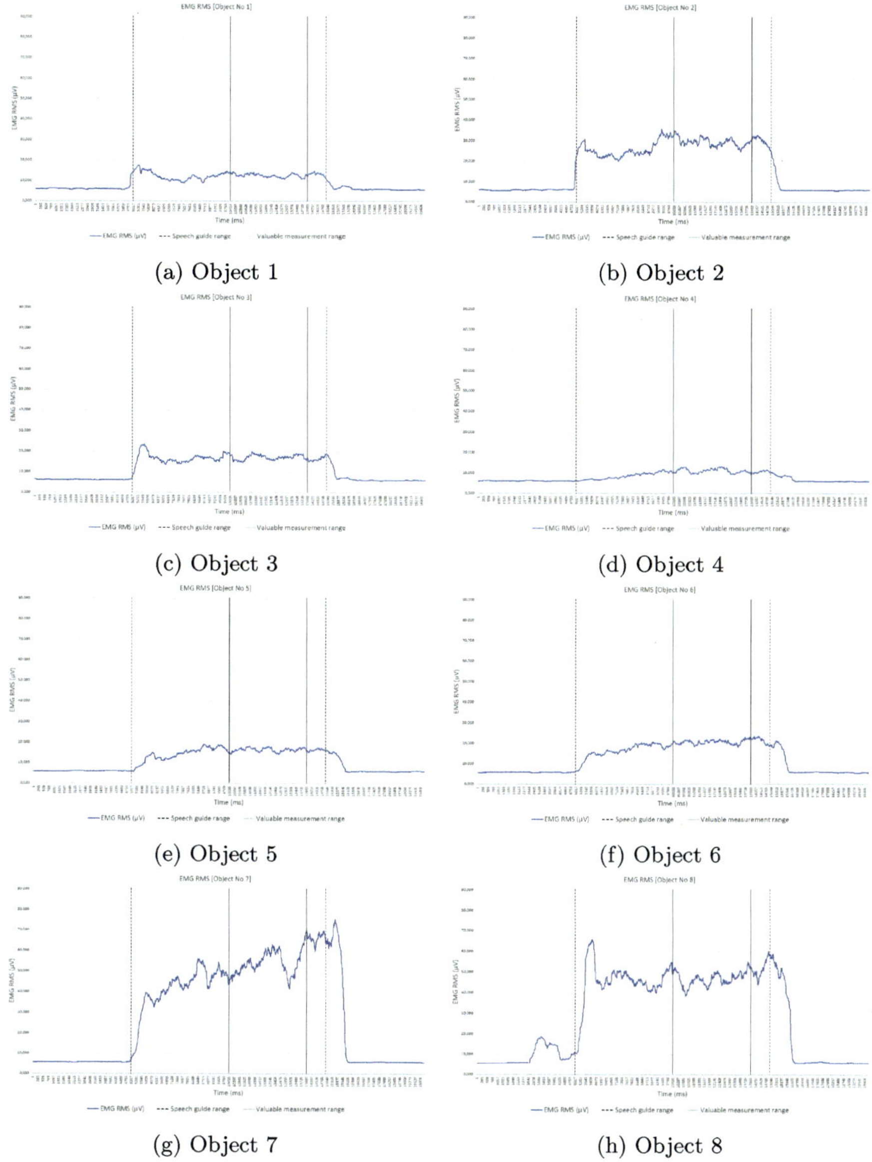

(a) Object 1 (b) Object 2

(c) Object 3 (d) Object 4

(e) Object 5 (f) Object 6

(g) Object 7 (h) Object 8

Fig. 9 sEMG RMS values for objects listed in Table 2. Each plot represents values obtained for a single object

4 Conclusions

Quantitative data describing forces needed for daily-life operations on particular objects is necessary in rehabilitation. The standard approach would require elaborate equipment and tools, such as special gloves [10] and dedicated grip measurement tools (such as load cells) and special algorithms [11]. Moreover, much attention must be put to providing appropriate contact surface or surface finish (e.g. by using silicone overlays with deformations similar to the human hand). The method proposed in this paper significantly simplifies the task of collecting data on scissor grip force. It employs the same device and signal acquisition unit that is subsequently used in rehabilitation exercises.

The proposed calibration workstation, consisting of a simple force measurement module with a trapezoidal grip overlay provides sufficient hand contact for a scissor grip. Though data collected in this work is by no means complete, it clearly indicates relevance of the proposed approach. To our knowledge, there are no comprehensive databases that could be used in calibrating rehabilitation equipment for the goals stated in this paper, nor literature devoted to it. Therefore, creating it might significantly enhance rehabilitation procedures. Based on the authors' experience, the developed method requires approximately 15 min to conduct tests for 8 different objects.

Further test will be performed to incorporate objects kept in different positions and to check if the calibration plot is sex-dependent.

Acknowledgements This study was supported by the National Centre for Research and Development of Poland under Grant No. POIR.01.01.01-00-1859/20 "Robot for neurorehabilitation and occupational theraphy of the hand using electromyography" (MS, MM) and SUT internal grant 02/040/BK_23/1035 (JS).

References

1. Jarque-Bou, N.J., Sancho-Bru, J.L., Vergara, M.: A systematic review of EMG applications for the characterization of forearm and hand muscle activity during activities of daily living: results, challenges, and open issues. Sensors **21**, 3035 (2021)
2. Merletti, R., Botter, A.: Electromyography: Physiology, Engineering, and Non-invasive Applications. Springer (2009)
3. Winter, D.A.: Biomechanics and Motor Control of Human Movement. Wiley-Interscience (1990)
4. Farina, D., Merletti, R.: Introduction to Surface Electromyography. IEEE Press (2000)
5. Bressi, F., Santacaterina, F., Cricenti, L., Campagnola, B., Nasto, F., Assenza, C., Morelli, D., Cordella, F., Lapresa, M., Zollo, L., et al.: Robotic-assisted hand therapy with gloreha sinfonia for the improvement of hand function after pediatric stroke: a case report. Appl. Sci. **12**, 4206 (2022)
6. Delph, M., Fischer, S., Gauthier, P., Martinez-Luna, C., Clancy, E., Fischer, G.: A soft robotic exomusculature glove with integrated sEMG sensing for hand rehabilitation. In: Proceedings of IEEE International Conference on Rehabilitation Robotics, pp. 1–7 (2013)

7. Xie, C., Yang, Q., Huang, Y., Su, S.W., Xu, T., Song, R.: A hybrid arm-hand rehabilitation robot with EMG-based admittance controller. IEEE Trans. Biomed. Circuits Syst. **15**(6), 1332–1342 (2021)
8. Lee, B.J.B., Williams, A., Ben-Tzvi, P.: Intelligent object grasping with sensor fusion for rehabilitation and assistive applications. IEEE Trans. Neural Syst. Rehabil. Eng. **26**(8), 1556–1565 (2018)
9. Rupasov, V.I., Lebedev, M.A., Erlichman, J.S., Lee, S.L., Leiter, J.C., Linderman, M.: Time-dependent statistical and correlation properties of neural signals during handwriting. PLoS One (2012)
10. Lapresa, M., Lauretti, C., Scotto di Luzio, F., Bressi, F., Santacaterina, F., Bravi, M., Guglielmelli, E., Zollo, L., Cordella, F.: Development and validation of a system for the assessment and recovery of grip force control. Bioengineering **10**, 63 (2023)
11. Romeo, R.A., Lauretti, C., Gentile, C., Guglielmelli, E., Zollo, L.: Method for automatic slippage detection with tactile sensors embedded in prosthetic hands. IEEE Trans. Med. Robot. Bionics **3**, 485–497 (2021)

Data Glove for the Recognition of the Letters of the Polish Sign Language Alphabet

Jakub Piskozub and Paweł Strumiłło

Abstract This article presents the results of a study on the classification of 36 letters of the Polish Sign Alphabet (PSL) using a data glove. The data glove includes a Raspberry Pi, a PCB with four 16-bit ADS1115 ADCs, and ten piezoresistive sensors. The sensors function as potentiometers that respond to finger bending. A neural network consisting of LSTM and convolutional layers was used for classification, achieving an efficiency of 99% on data from a single subject, which underwent prior augmentation. The study suggests that the device could potentially be calibrated solely from the target user. This is the first study of its kind on PSL.

Keywords Data glove · Sign alphabet · Deep neural networks · Sign language recognition

1 Introduction

There are more than 300 sign languages worldwide, used by a total of about 70 million people [1]. It should be noted that this is not the only form of communication for all who use them. There is no accurate data on the number of people who speak sign languages only but basing on the British Sign Language user data [2], where over a half of its users are deaf, it can be assumed that for around half of sign language users globally treats them as their native language. And while there are times when mutual unfamiliarity with verbal and visual languages by both sides of every-day conversation is a barrier, it is not uncommon for sign language users to have a functioning speech system and be able to communicate verbally or with an aid of writing. However, one can imagine cases in which such communication is further hampered. Lack of knowledge of the local language or its grammar, inoperative speech apparatus or problems with its use [3], or the general communication difficulties experienced by autistic people who also sometimes use sign language,

J. Piskozub (✉) · P. Strumiłło
Lodz University of Technology, Lodz, Poland
e-mail: Jakub.Piskozub@dokt.p.lodz.pl

P. Strumiłło et al. (eds.), *The Latest Developments and Challenges in Biomedical Engineering*, Lecture Notes in Networks and Systems 746,
https://doi.org/10.1007/978-3-031-38430-1_27

or the stress of an emergency can often make communication with a hearing person significantly difficult or impossible. Several services and devices have also been developed to make daily life easier for sign languages users [4]. However, none of them allow sign language to be translated into the vernacular directly, as modern translators do.

1.1 Sign Language Translators—Requirements

Due to the complex nature of sign languages, potential sign language translators face challenges far beyond those that accompany the translation of verbal languages. This is because the input data of an ordinary translator is only the typed text or utterance, while to correctly translate sign language, three inputs need to be recorded and interpreted simultaneously: hand movement, facial expression and lip movement, and body language. When considering devices to serve the task of sign language translation, the ergonomics of such a device should be a primary consideration. In addition, sign languages undergo dialect formation just like verbal languages. However, due to the much smaller, dispersed, and closed clusters of its users, these dialects are much less likely to penetrate the standard sign language, often being unintelligible even to other user groups even within a single localization [5]. It would therefore be necessary for any such device to be able to adapt not only to the user's physiology, but also to his unique vocabulary and movements.

As with hand movement, body pose, and facial expressions can be tracked by direct methods by placing sensors directly on the part of the body under study. However, given the current state of sensor technology and the priority ergonomics, this type of method should be ruled out. Today's vision sensor technology makes it possible to put sensors in smartphones that are sophisticated and powerful enough to equip every owner of a mobile device with a computer vision tool. Combined with the upcoming capabilities of 5G technology, this will allow the creation of an application to track and interpret facial expressions and poses. There are also vision-based methods for tracking movements and interpreting their meaning in sign language. However, their significant drawback stems from the very nature of this method. Namely, it is the lack of a perfect perspective to observe every hand movement or finger positioning. The slight rotation of the hand in relation to the camera, obscuring the positioning of the fingers, effectively makes it difficult or even impossible to track and interpret them [6]. A suitable alternative to vision systems in this case are direct, sensor-based methods, a breakdown of which can be seen in Fig. 1.

Since it is necessary to track the movement of fingers, devices for direct measurement of their position most often take the form of a glove. Such a device is called a data glove or smart glove. And it is currently one of the most effective methods of tracking finger movement, resistant to obscurity or poor lighting conditions.

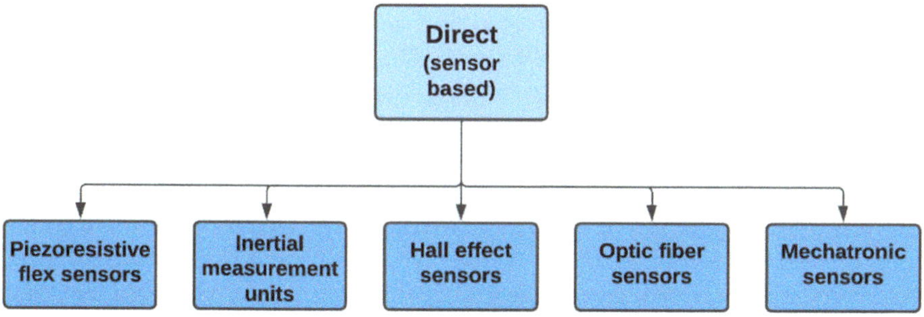

Fig. 1 Breakdown of devices by type of hand tracking method

1.2 *Data Glove as a Sign Language Translator*

Because of the advantages, the data glove is often adapted to the role of a "sign language interpreter." This is an extremely misleading statement often used by the authors of such devices. It comes from the oft-repeated myth that sign language is just hand movements. The erroneous conception of the device, which neglects the basic components of sign language communication is the cause of the reluctance of deaf people to accept such solutions [7]. And these are not unfounded arguments. It is not difficult to imagine a scenario where the malfunction of a communication device, which is also susceptible to damage in transit, only adds to the difficulty of the conversation by causing misunderstandings through misinterpretation.

Therefore, it is necessary to look for methods to increase the effectiveness of such devices while reducing their size and increasing their adaptability to potential users without burdening them with a monotonous learning process. This article describes the results of work on a glove for recognizing the letters of the Polish Sign Language alphabet. The work investigated the prospects for re-reducing the time spent on creating training sets for a neuro-network that recognizes individual letters by using data augmentation.

2 Related Work

Authors of [8] have collected in summary dozens of articles describing devices based on direct measurement methods, mostly in the from of a glove, and classification and segmentation methods. The percentage of articles presented by the authors clearly shows the general level of development of data glove devices in gesture recognition and sign language applications. Only 5% of the presented solutions for gesture or sign language recognition dealt with recognition of movements performed continuously. The others operate on artificially separated signals. Mostly because of the data acquisition method itself. Nearly half also involve static gestures, both understood as signal and spatiotemporal. This shows that such technologies are currently in the early stages of development. However, several works describing significant achievements in the development of the field should be mentioned.

Lee et al. in [9] have proposed an effective solution to the problem of segmenting gestures from a continuous sequence. The authors developed a method based on the dynamics of signal changes from 10 piezoresistive sensors placed on the top of the hand and an inertial sensor. Combined with a neural network consisting of LSTM units and dense layers, this allowed them to accurately determine when the correct phase of a gesture begins and ends. In addition, the authors presented the results of classifying the 11 gestures they studied using a neural network with a similar architecture to that of gesture stage detection. In Ref. [10] a further stage of the authors' research is presented, involving the application of previously developed methods together with a new method for compressing signals from piezoresistive sensors. The proposed compression method can speed up signal processing by up to 80%, while the authors' claimed efficiency in recognizing gestures in an uninterrupted sequence is 94%. The entire process of detection, compression and classification takes only 28 ms.

As indicated by the data collected by the authors [8], systems based on artificial neural networks require training data acquired from a dozen people repeating a set of test letters or gestures up to a hundred times. This implies significant amounts of time spent on data acquisition alone. To the best of our knowledge, the possibility of acquiring data by augmenting data obtained from just one person has not yet been explored.

3 Materials and Methods

3.1 Glove Construction

The glove with which the measurements discussed in this article were obtained was based on a popular System On Chip (SOC)—Raspberry Pi Zero 2 W, equipped with a Broadcomm BCM2835. The SOC is the core of the measurement and control unit, complemented by a PCB containing four, 16-bit ADS1115 ADCs, operating at a rate of 860 samples per second. The sensors function as potentiometers that respond to the bending of a finger, decreasing their resistance and thus changing the voltage at the input of the converter in proportion to the change in the angle of bending of the finger. The sensors exhibit linear resistance characteristics, regardless of the location and number of bending points as proven earlier research [11]. Connections between the sensors and the measurement and control unit were made using electrically conductive thread. This made it possible to eliminate connections using standard wire. Relative to the previous stage of testing, the glove was expanded to include an inertial sensor with 9 degrees of freedom BNO055, located in the center of the top layer of the hand. The symbolic arrangement of the components and their block diagram can be seen in Fig. 2.

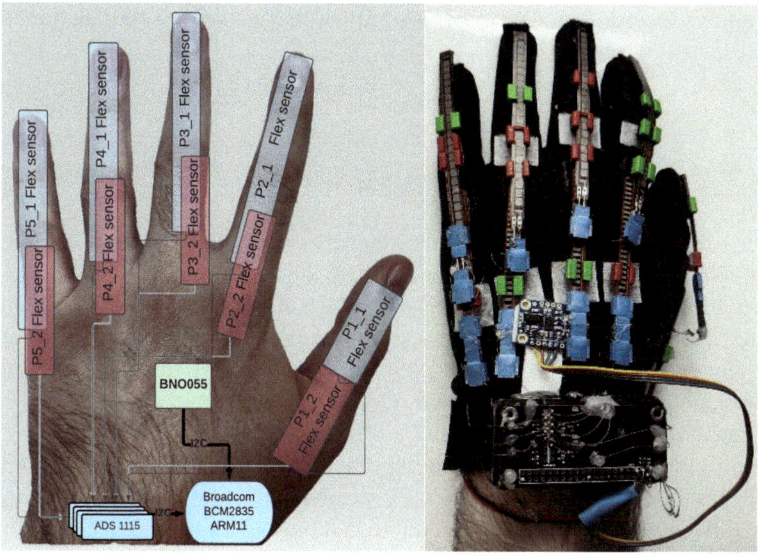

Fig. 2 Block diagram of the device and a picture of a device

Previous versions of the glove have shown the need for a proper sensor mounting system, as improperly made, significantly affects the effectiveness of the glove. To allow the sensors to move freely along the finger with its flexion and at the same time to be rigidly mounted on its tip, a Velcro attachment was used. The sensor is attached to the tip of the finger with Velcro glued to its end, while the path of its movement is determined by guides made by 3D printing, attached to the surface of the glove also with Velcro. This allowed to achieve the necessary freedom for the sensor to work properly. An additional advantage of this system is the ability to adjust the placement of the sensor to the anatomy of the subject's hand, so that the sensor always moves parallel to the finger.

3.2 *Software*

To ensure the multitasking of the glove and the mobility of the test bench, a separate server and router were used which connected all the devices in one network. The server was built based on another SOC—Raspberry Pi 4 version with 4 GB RAM. It acts as a Postgres database server, a Grafana visualization server or an MQTT communication broker depending on the glove software task currently being performed. The glove's control scripts were written in Python using the circuit-Python library to handle individual control tasks. The glove can operate in two modes: continuous operation—which involves sending a continuous stream of data at the maximum possible speed via MQTT communication, or measurement mode, which allows recording at user-controlled intervals and writing all data directly to a database.

3.3 Data Acquisition

Programs for processing, visualization and data analysis were also written using Python. The study was conducted on a group of 15 volunteers, men and women between the ages of 21 and 63. Prior to the survey, the size of the hand was measured from the base of the palm to the tip of the index finger. The range of these values was 16.1 mm to 20.8 mm. None of the test subjects suffered from musculoskeletal disorders of the hand. One of the test subjects is certified in sign language communication.

The process of data acquisition during the measurement session began after the subject was familiarized with the test procedure and the Polish Sign Language alphabet. After activating the entire system and putting on the glove, the test person assumed a comfortable sitting position with the ability to support the hand at the elbow. The test consisted of the subject performing each of the 36 PSL letters ten times. A 3.3 s period was allowed for the execution of a single letter, during which the measurement system recorded 75 vectors containing data from all 10 piezoresistive sensors, as well as from 3 axes of the accelerometer and 3 of the gyroscope. The duration of the measurement window was adjusted empirically. The subject was informed of the recording advancement by a progress bar displayed on the monitor screen. Data viewing immediately after recording was provided to the test person by the Grafana service dashboard. A single examination lasted about 70 min. The test method was approved by the Research Ethics Committee of the Technical University of Lodz.

3.4 Data Augmentation

Using the tsaug package for time series augmentation [12] a method for augmenting the acquired database was developed, that over time the data from one, two or all three groups: piezoresistive sensors, accelerometer, with a fixed number of 75 samples per single time series. The augmentation was random and occurred a maximum twice per series and lasted no longer than 1 s in total. Noise was also added to the signal, amounting to no more than 3% of the average signal value, and random drift. Drift consists of a linear change in the signal value to a value of no more than 50% of the original signal value over a given time range. Example results of the signal augmentation are shown in Fig. 3.

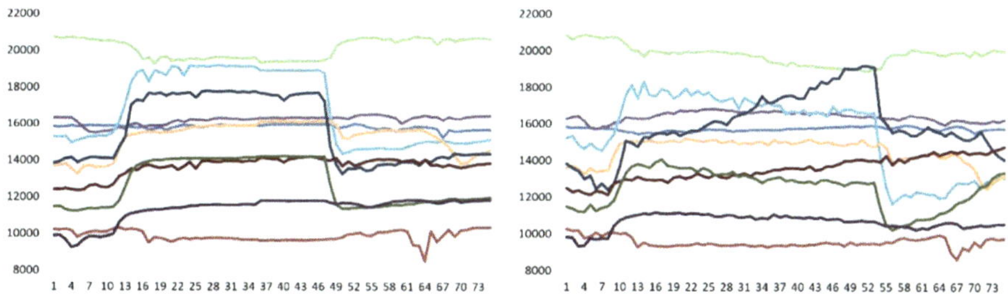

Fig. 3 Example of flex senor data augmentation (original signal on the left)

3.5 *Signal Preprocessing*

Due to the lack of isolation of the conductive threads used, there were occasional short circuits between the signal thread and the power thread. This resulted in the appearance of a supply voltage at the input of the transmitter and thus the appearance of false readouts. To eliminate to eliminate noise and outliers it was necessary to filter signals from all 16 channels using a moving median filter with a window width of 7 samples of the piezoresistive sensor data channels.

3.6 *Data Classification Methods*

We used recurrent neural networks (RNNs) for time series classification. However, simple recurrent layers cannot cope with the classification of every more complex time series. This is due to the problem of fading gradient—a phenomenon involving the inability to train the network [13]. The answer to this problem is the Long short-term memory (LSTM) layers [14] and gated recurrent unit (GRU) [15] which solve this problem by adding a "gate" to each neuron. The LSTM layer is based on memory cells, which can store information for long periods of time, thus LSTM layer can learn to extract and preserve long-term relationships between data. The GRU layer works on a similar principle to the LSTM layer It consists of two gates: a reset gate and an update gate. The reset gate determines which data should be forgotten, and the update gate determines which data should be included in memory state updates. The GRU has fewer parameters than LSTM and is usually faster to learn.

4 Results

4.1 Classification Process

The classification experiments were conducted in parallel on a personal computer and a StandardDS12 computing unit of the Azure platform. The PC is equipped with an Intel i9-13900 CPU, a RTX 2070 SUPER GPU and 32 GB of RAM. StandardDS12 is a platform equipped with a quad-core processor and 28 GB of RAM.

The best performing network architecture, empirically selected through experiments. It consists of convolution values alternating with GRU layers. The deep layers use the SELU activation function.

The effectiveness of the network was measured by the accuracy (1) index defined by the formula:

$$\text{Accuracy} = \frac{\text{Number of correct predictions}}{\text{Total number of predictions}} \tag{1}$$

Three additional values defining the effectiveness of the classification were also determined: specificity (2) sensitivity (3) and precision (4) defined as follows:

$$\text{Specificity} = \frac{\text{True Negatives}}{\text{True Negatives} + \text{False Positives}} \tag{2}$$

$$\text{Sensitivity} = \frac{\text{True Positives}}{\text{True Positives} + \text{False Negatives}} \tag{3}$$

$$\text{Precision} = \frac{\text{True Positives}}{\text{True Positives} + \text{False Positives}} \tag{4}$$

Three variants of the experiment were carried out to classify the signals recorded with the glove: with Leave One Out Cross-Validation (LOOCV) by individual test subjects, with fivefold cross-validation based on data from all test subjects, and with fivefold cross-validation using data from only one test subject multiplied dozens of times using an augmenter. In the last two cases, the data was divided into three sets: training, validation, and test. The validation set accounted for 20% of the total samples, while the test set accounted for 10%. In the case of LOOCV, the test data represented readings from one of the subjects, while the 20% validation set was selected from the remaining data. The performance of these three variants was designed to test the glove's effectiveness in three end-user matching scenarios: no network tuning, network tuning, and network training based on calibration data. 100 epochs were allocated for training the network in each variant. The input of the network is 1/100 of the total number of time series.

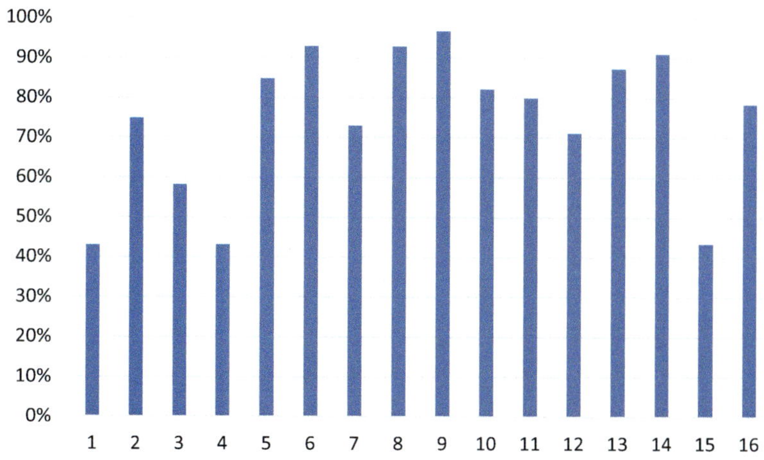

Fig. 4 Recognition accuracy of the signs when testing with the data from an individual subject

4.2 *Classification Results*

As a result of an experiment using data from all subjects, the proposed network architecture achieved a classification accuracy of 76% for the 36 letters of the Polish Sign Alphabet. The network reaches its highest accuracy after 50 epochs. One epoch lasts about 40 s, the entire training takes about 33 min.

The experiment using LOOCV achieved an average efficiency of 70%. In the type case, the network also reached its highest efficiency around 40 epochs. The individual efficiencies for the test subjects are shown in Fig. 4.

To test the effect of augmentation on the effectiveness of gesture classification, training of the network was carried out twice using augmented data from individuals for whom, in the case of LOOCV verification, the network achieved the lowest and highest efficiencies, i.e., from person #1 and 9. For these two cases, the classification with the help of the described network was accurate in 89% and 99% respectively. For a classifier achieving 99% classification accuracy sensitivity was: 97,6%, specificity: 99,9% and precision: 97,7%. The confusion matrix, determined for the test set, is shown in Fig. 5

5 Discussion

Among the three presented methods of training the proposed network architecture for the classification of the letters of the Polish Sign Alphabet, the best results were achieved by the method based on data only from a single surveyed person using data augmentation. The developed augmenter allowed to multiply the data acquired through measurements several times. This means that in the case of a hypothetical

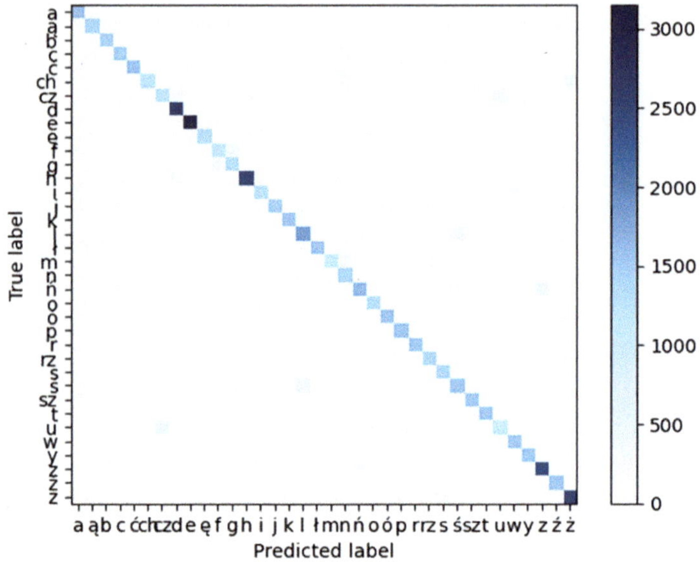

Fig. 5 Confusion matrix

PSL letter recognition device, it will not be necessary to train the network based on data collected from multiple users, but data only from the target user will suffice. It is also possible that there will be no need for the user to perform each letter as many as 10 times in the calibration process.

Reducing this number to 5 repetitions will proportionally reduce the duration of the calibration procedure, thereby making it less burdensome for the user.

The results of the variant with LOO cross-validation show that it is not possible to create an effective universal network without the need for retraining or transfer learning. This is because there is no correlation between the low effectiveness of the network and the parameters of the test, as these were always the same, or the features of the subject's anatomy. The reasons for the low effectiveness are therefore to be found in the mannerisms of some test subjects.

The confusion matrix (Fig. 5) shows that the training network based on the multiplied data of only one person does not have this problem. In addition, it also shows no problem recognizing any of the letters. Even similar letters like "O" and "S" or "Z", "Ź" and "Ż" are successfully recognized by the network.

Comparing the raw efficiency of the presented solution with others collected in [8, 9], it can be stated that it ranks at the top of the efficiency of starting to recognize isolated letters of the sign alphabet. And although many other solutions allowed to achieve a recognition efficiency of 95% one should consider the usability aspect of the proposed solution. The real advantage of the glove is revealed in the fact of conducting tests with its help in many remote locations, in different rooms and times of the day, as opposed to vision solutions checked only in one frame and lighting, or gloves using magnetic field sensors, working in a specific field space. In addition, the design based on the concept of resistive sensors acting as potentiometers allowed simpler implementation and data pre-processing compared to other direct methods.

However, it should be noted again that the presented system is not a full-fledged and independent sign language interpreter. It is only a part designed to function in a system that would additionally recognize other parts of sign language syntax. At present, the glove and the accompanying software allows the examination and recognition of letters performed in a specific period, and thus artificially separated. The next step in the development of the described device will be to enable the examination and recognition of letters/gestures performed in an uninterrupted sequence, which is currently one of the biggest challenges of technologies interpreting sign languages.

6 Conclusions

Presented in this article are the results of signal classification of 36 letters of the Polish Sign Alphabet collected using the author's data glove. To the best of our knowledge, no similar studies and PSL have been conducted to date. A neural network consisting of LSTM layers interleaved with convolutional layers was used for classification. The effectiveness of the proposed architecture, and of the device, was confirmed based on fivefold cross-validation training of the network on three different data sets. The best efficiency of 99% was achieved based on data from a single subject, multiplied using the developed database augmenter. Such high efficiency, relative to the other variants, leads us to believe that in the hypothetical case of the device's introduction into consumer use, a calibration conducted solely from the target user will be sufficient for the device to function properly.

The previous stage of research established a hierarchy of sensors in the process of classifying static PSL-like gestures. The next planned stage of research on the current version of the glove is to train the network again, this time using a reduced number of features, starting with those selected in the previous stage. Confirming the feasibility of such a reduction will make it possible to significantly reduce the size of devices that could potentially serve as a part of sign language interpreters systems.

References

1. U. Nations, "International Day of Sign Languages | United Nations," *Nations, United*. 2021. [Online]. Available: https://www.un.org/en/observances/sign-languages-day
2. "What are the different types of sign language?," *Sign Solutions*. 2021. [Online]. Available: https://www.signsolutions.uk.com/what-are-the-different-types-of-sign-language/
3. Geers, A.E., Mitchell, C.M., Warner-Czyz, A., Wang, N.-Y., Eisenberg, L.S. (2017) Early sign language exposure and cochlear implantation benefits. Pediatrics 140(1). doi: https://doi.org/10.1542/peds.2016-3489
4. NCDHHS (2021) Assistive Technology for the Deaf and Hard of Hearing
5. What Is American Sign Language (ASL)? | NIDCD (2023) Mar. 2019. https://www.nidcd.nih.gov/health/american-sign-language (accessed Feb. 21, 2023)

6. Do, N. T., Kim, S. H., Yang, H. J., Lee, G. S.: Robust hand shape features for dynamic hand gesture recognition using multi-level feature LSTM. Appl. Sci. 10(18). doi: https://doi.org/10.3390/APP10186293
7. Hill, J.: Do deaf communities actually want sign language gloves? Nat. Electron. **3**(9), 512–513 (2020). https://doi.org/10.1038/s41928-020-0451-7
8. Saeed, Z. R., Zainol, Z. B., Zaidan, B. B., Alamoodi, A. H.: A systematic review on systems-based sensory gloves for sign language pattern recognition: an update from 2017 to 2022. IEEE Access 1–1, 2022, doi: https://doi.org/10.1109/ACCESS.2022.3219430
9. Lee, M., Bae, J.: Deep learning based real-time recognition of dynamic finger gestures using a data glove. IEEE Access **8**, 219923–219933 (2020). https://doi.org/10.1109/ACCESS.2020.3039401
10. Lee, M., Bae, J.: real-time gesture recognition in the view of repeating characteristics of sign languages. IEEE Trans. Ind. Informatics **18**(12), 8818–8828 (2022). https://doi.org/10.1109/TII.2022.3152214
11. Piskozub, J.: Rozpoznawanie gestów statycznych ręki za pomocą rękawicy sensorowej. PRZEGLĄD ELEKTROTECHNICZNY **1**(9), 144–147 (2019). https://doi.org/10.15199/48.2019.09.29
12. Wen, T., Keyes, R.: Time series anomaly detection using convolutional neural networks and transfer learning. Accessed: Mar. 06, 2023. [Online]. Available: https://www.fast.ai
13. Bengio, Y., Simard, P., Frasconi, P.: Learning long-term dependencies with gradient descent is difficult. IEEE Trans. Neural Networks **5**(2), 157–166 (1994). https://doi.org/10.1109/72.279181
14. Hochreiter, S., Schmidhuber, J.: Long short-term memory. Neural Comput. **9**(8), 1735–1780 (1997). https://doi.org/10.1162/neco.1997.9.8.1735
15. Chung, J., Gulcehre, C., Cho, K., Bengio, Y.: Empirical evaluation of gated recurrent neural networks on sequence modeling (2014). doi: https://doi.org/10.48550/arXiv.1412.3555

Telemonitoring & Measurement

Smart Pillcase System to Support the Elderly and the Disabled

Michał Śniady and Aleksandra Królak

Abstract This paper analyzes the data concerning the elderly and the aspects of their lives that include taking medications and then proposes a device that was created in order to help elderly organize their medications. A solution consisting of mobile application and a 3D-printed box with Arduino board is proposed. Mobile application is responsible for gathering and processing all the information received from the Arduino, like status of the box (opened or closed), temperature, humidity, and pressure. It is also responsible for saving all new user's medications and displaying them in a form of calendar and list of medications for given day. When the box gets opened, the right medication gets automatically set as taken. The tests made by users showed that the whole product was received really well. Most suggestions concern the software of the product, and they can be easily implemented with software update.

Keywords Pillcase · Arduino · Mobile application · Calendar reminder

1 Introduction

Medicine and its development have enhanced our lives greatly. Diseases that were once lethal can now be controlled and cured with a help of something as little as a pill. Compared to a surgical procedure, downsides of taking pill are almost negligible, the biggest one of them being the fact that patient has to remember about them. Even the most advanced medication will not be effective if patient does not take it with accordance to recommendations of doctor or manufacturer. The problem becomes even bigger when it comes to elderly people. On average in recent years the number of medications taken per day among adults aged 65 and older doubled, and more than a third of them take more than 5 medicaments each day [1]. What is more, according to research from 2015 around a third of the EU elderly live alone [2] which means that there is no one around to remind them and monitor whether correct pills are taken at correct time.

M. Śniady · A. Królak (✉)
Institute of Electronis, Lodz University of Technology, Lodz, Poland
e-mail: aleksandra.krolak@p.lodz.pl

P. Strumiłło et al. (eds.), *The Latest Developments and Challenges in Biomedical Engineering*, Lecture Notes in Networks and Systems 746,
https://doi.org/10.1007/978-3-031-38430-1_28

In the years 2015–2018 in USA, 66.4% of people aged 65 and over were taking three or more prescribed drugs and 41.9% of people aged 65 and over were taking five or more prescription drugs in a time span of one month. Even though the percentages are decreasing, last number is extremely important. Around 50% of elderly take at least one OTC and at least one dietary supplement. It can be safely assumed that almost 50% of elderly are taking around 7 medications of different types during a time span of a month. That number is extremely high considering that most of these medications might have completely different schedule for taking them. It can be clearly seen that when increasing the number of medications that person takes, the percentage of people that meet this criterion decreases quite significantly [3]. Taking a closer look at these statistics but with Europe in mind, it can be seen almost immediately that the trends are nearly identical. While the raw values may be couple percent smaller than the values for US population, they still show that among people aged 65–74 years, 78.1% of them take prescribed medications and this value increases to 87.1% when we change the age criterion to 75 years and over [4]. Aforementioned statistics depict how many different types of medications people, especially elderly, have to administer on a daily basis. It is without a doubt that sooner or later some mistake with taking these medications is going to be made. Study presented in [5] shows that during the 13-year period error rate increased by 100%. Inadvertently taking the wrong drug, administering the incorrect dose, or taking the medication twice are all examples of these errors. Estimations were made by the FDA that around 1.3 million people are injured every year in the US by medication errors [6].

Recent studies show that over 60% of elderly use smartphones [7] and this number was rising at fast pace. This information opens a possibility of a totally different approach to the control of taking pills regularly. With the use of smartphones, the elderly can have an easy access to a mobile application that will act as a virtual assistant and will remind them when it is time to take their medications.

Main goal of this project is to give users a product that will allow them to monitor every aspect of taking their medications with the use of a mobile application and a smart box for pills. On the app user will be able to create its own profile where all the information will be stored. App will be divided into 3 parts: overview of profile with listed medications; detail calendar view with information concerning the time of taking the pills and if the pill was taken; detail view of conditions where the smart box is kept. The second part of the product, the box, will be monitoring and sending data to the app about the conditions and time when it was opened. Box will also serve as a place to store whole blister packs of user's pills.

2 Related Works

With advancing populations's age (especially in western countries) and diseases, technology also advances and produces new solutions that can help people to manage their pill-taking schedule. There are numerous devices like pill organizers, pill dispensers or reminder applications that assist people with their medications. To begin with there are numerous mobile applications that help with managing medications [8–10]. Medisafe Pill Reminder [11] is known for being user-friendly and having sleek design. It is considered to be the best medical reminder software. The app includes personalized daily reminders, essential drug interactions warnings, missed medicine alarms, refill reminders when user is running low and family scheduling features. Thanks to that users and their caregivers receive notifications. MyTherapy [12] is a health management tool that helps user stay in control of their health. Pill reminders, measurement tracking and health and exercise alerts are all included in the app. Users can track patterns and create goals for future with the help of overview charts. Users of this app can manage their medications, receive alerts to take their tablets, and create reminders for medical visits. They can also email reports to doctors directly from the app. Interesting project is My MediHealth [13], medication management application addressed to the pediatric population, mainly to children suffering from asthma. There is also a solution dedicated for women undergoing In Vitro Fertilization (IVF), called Saathi, which works as a reminder of the daily medications and injections schedule [14]. Aside from software solutions, there are also hardware, online pharmacy services and complete products that are created with elderly and pill-taking in mind. Among the commercial products one of the useful devices is TimeCap & Bottle from e-Pill [15]. It displays time and day of the week when the bottle was opened the last time. It allows the user to set up to 24 alarms. When it is the time to take the medicine, the alarm sounds for 10 s and the display flashes until the cap is removed. The alarms repeat every day. If the medication was not taken, the display keeps on flashing. Another product is TabTime Timer [16] which is a solution for people who do not require a pillbox but who want a remainder to take their medications. It is small so it can fit to user's pocket. The TabTime Timer can be used in a variety of situations: as a pill or medication reminder, alarm clock or as a nursing aid on the ward. Scientific projects are mostly based on hardware only solutions, including pill box and microcontroller-based system with RTC (real time clock), LCD display and an alert module composed of a buzzer and LED [17–19]. There are many systems based on various models of Arduino board, with their functionality limited to dispense the pills at the given time and generate the alert [20–24], some of them including also Skype message [18] or SMS to the users and caregivers about the time of taking the medication [25–28]. In recent years the smart pillboxes are developed together with the mobile applications allowing for receiving the notifications and viewing the history of medicine taking [26, 29–32]. There are several projects offering additional functionalities apart from reminding about taking the pills and dispensing them. System described in [32] offers the user interface with Chinese symbols. IR sensor, camera and RFID tags are used in [27] to count the pills in the dispenser. Recognition of the pills, based on CNN, is implemented in the device built of 3D printed components described in [30]. Additional functions, such as monitoring the temperature and humidity of the surrounding, are included in the project presented in [31]. A set of useful sensors, including glucometer, heart rate and temperature measurements, as well as CPR patch, are inplemented in a system described in [25].

3 Materials and Methods

Smart pill boxes should have sensors that can track when the box is opened, allowing the user to determine whether or not the pill was consumed. Sensors in the box should be able to connect with the user's phone through a mobile application. Users of mobile applications should be able to schedule different medications, check if they were taken, and view information about the environment in which the medication box and pills are kept. That was the project's goal. Three fundamental pillars were required in order to achieve the goals:

- A smartphone application
- Smart pill box
- Arduino micro-controller.

3.1 Arduino

Following criteria were utilized to select an Arduino board that was appropriate for this project:

- It must be small in size
- It must be able to transfer data via Bluetooth or the internet
- It must work with motion sensor(s) to determine whether the board has been moved
- It must work with sensor(s) to determine the desired environmental conditions, such as temperature and humidity.

The Arduino Nano 33 BLE Sensing board was selected for this project keeping in mind these specifications. It had all the requested characteristics, and it also had all the required sensors incorporated onto the board, enabling it to operate as a system with just one cable needed to connect it to a power source. Considering that this board has a width of 18 mm and a length of 45 mm, the first condition was easily met. Bluetooth Low Energy connection was utilized to send data from the board to the mobile application. A built-in Inertial Measurement Unit (IMU) was used to monitor whether the box was opened. IMU unit built into this board is a LSM9DS1 sensor. This board has also HTS221 sensor that can check temperature and humidity. Monitoring of air pressure is another characteristic that was used as a result of this board's selection. This was possible because to the Arduino Nano 33 BLE Sense's integrated LPS22HB barometric pressure sensor.

3.2 Mobile Programming Environment

A source code editor by the name of Microsoft Studio Code was used to create the mobile application. The application was created using Dart, version 3.34.0. Version

2.5.3 of the Flutter development kit was used to make the app compatible with iOS and Android devices. Early on in the development process, an application was uploaded by cable to a single iOS phone. Afterwards, a personal Apple Developer account was utilized to collect feedback and test the app on as many devices as feasible.

3.3 CAD Software

A 3D model of a box was created with the help of Creo Parametric Software. It is a computer-aided design program that enables users to create items that can then be 3D printed. The pill box was designed, visualized, and printed using three different types of files:

- *.prt—file generated when a single component of the object is designed. When viewed in Creo Parametric, the user can examine closely at a particular section, change its proportions, and alter its design.
- *.asm—file generated when the user wants to see how an object with numerous parts is combined from several prt files.
- *.stl—native file format for stereolithography. When a part is saved using this extension, the item is converted into a triangle-based object, allowing for 3D printing of the part.

3.4 3D Printing

Model of a pill box was printed using 3D printer called Raise3D E2. It has a build volume of 330 × 240 × 240 mm and the size of the nozzles is 0.4 mm [33]. The box was printed using a PLA material.

4 Project Implementation

4.1 Developed Algorithm for Arduino

The Arduino Integrated Development Environment is the company's proprietary IDE. This IDE is used to write the project's Arduino-specific code. To enable a mobile app to receive, show, or otherwise make use of this data, it must collect and publicize all relevant data. IMU must continuously broadcast temperature, humidity, and pressure even when it is not being used. Only when the IMU is turned on does it need to disclose extra information that the mobile app can turn into proof that the box was opened and medicine was taken. In the meantime, some simple information

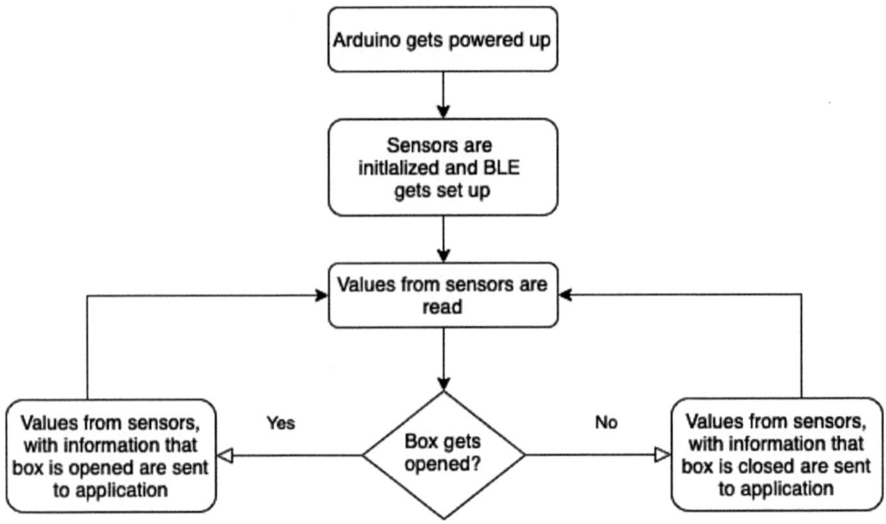

Fig. 1 Diagram of Arduino operation

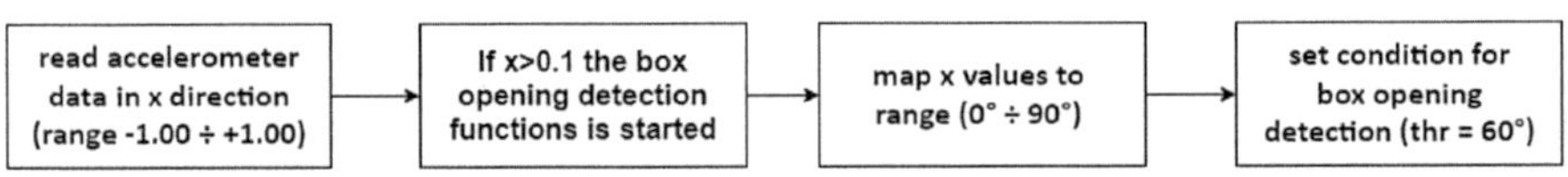

Fig. 2 Box opening detection algorithm

about the state of the Arduino and about the state of the box (if it was opened) are displayed on the serial monitor.

The operation of the Arduino board is presented in Fig. 1.

The flowchart of an algorithm for monitoring if the box was opened is presented in Fig. 2. It is based on the reading from the IMU unit integrated with the Arduino board.

The first step of the algorithm is to access the values of the LSM9DS1 accelerometer readings in X axis. These values range from –1.00 to 1.00 and the value is 0.00 while the Arduino board is stationary. Therefore, the threshold of 0.1 is set for function to be used as it implies that the cover of the box was moved. The next thing to do is to map the readings to a different range, from 0° (the box is closed) to 90° (the box is opened and creates a right angle). The threshold of 60° was set empirically for determining if the box was opened. This algorithm has certain limitations as it may detect the box opening in case the pox is lifted and moved, therefore in the next version of the application data from Y and Z accelerometer axes, as well as from the gyroscope, will be used to prevent false box opening detections.

4.2 Mobile Application

The project's Arduino component was in charge of collecting and sending all the required data. Yet, the user needs to see the data in an understandable and graphical format. The project's mobile application component came in handy in this situation. The Flutter framework was used to construct the app. A single project that can subsequently be compiled for both the iOS and Android platforms was written in the Dart programming language. TakeYourPills is the name of the app, which was created in such a way that it already has all the required files and certifications to be uploaded and made available on Google Play and the AppStore. It has an icon and is connected to a Firebase database.
The operation of the TakeYourPills app is presented in Fig. 3.

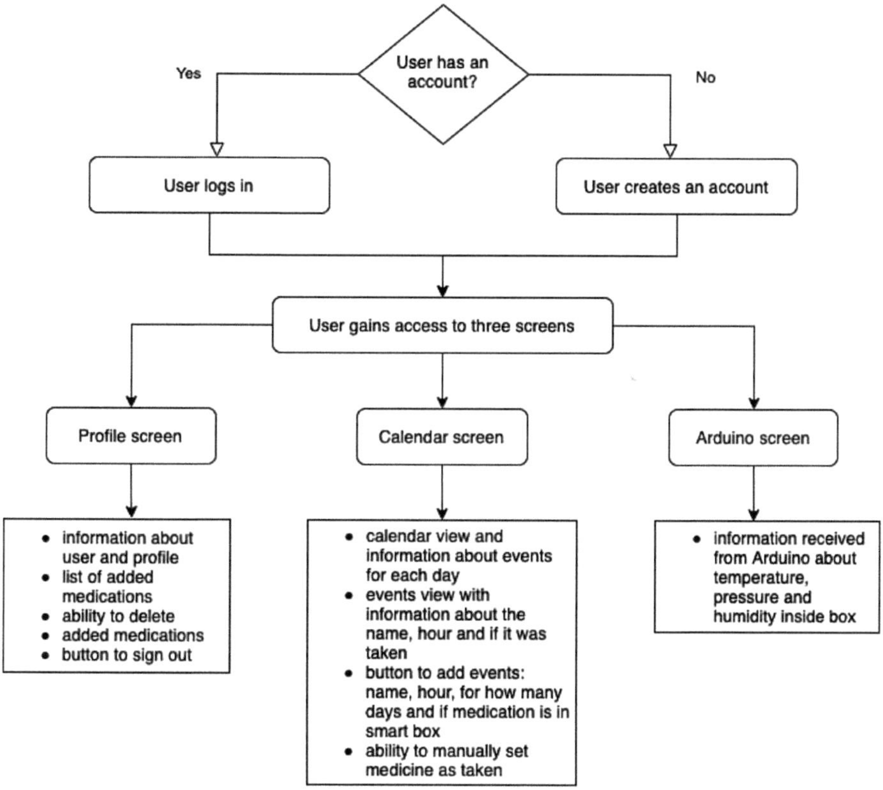

Fig. 3 Diagram of Arduino operation

Application uses several dependencies which allow app to function properly:

– flutter_blue—flutter Bluetooth plugin that enables phone to be used as a central device in Bluetooth low energy connection

- flutter_auth—plugin that enables the app to use Firebase Authentication API i.e., for creating accounts and logging in
- firebase_core—plugin that enables the app to use Firebase Core API, which allows connecting to multiple Firebase apps
- cloud_firestore—plugin used to add database functionality to app which allows user to have their data stored in cloud
- flutter_clean_calendar—plugin based on flutter_calendar package. It adds a simple layout calendar and events functionality to the app
- intl—plugin providing internationalization and localization facilities, including message translation, plurals and genders, date/number formatting and parsing, and bidirectional text
- modal_bottom_sheet—plugin used often i.e., to display additional information or confirmation messages that appear from the bottom of the screen
- flutter_spinkit—plugin enabling the app to use different types of loading indicator animations
- flutter_local_notifications—this plugin is used for displaying local notifications
- rxdart—plugin extending the capabilities of Streams and StreamControllers, which in this app are used i.e., to monitor data advertised by Arduino and to monitor if, and which user is currently logged in.

Screenshots of the mobile aplication are depicted in Figs. 4 and 5.
Current version of the application is created for the households. Users can add, edit and delete medications by themselves. There is no doctor module in current version of the app.

4.3 Pillcase Design

Pill box part of the project is responsible for storing pills and Arduino board. Thanks to the movement of this box, Arduino is able to detect its movement and advertise the data concerning temperature, pressure, humidity and if it is opened or not. The box is composed of three parts. The design of the pillcase is shown in Fig. 6.

First part of the box is the bottom piece. It is created as a base for the whole box. Its main function is to store blisters of tablets that user sets as stored in the pill box when they are adding given medication tin the mobile application. The protrusion was designed to work with the upper piece of the box creating a latch, which ensures, that the box does not get opened unwillingly. Two hinge elements of bottom piece create with hinge element of the upper piece, a tunnel. A bolt is used to connect bottom and upper pieces of the box with the use of these hinge parts. The dimensions of the bottom piece (excluding the protrusion and hinge parts) are 16 cm length, 8 cm width and 4 cm height. These dimensions ensure that many blister packs of different form can be stored in the box. The upper piece of the box is designated to be an opening and closing part. Width and length dimensions (without the additional elements on the front and back of this part) are the same as the bottom piece however,

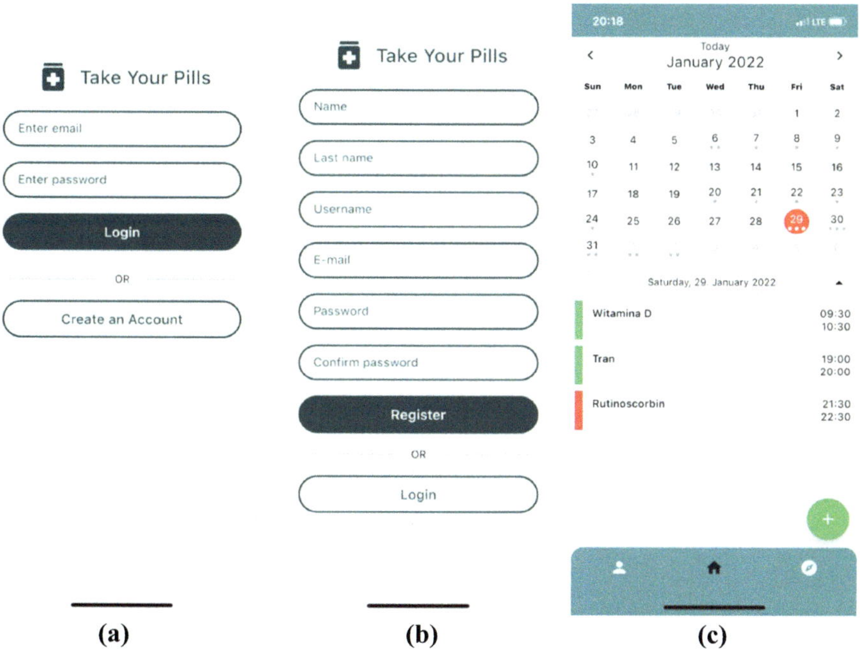

Fig. 4 Example screens from moblie appplication: **a** Login screen, **b** Registration screen, **c** Calendar screen

the height is 2 cm. This part is mostly hollow however, in the middle there is a special tunnel designed to house Arduino board. Because of that, Arduino is able to detect if the box is opened or not. A latch in the upper piece together with a protruding part in the bottom piece create a locking structure. Because of that, in case where the whole box is being moved the chances of it opening unexpectedly are significantly lower. A third part of the hinge mechanism goes between the two elements in the bottom piece, creating a sturdy hinge mechanism that connects these two pieces, and allows the upper one to open and close. Additional elements are two arms that work together with a third element of the box, the latch. The last part of the box is the latch. It is used to cover the entrance to the tunnel where Arduino board is stored. This prevents the board from falling out of the box or from dust or water entering this tunnel and damaging the board. The narrow part of the latch is designed to fit into the tunnel, while the wider parts are designed in a way to fit between the wall of the upper piece and special arms that are next to the tunnel. Such a solution results in that this latch can only be moved in upwards and downwards directions, thanks to the arms it cannot accidentally fall out of its place. The placement of Arduino board (upper part of the box, tunnel in the middle) is presented in Fig. 7.

There is an empty area in the middle of the latch to allow the Arduino to connect to some power source with a wire. In the future version of this box a power bank can be placed in the hollow space in the upper part of the box. Assuming that we

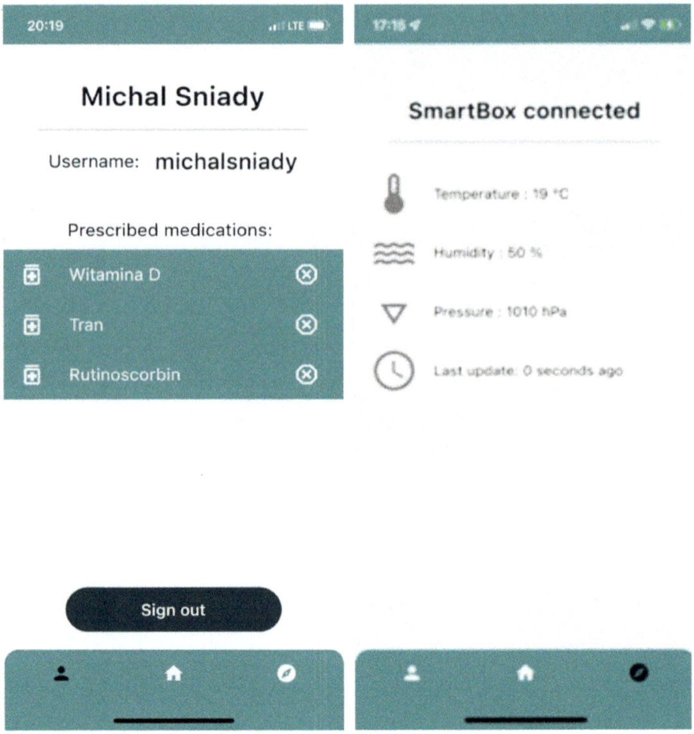

(a) **(b)**

Fig. 5 Example screens from mobile application: **a** Profile, **b** Environment information

(a) (b) (c)

Fig. 6 Pillbox design: **a** front view, **b** back view, **c** front view opened

would be able to fit a 2000 mAh powerbank into the box, and that the average power consumption of our board (while having a delay of 0.1 s set every time it sends an information) is 22 mW, it can be calculated that the board could have been powered with this powerbank for 2 weeks.

5 Results

Qualitative and quantitative results were obtained after performed tests with the prototype. In order to get these results 10 participants tested the product—application and the smart box. There was one criterion that had to be met in order to be included in this test—the tester had to be an elderly. According to theoretical part of this thesis, that meant, that tester had to be around 60 years old or above. During this test, the participant was asked to create an account, add one medicine which is placed in smart box and one which is not, set medications as taken, delete medicine from the list and read values of temperature, humidity, and pressure from Arduino. When it comes to creating an account, this process went smoothly in all cases thanks to an implemented feedback system. A way to access modal bottom sheet to add a medicine was also intuitive. However, a way to manually set medicine as taken turned out not to be that obvious. Suggestions were made that a button that would be visible next to medications that were not taken would be a better way instead of the whole row working as a button. Another suggestion was made concerning the information when setting the medication as taken and when deleting it. Suggestion was to add a name of the medication to the text so that the user can see which medicine will be updated or deleted. When the function of 'Medicine is placed in SmartBox' checkbox was explained to testers, they were also asked to put a blister of tablets into the box and to open it, just to take this blister out. In most of the cases this movement was read correctly, and the application received the information that the box was opened however, in two cases testers took the box into their hands and then opened. Because of the fact that the box was tilted then kept in hands, in one case application received the information that the box was opened because the whole box was tilted, and in the second case the movement of opening the box was not registered at all. Explaining that the box needs to stay on flat surface during opening in order to send correct information solved this problem and in the next try movement was detected correctly. When it comes to layout, the reviews were positive. The factors that were highlighted the most were three easy to navigate screens, green and red indications next to the medications, lack of hidden menus. There was one suggestion made however, about the text size. Tester noted that a slider in profile screen similar

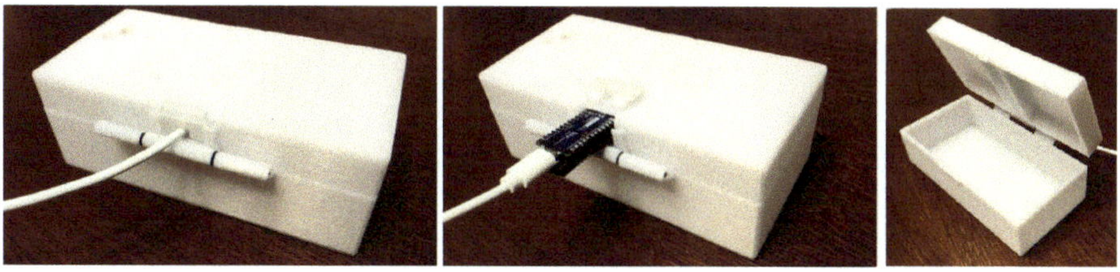

(a) (b) (c)

Fig. 7 Printed box: **a** back view with latch closed, **b** back view with Arduino board visible, **c** front view opened

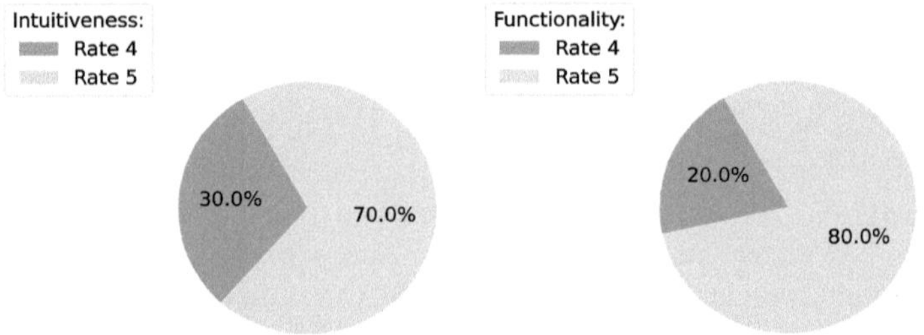

Fig. 8 Results of users questionnaires

to the one in iOS setting with even two text sizes, the default and larger one, would be useful, because they find it hard to read the time span of when to take the medication. Local notifications were also received well however, a request was made to add a third notification exactly at a time when the medication was set. Aside from couple suggestions listed above, the application received positive feedback.

Testers were also asked to assess the app in three aspects: intuitiveness, functionality, and aesthetics on a scale of 1–5. Aesthetics part received the highest mark from all of testers. Intuitiveness (Fig. 8) received the highest mark from 70% of testers, while the functionality received the highest mark from 80% of testers. It is worth noting that no mark below 4 was given for any aspect of the app.

Quantitative research proved that the Arduino code works as planned and that the opening movement gets detected correctly. During this research however, it was found that in order for box to work correctly, during compilation of the code it needs to stay on a flat surface. If it does not, the accelerometer may not be calibrated correctly, therefore not detecting the opening or giving false positive information of the opened box. The values of temperature, humidity, and pressure that were presented to the user in the mobile application were also compared with values received on other sensor in three different rooms. The differences between Arduino and other sensors were in a range of ± 0 to 2 °C for temperature, ± 0 to 4% for humidity and ± 0 to 5 hPa for pressure.

Quantitative and qualitative results proved that testers were happy with how the application works, with only couple suggestions made, and that the product works correctly, giving correct feedback and information to users.

6 Conclusions

The goal of this project was to create a device and an app that will aid elderly in taking their medications. After the implementation of all three parts, a ready to use product was created. All desired features were available to the user of the box and the application. The application is able to create a profile for each user, save in cloud all the medications that user has to take, display all medications for each day with the help of calendar and list of medications, set medications as taken manually and with the help of the SmartBox, send notifications considering medications that should be taken at given time, receive environment information from the box and more. Application works smoothly and is ready to be uploaded to AppStore and Google Play as it was tested and verified with TestFlight tool. The box also meets all the requirements for this project. It can store a lot of blisters of tablets, it constantly checks if it was opened and it sends information about the temperature, humidity and pressure which gives not only a good overview of the conditions that pills are stored under, but also an overall information about factors that are important for user's health and well-being. What is more, it is a one-time setup device. Whenever user's phone is close enough to receive BLE advertised information, the app will automatically connect with the box. Qualitative testing showed that even though there were some suggestions received, the implementation of these suggestions can be done with just one update of the application.

When it comes to improvements in the following versions of this project, aside from adding additional features in application, there is one big change that should be implemented. A microcontroller should be designed and created specifically for this project. Because of that the device could be much more energy efficient, allowing it to operate on batteries instead of having it always plugged to power source. This would also mean that the box could be redesigned.

References

1. Charlesworth, C.J., Smit, E., Lee, D.S., Alramadhan, F., Odden, M.C.: Polypharmacy among adults aged 65 years and older in the United States: 1988–2010. J.S Gerontol. Ser. A, Biol. Sci. Med. Sci. **70**(8), 989–995 (2015)
2. Eurostat: a look at the lives of the elderly in the EU today (2017). https://ec.europa.eu/eurostat/cache/infographs/elderly/index.html
3. CDC: use of multiple prescription drugs (2019). https://www.cdc.gov/nchs/data/hus/2019/039-508.pdf
4. Eurostat: medicine use statistics (2022). https://ec.europa.eu/eurostat/statistics-explained/index.php?title=Medicine_use_statistics#Prescribed_medicines
5. Hodges, N.L., Spiller, H.A., Casavant, M.J., Chounthirath, T., Smith, G.A.: Non-health care facility medication errors resulting in serious medical outcomes. Clin. Toxicol. **56**(1), 43–50 (2018)
6. Columbus, C.: More People Are Making Mistakes With Medicines At Home, Shots-Health News from NPR (2017). https://www.npr.org/sections/health-shots/2017/07/12/536519077/took-the-wrong-medicine-by-mistake-study-finds-such-errors-are-on-the-rise?t=1642521159331
7. Perrin, A.: Mobile Technology and Home Broadband 2021. Pew Research Center (2021)
8. Teixeira, A., Ferreira, F., Almeida, N., et al.: Design and development of Medication Assistant: older adults centred design to go beyond simple medication reminders. Univers. Access Inf. Soc. **16**, 545–560 (2017). https://doi.org/10.1007/s10209-016-0487-7
9. Ameta, D., Mudaliar, K., Patel, P.: Medication reminder and healthcare-an android application. Int. J. Manag. Public Sect. Inf. Commun. Technol. (IJMPICT) **6**(2) (2015)
10. Oprea, F., Rosner, D., Popescu, F., Scrab, A.: Mediminder-medication management and reminder application. In: 2021 20th RoEduNet Conference: Networking in Educa-

tion and Research (RoEduNet), Iasi, Romania, pp. 1–6 (2021). https://doi.org/10.1109/RoEduNet54112.2021.9637712
11. Medisafe App. https://www.mymedisafe.com/
12. MyTherapy Pill Reminder App. https://www.mytherapyapp.com/
13. Slagle, J.M., Gordon, J.S., Harris, C.E., Davison, C.L., Culpepper, D.K., Scott P., Johnson, K.B.: MyMediHealth-designing a next generation system for child-centered medication management. J. Biomed. Inform. **43**(5) (2011)
14. Wadibhasme, P., Amin, A., Choudhary, P., Saindane, P.: Saathi-a smart iot-based pill reminder for IVF patients. In: Senjyu, T., Mahalle, P.N., Perumal, T., Joshi, A. (eds.) Information and Communication Technology for Intelligent Systems. ICTIS 2020. Smart Innovation, Systems and Technologies, vol. 196. Springer, Singapore (2021). https://doi.org/10.1007/978-981-15-7062-9_70
15. TimeCap & Bottle Last Opened Time Stamp with Reminder. https://www.epill.com/timecap.html
16. TabTime Timer. https://tabtime.com/
17. Kuo, G.H.-W.: Research and Implementation of Intelligent MedicalBox. Department of Electrical Engineering, I-Shou University, Kaohsiung, TW. M.S. Thesis (2009)
18. Huang, S.-C., Chang, H.-Y., Jhu, Y.-C., Chen, G.-Y.: The intelligent pill box-design and implementation. In: 2014 IEEE International Conference on Consumer Electronics-Taiwan, Taipei, Taiwan, pp. 235–236 (2014). https://doi.org/10.1109/ICCE-TW.2014.6904076
19. Hayes, T.L., Hunt, J.M., Adami, A., Kaye, J.A.: An electronic pillbox for continuous monitoring of medication adherence. In: Proceedings of the 28th IEEE EMBS Annual International Conference, Aug. 30–Sept. 3 (2006)
20. Minaam, D.S.A., Abd-ELfattah, M.: Smart drugs: improving healthcare using smart pill box for medicine reminder and monitoring system. Futur. Comput. Inform. J. **3**, 443–456 (2018)
21. Shashank, S., Tejas, K., Pushpak, P., Rohit, B.: A smart pill box with remind and consumption using IOT. Int. Res. J. Eng. Technol. **4**(12), 152e4 (2017)
22. Goh, M.L.I., Garcia, M.B., Lalata, J.-A.P., Lagman, A.C., Vicente, H.N., De Angel, R.M.: A pocket-sized interactive pillbox device: design and development of a microcontroller-based system for medicine intake adherence. In: 2019 International Conference on Computational Intelligence and Knowledge Economy (ICCIKE), Dubai, United Arab Emirates, pp. 718–723 (2019). https://doi.org/10.1109/ICCIKE47802.2019.9004276
23. Mohammad, S., Bhowmick, T., Siddique, M.S.U.Z., Khan, M.M., Bhattacharyya, S.: Research and development of a artificial intelligence based smart medicine box. In: 2022 6th International Conference on Computing Methodologies and Communication (ICCMC), Erode, India, pp. 407–412 (2022). https://doi.org/10.1109/ICCMC53470.2022.9753751
24. Bhati, S., Soni, H., Zala, V., Vyas, P.: Smart medicine reminder box. IJSTE-Int. J. Sci. Technol. Eng. **3**(10) (2017). , ISSN: 2349-784X (online)
25. Nalini, M., Abirami, V., Aishwarya Lakshmi, G., Harini, D.: IoT based smart medicine kit. Mater. Today: Proc. **46**, 4125–4127 (2021)
26. Srinivas, M., Durgaprasadarao, P., Raj, V. N. P.: Intelligent medicine box for medication management using IoT. In: 2018 2nd International Conference on Inventive Systems and Control (ICISC), Coimbatore, India, pp. 32–34 (2018). https://doi.org/10.1109/ICISC.2018.8399097
27. Harshitha, V., Sandeep, K., Swasthika Jain, T.J.: An interactive pill box using IoT. Int. J. Eng. Adv. Technol. (IJEAT) **9**(3) (2020). ISSN: 2249-8958 (Online)
28. Deepthi, A.S., Satish Kumar, G.A.E.: Smart pill notification device using Arduino. In: 2022 IEEE International Conference on Distributed Computing and Electrical Circuits and Electronics (ICDCECE), Ballari, India, pp. 1–5 (2022). https://doi.org/10.1109/ICDCECE53908.2022.9792842
29. Ahmad, S., Hasan, M., Mohammed, G.P., Shahabuddin, M., Tabassum, T., Allvi, M.W.: IoT based pill reminder and monitoring system. IJCSNS Int. J. Comput. Sci. Netw. Secur. **20**(7) (2020)
30. Tsai, K.-L., Liau, B.-Y., Hung, Y.-M., Yu, G.-J., Wang, Y.-C.: Development of smart pillbox using 3d printing technology and convolutional neural network image recognition. Sens. Mater. **32**(5), 1907–1912 (2020)

31. Pradheep, B., Rajan, T., Brightny, A., Rebeak, C. D., Poojasen, S., Vijay, P., Suresh, M., Rohith Alex, R.: Smart pill box with reminder to consume and auto-filling process using IoT. In: Proceedings of the 5th International Conference on I-SMAC; IoT in Social, Mobile, Analytics and Cloud (2021)
32. Tsai, H.-L., Tseng, C.H., Wang, L.-C., Juang, F.-S.: Bidirectional smart pill box monitored through internet and receiving reminding message from remote relatives. In: 2017 IEEE International Conference on Consumer Electronics-Taiwan (ICCE-TW), Taipei, Taiwan, pp. 393–394 (2017). https://doi.org/10.1109/ICCE-China.2017.7991161
33. Raise 3D E2. https://www.techradar.com/reviews/raise3d-e2

Opportunities of Data Medicine: Telemonitoring of Multimodal Medical Data in Outpatient Care

Alexander Keil, Nick Brombach, Olaf Gaus, Rainer Brück, and Kai Hahn

Abstract Telemonitoring of vital signs combined with other multimodal health data can provide a basic foundation for data medicine. The collected data can then be analyzed semi-automatically and made available to the physician for better assessment of the state of health. This procedure was tested in use-cases in a nursing home and with elderly people at home. In addition to several vital signs, patients were able to submit details of their health status. To enable a high number of valid measurements, support measures were established that are adapted to residents in nursing homes, in the home, and caregivers.

Keywords Telemonitoring · Medical data · Data driven medicine · Vital signs · Remote patient monitoring

1 Introduction

Demographic change confronts society with a variety of challenges [1]. This applies in particular to medical care, as demand increases with age [2, 3]. Technical processes can help here and open the way to data driven medicine. In this, multimodal medical data is linked to semi-automatic evaluation to enable an assessment of the patient's state of health [4]. An important part of this data can be collected by telemonitoring, where the patient regularly records and transmits his/her vital signs. This data can

---

A. Keil · R. Brück · K. Hahn
University of Siegen, Medical Informatics and Microsystem Engineering, Am Eichenhang 50, 57068 Siegen, Germany

N. Brombach
University of Siegen, Cyber-Physical Systems, Kohlbettstraße 15, 57072 Siegen, Germany

A. Keil (✉) · O. Gaus
University of Siegen, Institute Digitale Modellregion Gesundheit Dreiländereck, Weidenauer Straße 167, 57076 Siegen, Germany
e-mail: alexander.keil@uni-siegen.de

P. Strumiłło et al. (eds.), *The Latest Developments and Challenges in Biomedical Engineering*, Lecture Notes in Networks and Systems 746,
https://doi.org/10.1007/978-3-031-38430-1_29

make a valuable contribution not only to the treatment of the patient but also to prevention from diseases [5].

In the following, use-cases for telemonitoring of multimodal data will be presented. The data was processed and made available to the physician for evaluation. Subsequently, the experiences and acceptance of patients, physicians and nursing staff are discussed. Finally, technical and non-technical support options for conducting an RPM study are discussed and their effectiveness evaluated. The purpose of the work is to help to pave the way to an efficient data based health system where data mobility could partly replace patient mobility taking into account the specific German lack of GPs in rural areas.

2 Related Works

Remote Patient Monitoring (RPM) have already been practiced for a long time but in the last decades it has improved dramatically [6, 7]. Several studies have been performed about the topic with different focus. According to [8], the number of publications about RPM has nearly doubled from 2009–2011 to 2012–2014, and again increased by 64% in 2015–2018.

RPM has been shown to have the potential to support many common clinical pictures. In [8] it was examined what type of diseases were present in the studies. From the 272 selected articles, 47.8% were about cardivascular diseases, 18.0% about endocrinological diseases, 13.6% about elderly patient care and 11.3% about pulmonary diseases [8]. Another review showed that RPM has a positive treatment effect for patients with hypertension [9].

In another review, the focus was about the evaluation criteria of RPM projects with heart failure. 128 publications have been analyzed by 52 evaluation criteria in the following six dimensions: clinical, economic, user perspective, educational, organizational and technical. 97% of the studies considered only the three of these dimensions, none all of them [10].

The study in [11] is very similar to the one presented below. It has also taken place in Germany. Multimorbid patients older than 65 years were selected. They were asked to measure their vital signs independently using provided measuring devices. The data was transmitted to the physicians via a tablet with a corresponding app. The measurement intervals and vital signs were specified by the physician. Measuring devices for blood pressure, heart rate, oxygen and weight were available. There were a total of 117 study participants, of whom 116 completed the study. The main reason for dropout was no perceived (informational) health benefit. For a smooth running of such a study it was recommended to include the users in the development process and to provide technical support [11].

Also the field of wearable vital sign monitors is growing allowing continuous monitoring. These can be patches, in part of the clothes, chest straps of bands for arm of wrist [12].

3 Methods and their Applications in Use-Cases

Telemontoring consists of gaining medical data from patients that are not under direct supervision of medical personnel. Different scenarios are possible in this context. The focus of this paper is on outpatient care where vital sign measurements take place at the patients' homes or in nursing homes. The vital data acquisition is performed without physicians either by the patients themselves or by caregivers. General practitioners (GP) have remote access to the measured data. The required methods were developed in real-life use-cases. The technical infrastructure as well as the medical procedures were subject of a research project carried out in a rural area of North Rhine-Westphalia in Germany in the years 2021/2022.

As vital signs blood pressure, heart rate, blood oxygen saturation, weight, ECG and blood glucose came into consideration. These vital signs can easily be measured with inexpensive and commercially available devices and are important to monitor several common diseases such as heart failure, arterial fibrillation and hypertension, as well as cases of multimorbidity [13]. GPs prescribed individual vital sign measurements for their patients. These prescriptions describe which vital sign to be measured, at which time of day, and how often. The goal of the project was to test the transfer of self-measured vital data of elderly people from the patient to the attending GP. This way, the physician has a better foundation for possible modifications of the treatment or the medication.

An overview of the developed technical infrastructure is shown in Fig. 1. The measuring devices used were from Beurer (a leading producer of such devices). All of them were equipped with Bluetooth and transmitted their data wireless to a smartphone.

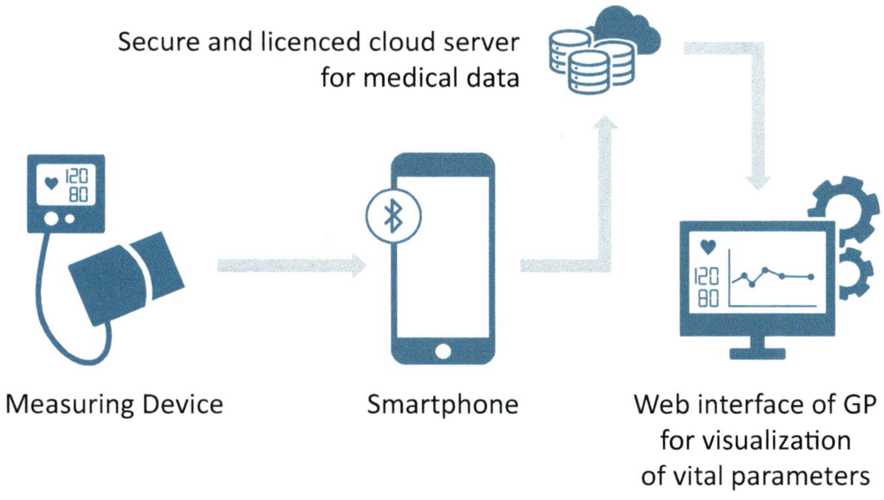

Fig. 1 Technical infrastructure of the use-case project

3.1 Methods for Secure Data Transfer to the Cloud

As smartphones are used more and more [14] and even in the older generations [15], they were chosen to serve as a gateway to the internet. Receiving the Bluetooth transmission from the devices, an app from Beurer forwards it internally to Apple Health environment, which serves as a hub for exchanging vital data. The crucial transfer software for cloud access was developed in the project and realised in an iOS app. It fetched the data from Apple Health and forwarded them to AWS (Amazon Web Services) cloud environment. Additionally, it displayed the medical prescription, the history of all vital data and a feedback about successful data transmission [16].

AWS was used as a certified cloud service to store medical data. DynamoDB was chosen as database and Cognito for user authentication. In future implementations, Medical Data Spaces can be used instead of classical clouds like proposed by the International Data Spaces Association (IDSA). In this way, it is possible to reduce data silos by using federated data systems and enable interoperable exchange of data between systems from different providers. This opens up new perspectives with regard to the patient's data sovereignty [17].

3.2 Secure Web Interface for the Physician

For the GPs, a web interface was provided to access the cloud data and visualize them. Implemented in C# with ASP.NET it was hosted on a virtual server of the university's data center. It was protected with individual credentials for each physician. Two general practitioners (GP) and a nursing home participated in the project. Initially GPs recruited their patients. In a first phase, patients measured their vital signs with or without the help of the nursing staff in a nursing home and sent them to the cloud. In a second phase patients living at home sent their vital data. GPs only needed a web browser to get access to all patients' data [18].

4 Multimodal Data Acquisition

In the use-case studies, data was collected from multiple sources as shown in Table 1.

4.1 Vital Data

Vital data was recorded from different measurement devices, mostly from Beurer. Table 2 shows an overview. The BM 85 was used to measure blood pressure and

Table 1 Collected data and their sources

Data type	Data source	Data input
Blood pressure, heart rate	Beurer BM 85	Bluetooth
Blood oxygen saturation	Beurer PO 60	Bluetooth
Weight	Beurer BF 700	Manual input
ECG	Apple Watch Series 6	Bluetooth
Blood glucose	Patient's own devices	Manual input
Health complains	Input form in app	Manual input
Metadata	Smartphone	Automatically generated
Physicians comments	Physician's web frontend	Manual input
"Read" mark for vital data	Physician's web frontend	Manual input

Table 2 Overview of the used vital sign meters

Device name	Used vital values	Comment
Beurer BM 85	Blood pressure, heart rate	–
Beurer PO 60	Blood oxygen saturation	–
Beurer BF 700	Weight	Bluetooth not used
Apple watch series 6	ECG	Heart rate and oxygen saturation not used

heart rate, the PO 60 measured blood oxygen saturation. As scale the BF 700 was used but it's integrated bluetooth functionality was not used. When the scale was connected to the smartphone together with other Bluetooth devices, there were permanent disturbances in the data transfer of all devices. Therefore, it was decided to enter the weight manually in the app instead.

For measuring ECGs, Apple Watch Series 6 were used. For Apple Health, there is currently no interface available to import ECG data from third party apps. Therefore the only way to add ECG data is to measure them with an Apple Watch. The Apple watch is also capable of measuring the heart rate and the oxygen saturation, but these functions were not used. The heart rate is measured continuously, but without any context about the activity of the wearer, these large numbers of values cannot be interpreted by the physician. The measurement of oxygen saturation is not medically certified and was therefore not used in the project.

In addition to the vital data, meta data was collected about date and time of the measurement and the data transmission, and the source of the data.

4.2 Manual Input from Patients

Measuring blood glucose was not included in the original project concept, therefore no measurement devices were acquired. However, since the physicians involved in the project have repeatedly emphasized the relevance of blood glucose measurements especially for their diabetes type 2 patients, it was decided to include it in the project. Since patients used their own devices, an automatic transmission of vital signs could not be realized. A manual interface for patients with appropriate measuring devices have been created to enter their values manually and transmit them to the cloud.

An option for patients to manually add health complains has also been provided. For this purpose, a list of common health complaints was developed together with the physicians participating in the project. This includes e.g. diarrhea, fever, and various different pains. In addition to the predefined complaints, "Others" can be selected and a text field can be used to explain the complaint further.

4.3 Data Input by Physicians

Using the web interface, the physicians were able to add comments for internal use to every vital value. They were also able to mark vital values as "read". This information was communicated to patients via the app so they could be sure their vital signs were received and seen by their GP.

4.4 Data-Based Physician-Patient Consultation

The physician could access all vital data and metadata of their patients. This information was used during consultations with the patients to allow a better assessment of the state of health. Here, the prescribed measurements could be changed to have a closer look on certain values, however, this option was not taken up by the physicians involved. Furthermore, the physician could adjust the medication of their patients at any time based on the collected data. This option was used several times, resulting in improved treatment for the patients.

5 Semi-automatic Data Analysis and Approach for Automation

Since information is defined by data with semantics, the data must be interpreted to be of use to the physician. Data preparation and data preprocessing is necessary as physicians do not have time trying to assess long list of data in order to identify anomalies. Several options for semi-automatic data preparation were considered and are explained in the following sections.

5.1 *Graphical Data Preparation*

In order to have an overview of the patient's data set at a glance, it is necessary to prepare it graphically. For the use-cases described above, a graphical overview page was added to the web interface for the physician. On this page all recorded vital data of the selected patient of the last 14 days are displayed using a common time axis (see Fig. 2).

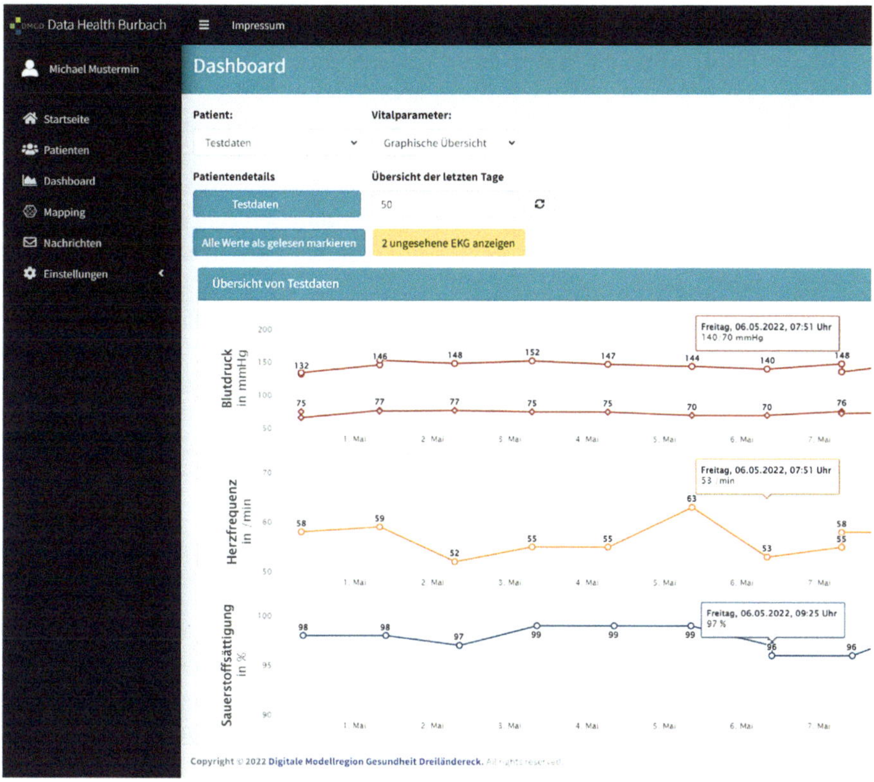

Fig. 2 Graphical overview of all vital values of a selected patient in the physician's web interface

5.2 Thresholds and Early Warning Scores

In addition, patient individual thresholds are in use for every vital parameter. If a value is not within a certain value range, it is marked as an outlier and highlighted accordingly. This is similar to the British National Early Warning Score (NEWS2) which was introduced in 2012 as a guide used by medical services to quickly determine the degree of illness of a patient [19]. However, because only few or single vital parameter are available, a useful summarized score as proposed in NEWS2 cannot be created in this approach.

5.3 Use of Artificial Intelligence

In future, the integration of data analysis based on artificial intelligence (AI) is conceivable and will be feasible. AI-based data evaluation automatically detects anomalies within value sets and highlight them appropriately for the treating physician [20]. AI can also be able to use vital data as well as data from other possible sources like an electronic health record to detect impending illnesses at an early stage. This could possibly prevent the outbreak of the disease completely or at least mitigate it [4]. However, it is important to note that an AI can only serve as a decision support system. The final decision regarding diagnosis and treatment must always be made by a physician.

In perspective, the vital data generated by telemonitoring together with further examinations and medical diagnoses can serve as a foundation of a Digital Twin. This digital representation of the patient's body (or a least of single human organs) can be used as simulation model that enables diagnoses resulting in preventive action to be taken at an early stage in the event of impending illnesses [21].

6 User Experience and Acceptance

Such a digital health solution can only be widely deployed if it is supported by all steakholders and meets with widespread acceptance. For this reason, a parallel study was conducted to assess the experiences of patients, physicians and caregivers using the system. At the same time, the technical infrastructure was permanently enhanced on the basis of feedback from the stakeholders and support structures were set up.

6.1 By Patients

With regard to acceptance, a mixed picture has emerged among patients. The patients living at home were on average 69.4 years old and most of them were relatively fit, both physically and mentally. The patients in the nursing home however were on average about 8 years older. Furthermore, some of them were physically as well as mentally impaired.

Nursing home Some of the patients in the nursing home had a relatively high affinity for technology and were also in possession of computers, smartphones or voice assistants. These persons did not have any major problems in measuring their vital signs by themselves. Some of the residents were no longer mentally capable of familiarizing themselves with the devices due to their age. Among the other residents, there were some who took up the new technical challenge and found out how easy it was to use after an initial familiarization phase. However, there were also residents who rejected the new technology from the beginning and did not even want to try to operate it. In the end, 8 of the 19 residents of the nursing home measured their vital signs themselves, the rest had to require help from the nursing staff.

Figure 3 shows an overview of the measurements done in the nursing home. The blue area represents the number of incoming measurements of every day from the patients who measure their vital signs themselves. The green add-on area above shows measurements by the nursing staff. The grey vertical lines represent visits of the project team at the nursing facility. At the end of the project phase, there was an outbreak of COVID-19 which prohibited further visits marked in red.

At the end of the right end of Fig. 3, the number of measurements is suddenly dropping. At this time, interviews with several participants were conducted who then thought that the project was over now and no further measurements have to be done. A few days later, all the equipment was collected from the nursing home and this project phase ended.

It should be mentioned that within this project neither the use of the system in the nursing home nor at patients' home were intended to react on alarm events such as vital data requiring immediate help from emergency units. Since the data was only made available to the doctor by intentionally accessing the data base, emergency scenarios had been provided conceptionally but as the legal and organisational requirements are very high, they were not pursued further in these use cases.

Patient's home Due to the comparatively younger age of the patients at home, the acceptance was completely different. All project participants were used to smartphones, so that basic operations were not a problem. Since the participants were spread over a large rural area, it was no longer possible to visit them personally in the event of technical problems. Instead, a support helpline was set up so that project participants could contact technical project staff at any time.

Figure 4 shows an overview of the measurements at patient's home. The blue area is the number of measurements per day. The yellow lines mark visits at the patients home while the brown lines marks calls at the support helpline.

6.2 By Caregivers

The organization of measurements by the nursing staff has proven to be difficult. There is about the same number of caregivers for every resident (out of 120 in total) in the facility. They are organized in three shifts and most of them work only part-time. Therefore, there is always a mix of different caregivers at the stations and it is virtually impossible to inform everyone personally about the project implementation.

Furthermore, it was unclear to the caregivers which vital signs had to be measured when and for which resident. As part of another digitization project in the nursing home, information screens for the care givers were installed during the project period to provide current information. They were operational a little after half the project phase, and were used to indicate upcoming measurements at each station. Therefore, as can be seen in Fig. 3, the number of vital data measured by the nursing staff increase sharply from Feb 7. At the end of the project phase, it is dropping again because of an outbreak of COVID-19.

Although the nursing staff recognized potential in the specific measurement of vital signs, the evaluation was still rather negative. The performance of the measurements was perceived as an additional burden. Furthermore, an integration of the vital data into the nursing documentation was desired, as some measurements would otherwise have to be entered here additionally in the course of the nursing routine.

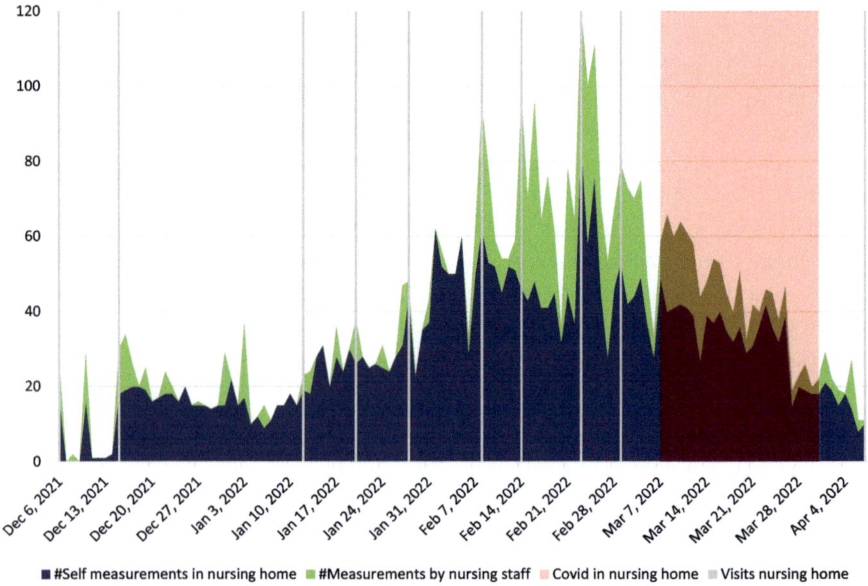

Fig. 3 Measurements in the nursing home during the project

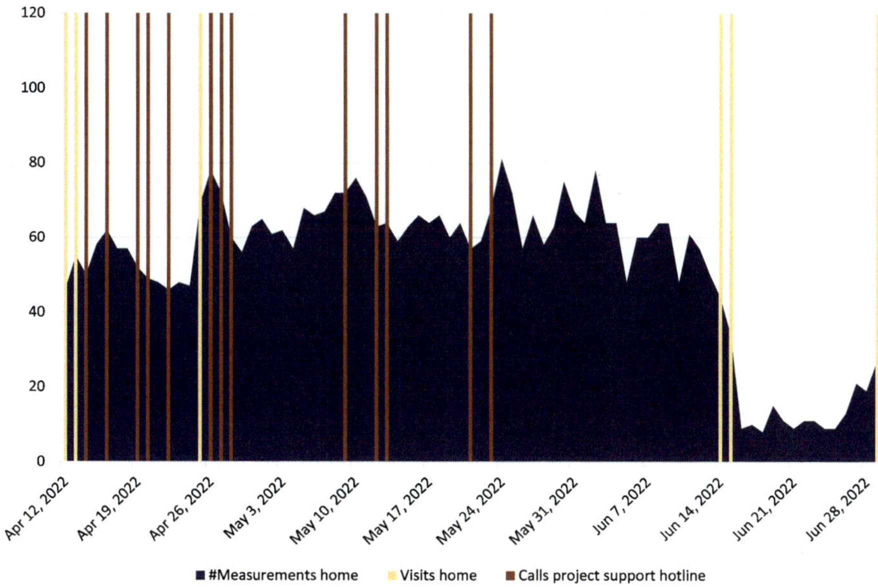

Fig. 4 Measurements at patient's home during the project

6.3 By Physicians

The GPs involved saw great potential in data-based medicine, as it can provide better insight into the patient's state of health. Based on the vital data, several medications were also adjusted during the course of the project. The range of functions and the visual design of the web interface were optimized several times in dialog with the physicians.

However, in order to be used practically in future, it is not sufficient to have the vital data on a separate web interface. It must be integrated into the patient's electronic health record within the practice management system, as it is already done for example in Denmark [22], and practice management systems so that this can be easily accessed.

6.4 Support Measures

During the project, many support measures were implemented to ensure smooth transmission of vital data.

Regular visits have proven to be the most effective measure in the nursing home. These were intended in particular to support residents who measure their vital signs themselves. Before each visit, the project team was able to use the physician's website with an separate access to determine which patients had no or incomplete new vital data. They were then visited in a targeted manner. In the home, this measure proved to be impractical due to the large distance, as it would have taken several days to visit all participants. However, due to the better technical affinity, this was also not necessary to the same extent.

One of the first additional measures was the establishment of a telephone helpline. Via this helpline, contact could be established with a project staff member at any time. The helpline was already available most of the time in the nursing home, but was only used by the residents in rare cases. In the home, however, the helpline proved to be a very effective and was frequently used. A large number of problems could be solved efficiently without time-consuming on-site visits.

An explanatory video was created to introduce the project, the measuring devices and the app. This was mainly used to inform the nursing staff. However, it was also integrated into the app so that every study participant could watch it at any time.

6.5 Additional Challenges

There were some additional challenges, especially in the nursing home. In addition to the aforementioned COVID-19 outbreak, there were also cases of scabies, which is why one visit had to be terminated prematurely. Furthermore, one patient in the nursing home died shortly after the start of the intervention. Another patient became a palliative case and therefore dropped out of the project. Because patients in the home were much more mobile, several went on vacation during the project phase. Due to the resulting changes in daily routine or the lack of an Internet connection, there were gaps in the data, sometimes lasting several days.

7 Conclusion

Telemonitoring of patients has high potentials in the future. It can make a valuable contribution to the treatment, but especially to the prevention and early detection of diseases. In two use-case scenarios, multimodal medical data was obtained first in a nursing home and second with elderly people living at home. The vital data came from several measuring devices. These data were supplemented by manual inputs from both the patient and the physician and stored together in a cloud. Metadata was collected in parallel.

This data was visually processed for analysis by the physician and provided with threshold values similar to early warning scores. In the future, the data analysis could be supplemented by AI components and additional data sources like an electronic health record.

However, the realization of such a telemonitoring project also involves a number of challenges that have to be met, both humanly and technically. For self-measuring residents in the nursing home, regular visits are the best measure to ensure continuous data. Decentralized information systems, such as information in the individual wards and explanatory videos, are a suitable measure to support the nursing staff during the measurements. In the case of patients at home, who are clearly fitter, most usage problems can already be solved by advice via telephone. Physicians see the potential, but the data must be included in their IT systems.

References

1. van den Bussche, H.: Die Zukunftsprobleme der hausärztlichen Versorgung in Deutschland: Aktuelle Trends und notwendige Maßnahmen In: Bundesgesundheitsblatt 62, pp. 1129–1137 (2019). https://doi.org/10.1007/s00103-019-02997-9
2. Ulrich, R.E.: Demographic change in Germany and implications for the health system. J. Publ. Health **13**, 10–15 (2005). https://doi.org/10.1007/s10389-004-0084-8
3. Denton, F.T., Spencer, B.G.: Chronic health conditions: changing prevalence in an aging population and some implications for the delivery of health care services. Can. J. Aging/La Revue canadienne du vieillissement **29**(1), 11–21 (2010). https://doi.org/10.1017/S0714980809990390
4. Tawalbeh, L.A., Mehmood, R., Benkhlifa, E., Song, H.: Mobile cloud computing model and big data analysis for healthcare applications. IEEE Access **4**, 6171–6180 (2016). https://doi.org/10.1109/ACCESS.2016.2613278
5. El-Rashidy, N., El-Sappagh, S., Islam, S.M.R., M. El-Bakry, H., Abdelrazek, S.: Mobile health in remote patient monitoring for chronic diseases: principles, trends, and challenges. Diagnostics **11**, 607 (2021). https://doi.org/10.3390/diagnostics11040607
6. Nangalia, V., Prytherch, DR., Smith, GB.: Health technology assessment review: remote monitoring of vital signs—Current status and future challenges. Crit. Care. **14**, 233 (2010). https://doi.org/10.1186/cc9208
7. Zhao, F., Li, M., Tsien, J.Z.: Technology platforms for remote monitoring of vital signs in the new era of telemedicine. Expert Rev. Med. Dev. **12**(4), 411–429 (2015). https://doi.org/10.1586/17434440.2015.1050957
8. Frederico Arriaga Criscuoli de Farias, F.A.C., Carolina Matté Dagostini, C.M., Yan de Assunção Bicca, Y.A., Vincenzo Fin Falavigna, V.F., Asdrubal Falavigna, A.: Remote patient monitoring: a systematic review. Telemed. e-Health **26**(5), 576–583 (2020). https://doi.org/10.1089/tmj.2019.0066
9. Kalagara, R., Chennareddy, S., Scaggiante, J., Matsoukas, S., Bhimani, A., Smith, C., Putrino, D., Dangayach, N.S., Mocco, J., Schnipper, J.L., Auerbach, A.D., Kellner, C.P.: Blood pressure management through application-based telehealth platforms: a systematic review and meta-analysis. J. Hypertens. **40**(7), 1249–1256 (2022). https://doi.org/10.1097/HJH.0000000000003164
10. Farnia, T., Jaulent, M., Steichen, O.: Evaluation criteria of noninvasive telemonitoring for patients with heart failure: systematic review. J. Med. Int. Res. **20**(1) (2018). https://doi.org/10.2196/jmir.7873
11. Lang, C., Voigt, K., Neumann, R., Bergmann, A., Holthoff-Detto, V.: Adherence and acceptance of a home-based telemonitoring application used by multi-morbid patients aged 65 years and older. J. Telemed. Telecare **28**(1), 37–51 (2022). https://doi.org/10.1177/1357633X20901400
12. Soon, S., Svavarsdottir, H., Downey, C., Jayne, DG.: Wearable devices for remote vital signs monitoring in the outpatient setting: an overview of the field. BMJ Innov. **6**, 55–71 (2020). https://doi.org/10.1136/bmjinnov-2019-000354

13. Keil, A., Hahn, K., Farhan, N., Hillerkuss, D., Brück, R., Gaus, O.: Self-measurement of vital signs to prevent and treat common disease patterns. In: E-Health and Bioengineering Conference (EHB), pp. 01–04, Iasi, Romania (2022). https://doi.org/10.1109/EHB55594.2022.9991709
14. Anteil der Smartphone-Nutzer* in Deutschland in den Jahren 2012 bis 2021 [Graph]. VuMA. https://de.statista.com/statistik/daten/studie/585883/umfrage/anteil-der-smartphone-nutzer-in-deutschland/
15. Anteil der Smartphone-Nutzer in Deutschland nach Altersgruppe im Jahr 2021 [Graph]. VuMA. https://de.statista.com/statistik/daten/studie/459963/umfrage/anteil-der-smartphone-nutzer-in-deutschland-nach-altersgruppe/
16. Keil, A., Hahn, K., Brombach, N., Farhan, N., Brück, R., Gaus, O.: Acquisition and processing of biomedical data for outpatient care in rural areas. Curr. Direct. Biomed. Eng. **8**(2), 81–84. https://doi.org/10.1515/cdbme-2022-1022
17. Nagel, L., Lycklama, D.: Design principles for data spaces (1.0). Zenodo Dortmund (2021). https://doi.org/10.5281/zenodo.5244997
18. Keil, A., Hahn, K., Brück, R., Brombach, N., Farhan, N., Gaus, O.: Cloud-based system for vital data recording at patients' home. In: Information Technology in Biomedicine. ITIB 2022. Advances in Intelligent Systems and Computing, vol. 1429. Springer, Cham. https://doi.org/10.1007/978-3-031-09135-3_2
19. National Early Warning Score (NEWS) 2: Standardising the assessment of acute illness severity in the NHS. Updated report of a working party. Royal College of Physicians, London (2017)
20. Kahn, Z.F., Alotaibi, S.R.: Applications of artificial intelligence and big data analytics in m-health: a healthcare system perspective. J. Healthc. Eng. Hindawi (2020). https://doi.org/10.1155/2020/8894694
21. Bruynseels, K., Santoni de Sio, F., van den Hoven, J.: Health care in the cyberspace: medical cyber-physical system and digital twin challenges. Front. Genet. **9** (2018). https://doi.org/10.3389/fgene.2018.00031
22. Fischer, P.: Digital Health: Untersuchung zur Akzeptanz der elektronischen Gesundheitsdatenspeicherung in Form der elektronischen Patientenakte (ePA) in Deutschland. In: ifgs Schriftenreihe der FOM 17, MA Akademie Verlags- und Druck-Gesellschaft mbH, Essen (2019). ISBN: 978-3-89275-116-8

Measurement of Blood Flow in the Carotid Artery as one of the Elements of Assessing the Ability for Pilots in the Gravitational Force Conditions–Review of Available Solutions

Ewelina Sobotnicka, Jan Mocha, Aleksander Sobotnicki, Jerzy Gałecka, and Adam Gacek

Abstract Continuous monitoring of cardiac function is highly desirable for long-term assessment of cardiovascular health, detection of acute cardiac dysfunction. Information from volume flow rate may be important in a variety of different clinical circumstances, such as stroke, arteriovenous malformation, cardiac failure. The main aim of the work was review of non-invasive measurements of carotid blood flow methods and devices. There are various types of methods and devices for monitoring blood flow in the carotid artery, however, so far no mobile recorder of the appropriate quality of registration has been developed. The most promising solutions are wireless wearable patches that, when applied to the skin of the neck, will not be a burden during measurements. This is important when measurements are to be made, during extreme activities, such as in pilots. As a result of the conducted research, there is a presumption that carotid blood flow measurements methods has great potential to be a useful tool in the diagnosis of the pilot's predisposition, at the same time performing a warning function, the so-called alerters, before the imminent overload loss of consciousness.

Keywords Carotid artery blood flow · Doppler ultrasound · Wearable ultrasound · Non-invasive hemodynamic monitoring

E. Sobotnicka (✉) · J. Mocha · A. Sobotnicki · J. Gałecka · A. Gacek
Lukasiewicz Research Network, Krakow Institute of Technology, The Centre for Biomedical Engineering, Zakopianska Str. 73, 30-418 Krakow, Poland
e-mail: ewelina.sobotnicka@kit.lukasiewicz.gov.pl

J. Mocha
PhD School, Silesian University of Technology, Gliwice, Poland

P. Strumiłło et al. (eds.), *The Latest Developments and Challenges in Biomedical Engineering*, Lecture Notes in Networks and Systems 746,
https://doi.org/10.1007/978-3-031-38430-1_30

1 Introduction

The carotid arteries are major blood vessels in the neck that supply blood to the brain, neck, and face [3]. There are two carotid arteries, one on the right and one on the left. In the neck, each carotid artery branches into two divisions: the internal carotid artery supplies blood to the brain and the external carotid artery supplies blood to the face and neck [3]. Common carotid artery (CCA) volume flow rate (VFR) is clinically useful for study of cerebrovascular disease [2, 4]. Continuous monitoring of cardiac function is highly desirable for long-term assessment of cardiovascular health, detection of acute cardiac dysfunction, and clinical management of critically ill patients or patients undergoing surgery [1]. Information from VFR may be important in a variety of different clinical circumstances, such as stroke, arteriovenous malformation, cardiac failure [4, 5]. In patients with stenosis in the internal carotid artery (ICA), there are significant changes in volume flow rate of CCA with high-grade stenosis or occlusion [4]. These hemodynamic changes can provide valuable diagnostic information to assist with interpretation of carotid ultrasound results, especially in pilots during flight or flight simulation with gravitational force conditions where changes in flows are significant [4]. Various methods can be used to measure blood flow. For example, Carotid Doppler ultrasonography (CDU) has been recently investigated as a non-invasive hemodynamic monitoring modality [6]. Due to its superficial location and easy accessibility of the carotid artery, CDU is an attractive tool [6, 7].

2 Carotid Artery Blood Flow Measure Methods

From the state of the art, there are many methods that can be used to measure blood flow in the body. These include for example: Doppler based ultrasound, Laser-Doppler, Duplex Doppler, Color Doppler, bioimpedance, thermal method, Near Infrared Spectroscopy (NIRS), optical and acousto-optic methods. However, not all methods have sufficient tissue penetration capability to measure blood flow in the carotid artery. Due to the review character of this paper, the following are briefly discussed methods that were considered by the authors to be feasible use for measuring carotid artery blood flow in pilots during flight or flight simulation with gravitational force conditions.

2.1 Carotid Artery Doppler Ultrasound

A carotid artery Doppler ultrasound is a diagnostic test used to check the circulation in the large arteries in the neck. This method shows any blockage in the vessels by a blood clot or thrombus formation [9]. CDU has been explored as a non-invasive and inexpensive alternative for long-established invasive hemodynamic monitoring techniques [8]. According to the authors of the work [8, 9] for proper operation of the CDU a linear ultrasound converter (5–18 MHz) with B-mode and pulsed Wave Doppler is required. B-mode imaging is used to estimate the cross-sectional area and Doppler angle required for absolute flow measurements [9]. Pulsed Wave Doppler is used to measure the velocity waveform at a selected sample volume in a vessel. Alternatively, a continuous wave Doppler transducer may be used to acquire velocity information from the entire field of view of the transducer [10, 11].

2.2 *Laser Doppler*

Laser Doppler enables optical assessment of perfusion [12, 13]. Laser Doppler probe emits laser light that penetrates the surface layers of the tissue [12]. This light scatters on moving blood cells with a change in light frequency [13]. The microcirculatory vascular system within the skin affects the laser light. The angles between the paired light beam and the vessels are unknown. This method allows measurement of the average flow velocity. The signal strength is proportional to the product of the number of blood cells within the examined tissue fragment and the speed of blood cell movement [12]. Using this technique, the state of skin microcirculation can be assessed. Assessment of skin perfusion using the Laser Doppler technique is difficult to calibrate, the results are individually variable and dependent on the place of measurement (even a slight shift of the measuring probe on the skin can cause is totally change the results) [12]. Despite these limitations, due to its non-invasiveness and simplicity of conducting the measurement, the method has gained wide acceptance especially for assessing the acute response of the microcirculation to ischemia, cold or the effects of drugs [12].

2.3 *Duplex Doppler*

Duplex Doppler ultrasonography involves imaging the arteries with simultaneous flow measurements [12]. It combines the anatomical image of the tissues in a two dimensional B-presentation with the flow image of the area under test [14]. Ultrasonography with dual imaging of the vessels includes their visualization in longitudinal and transverse sections, evaluation of their morphology and the selected site of blood flow. A record of the flow curve is obtained from the accurately marked sampling location. Once the flow line is determined with a suitable marker and the angle between the direction of the ultrasound wave and the direction of blood flow is calculated, numerical data on various flow parameters can be obtained [15].

2.4 Color Doppler

Ultrasound color mapping is an extension of the duplex technique and is widely used in clinical diagnostics [12, 16]. In this method, the occurrence of the Doppler phenomenon is analyzed in the area selected on the morphological image, along the individual lines. Color Doppler is used to detect laminar flow, which is widespread in the human circulatory system [16]. In color Doppler ultrasound machines, image pixels corresponding to echoes reflected from tissues are encoded in grayscale, while echoes reflected from moving elements (blood cells), are assigned a different color, depending on the speed and direction of movement. The higher speed, the more intense colors [15]. Color Doppler ultrasound enables precise localization of even small vessels supplying individual structures as well as precise evaluation of the vascular system [15]. Color Doppler allows detection and evaluation of the severity of backflow in the superficial and deep venous system of the lower limbs, the presence and extent of atherosclerotic lesions in the peripheral arteries and aorta, the presence and structure of aneurysms of the abdominal aorta and other arteries [15].

2.5 Bioimpedance Methods

Bioimpedance methods measures the tissue impedance and captures important physiological parameters including blood flow, muscle contractions, body fluid shifts, and other cardiovascular parameters [21]. Bioimpedance methods are a promising tool, leveraging deep penetration of small electrical signals into the tissue to extract information that is modulated with the blood flow [21].

2.6 Calorimeter Measurements

Calorimetry is an accurate technique to determine energy expenditure [18]. It mainly relies on thermal correlation of tissue blood flow with heat transfer to a specially adapted calorimeter [12]. Blood is generally considered to be a non-Newtonian fluid, so that reason many related problems may not have the exact solutions. Therefore, numerical simulations are suggested as a useful tool to understand thermal behavior of blood flows [20]. All techniques for measuring thermal blood flow have in common a small heat source and two temperature difference sensors placed in or on the tissue [19]. Conversion of the heat transport coefficient into specific blood flow is possible only if the vessels in the temperature field are predominantly capillaries [19].

3 Material and Methods

The main aim of the paper is to discuss methods and innovative devices enabling measure carotid artery blood flow. Work has been carried out resulting in scientific evidence endorsing the claim that it is possible use of methods for measurement blood flow in the carotid artery in difficult environmental conditions, for example in pilots during gravitational force. In order to find publications, a certain search strategy and a certain method of data selection were used.

3.1 Search Strategy

A review of the literature was conducted to obtain reliable information on the possibility methods measurement of carotid artery blood flow. The PubMed, Cochrane medical database and internet search engines was conducted. The search range included dates between 01.01.2015 and 03.03.2023. Keywords such as: "carotid artery blood flow", "blood flow", "carotid blood flow", "Doppler ultrasound", "Doppler", "optical flow", "bioimpedance", "thermal carotid artery blood flow" was used to search medical database and internet.

3.2 Data Selection

The inclusion and exclusion criteria relate to the characteristics of the literature sought and indicate the type of intervention used. For the analysis of publications related to the carotid artery blood flow methods, the following were indicated as main inclusion criteria: ultrasound method, blood flow measure method, non-invasive hemodynamic monitoring. The review was conducted in two stages. In the first stage, literature was rejected unrelated to the topic after reviewing the title of the paper. Next, publications were rejected that proved to be unhelpful after reviewing the abstract. After applying a search strategy with inclusion and exclusion criteria, publications were obtained and used as a reliable source of information to endorse the claim that it is possible to use methods to measure blood flow in the carotid artery in pilots during gravitational force. Review obtained information on existing solutions (wireless, wearable) for measuring blood flow in the carotid artery and their potential capabilities.

4 Results

As a result of the selection of identified scientific reports, publications were obtained, which are presented in Table 1.

Table 1 Articles included in the review associated with carotid artery blood flow methods

Study	Applied method	Short description	Transducer frequency
Hu et al. [1]	B-mode ultrasound imaging strategies	Wearable ultrasonic patch for continuous, real-time cardiac function assessment	3 MHz
Bu et al. [22]	Carotid doppler ultrasound	Comparison of carotid blood flow measured by ultrasound and cardiac output in patients undergoing cardiac surgery	1–5 MHz
Cheong et al. [23]	Carotid doppler ultrasound	Comparison of carotid blood flow measured by ultrasound and cardiac output	2–4 MHz
Du et al. [16]	Ultrasound color flow mapping	New generation device which can dynamically visualize the blood flow in carotid artery	15 kHz
Du et al. [24]	V Flow (high-frame-rate ultrasound vector flow imaging)	Comparison of V Flow and pulsed wave Doppler and evaluate the accuracy of the V Flow technique	3–10 MHz
Kenny [25]	Doppler ultrasound	Wearable Doppler ultrasound for carotid blood flow measured	4 MHz
Kenny et al. [26]	Carotid Doppler ultrasound	Wearable Doppler ultrasound system that continuously measures the carotid artery flow and detects central hypovolemia during lower body negative pressure	4 MHz
Kenny et al. [27]	Carotid Doppler ultrasound	Wearable Doppler ultrasound device adheres to the neck which measure beat to beat changes across multiple cardiorespiratory cycles	4 MHz
Suriani et al. [8]	Carotid Doppler ultrasound	The use of CDU for non-invasive hemodynamic monitoring	5–18 MHz
Kenny et al. [28]	Carotid Doppler ultrasound	Novel ultrasound patch for detecting blood flow quickly in the carotid	4 MHz
Wang et al. [29]	Color Doppler	Stretchable wearable devices for the continuous carotid blood flow monitoring	7,5 MHz
Wang et al. [30]	Carotid Doppler ultrasound	Flexible Doppler ultrasound device for the continuous monitoring blood flow	5 MHz
Hu et al. [31]	Ultrasound imaging technologies	Stretchable ultrasonic transducer array which can conform	2–6 MHz
Shen et al. [32]	Bioimpedance method	Mathematical models to describe the relationship between the center-line velocity, the arterial radius, blood conductivity, and blood flow	2.5–5 MHz
Campen et al. [33]	Doppler imaging	Carotid blood flow measure during orthostatic stress	6–3 MHz
Jalil et al. [34]	Pulsed wave doppler	Comparison of changes in carotid flow time and stroke volume induced by passive leg raising	1.5–3.5 MHz

(continued)

Table 1 (continued)

Study	Applied method	Short description	Transducer frequency
Webb et al. [35]	Thermal method	Sensor technology for thermal measurements of blood flow	2 Hz

5 Discussion

The development of material technology and mainly measurement methods has contributed to the development of innovative solutions for the detection of blood flow in the carotid artery. The most promising solutions are wireless wearable patches that, when applied to the skin of the neck, will not be a burden during measurements. This is important when measurements are to be taken while person are performing certain extreme activities, such as in pilots during fly or simulated gravitational force (g-force). For example, in the paper [1], the authors described an innovative design of a device for continuous monitoring of flows in the carotid artery. Developed technology enables dynamic, portable cardiac performance monitoring with significantly improved accuracy in a variety of environments. The authors [1] have developed flexible composite electrodes that can withstand various deformations so that they can be placed directly on the skin of the patient. The paper [8] is a review discussing the aspect of using carotid Doppler for noninvasive hemodynamic monitoring. Accurate hemodynamic monitoring is the cornerstone of critical patient treatment. Provides optimization of tissue and organ perfusion may prevent multi-organ failure [8]. An overview of the equipment required for CDU testing, the limitations of this method and technical challenges potentially facing CDU ultrasound, and the cerebrovascular mechanisms that may affect CDU measurement results is also provided in [8]. The paper also describes alternative techniques for non-invasive hemodynamic monitoring (cardiac output, volume status, and fluid responsiveness assessment), detailing the advantages and limitations compared to CDU. On the other hand, the paper [22] conducted a study measuring cerebral blood flow during different cardiac cycles using ultrasound. Carotid artery systolic blood flow (SCF), carotid artery diastolic blood flow (DCF) and total (systolic and diastolic) carotid artery blood flow (TCF) were studied as well as the direct correlation between CBF and cardiac output (CO). The authors in the paper [22] suggested that SCF should be used as an indicator to replace CO. In the study [23], analyzes were carried out consisting in examining the new position of the transducer during ultrasound measurements. Carotid flow was measured by placing the probe at the level of the left supraclavicular fossa, directing the probe towards the foot. This placement made it possible to assess flow in the carotid artery near its exit in the aorta [23]. Moreover, with this maneuver, it was possible to achieve a perfect match between the direction of flow and the ultrasound wave. In the work of [16] and [24], attention was focused on ultrasound color mapping using the V Flow technology (Vector Flow Imaging), and the development of a device to enable dynamic visualization of blood flow. In papers [25–27] and

[28], the authors focused on developing and presenting a novel wireless wearable Doppler ultrasound device. The developed device detected reduced stroke volume corresponding with hypovolemia and was able to continuously measure carotid artery heart rate. In summary, Kenny and co-authors [25, 26] have developed a novel ultrasound patch for detecting blood flow quickly in the carotid. Wang and co-authors [29] developed a flexible device for continuous monitoring of blood flow velocity in deep tissues. The developed device is thin (1 mm), lightweight (0.75 g) and adapts to the skin and can be used to monitor blood flow velocity after vascular reconstruction procedures or preventive testing for certain chronic vascular diseases [29]. The authors [30] also developed stretchable ultrasonic transducer arrays. The developed device can obtain three-dimensional images of complex skin surface (under flat, concave and convex surfaces) [30]. In the paper [32], the authors proposed a simplified mathematical model describing the relationship between centerline velocity and radius and blood conductivity.

The measurement procedures described in papers [33] and [34] provide valuable information on the feasibility and validity of performing carotid artery flow measurements during an orthostatic stress test. Flow changes that occur during tilt testing can be measured by Doppler imaging of the right internal carotid artery and vertebral artery [33]. Similar in the paper [34], the authors studied flows during tilting, particularly during leg raising. During testing, changes over time in carotid artery flow and stroke volume induced by passive leg elevation were compared.

An equally interesting approach to measuring blood flow is the method and device described in [35]. The authors [35] have developed a thermal sensor that has the ability to non-invasively continuously monitor blood flow. Tests conducted with the thermal sensor have helped establish a number of guidelines for its use, such as the optimal placement of the sensor and how deep below the skin's surface the device can assess blood flow [35]. This device promises to be a low-cost, easy-to-manufacture sensor for use in outpatient or hospital settings [35].

6 Summary and Conclusions

A literature review of the available solutions to study blood flow in the carotid artery was aimed at finding out the current state of knowledge. Found publications, are a source of information that can will be used to develop an innovative medical device that allows blood flow measurements under the extreme conditions associated with the examination and training flying personnel. Studies with flight attendants and pilots will be designed to determine the undershooting changes in carotid artery blood flow as a result of low-body negative pressure (LBNP). LBNP is a relative simple method that simulates the effect of g-force on the human body [36–38]. Determination of qualifying thresholds for assessing the blood flow capacity of the carotid arteries can be a useful tool in the diagnosis of pilot predisposition, while acting as a warning, so-called alerters, of impending overload loss of consciousness (G-LOC). A simulator testing methodology designed to assess the ability of flight personnel to serve in the air could also be useful for assessing the predisposition of individuals to work in hypoxic conditions.

The device, which will be developed based on the guidelines obtained from the review, will have to meet safety requirements for medical devices and should be immunity to motion artifacts related to the nature and conditions of the tests for which it will be used as well as measure and present actual data continuously. In the future, tests with the device for carotid blood flow measured may contribute to the effective selection of pilots with predisposition to good tolerance the g-force.

Funding This work was supported by the resources of the National Centre of Research and Development, Poland, within the framework of the Program for National Defense and Security, Project DOB-BIO-12–05-001–2022.

References

1. Hu, H., Huang, H., Li, M., et al.: A wearable cardiac ultrasound imager. Nature **613**, 667–675 (2023)
2. Houte, J., Mooi, F.J., Montenij, L.J., Meijs, L.P.B., Suriani, I., Conjaerts, B.C.M., Houterman, S., Bouwman, A.R.: Correlation of carotid doppler blood flow with invasive cardiac output measurements in cardiac surgery patients. J Cardiothorac Vasc Anesth., **36**(4), 1081–1091 (2022)
3. Goldman, L.: Cecil medicine. Saunders Elsevier, (2008)
4. Likittanasombut, P., Reynolds, P., Meads, D., Tegeler, C.: Volume flow rate of common carotid artery measured by Doppler method and Color Velocity Imaging Quantification (CVI-Q). J. Neuroimaging **16**(1), 34–38 (2006)
5. Eicke, B.M., Tegeler, C.H.: Ultrasonic quantification of blood flow volume. In: Tegeler, C.H., Babikian, V.L., Gomez, C.R. (eds.) Neurosonology. Missouri: Mosby-Year Book, Inc., St. Louis, MO, pp. 101 – 110 (1996)
6. Houte, J., Raaijmaakers, A.E., Mooi, F.J., Meijs, L.P.B., Boer, E.C., Suriani, I., Houterman, S., Montenij, L.J., Bouwman, A.R.: Evaluating corrected carotid flow time as a non-invasive parameter for trending cardiac output and stroke volume in cardiac surgery patients. J. Ultrasound, (2022)
7. Girotto, V., Teboul, J.L., Beurton, A., et al.: Carotid and femoral Doppler do not allow the assessment of passive leg raising effects. Ann Intensive Care **8**, 67 (2018)
8. Suriani, I., Houte, J., Boer, E.C., Knippenberg, L., Manzari, S., Mischi, M., Bouwman, R.A.: Carotid Doppler ultrasound for non-invasive haemodynamic monitoring: a narrative review. Physiol Meas. **43**(10), (2023)
9. Brandt, A.H. et al.: A comparison study of vector velocity, spectral Doppler and magnetic resonance of blood flow in the common carotid artery Ultrasound Med. Biol. **44**, 1751–61 (2018)
10. Bussmann, B.M., Sharma, S., McGregor, D., Hulme, W., Harris, T.: Observational study in healthy volunteers to define interobserver reliability of ultrasound haemodynamic monitoring techniques performed by trainee doctors Eur. J. Emerg. Med. **26**, 217–223 (2019)
11. Cencetti, S., Lagi, A., Cipriani, M., Fattorini, L., Bandinelli, G., Bernardi, L.: Autonomic control of the cerebral circulation during normal and impaired peripheral circulatory control. Heart **82**, 365–372 (1999)
12. Mnich, K.: The methods of survey blood fl ow in human body, Works of the Strata Mechanics Research Institute PAN, Tom 6, nr 3–4, s. 171–184 (2004), in Polish

13. Kuwabara, K., Higuchi, Y., Ogasawara, T., Koizumi, H., Haga, T.: Wearable blood flowmeter appcessory with low-power laser Doppler signal processing for daily-life healthcare monitoring. Annu. Int. Conf. IEEE. Eng. Med. Biol. Soc. **2014**, 6274–6277 (2014)
14. Traczyk, W.: Human functional diagnostics. Medical Publisher, Warszawa (1999). in Polish
15. Noszczyk, W.: Surgery of peripheral arteries and veins. Medical Publisher, Warszawa (1998). in Polish
16. Du, Y., Shen, Y.: V Flow a novel visualization of blood flow. Mindray Healthc. Reach., (2016)
17. Filipczyński, L.: Blood flows, Polish Academy of Sciences PAN. Warszawa- Poznań (1980), in Polish
18. Steinmetz, M.P.: Nutritional care of the spinal cord–injured patient, calorimetry: measurement of energy expenditures. Benzel's Spine Surg., (2017)
19. Priebe, L.: Methods of thermal blood flow measurement. Bibl Radiol. **6**, 33–44 (1975)
20. Xingting, L., Xingyu, C., et al.: The thermal behavior of blood flow in the arteries with various radii and various stenosis angles using non-Newtonian Sisko model. Alex. Eng. J. **61**(9), 7195–7201 (2022)
21. Sel, K., Osman, D., Jafari, R.: Non-invasive cardiac and respiratory activity assessment from various human body locations using bioimpedance. IEEE Open J. Eng. Med. Biol. **2**, 210–217 (2021)
22. Bu, X., Zhang, Y., Chen, L., et al.: Comparison of carotid blood flow measured by ultrasound and cardiac output in patients undergoing cardiac surgery. Research Square, (2022)
23. Cheong, I., Otero, C.V., Sosa, F.A., Tort, O.B., Merlo, P.M., Tamagnone, F.M.: Carotid flow as a surrogate of the left ventricular stroke volume. J. Clin. Monit. Comput, (2022)
24. Du, Y,. Ding, H., He, L., Yiu, B.Y.S., Deng, L., Yu, A.C.H., Zhu, L.: Quantitative blood flow measurements in the common carotid artery: A comparative study of high-frame-rate ultrasound vector flow imaging, pulsed wave doppler, and phase contrast magnetic resonance imaging. Diagnostics (Basel), **12**(3), 690 (2022)
25. Kenny, J.S.: Assessing fluid intolerance with doppler ultrasonography: a physiological framework. Med Sci (Basel) **10**(1), 12 (2022)
26. Kenny, J.S., Elfarnawany, M., Yang, Z., Eibl, A.M., Eibl, J.K., Kim, C.H., Johnson, B.D.: A wireless ultrasound patch detects mild-to-moderate central hypovolemia during lower body negative pressure. J. Trauma. Acute. Care. Surg, **93**(2S Suppl 1), S35-S40 (2022).
27. Kenny, J.S., Barjaktarevic, I., Mackenzie, D.C., Elfarnawany, M., Yang, Z., Eibl, A.M., Eibl, J.K., Kim, C.H., Johnson, B.D.: Carotid artery velocity time integral and corrected flow time measured by a wearable Doppler ultrasound detect stroke volume rise from simulated hemorrhage to transfusion. BMC Res Notes **15**(1), 7 (2022)
28. Kenny, J.S., Munding, C.E., Eibl, J.K., Eibl, A.M., Long, B.F., Boyes, A., Yin, J., Verrecchia, P., Parrotta, M., Gatzke, R., Magnin, P.A., Burns, P.N., Foster, F.S., Demore, C.E.M.: A novel, hands-free ultrasound patch for continuous monitoring of quantitative Doppler in the carotid artery. Sci Rep. **11**(1), 7780 (2021).
29. Wang, C., Qi, B., Lin, M., Zhang, Z., Makihata, M., Liu, B., Zhou, S., Huang, Y.H., Hu, H., Gu, Y., Chen, Y., Lei, Y., Lee, T., Chien, S., Jang, K.I., Kistler, E.B., Xu, S.: Continuous monitoring of deep-tissue hemodynamics with stretchable ultrasonic phased arrays. Nat. Biomed. Eng. **5**(7), 749–758 (2021)
30. Wang, F., Jin, P., Feng, Y., Fu, J., Wang, P., Liu, X., Zhang, Y., Ma, Y., Yang, Y., Yang, A., Feng, X.: Flexible doppler ultrasound device for the monitoring of blood flow velocity. Sci Adv. **7**(44), (2021).
31. Hu, H., Zhu, X., Wang, C., Zhang, L., Li, X., Lee, S., Huang, Z., Chen, R., Chen, Z., Wang, C., Gu, Y., Chen, Y., Lei, Y., Zhang, T., Kim, N., Guo, Y., Teng, Y., Zhou, W., Li, Y., Nomoto, A., Sternini, S., Zhou, Q., Pharr, M., Scalea, FL., Xu, S.: Stretchable ultrasonic transducer arrays for three-dimensional imaging on complex surfaces. Sci Adv., **4**(3), (2018)
32. Shen, H., Li, S., Wang, Y., Qin, K.R.: Effects of the arterial radius and the center-line velocity on the conductivity and electrical impedance of pulsatile flow in the human common carotid artery. Med. Biol. Eng. Comput. **57**(2), 441–451 (2019)

33. Campen, C.L.M.C., Verheugt, F.W.A., Visser, F.C.: Cerebral blood flow changes during tilt table testing in healthy volunteers, as assessed by Doppler imaging of the carotid and vertebral arteries. Clin. Neurophysiol. Pract., **3**, 91–95 (2018)
34. Jalil, B., Thompson, P., Cavallazzi, R., Marik, P., Mann, J., El-Kersh, K., Guardiola, J., Saad, M.: Comparing changes in carotid flow time and stroke volume induced by passive leg raising. Am. J. Med. Sci. **355**(2), 168–173 (2018)
35. Webb, R.C., Ma, Y., Krishnan, S., Li, Y., Yoon, S., Guo, X., Feng, X., Shi, Y., Seidel, M., Cho, N.H., Kurniawan, J., Ahad, J., Sheth, N., Kim, J., Taylor, J.G., Darlington, T., Chang, K., Huang, W., Ayers, J., Gruebele, A., Pielak, R.M., Slepian, M.J., Huang, Y., Gorbach, A.M., Rogers, J.A.: Epidermal devices for noninvasive, precise, and continuous mapping of macrovascular and microvascular blood flow. Sci Adv. **1**(9), (2015)
36. Sobotnicka, E., Sobotnicki, A., Czerw, M., Badura, G., Sobiech, M., Krej, M., Puchalska, L., Dziuda, Ł.: A multiparameter examination system to assess self-regulatory mechanisms of the cardiovascular system under simulated hypergravity conditions. In: 2018 Baltic URSI Symposium (URSI), pp. 87–90 (2018). https://doi.org/10.23919/URSI.2018.8406698
37. Sobotnicka, E., Sobotnicki, A., Czerw, M., Badura, G., Sobiech, M., Krej, M., Puchalska, L., Dziuda L.:Tests for pilots under simulated hypergravity conditions—technological challenges and research methodology. In: 2018 25th International conference "mixed design of integrated circuits and system" (MIXDES), pp. 465–470 (2018). https://doi.org/10.23919/MIXDES.2018.8436651
38. Sobotnicka, E., Sobotnicki, A., Czerw, M., Badura, G., Krej, M., Puchalska, L., Kowalczuk, K., Gaździński, S., Dziuda Ł.: Methods to assess self-regulatory mechanisms of the cardiovascular system under simulated hypergravity conditions. In: 2019 MIXDES—26th International conference mixed design of integrated circuits and systems. pp. 407–412(2019). https://doi.org/10.23919/MIXDES.2019.87871529

Application of Unsuppressed Water Peaks for MRS Thermometry

Marcin Sińczuk, Jacek Rogala, Ewa Piątkowska-Janko, and Piotr Bogorodzki

Abstract This study explores the feasibility of unsuppressed water (WU) peaks acquired as a part of prescanning in conjunction with metabolites from suppressed water (WS) spectra for temperature measurements. Calibrations and in vivo single-voxel 1H MRS was conducted on a 3-T GE scanner. Calibration constants for the water-chemical shift were obtained by using a temperature-controlled phantom containing an aqueous solution of N-acetyl aspartate (NAA), Creatine (Cr), and Choline (Cho). Estimations of absolute human brain temperature were performed utilizing the correlation of temperature to the water-chemical shift for the NAA, Cr, and Cho resonances. Commercially available single-voxel point-resolved spectroscopy (PRESS) sequences (repetition/echo time = 1500 ms/30 ms; voxel size $2 \times 2 \times 2$ cm^3) were used to acquire data for calibration of the metabolites' chemical shift differences and in vivo temperature estimations. Each sequence included 16 averages without water suppression and 96 averages with water partially suppressed. In vivo study consisted of 2 PRESS sequences, one before and one after extensive 30-min fMRI task acquisition. The mean brain temperatures calculated using NAA, Cr, and Cho with WS were 37.49 $\mp$ 0.40 °C, 37.09 $\mp$ 0.34 °C, and 38.18 $\mp$ 0.37 °C before fMRI and 37.19 $\mp$ 0.45 °C, 36.81 $\mp$ 0.39 °C, and 37.88 $\mp$ 0.41 °C after fMRI, with a difference of approximately 0.3°C. When using WU with NAA, Cr, and Cho, the mean brain temperature was 38.13 $\mp$ 0.49 °C, 37.67 $\mp$ 0.44°C, and 38.86 $\mp$ 0.52 °C before fMRI and 37.89 $\mp$ 0.52 °C, 37.45 $\mp$ 0.50 °C, and 38.62 $\mp$ 0.52°C after fMRI, with a difference of approximately 0.24°C. Similar differences reported by using both methods for temperature calculation suggest that using unsuppressed water for temperature measurement in absence of its suppressed counterpart is feasible and may add another angle to already concluded past MRS studies which is temperature.

Keywords MR spectroscopy · Thermometry · Partially suppressed water · Separately acquired water · Calibration · Brain temperature · Human · Noninvasive

---

M. Sińczuk (✉) · E. Piątkowska-Janko · P. Bogorodzki
Nalecz Institute of Biocybernetics and Biomedical Engineering PAS, Warsaw, Poland
e-mail: msinczuk@ibib.waw.pl

J. Rogala
Institute of Experimental Physics, Faculty of Physics, University of Warsaw, Warsaw, Poland

P. Strumiłło et al. (eds.), *The Latest Developments and Challenges in Biomedical Engineering*, Lecture Notes in Networks and Systems 746,
https://doi.org/10.1007/978-3-031-38430-1_31

1 Introduction

Temperature plays an important role in the regular functioning of the brain. Measuring it is crucial in potentially diagnosing traumatic brain injury, stroke, hypothermia, epilepsy, or even tumors [1–5]. However, conventional temperature measurement methods, such as tympanic and rectal thermometry, reflect poorly on actual brain temperature [6, 7]. Invasive methods used in clinical routines, such as implanted sensors provide precise measurements; however neurosurgical intervention associated with probe insertion poses significant risks and therefore is difficult to justify.

There are currently several methods of non-invasive temperature measuring using magnetic resonance imaging (MRI). Among them, there are techniques which are utilizing the temperature dependence of such parameters as proton density [8, 9], T1 [10, 11], and T2 [12] relaxation times, the diffusion coefficient [13, 14], magnetization transfer [15, 16], and the proton frequency shift divided into Phase Mapping [17] and chemical shift [18].

The method using waters chemical shift namely 1H Magnetic Resonance Spectroscopy Thermometry (1H MRST) has been demonstrated as the most feasible method for the estimation of absolute temperature when combined with high-resolution MRI [19–27]. It is based on the principle that water chemical shift correlates linearly with temperature by approximately −0.01 ppm/°C [18], which can be used to provide actual temperature measurements using appropriate calibration. 1H MRST is based on measuring chemical shift difference between the spectral water peak and one or more temperature-independent metabolite peaks, such as creatine, N-acetyl aspartate [19, 28, 29], or choline-containing compounds[30]. The uncertainty of the technique has been reported to be ± 0.2 °C in phantoms using single-voxel MRS at 3 T and a set temperature calibration and ± 0.8 °C by 1H MRS imaging [22] but it is discussed that in brain uncertainty of less than ± 0.5 °C is questionable[27].

Currently, there are two major techniques for 1H MRST. One of them is the two-spectra approach, which measures temperature using water frequency from unsuppressed spectra in conjunction with reference frequency from suppressed spectra [31]. Problems with such a method may arise from potential field drifts and subject motion between the two MRS sessions[28] The other method uses only partial suppression on the water peak, therefore, providing a way to simultaneously acquire reference and partially suppressed water signals [19, 27, 32]. This approach is founded on the reasoning that partially suppressing the water can mitigate the artifacts of unsuppressed water acquisition, while still enabling concurrent measurements of water and reference signal frequencies [28, 31].

There is an understanding that non-suppressed and partially suppressed 1H MR spectra yield comparable results in estimating brain temperatures [19, 22], although recent research concludes that using partially suppressed water signals in 1H MRST may result in low accuracy where using non-suppressed signals in conjunction with temperature-independent metabolite peaks suffers from low precision resulting in difference of even 1 °C between both methods [33]. Currently, numerous MRS studies have concluded, in which temperature could play significant role in aiding researchers in understanding various biological processes and pathologies therefore gathering new conclusions based on that new information. These studies typically focus on metabolites as the main point of interest, causing the water peak in the suppressed spectra to be mostly invisible. Consequently, the viability of utilizing the unsuppressed water peak in combination with metabolites from these suppressed spectra for temperature measurements should be investigated further.

At present, GE scanners employ commercially available methods of spectral collection that involve acquiring brief prescan spectra with unsuppressed water, rendering metabolites non-detectable, followed by partially suppressed spectra with fully visible metabolites (Fig. 1). With the ability to use both methods in the same scanner and access to a large sample size, we have a unique opportunity to make a reliable comparison between the two methods. Our study aimed to evaluate the feasibility of using these unsuppressed spectra to measure brain temperature. Additionally, we aimed to investigate the effect of prolonged MRI acquisition on brain temperature using these methods.

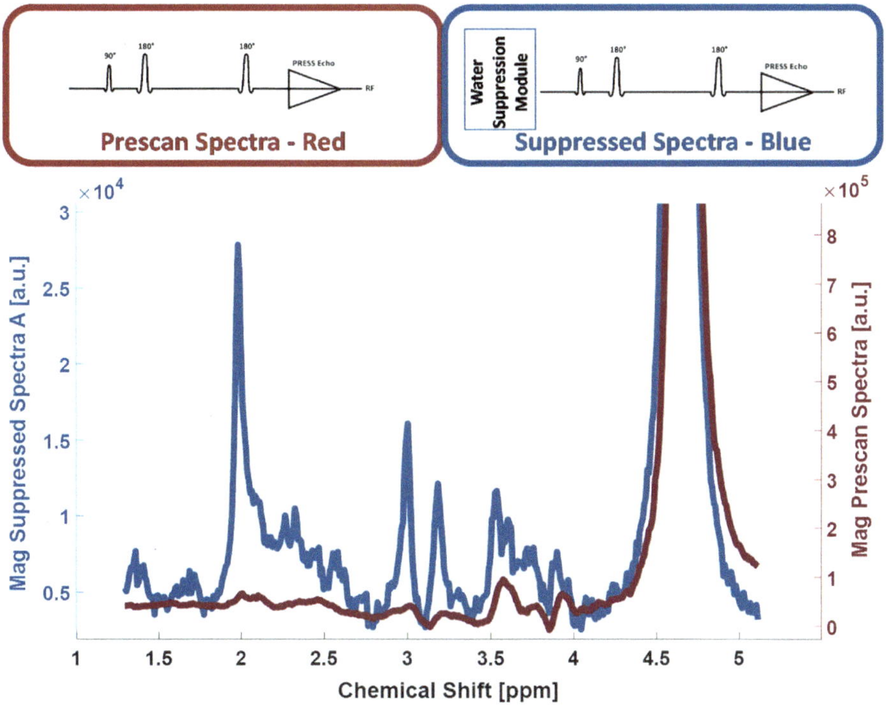

Fig. 1 Suppressed (Blue) and Unsuppressed Prescan (Red) spectra, Suppressed spectra are characterized by noticeable metabolite peaks, lower amplitude and partially or entirely suppressed water peak (at approx. 4.7 ppm), Unsuppressed prescan spectra are characterized by imperceptible metabolite peaks, bigger amplitude and seldom water peak (at approx. 4.7 ppm)

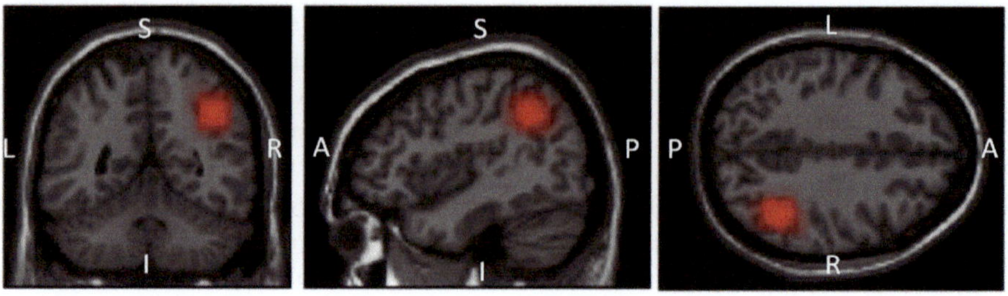

Fig. 2 Voxel placement of in vivo subject located in right inferior parietal lobule (rIPL)

2 Materials and Methods

2.1 MRI Data Scanning Protocol

Data was acquired using a 3-T GE scanner (GE Discovery 750w) with a standard transmit/receive 8-channel head coil. The room temperature and lighting were kept constant at 20.5 °C (standard deviation [SD] ± 0.5 °C). A high-resolution structural MRI sequence (TR/TE = 6,936 ms/2,968 ms, matrix 256 × 256, voxel 1,0547 × 1,0547 × 1,2 mm^3) was obtained after standard shimming for improved voxel localization. Single-voxel 1H-MRS spectra (PRESS) were obtained from a VOI (20 × 20 × 20 mm^3) located in the right inferior parietal lobule (rIPL) (Fig. 2). This large VOI ensured a reasonable signal-to-noise ratio to obtain a sufficient 1H-MRS signal for further data processing. The scan parameters for PRESS were TR = 1500 ms, TE = 30 ms, bandwidth 5 kHz, 4096 points, NEX = 8, with 96 averages and a scan time of 3 min. Water suppression was realized by an automated CHESS [34] module embedded in the commercially available GE PRESS sequence, which consisted of ten Very Selective Saturation (VSS) RF suppression pulses. Before each MRS scanning sequence, a non-suppressed scan with parameters identical to a water-suppressed scan was performed (with the only change being 16 averages instead of 96). During in vivo studies Six functional MRI scans were collected after the MRS scanning sequence, using a t2*−weighted gradient echo planar imaging sequence (TR = 2500 ms, TE−25 ms, voxel 3 × 3 × 3 mm^3). During the first two scans that lasted 5′50", subjects performed an Anticipation Attention Task (ATT) [35], while in the last four scans, each lasting 4′25", the N-Back task [36] was carried out. Following the functional scans, a second PRESS scan was conducted with the same parameters.

2.2 Phantom Measurements

For the calibration measurements sulphuration flask with a volume of 500 ml was fitted as a water jacket for the phantom 50 mL test tube. Phantom contained an aqueous solution to which we added the metabolites with the following concentrations 12.5 mM NAA, 12.5 mM Glu, 10 mM Cr, 7.5 mM mIns, 5 mM Lac, and 3 mM Cho. This aqueous solution contained an additional 72 mM K2HPO4 (dipotassium phosphate), 28 mM KH2PO4 (monopotassium phosphate), 20 mM NaFor (sodium formate), and 1.00 g/L of NaN3 (sodium acide) to provide microbiological and pH stability (Covaciu et al., 2010). To mimic an intracellular environment the aqueous solution pH was set to 7.0. The Thermo Scientific SC100 circulation thermostat with the S5P bath was positioned outside the MR examination room and used to control and stabilize the temperature. This device is equipped with a heater, and the controller immersed in a water tank allows you to adjust the temperature with an accuracy of 0.1 °C. Polyvinylchloride tubes were connected from the bath to the water jacket around the phantom to ensure continuous circulation and control over the thermal state inside the phantom. Inside the scanner, the measurement in the phantom was made using two optical fiber OSENSA PRB-G40-2.0 M-STM-MR temperature sensors placed inside. The sensors were connected to the FTX-200-lx + converter. The temperature monitoring system was connected with a USB cable to a computer where the temperature was read using a dedicated application provided by OSENSA. The accuracy of the sensors according to the manufacturer's data is (+−0.1 °C). Spectroscopic measurements were performed while the aqueous solution in the phantom reached thermal equilibrium at eleven different temperature levels ranging from 31.5 to 39.5 °C. A minimum of 20 min was devoted to stabilizing the temperature at the level we selected. The average temperature fluctuation during each measurement was always no more than 0.02°C. At each temperature level, five spectroscopic measurements were made with a break for auto-prescan performed by the scanner before each measurement (For PRESS sequence protocol setting please refer to "MRI data scanning protocol"). The tested voxel was placed in the center of the test tube close to the fiber optic temperature sensors. To reduce the temperature drift as much as possible, water circulation in the phantom around the tube was maintained during the acquisition of MR signals. After signal acquisition, the spectroscopic data was transferred to a local workstation for further data analysis using in-house-developed automated software based on FID-A [37].

2.3 Subjects

This study involved the participation of 164 healthy adults who were right-handed. Their mean age was 28 years with a standard deviation (SD) of 6 years and the age range was 24–45 years. Of these participants, 87 were female. The study was conducted in compliance with ethical principles for medical research involving human subjects, and approval from the local ethics committee was obtained. All volunteers provided informed consent after a detailed explanation of the study procedures and risks. Prior to participation, potential contraindications to MRI examination such as claustrophobia, cochlear implants, metal fragments in the eye or head, or pacemakers were verified to ensure the safety of the participants.

2.4 In Vivo Measurements

The measurements were conducted between 8:00 am and 4:00 pm. Prior to and after the examination, the subjects' temperature was recorded three times in the right ear using an infrared ear thermometer (Braun ThermoScan® 3). During the examination, the subjects were comfortably positioned inside the scanner, with their head centrally located within the coil and supported by padding and foam to minimize movement in any plane. Earplugs were provided for the subjects, and MR data was acquired (please refer to the "**MRI data scanning protocol**" section for a detailed description). Following the scan, temperature measurements were taken again in the right ear. The entire experiment was scheduled to take approximately 60 min per subject, with 15 min allocated for subject preparation and 45 min for MR acquisition. A total of 328 MR measurements were performed, with 164 subjects undergoing 2 PRESS scans, each of which consisted of water-suppressed and unsuppressed data. Data collected during fMRI sessions were not utilized in this study.

2.5 Data Analysis

The preprocessing, line shape fitting, and post-processing were performed with in-house-developed automated software based on FID-A [37]. Following the latest consensus on preprocessing of MRS data [38], the preprocessing process consisted of Fourier transformation of the data, RF coil combination, signal averaging, and phase correction.

2.6 Post-processing

Finally, the signals of metabolites NAA, Cr, and Cho and water peaks from suppressed and unsuppressed spectra were processed using Lorentzian lineshape modeling to acquire Amplitudes, Linewidth, and Center of the investigated peaks. Using the standard linear regression we estimated the correlation between the chemical shift of each metabolite with the measured temperature inside the phantom gaining us six regression curves. Three of these curves were generated by combining the peaks of NAA, Cr, and Cho from suppressed spectra with the water peak from the same suppressed spectra. The other three curves were generated by combining the same metabolite peaks from suppressed spectra with the water peak from unsuppressed spectra.

The general equation we were using for temperature measurement was as follows:

$$T(^{\circ}\mathrm{C}) = A \times \Delta_{W-M} + B \tag{1}$$

Table 1 Calibration results of obtained regression line parameters

Metabolite	Slope (°C/ppm)	Intersection (°C)	R^2
WS-NAA	−101.088	305.8945	0.99
WS-Cr	−100.3	201.5	0.99
WS-Cho	−105	192.8	0.99
WU-NAA	−99	300.4	0.99
WU-Cr	−98.53	198.6	0.99
WU-Cho	−102.5	189.2	0.99

*Water-suppressed spectra (WS), unsuppressed spectra (WU)

where A is the slope and B is an intersection of the regression line measured with conjunctions of metabolites and water peaks and Δ_{W-M} is a difference between frequencies of water and referenced metabolites peaks. Our study used NAA, Cr, and Cho as reference metabolites.

For correlation analyses, Pearson correlation and regression analyses were applied. For comparison reasons, a two-sided Student's t-test was performed. $p < 0.001$ was considered statistically significant. JASP Version 0.17.1 (JASP Team (2023)) [39] and MATLAB 2020b (Mathworks, Natick, MA, USA). were used to analyze data; for data represented as graphs or tables, we used Microsoft Office 2016 Package (Microsoft Corp., Redmond, WA) and MATLAB 2020b.

3 Results

3.1 Phantom Calibration Measurements

Eleven sets of fluoroptic temperature measurements were performed from 31.5 °C to 39.5 °C. The relation between temperature and the chemical shift of the metabolites was estimated by regression line analysis and yielded six regression coefficients: three of which use water suppressed in conjunction with one of metabolites NAA, Cre or Cho; three of which use water unsuppressed with one of metabolites NAA, Cre or Cho. Each regression yielded excellent correlation coefficients (Table 1).

3.2 Water Peak Comparisons Phantom

The in vivo phantom data showed an average water suppression of 95.72%. The average linewidth of the water-suppressed spectra (WS) was significantly higher than that of the unsuppressed spectra (WU) (mean WS = 0.0408 Hz, SD = 0.0119 Hz, compared to WU = 0.0393 Hz, SD = 0.0112 Hz, $p < 0.005$), and the data were strongly correlated ($r = 0.96$, $p < 0.001$). Although the average center frequency

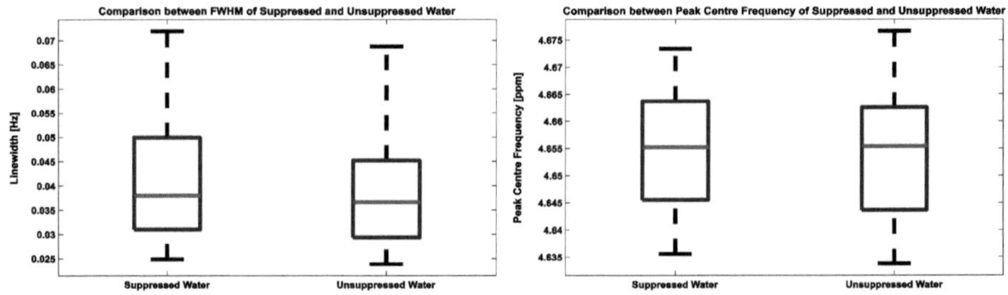

Fig. 3 Comparisons between Linewidth in Water Suppressed and Unsuppressed water (left) and comparisons of peak center frequencies between WS and WU (right), in the phantom data

peaks of the WS were not significantly different from those of the WU (mean WS = 4.6538 ppm, SD = 0.0106 ppm, compared to WU = 4.6541 ppm, SD = 0.0114 ppm, p = 0.53), they were highly correlated (r = 0.97, p < 0.001) (see Fig. 3).

3.3 Water Peak Comparisons In Vivo

The in vivo data showed an average water suppression of 96.01%. The average linewidth of the water-suppressed spectra (WS) was significantly higher than that of the unsuppressed spectra (WU) (mean WS = 0.0866 Hz, SD = 0.0068 Hz, compared to WU = 0.0835 Hz, SD = 0.0058 Hz, p < 0.0000), and the data were strongly correlated (r = 0.87, p < 0.0000). The average center frequency peaks of the WS were significantly higher than those of the WU (mean WS = 4.6129 ppm, SD = 0.0489 ppm, compared to WU = 4.6071 ppm, SD = 0.0486 ppm, p < 0.0000), and they were highly correlated (r = 0.99, p < 0.0000) (see Fig. 4).

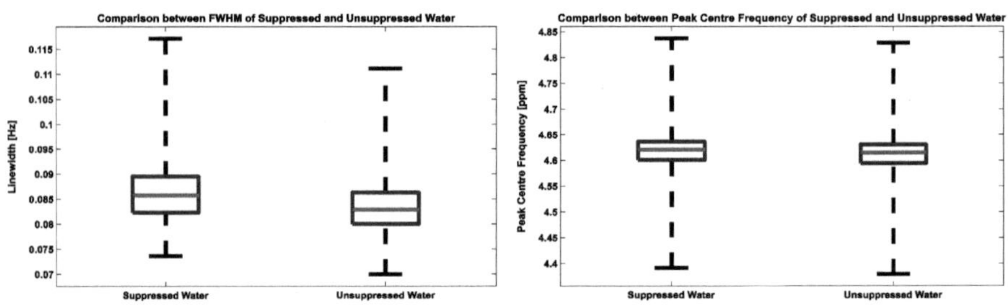

Fig. 4 Comparisons between Linewidth in Water Suppressed and Unsuppressed water (left) and comparisons of peak center frequencies between WS and WU (right), in the in vivo data

Table 2 In vivo measurements: Comparisons of mean temperatures calculated with standard deviations on the basis of six regression curves with Pearson correlation between before and after measurements

Model	T NAA WS	T NAA WU	T Cr WS	T Cr WU	T Cho WS	T Cho WU
PRE fMRI (°C)	37.49(±0.40)	38.13(±0.49)	37.09(±0.34)	37.67(±0.44)	38.18(±0.37)	38.86(±0.52)
POST fMRI (°C)	37.19(±0.45)	37.89(±0.52)	36.81(±0.39)	37.45(±0.50)	37.88(±0.41)	38.62(±0.52)
Change (°C)	-0.3	−0.24	−0.28	−0.22	−0.3	−0.24
Pearson r	0.71	0.34	0.66	0.25	0.59	0.2

*NAA−N-acetylaspartate; Cr−creatine; Cho−Choline; WS−water suppressed, WU−water unsuppressed

3.4 Measurement of Subject Ear Temperature

After scanning, the mean temperature of subjects measured with an infrared ear thermometer showed a significant increase (mean temperature before scanning = 36.86 °C, SD = 0.23, compared to after scanning = 36.95 °C, SD = 0.22, $p < 0.0000$). There were no significant differences in temperature changes between male and female subjects.

3.5 Subject Single-Reference Temperatures using WS

Temperatures were calculated using single-reference metabolite with WS. These calculations showed significant differences between absolute temperature measurements. The smallest SD was observed for TCRE, followed by TCHO, and then TNAA (see Table 2 and Fig. 5).

3.6 Subject Single-Reference Temperatures using WU

Temperatures were also calculated using single-reference metabolite with WU. These calculations showed significant differences between absolute temperature measurements. The smallest SD was observed for TCRE, and the SDs of TCHO and TNAA were similar (see Table 2 and Fig. 5).

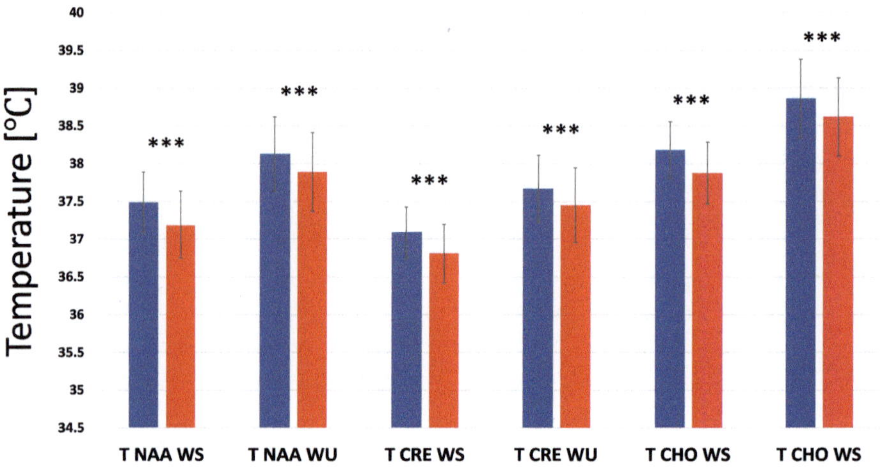

Fig. 5 Comparisons of mean temperature calculations using Suppressed (WS) or Unsuppressed (WU) water peaks with chosen reference peak NAA, Cr or Cho–before (Blue) and after (Red) fMRI acquisition. Asterisks in the picture represent statistical significance

3.7 Subject Single-Reference Temperatures–Differences between WS and WU

Temperatures calculated using WS and WU peaks showed a significant increase of approximately 0.64 °C. The SD of WU temperatures was higher than that of temperatures measured using WS (see Table 2 and Fig. 5).

3.8 Changes Before and After fMRI Scan

All methods of calculation showed a significant decrease in temperatures of approximately −0.3 °C for methods using WS and −0.23 °C when using WU (see Table 2 and Fig. 5).

4 Discussion

The present study was set to evaluate temperature readings acquired using unsuppressed water spectra and their potential use in other spectroscopic studies. To this end, we used a commercially available magnetic resonance spectroscopy sequence in a GE scanner to acquire unsuppressed water spectra simultaneously with those partially suppressed. By performing phantom measurements we subsequently generated scanner-specific calibration curves which are paramount to absolute temperature measurements as in agreement with previous literature [25, 27, 30]. SVS measurements examined 167 healthy volunteers. To this date, it is one of the biggest data sets used for MRST measurements. Looking at only water peaks originating from in vivo and phantom data we acquired information about how the effect of CHESS sequence suppression on those water peaks—overall peak suppression was at approximately 95%, and in both sets of data there was a small yet significant change in Linewidth between suppressed and unsuppressed water peaks which is consistent with recently reported data [33]. Interestingly within in vivo data we observed a significant change between the frequency of the peak of suppressed water and unsuppressed water—which was not observable in phantom data. This change should explain the rise in temperature calculated on the basis of unsuppressed spectra compared to those calculated by its suppressed counterpart. Ear temperature of in vivo subjects reported an overall rise. Still, it could be explained by using earplugs inside the scanner—in later studies, we think that using 2 types of body temperature (tympanic and core) measurement could prove most beneficial.

MRS thermometry as a relatively new technique has been validated against invasive methods. Studies on patients with acute brain injury have shown that brain temperature can vary from 32.6 °C to 45.0 °C, with daily and hourly fluctuations of around one degree [40–42]. However, individual variations over a 72-h monitoring period can be as high as four degrees, depending on the patient's behavior [42].

In this study, we aimed to assess the reliability of the MRS thermometry method for measuring brain temperature. Our combined measurements using suppressed water peaks with all metabolites fall within the range of invasive results, with a range of 35.82–39.58 °C and an amplitude of 0.3°. Similarly, combined measurements of unsuppressed water peaks with all metabolites fall within a range of 36.45–40.59 °C with an amplitude of 0.23. These findings also fall within the range of other MRS-t measurements reported in the literature (Zhang et al. 2020: 35.4–37.4 °C; Rzechorzek et al. 2022: 36.1–40.9 °C; Sung et al. 2022: 35.5–40.1 °C).

Upon comparing suppressed and unsuppressed data, we observed a noteworthy increase of 0.64 °C in the absolute temperature calculated using unsuppressed water peak combined with metabolite peaks from the suppressed spectrum. This rise could be attributed to the difference in the water frequency peak seen only observed in in vivo data. It could be explained due to how suppression of the water peak can result in systematic errors in the estimations of absolute temperature values as per recent studies [33]. However, despite this disparity, we noted a similar temperature change before and after the fMRI session regardless of the method used. Those findings suggest that utilizing an unsuppressed water peak could be advantageous to the field of MRS thermometry, as it would enable the inclusion of temperature as an additional covariate in studies that utilize old spectroscopic data, as recently demonstrated by Mintzopoulos [43]. As in many spectroscopic studies water signal is suppressed along with other unwanted signals. Therefore the water signal may not be detectable in spectra with weak signals or low signal-to-noise ratios. Using unsuppressed water peaks for referencing allows for the detection of the water signal even in spectra with weak signals, increasing the sensitivity of analysis.

Although our study used commercially available sequences, it is important to acknowledge the clear limitations. We did not test our findings on other scanners or with sequences embedded with different water-suppression modules. However, we believe that our results may be applicable to other scanners since their fundamental basis is similar. It is worth noting that our study was conducted on healthy subjects, and therefore, further clinical studies would be necessary to fully understand the implications of our findings.

5 Conclusion

In conclusion, this study validates the use of MRS thermometry as a reliable and non-invasive method for measuring brain temperature. The comparison of suppressed and unsuppressed water peaks in both phantom and in vivo data highlights the potential advantage of utilizing unsuppressed water peaks in future or past MRS studies. Negative change in temperature measured by both techniques will need further research as understanding how prolonged stay inside the MR scanner may prove beneficial to all further MRI studies. The findings also suggest that the use of two types of body temperature measurements, tympanic and core, could be beneficial in future studies. This study adds to the growing body of literature on MRS thermometry and its potential applications in various fields such as brain injury and other spectroscopic studies.

Acknowledgements Project was financed by funding provided by National Centre for Research and Development; grant POIR–01.01.01–00–178/15

References

1. Corbett, R.J., Purdy, P.D., Laptook, A.R., Chaney, C., Garcia, D.: Noninvasive measurement of brain temperature after stroke. Am. J. Neuroradiol. **20**(10), 1851–1857 (1999)
2. Weis, J., Covaciu, L., Rubertsson, S., Allers, M., Lunderquist, A., Ahlström, H.: Noninvasive monitoring of brain temperature during mild hypothermia. Magn. Reson. Imaging **27**(7), 923–932 (2009)
3. Soukup, J., Zauner, A., Doppenberg, E.M., Menzel, M., Gilman, C., Young, H.F., Bullock, R.: The importance of brain temperature in patients after severe head injury: relationship to intracranial pressure, cerebral perfusion pressure, cerebral blood flow, and outcome. J. Neurotrauma **19**(5), 559–571 (2002)
4. Babourina-Brooks, B., Wilson, M., Arvanitis, T.N., Peet, A.C., Davies, N.P.: MRS water resonance frequency in childhood brain tumours: a novel potential biomarker of temperature and tumour environment. NMR Biomed. **27**(10), 1222–1229 (2014)
5. Sone, D., Ikegaya, N., Takahashi, A., Sumida, K., Ota, M., Saito, T., Sato, N.: Noninvasive detection of focal brain hyperthermia related to continuous epileptic activities using proton MR spectroscopy. Epilepsy Res. **138**, 1–4 (2017)
6. Childs, C., Machin, G.: Reliability issues in human brain temperature measurement. Crit. Care **13**(4), 1–4 (2009)

7. Rango, M., Arighi, A., Bresolin, N.: Brain temperature: what do we know? NeuroReport **23**(8), 483–487 (2012)
8. Johnson, F.H., Johnson, S., Eyring, H., Stover, B.J.: The theory of rate processes in biology and medicine. Wiley-Interscience (1974)
9. Chen, J., Daniel, B.L., Pauly, K.B.: Investigation of proton density for measuring tissue temperature. J. Magn. Reson. Imaging **23**(3), 430–434 (2006)
10. Bloembergen, N., Purcell, E., Pound, R.: Relaxation effects in nuclear magnetic resonance absorption. Phys. Rev. **73**(7), 679–712 (1948)
11. Parker, D.L.: Applications of NMR imaging in hyperthermia: an evaluation of the potential for localized tissue heating and noninvasive temperature monitoring. IEEE. Trans. Biomed. Eng. **31**(1), 161–167 (1984)
12. Nelson, T.R., Tung, S.M.: Temperature dependence of proton relaxation times in vitro. Magn. Reson. Imaging **5**(3), 189–199 (1987)
13. Bleier, A.R., Jolesz, F.A., Cohen, M.S., Weisskoff, R.M., Dalcanton, J.J., Higuchi, N., Feinberg, D.A., Rosen, B.R., McKinstry, R.C., Hushek, S.G.: Real-time magnetic resonance imaging of laser heat deposition in tissue. Magn. Reson. Med **21**(1), 132–137 (1991)
14. MacFall, J., Prescott, D.M., Fullar, E., Samulski, T.V.: Temperature dependence of canine brain tissue diffusion coefficient measured in vivo with magnetic resonance echo-planar imaging. Int. J. Hyperthermia. **11**(1), 73–86 (1995)
15. Young, I., Hand, J., Oatridge, A.: Modeling and observation of temperature changes in vivo using MRI. Magn. Reson. Med. **32**(3), 358–369 (1994)
16. Graham, S.J., Stanisz, G.J., Kecojevic, A., Bronskill, M.J., Henkelman, R.M.: Analysis of changes in MR properties of tissues after heat treatment. Magn. Reson. Med **42**(6), 1061–1071 (1999)
17. Ishihara, Y., Calderon, A., Watanabe, H., Okamoto, K., Suzuki, Y., Kuroda, K., Suzuki, Y.: A precise and fast temperature mapping using water proton chemical shift. Magn. Reson. Med. **34**(6), 814–823 (1995)
18. Hindman, J.C.: Proton resonance shift of water in the gas and liquid states. J. Chem. Phys. **44**(12), 4582 (1966)
19. Cady, E.B., D'Souza, P.C., Penrice, J., Lorek, A.: The estimation of local brain temperature by in vivo 1H magnetic resonance spectroscopy. Magn. Reson. Med. **33**(6), 862–867 (1995)
20. Kuroda, K.: Non-invasive MR thermography using the water proton chemical shift. Int. J. Hyperth. **21**(6), 547–560 (2005)
21. McDannold, N.: Quantitative MRI-based temperature mapping based on the proton resonant frequency shift: review of validation studies. Int. J. Hyperth. **21**(6), 533–546 (2005)
22. Childs, C., Hiltunen, Y., Vidyasagar, R., Kauppinen, R.A.: Determination of regional brain temperature using proton magnetic resonance spectroscopy to assess brain–body temperature differences in healthy human subjects. Magn. Reson. Med.: Off. J. Int. Soc. Magn. Reson. Med. **57**(1), 59–66 (2007)
23. Pauly, K.B.: MR thermometry. J. Magn. Reson. Imaging: Off. J. Int. Soc. Magn. Res-Onance Med. **27**(2), 376–390 (2008)
24. Bainbridge, A., Kendall, G.S., Vita, E.D., Hagmann, C., Kapetanakis, A., Cady, E.B., Robertson, N.J.: Regional neonatal brain absolute thermometry by 1H MRS. NMR Biomed. **26**(4), 416–423 (2013)
25. Babourina-Brooks, B., Simpson, R., Arvanitis, T.N., Machin, G., Peet, A.C., Davies, N.P.: MRS thermometry calibration at 3 T: effects of protein, ionic concentration and magnetic field strength. NMR Biomed. **28**(7), 792–800 (2015)
26. Rango, M., Bonifati, C., Bresolin, N.: Post-activation brain warming: a 1-H MRS thermometry study. PLoS ONE **10**(5), e0127314 (2015)
27. Verius, M., Frank, F., Gizewski, E., Broessner, G.: Magnetic resonance spectroscopy thermometry at 3 tesla: importance of calibration measurements. Ther. Hypothermia Temp. Manag. **9**(2), 146–155 (2019)
28. Corbett, R., Laptook, A., Weatherall, P.: Noninvasive measurements of human brain temperature using volume-localized proton magnetic resonance spectroscopy. J. Cereb. Blood Flow Metab. **17**(4), 363–369 (1997). https://doi.org/10.1097/00004647-199704000-00001

29. Covaciu, L., Rubertsson, S., Ortiz-Nieto, F., Ahlström, H., Weis, J.: Human brain MR spectroscopy thermometry using metabolite aqueous-solution calibrations. J. Magn. Reson. Imaging: Off. J. Int. Soc. Magn. Res-Onance Med. **31**(4), 807–814 (2010)
30. Zhu, M., Bashir, A., Ackerman, J.J., Yablonskiy, D.A.: Improved calibration technique for in vivo proton MRS thermometry for brain temperature measurement. Magn. Reson. Med. **60**, 536–541 (2008). https://doi.org/10.1002/mrm.21699
31. Corbett, R.J.T., Laptook, A.R., Tollefsbol, G., Kim, B.: Validation of a noninvasive method to measure brain temperature in vivo using 1H NMR spectroscopy. J. Neurochem. **64**, 1224–1230 (1995). https://doi.org/10.1046/j.1471-4159.1995.64031224.x
32. Thrippleton, M.J., Parikh, J., Harris, B.A., Hammer, S.J., Semple, S.I.K., Andrews, P.J.D., Wardlaw, J.M., Marshall, I.: Reliability of MRSI brain temperature mapping at 1.5 and 3 T. NMR Biomed. **27**, 183–190 (2014). https://doi.org/10.1002/nbm.3050
33. Dong, Z., Milak, M.S., Mann, J.J.: Proton magnetic resonance spectroscopy thermometry: Impact of separately acquired full water or partially suppressed water data on quantification and measurement error. NMR Biomed. **35**(6), e4681 (2022). https://doi.org/10.1002/nbm.4681
34. Haase, A., Frahm, J., Hanicke, W., Matthaei, D.: 1H NMR chemical shift selective (CHESS) imaging. Phys. Med. Biol. **30**(4), 341 (1985)
35. Jurewicz, K., Paluch, K., Kublik, E., Rogala, J., Mikicin, M., Wróbel, A.: EEG-neurofeedback training of beta band (12–22 Hz) affects alpha and beta frequencies–A controlled study of a healthy population. Neuropsychologia **108**, 13–24 (2018)
36. Kirchner, W.K.: Age differences in short-term retention of rapidly changing information. J. Exp. Psychol. **55**(4), 352–358 (1958). https://doi.org/10.1037/h0043688
37. Simpson, R., Devenyi, G.A., Jezzard, P., Hennessy, T.J., Near, J.: Advanced processing and simulation of MRS data using the FID appliance (FID-A)—An open source, MATLAB-based toolkit. Magn. Reson. Med. **77**, 23–33 (2017). https://doi.org/10.1002/mrm.26091
38. Near, J., Harris, A. D., Juchem, C., Kreis, R., Marjańska, M., Öz, G., Gasparovic, C.: Preprocessing, analysis and quantification in single-voxel magnetic resonance spectroscopy: experts' consensus recommendations. NMR Biomed., e4257(2020)
39. JASP Team.: JASP (Version 0.17.1)[Computer software]. (2023). https://jasp-stats.org/
40. Rossi, S, Zanier, E.R., Mauri, I., et al.: Brain temperature, body core temperature, and intracranial pressure in acute cerebral damage. J. Neurol., Neurosurg. & Psychiatry **71**, 448–454 (2001)
41. Lu, H.Y., Huang, A.P., Kuo, L.T.: Prognostic value of circadian brain temperature rhythm in basal ganglia hemorrhage after surgery. Neurol. Ther. **10**(2), 1045–1059 (2021). https://doi.org/10.1007/s40120-021-00283-y
42. Rzechorzek, N.M., Thrippleton, M.J., Chappell, F.M., Mair, G., Ercole, A., Cabeleira, M., Rhodes, J., Marshall, I.: A daily temperature rhythm in the human brain predicts survival after brain injury. Brain **145**(6), 2031–2048 (2022). https://doi.org/10.1093/brain/awab466
43. Mintzopoulos, D., Ratai, E.-M., He, J., Gonzalez, R.G., Kaufman, M.J.: Simian immunodeficiency virus transiently increases brain temperature in rhesus monkeys: detection with magnetic resonance spectroscopy thermometry. Magn Reson Med. **81**, 2896–2904 (2019). https://doi.org/10.1002/mrm.27635

Analyzing the Performance of Real-Coded Genetic Algorithm with Control Locations for Multi-Robot Path Planning

Karolina Wójcik and Adam Ciszkiewicz

Abstract The problem of navigating multi-robot systems through environments in a collision-free manner is becoming ever more important in the healthcare industry. An increasing interesting can be observed in optimization-based methods for path planning in multi robot systems. This study analyzes the performance of a Real-Coded Genetic Algorithm when planning paths for multi-robot systems with up to five robots. Our approach uses the control location method extended to multiple robot systems. The entire path-planning task was contained within the objective function, which meant that no custom modifications to the optimization algorithm were required, furthermore no constraints were placed on the robot displacements—the collision space was continuous. In total 625 simulations were performed over five custom collision maps and including five reruns of the procedure to minimize the effects of random number generator. Our findings indicated that the methods was viable in cases up to five robots, where the average success rate for path planning was more than 37% with a low number of generations and population size. Depending on the map the success rate could be as high as 80% for systems with 4 robots. Interestingly, avoiding robot-to-environment collisions was more difficult than robot-to-robot collisions, while the number of control locations proved to be more costly for the algorithm than the number of the robots. This might suggest that the algorithm is more suited to simpler collision spaces with higher number of robots.

Keywords Optimization · Continuous space · Collision check

1 Introduction

The problem of navigating robotic structures through varied environments in a collision-free manner is a complex one. Nevertheless, with the growing interest in automating common tasks in healthcare industry [1], such as: medicine and food

K. Wójcik · A. Ciszkiewicz (✉)
Faculty of Mechanical Engineering, Cracow University of Technology, Cracow, Poland
e-mail: adam.ciszkiewicz@pk.edu.pl

P. Strumiłło et al. (eds.), *The Latest Developments and Challenges in Biomedical Engineering*, Lecture Notes in Networks and Systems 746,
https://doi.org/10.1007/978-3-031-38430-1_32

delivery, cleaning and sanitization or computer-aided surgery [2, 3] and surgeon training system [4], the problem is becoming ever more important. These new requirements provide unique challenges, one of them being planning the paths for a multi-robot system.

Over the years many different solutions have been presented to path planning. These methods could be divided into single robot path planning and multi robot path planning. In terms of single robot path planning some of the more popular approaches include: sampling-based methods, such as Rapidly-Exploring Random Trees [5–8], graph-based approaches [4] and optimization [9–12]. Due to the nature of the path planning problem, most of these methods could be extended to multi-robot systems, a further summary of the approaches could be found in [13]. An increasing interest can be observed in methods based on optimization, as they tend be more efficient at finding high quality paths in complex systems with single robots [14]. Optimization has also been successfully employed in path planning for multi-robot systems [15–18]. Most of these studies use a modified version of Genetic Algorithm, specialized for the undertaken planning problem, while in some cases Genetic Algorithm is coupled with a different planning algorithm, which results in a hybrid solution [17].

The dynamic nature of multi-robot path planning makes the problem very complicated and nonlinearly scaling. This directly translates to the increased numerical complexity of the task and, while solutions are available, the problem remains open. Furthermore, new and efficient methods could provide opportunities for incorporating more features into path planning, such more complex representation of the collision space or higher-level planning, which takes into account the traversability of the terrain and more.

In this paper, the planning procedure was based on the control locations method [2, 9, 10, 19], which, in comparison to commonly used graph-based approaches, represented the collision space in a continuous manner, while only requiring a few control locations to define a path for a robot. The main aim of this study was to analyze the performance of a Real-Coded Genetic Algorithm (RCGA) with control locations method for path planning in multi-robot systems. The entire path-planning task was contained within the objective function, which meant that no custom modifications to the optimization algorithm were required, furthermore no constraints were placed on the robot displacements—the collision space was continuous.

2 Material and Methods

In this study RCGA was used to plan paths for up to five robots in multi-robot systems using the control locations method. The performance of the approach was tested with five custom collision maps over five reruns of the optimization to minimize the effects of the seed of the random number generator.

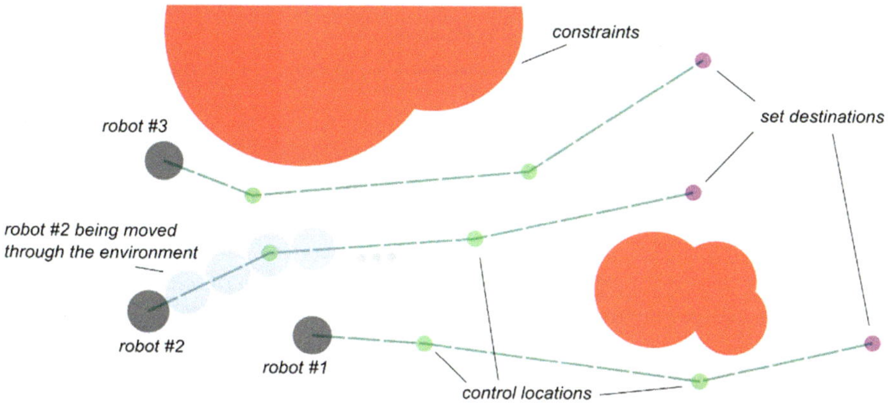

Fig. 1 The control locations method for multiple robots

2.1 *Control Locations Method for Multiple Robots*

The proposed approach was based on the control locations method for path planning, previously employed for single-robot path planning [2, 9, 10, 19]. Each robot had.

a path composed of a fixed number of connected linear segments. These segments were obtained from a fixed number of control locations (here: points)—see Fig. 1. Furthermore, the locations were optimized by Genetic Algorithm. The motivation for using the control locations method, was its relatively low numerical complexity (due to it only requiring a few locations to describe a path), while allowing for continuous representation collision space. These features could potentially prove useful in crowded healthcare centers.

The fitness function took into account the lengths of the paths traversed by the robots as well as the number of collisions, which occurred during the simulation. The use of the weighted sum of objectives allowed for transforming a multi-objective problem into a single-objective one. Low values of the fitness function meant more optimized the paths for the robots.

2.2 *The Fitness Function*

In order to assess the quality of the paths, the robots were moved along their paths concurrently with constant velocity in a step-by-step fashion. A check for robot-robot and robot-environment collision was performed after each step. Additionally, if a robot reached its destination, it remained stationary until all of the robots reached their destinations—in such cases the robot-to-robot collisions checks were still performed. After this simulation, the fitness function detailing the quality of the paths was computed as follows:

$$f(\boldsymbol{x}) = \mathrm{w}_1 \frac{\sum_{i=0}^{n} \sum_{j=0}^{m} \left\| \boldsymbol{p}_{i,j} - \boldsymbol{p}_{i,j+1} \right\|}{\sum_{j=0}^{n} \left\| \boldsymbol{p}_{i,ini} - \boldsymbol{p}_{i,fin} \right\|} + \mathrm{w}_2 col_{r-e} + \mathrm{w}_3 col_{r-r} \tag{1}$$

where: $\boldsymbol{x}$—the decision variable vector, which contained the control points for each robot: $\boldsymbol{x} = [x_{11}\ y_{11}\ x_{12}\ y_{12}\ \ldots\ x_{1n}\ y_{1n}\ x_{21}\ y_{21}\ x_{22}\ y_{22}\ \ldots\ x_{2n}\ y_{2n}\ \ldots\ x_{kn}\ y_{kn}]^T$, n—the number of robots in the system, m—the number of control locations (not counting the initial and final location), $col_{r\text{-}e}$ ($col_{r\text{-}r}$)—the sum of the collisions between any of the robots and the environment (the robots), w_k—the weight of the subobjective k (here: $w_1 = 8.$,
$w_2 = 1.$, $w_3 = 1.$), $\boldsymbol{p}_{i,j}$—the position vector of the control location j for robot i: $\boldsymbol{p}_{i,j} = [x_{ij}\ y_{ij}]^T$, ini (fin)—the initial (final) location of the robot.

2.3 Real-Coded Genetic Algorithm

The fitness function defined in the previous section was minimized using RCGA [20], a variant of Genetic Algorithm [21], which employs real-valued variables. RCGA performed its operations on a decision vector of real variables, which represented the coordinates of control locations, i.e. the points in the paths of the robots. The implementation of RCGA was based on [22]. The initial population of solutions was generated using normal distribution. This population was then transformed over multiple generations with roulette wheel selection [23], nonuniform mutation [24] and blx-α crossover ($\alpha = 0.5$) [25]. The crossover ratio was set to 0.7, while 10% of the best solutions were transferred to a new population through elitism. Additionally, the values of the fitness function were scaled using rank scaling. In all of the simulations the population size and the number of generations were set to 100. This number was deliberate as path planning solutions should be of low numerical complexity to be viable in dynamic environments.

2.4 Collision Maps and Robots

To test the algorithm five collision maps were randomly generated in Python. The initial locations of the robots were spaced over the y-axis with x set to 0. The destinations of the robots were manually set based on traversability of each map. The maps were presented in Fig. 2. In order to decrease the numerical complexity of the task, the collisive elements were represented with circles of varied radii (from 1. to 10.). The robots in the study were also represented with circular shapes, as seen in Fig. 2. This way the collision check between two objects could be performed by computing the distance between each two objects and verifying whether the distance was smaller

Fig. 2 One of the generated collision maps along with the initial (blue color) and final (green color) locations for the robots

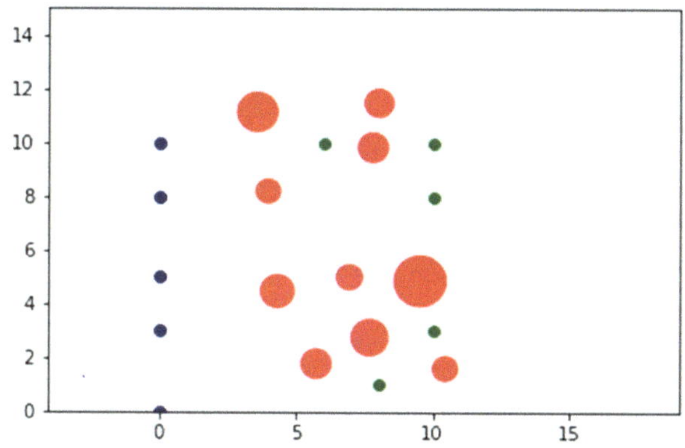

than the sum of their radii—for both robot-robot collisions and robot-environment collision. The code was vectorized using scipy [26]. It is worth mentioning, the robots and the collisive elements could also be represented with triangular meshes and geometry-trees, as seen in [2], without any changes to the algorithm, except for the collision check. Furthermore, more degrees of freedom could be introduced as needed.

2.5 The Simulations

The aforementioned algorithm was used to generate paths with up to five control locations (from one to five) for up to five robots (from one to five) within the five collision maps presented in Fig. 2. Each simulation was repeated five times under different values of the random number generator seed to minimize the effects of chance on the results. After each optimization, the best solution along with the values of its subobjectives (fitness values, path lengths, robot-to-robot and robot-to-environment collisions) were logged.

3 Results and Discussion

In total, 625 simulations were performed using the aformentioned approach. Tables 1 and 2 contained a summary of the obtained results in all of the performed simulations. It can be seen that in 392 out of 625 simulations, the algorithm was successful in finding a collision-free path for the robots. Interestingly, within the failed trials only 20 runs represented cases with robot-to-robot collisions—see Table 2.

With the assumed fitness function, it was easier to avoid robot-to-robot collisions than robot-to-environment collisions. It is worth mentioning that robot-to-robot collisions are dynamic due to the motion of the robots and thus more difficult. We believe

Table 1 The percentage of successful attempts of the proposed path planning procedure with regards to the number of robots, where: *n*—the number of robots in the system, *map*—the map number

n/map	*#1*	*#2*	*#3*	*#4*	*#5*	*Average*
1	88	80	80	92	80	84,0
2	72	92	72	76	76	77,6
3	60	60	52	56	88	63,2
4	40	80	36	40	60	51,2
5	32	44	28	36	48	37,6

that the obtained results are highly reliant on the weights of the fitness function and this should be studied further.

Based on the obtained results, the most difficult case solved with the assumed settings had 5 robots and 4 control locations. Nevertheless, as seen in Table 1, even with 5 robots, more than 37% of path planning attempts were successful. This number could be as high as 80% for 4 robots depending on the map with over 50% of average success rate over all maps. The obtained results were promising considering the settings of the algorithm.

As seen in Table 2, the algorithm performed very well with up to 5 robots and 3 control locations. These results were promising and showcased the viability of the approach. Selected successful attempts for 4 and 5 robots using different maps were presented in Fig. 3.

It is worth noting how much the paths changed, when the number of robots was increased from 4 to 5. Since the robots were assumed to move at the same, constant speed, the delay required for a safe passage within intersecting routes had to be incorporated into the paths themselves—some of the robots had to make a "detour" to allow others to safely pass them.

The efficiency of the algorithm dropped in higher dimensions. Interestingly, including more control locations was more taxing than increasing the number of robots. Based on our results, the algorithm was capable of finding paths for 5 robots with 2 control locations, while the case of 2 robots with 5 control locations had a significantly lower success rate. These results could potentially be improved with by increasing the number of generations. This was further evidenced in Fig. 4, where in 100 generations the algorithm could not simplify complex paths to fit simpler collision spaces and the resulting "jitter motion" was the main cause of collisions.

The results obtained in this paper prove that the control locations method coupled with Genetic Algorithm could be used for path planning in complex multi-robot systems. The main strength of the paper was that it showcased that RCGA can optimize paths for 5 robots moving concurrently, with reasonable effectiveness. Nevertheless, these findings were limited by the total number of simulations—625. This was especially evident in the cases with "jitter motion", which require a further study.

Table 2 A summary of collisions in the optimized paths based on 625 runs of the algorithm, where: *n*—the number of robots in the system, *m*—the number of control locations (not counting the initial and final location), *map* is the map number and *seed* is the seed number id

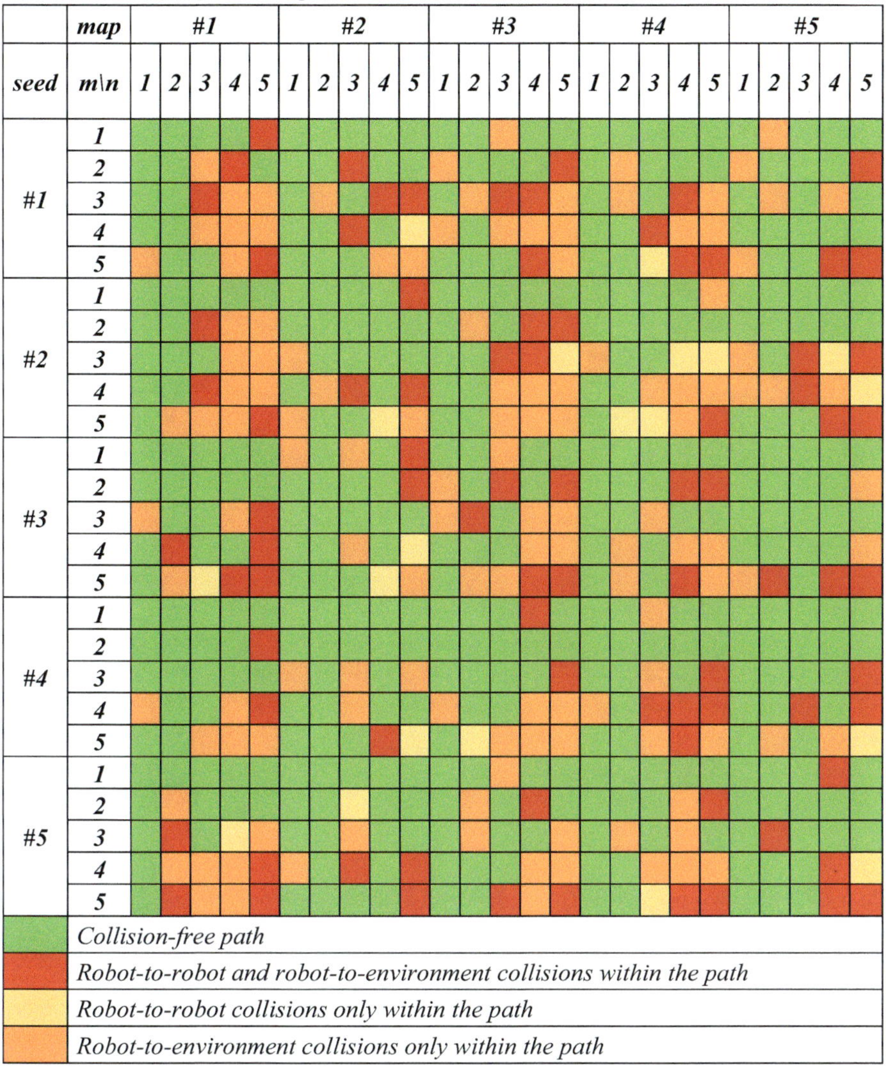

4 Conclusions

The analyzed path planning method for multiple robots proved to be viable in cases up to five robots, where the average success rate for path planning was more than 37% with a relatively low number of generations and population size—both at 100. Interestingly, depending on the map the success rate could be as high as 80% for 4

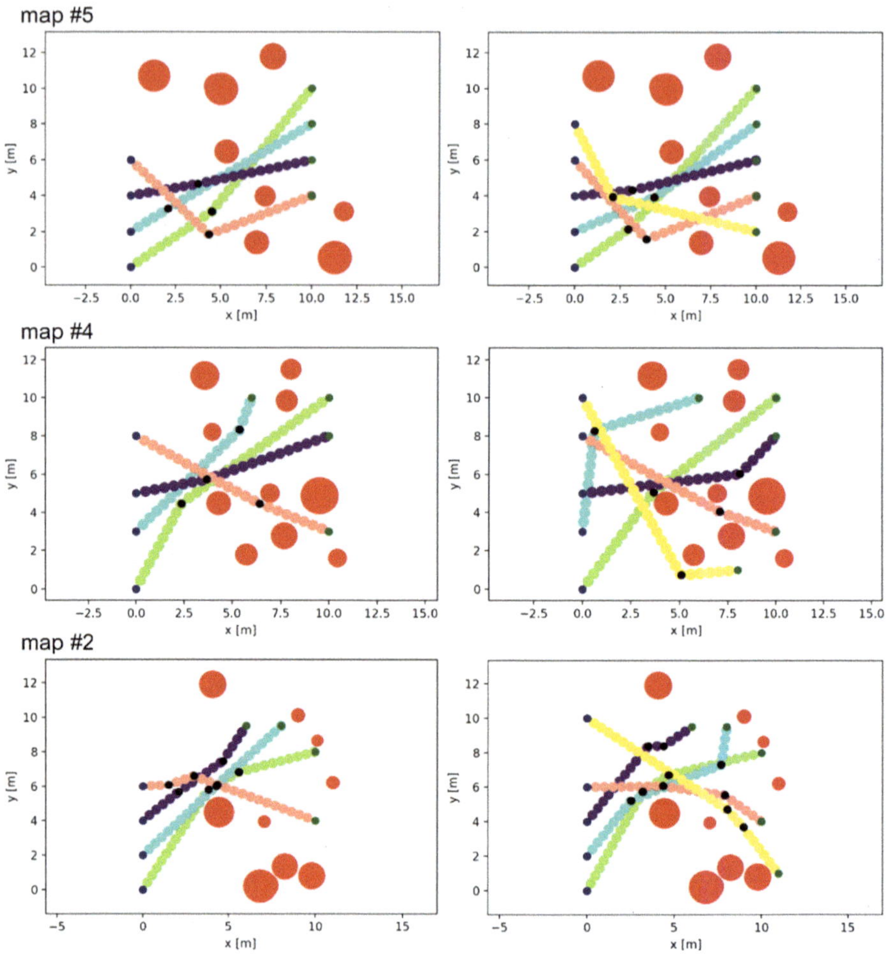

Fig. 3 Selected successful attempts for 4 (left) and 5 (right) robots on three different maps

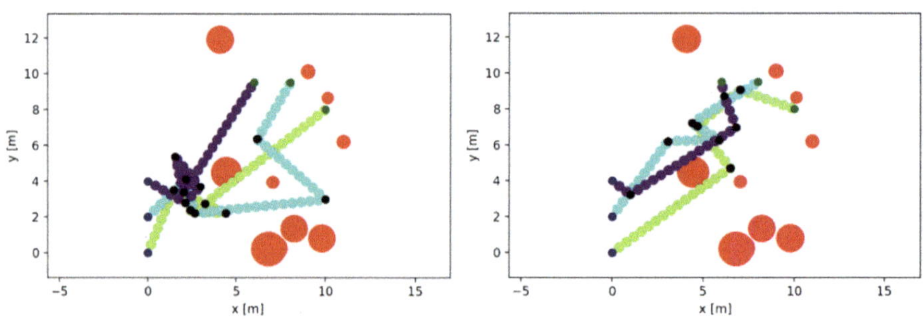

Fig. 4 Examples of "jitter motion", which caused collisions in some of the runs with higher number of control locations

robots. In total, the algorithm was successful in 392 out of 625 simulations. These results suggest that the algorithm could be viable in practical applications in the healthcare industry, for tasks such as automated medicine delivery in large medical centers. Due to the relatively low numerical complexity of the approach along with the continuous representation of the collision space, it is our conjecture that the approach could be useful for path planning in crowded healthcare centers.

It is worth mentioning that avoiding robot-to-environment collisions was more difficult than robot-to-robot collisions. The collisions between robots were dynamic in nature and since the constant speed was imposed on the robots, the paths had to take into account "detours" allowing for safe passages for the robots. In comparison, the collisions between the robots and environment seemed trivial making the results unexpected. We believe that the results might be connected to the weights of the fitness function—this requires a further study.

The number of control locations proved to be more costly for the algorithm. It was easier to incorporate more robots than to increase the number of control locations. We believe that this was mainly caused by the low number of generations, as in some of the solutions with 4 and 5 control locations, the robots moved in a chaotic way—the algorithm could not simplify the path efficiently and the resulting "jitter motion" caused collisions. Nonetheless, the control locations were more costly to incorporate. This might suggest that the algorithm was better suited to simpler collision spaces with higher number of robots.

References

1. Sharma, R., Singh, D., Gaur, P., Joshi, D.: Intelligent automated drug administration and therapy: future of healthcare. Drug Deliv. Transl. Res. **11**, 1878–1902 (2021). https://doi.org/10.1007/s13346-020-00876-4
2. Ciszkiewicz, A., Milewski, G.: Path planning for minimally-invasive knee surgery using a hybrid optimization procedure. Comput. Methods Biomech. Biomed. Engin. **21**, (2018). https://doi.org/10.1080/10255842.2017.1423289
3. Majak, M., Żuk, M., Świątek-Najwer, E., Popek, M., Pietruski, P.: Augmented reality visualization for aiding biopsy procedure according to computer tomography based virtual plan. Acta Bioeng. Biomech. **23**, 81–89 (2021). https://doi.org/10.37190/ABB-01811-2021-02
4. Napalkova, L., Rozenblit, J.W., Hwang, G., Hamilton, A.J., Suantak, L.: An optimal motion planning method for computer-assisted surgical training. Appl. Soft Comput. **24**, 889–899 (2014). https://doi.org/10.1016/j.asoc.2014.08.054
5. LaValle, S.M.: Rapidly-exploring random trees: a new tool for path planning. Res Report. **9811**, (1998) https://doi.org/10.1.1.35.1853
6. Kuffner, J.J., LaValle, S.M.: RRT-connect: An efficient approach to single-query path planning. In: Proceedings 2000 ICRA. Millennium Conference. In: IEEE International Conference on Robotics and Automation. Symposia Proceedings (Cat. No.00CH37065), pp. 995–1001. IEEE (2000). https://doi.org/10.1109/ROBOT.2000.844730
7. Yershova, A., LaValle, S.M.: Improving motion planning alorithms by efficient nearest-neighbor searching. IEEE Trans. Robot., 1–8 (2006). https://doi.org/10.1109/TRO.2006.886840

8. Qureshi, A.H., Ayaz, Y.: Intelligent bidirectional rapidly-exploring random trees for optimal motion planning in complex cluttered environments. Rob. Auton. Syst. **68**, 1–11 (2015). https://doi.org/10.1016/j.robot.2015.02.007
9. Chae, K.G., Park, J.H.: Trajectory optimization with GA and control for quadruped robots. J. Mech. Sci. Technol. **23**, 114–123 (2009). https://doi.org/10.1007/s12206-008-0920-9
10. Abu-Dakka, F.J., Rubio, F., Valero, F., Mata, V.: Evolutionary indirect approach to solving trajectory planning problem for industrial robots operating in workspaces with obstacles. Eur. J. Mech. A/Solids. **42**, 210–218 (2013). https://doi.org/10.1016/j.euromechsol.2013.05.007
11. Che, L., Kang, F.J.: Automatic global path generation for large-scale 3D scene exploration. Int. J. Model. Simulation, Sci. Comput. **11**, (2020)
12. Kundra, H., Khan, W., Malik, M., Rane, K.P., Neware, R., Jain, V.: Quantum-inspired firefly algorithm integrated with cuckoo search for optimal path planning. Int. J. Mod. Phys. C. **33**, (2022)
13. Madridano, Á., Al-Kaff, A., Martín, D., de la Escalera, A.: Trajectory planning for multi-robot systems: Methods and applications. Expert Syst. Appl. **173**, (2021). https://doi.org/10.1016/j.eswa.2021.114660
14. Chen, Z., Wu, H., Chen, Y., Cheng, L., Zhang, B.: Patrol robot path planning in nuclear power plant using an interval multi-objective particle swarm optimization algorithm. Appl. Soft Comput. **116**, 108192 (2022). https://doi.org/10.1016/j.asoc.2021.108192
15. Bai, X., Yan, W., Cao, M., Xue, D.: Distributed multi-vehicle task assignment in a time-invariant drift field with obstacles. IET Control Theory Appl. **13**, 2886–2893 (2019). https://doi.org/10.1049/iet-cta.2018.6125
16. Han, Z., Wang, D., Liu, F., Zhao, Z.: Multi-AGV path planning with double-path constraints by using an improved genetic algorithm. PLoS ONE **12**, 1–16 (2017). https://doi.org/10.1371/journal.pone.0181747
17. Nazarahari, M., Khanmirza, E., Doostie, S.: Multi-objective multi-robot path planning in continuous environment using an enhanced genetic algorithm. Expert Syst. Appl. **115**, 106–120 (2019). https://doi.org/10.1016/j.eswa.2018.08.008
18. Zhen, X., Enze, Z., Qingwei, C.: Rotary unmanned aerial vehicles path planning in rough terrain based on multi-objective particle swarm optimization. J. Syst. Eng. Electron. **31**, 130–141 (2020)
19. Hidalgo-Paniagua, A., Vega-Rodríguez, M.A., Ferruz, J., Pavón, N.: MOSFLA-MRPP: Multi-objective shuffled frog-leaping algorithm applied to mobile robot path planning. Eng. Appl. Artif. Intell. **44**, 123–136 (2015). https://doi.org/10.1016/j.engappai.2015.05.011
20. Herrera, F., Lozano, M., Verdegay, J.L.: Tackling real-coded genetic algorithms: operators and tools for behavioural analysis. Artif. Intell. Rev. **12**, 265–319 (1998). https://doi.org/10.1023/A:1006504901164
21. Goldberg, D.E.: Genetic algorithms in search. Optimization and machine learning. Addison-Wesley Longman Publishing Co., Inc, Boston (1989)
22. Ciszkiewicz, A.: Analyzing uncertainty of an ankle joint model with genetic algorithm. Materials (Basel). **13**, 1175 (2020). https://doi.org/10.3390/ma13051175
23. Baker, J.E.: Reducing bias and inefficiency in the selection algorithm. Proc ICGA. **2**, 14–21 (1987)
24. Michalewicz, Z.: Genetic algorithms + data structures = evolution programs. Springer-Verlag, New York (1992)
25. Eshelman, L.J., Schaffer, J.D.: Real-coded genetic algorithms and interval-schemata. Found. Genet. Algorithms **2**, 187–202 (1993)
26. van der Walt, S., Colbert, S.C., Varoquaux, G.: The NumPy array: A structure for efficient numerical computation. Comput. Sci. Eng. **13**, 22–30 (2011). https://doi.org/10.1109/MCSE.2011.37

Detection of People Swimming in Water Reservoirs with the Use of Multimodal Imaging and Machine Learning

Jakub Konert, Adam Dradrach, and Jacek Rumiński

Abstract Every year in many countries, there are fatal unintentional drownings in different water reservoirs like swimming pools, lakes, seas, or oceans. The existing threats of this type require creating a method that could automatically supervise such places to increase the safety of bathers. This work aimed to create methods and prototype solutions for detecting people bathing in water reservoirs using a multimodal imaging system and machine learning. Two types of cameras, RGB and thermal, were integrated and calibrated to form a multimodal imaging system. The system was designed and implemented to acquire real-world data for bathing people in swimming pools. The EfficientDet models were adapted and trained on collected data reaching at least 94% detection accuracy, with the highest result equal to 97.47%. The best accuracy obtained for the thermal data was lower and equal to 94.85%. However, thermal imaging allows observing scenes in low-light conditions or darkness. This could potentially highly improve the effectiveness of rescue missions, decreasing the death rates or improving the health of early rescued people. Thermal imaging could also be more acceptable regarding privacy, as high-frequency biometric features are not as easy to extract from thermal images as from high-resolution RGB images.

Keywords Machine learning · Multimodal vision · Object detection · Thermal imaging · Safety in water reservoirs · Human health

1 Introduction

It is estimated that drowning is responsible for approximately 320,000 deaths annually worldwide, of which approximately 25% occur in swimming pools [1, 2]. This is because, to a non-rescuer, a drowning person looks like they are having fun or is just swimming. During drowning, a person cannot call for help because he focuses on the

---

J. Konert · A. Dradrach · J. Rumiński (✉)
Faculty of Electronics, Telecommunications and Informatics, Gdańsk University of Technology, Gdańsk, Poland
e-mail: jacek.ruminski@pg.edu.pl

P. Strumiłło et al. (eds.), *The Latest Developments and Challenges in Biomedical Engineering*, Lecture Notes in Networks and Systems 746,
https://doi.org/10.1007/978-3-031-38430-1_33

chaotic oxygen intake into the lungs [2]. In 2020, in Poland alone, 483 people lost their lives due to drowning [3]. In the USA, there are over 4,000 fatal unintentional drownings, and more than 8,000 nonfatal drownings [4]. Therefore, a system providing additional support for bathing people could be very important. Such systems and related methods should work in real-time, or near-real-time [5]. Otherwise, it would be difficult to successfully detect bathers in video sequences [2]. Many possible hardware platforms enable low-power and efficient data processing for video data processing using a machine learning approach. In this work, the Coral Development Board was chosen [6]. Multimodal imaging allows recording devices with different modalities to complement each other. For example, a visible light camera can provide high-spatial-frequency details in normal light conditions. In contrast, a thermal camera could provide information in low light or even in darkness [7–9]. Due to technical reasons, such cameras are usually produced as separate devices and must be integrated. Therefore, such cameras should be calibrated to show images that represent similar scenes. This task is difficult since thermal images are smoother due to the heat flow in imaging objects [10, 11]. There are many methods for calibrating such cameras. The probably most popular method uses a calibration chart or board with squares arranged in a checkerboard pattern [12] for visible observations and as thermally different squares for thermal imaging. This arrangement ensures the presence of corners visible in the RGB and thermal images, which can be detected by the appropriate software and mapped to the appropriate coordinates. Different configurations of the cameras and their calibration were presented in [13, 14].

Limited related work has been published on detecting people in water reservoirs using multimodal or thermal imaging. In [15], authors developed multi-step algorithms (e.g., 20 steps) for intruder detection outside and inside swimming pool regions. They used a traditional image processing approach looking for human head patterns. The results obtained during experiments gave 95.58% accuracy for the region outside the swimming pool and 92.44 for the region inside the swimming pool. In another study, [16] authors collected thermal images of objects in water using Unmanned Aerial Systems (UAS) in maritime Search And Rescue (SAR) missions. The following categories of objects were labeled: Boats, Pallets, Humans (313), and Buoys. The neural network model with 2 Convolutional Layers was trained for the classification problem. The achieved average accuracy was 92.5%. Detection of inflatable boats with people using thermal imaging was proposed in [17]. The authors recorded video streams using a high-resolution thermal camera (640×480) and selected 2500 images for a dataset. Their dataset also contained thermal images of the people at night and day recorded in other scenarios (not related to water reservoirs). They trained Yolo detection models [18] and Faster R-CNN models using different CNNs for feature extraction. The best classification rate was 91% for Yolo and 95% for Faster R-CNN. Another, less related study on thermal imaging applied for swimming people was proposed in [19]. The authors captured thermal images before and after the swimming task. Each task consisted of a 7×200 m protocol in crawl or backstroke. The results showed no significant differences between both techniques analyzed for 10 subjects.

This preliminary study aims to analyze the potential role of thermal and visible light imaging in detecting people in water reservoirs.

Key contributions of this study include: (1) A data acquisition system was designed to collect visible light images and low-resolution thermal images. The cameras were calibrated using a specially designed calibration board. (2) A new dataset was collected with multimodal and annotated images of people in water reservoirs. (3) The EfficientDet models were adapted to detect people in water reservoirs, and the best model was converted for use in mobile devices. (4) The prediction speed and accuracy of two versions of the same architecture were analyzed and compared for visible and thermal images.

This paper is organized as follows: Sect. 2 describes the experimental setup, method, and created dataset. Section 3 describes the achieved results. Section 4 presents a discussion of the results. Section 5 summarizes this work.

2 Material and Methods

2.1 Multimodal Camera

A novel module that combines an RGB camera with a thermal camera has been designed and implemented to provide a cost-effective data-acquisition solution. The thermal camera was the FLIR Lepton 3.5 Mini Thermal Camera [20], while the RGB camera was the Logitech HD Webcam C270 [21]. Despite the large variations in their specifications, the problem was overcome via software calibration. Both cameras needed to be calibrated and required a standard reference board. To accomplish this, a chess board with 450 Ω resistors attached to the PCB was utilized to apply voltage to the alternating fields. The thermal section of the board was based on an earlier design, as shown in [22], while the visible light camera was painted with black-and-white squares.

The board was carefully positioned to avoid interference with the recording, and no reflections or other heat sources were allowed to impact it. The board was heated by applying a voltage to resistors and copper fields (Fig. 1). A simple Python script was used to collect frames simultaneously from the thermal and RGB cameras, which were positioned in three different locations relative to the calibration table. Approximately ten shots were taken, resulting in over 30 frames for each camera. The Matlab script was prepared to calibrate cameras using transformation parameters for distortion, translation, rotation, etc. The transformation parameters were imported by a Python script to perform the required pre-processing of corresponding frames. The cameras' pre-calibrated output images were modified, resulting in a resolution of 640×480 px for both modalities. Thermal images were upscaled from the default values, and visible light images were cropped and translated correspondingly.

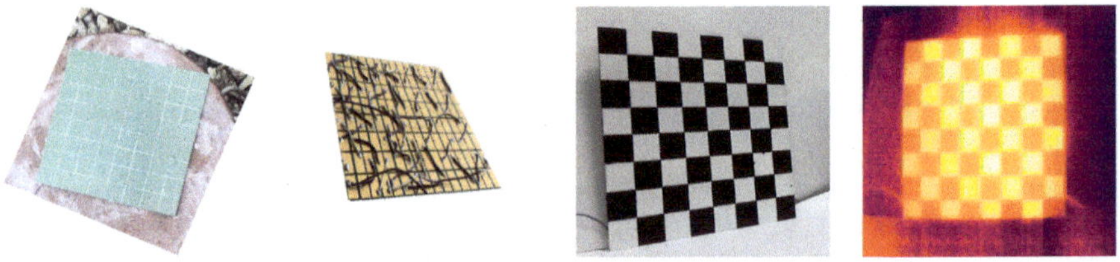

(a) Chessboard before painting (b) Back of the chessboard (c) Painted chessboard (d) Chessboard in thermal

Fig. 1 Calibration chessboard

2.2 *Data Acquisition Unit*

The set of cameras was enclosed in a specially placed housing that was printed on a 3D printer (Fig. 2). The housing has been additionally sealed to prevent the impact of environmental influences (e.g., water), which could damage the cameras used. In addition, additional lenses were used, one of ordinary silicon glass for the RGB camera and the other of germanium glass, which allows free penetration of far infrared radiation through its structure and thus allows the thermal camera to record the image from inside the housing. Inside the housing, the cameras were connected to a common USB hub, and the output cable led out. This allowed for the tightness of the housing to be maintained.

The second module of the system consisted of a computing unit, which was the Coral Dev Board as a single-board computer [23]. It was placed in another housing designed and printed on a 3D printer (Fig. 3). The connection of both modules was carried out using a USB cable. The USB hub inside the camera housing allowed us to avoid the problem with the limited USB inputs found on the Coral Dev Board.

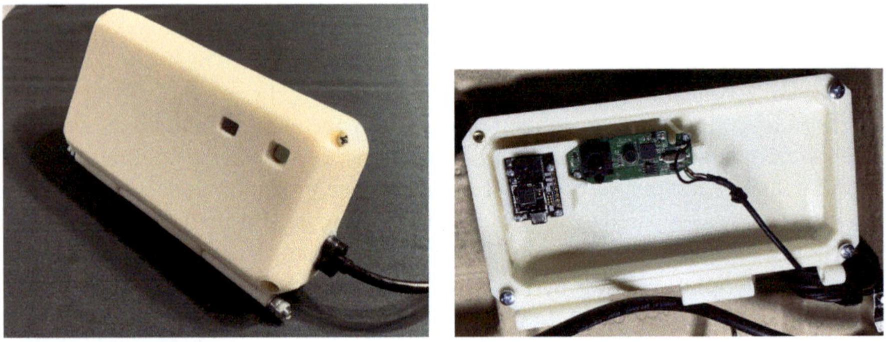

(a) Housing outside (b) Housing inside

Fig. 2 Cameras housing

(a) Housing outside (b) Housing inside

Fig. 3 Coral Dev Board housing

A division into separate software processes was used to synchronize both cameras. This method allows for avoiding the problem of global interpreter lock (GIL) possible to occur using synchronization across many threads. The data acquisition consisted of capturing a frame from one camera and blocking the process until the second camera captured the frame and unlocked the first process. This situation took place alternately. The whole solution was forced by the different frame rates of the cameras used. The frame rate of the RGB camera is 30 frames per second, while the thermal frame rate is only 9 frames per second.

2.3 Data Acquisition

A new dataset was collected using the designed system. For this purpose, the module with cameras has been moved to the private pool, on which tracks were reserved for this task. Only the volunteers who participated in the experimental study were in the pool during data collection. This allowed us to avoid the problem of registering the image of third parties without their consent. It is worth mentioning that the water in the pool was 26 °C, and despite this, after some time of bathing, the thermal images already showed a decrease in the surface body temperature of the participants. The module with the Coral Dev Board was located in the lifeguard's office, while the module with cameras was led out to the swimming pool via a USB cable. This procedure was for safety purposes and not to lead the power supply out of the 230 V socket, which was used to power the module from the Coral Dev Board. The data were collected after 30 min of swimming in the pool. Each camera collected data according to its frequency and standard image resolution. This allowed for the collection of more than 3000 RGB and thermal raw images. The example of collected RGB and thermal images is shown below in Fig. 4.

(a) RGB raw image (b) Thermal raw image

Fig. 4 Examples of collected images

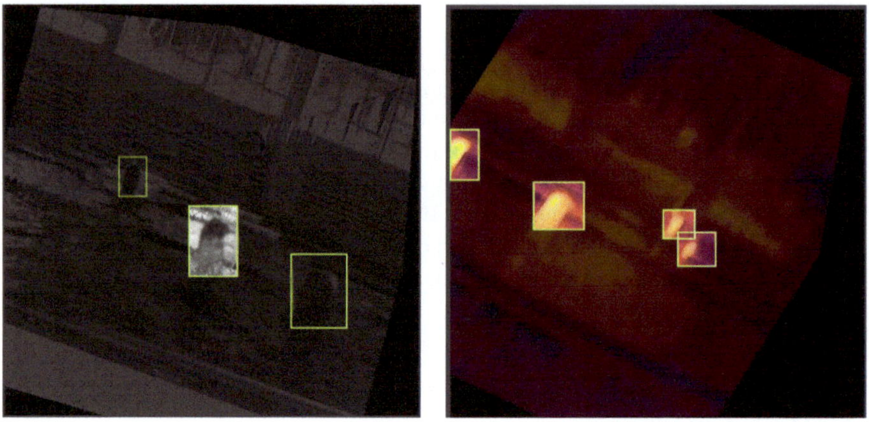

(a) Transformed and tagged RGB image (b) Transformed and tagged thermal image

Fig. 5 Examples of transformed and tagged images

Collected images were pre-processed and augmented (e.g. noise reduction, geometrical transformations) using tools from the Roboflow.com website. Finally, 7579 processed RGB images were obtained, and 7740 thermal images. Some of the data sets obtained in this way were also manually labeled, using tools from the Roboflow.com website. Only the processed and marked images were used to compose the final datasets, which were used to train the adapted CNN models. An example of processed and tagged images using Roboflow.com is shown in Fig. 5.

2.4 *Object Detection Methods*

The pre-trained models from the EfficientDet architecture versions D0 and D1 [24, 25] were adapted in this study. Separate models were trained for visible light images and thermal images. Four model versions were trained: two for the RGB images (EfficientDet D0 and D1), and two for the thermal images (EfficientDet D0 and D1). One of the objectives was to compare the results of the models trained for RGB and thermal images. All datasets were divided into training and test sets using 75% of training images and 25% of test images. Additionally, the Model Maker python library was used to train the models and immediately convert them to the .tflite format. The .tflite format represents the format of models created using the TensorFlow2 library, but for mobile devices, an example of this is Coral Dev Board. All the models were trained using the infrastructure available at the AI Living Lab of the Gdansk University of Technology (Nvidia A40 graphics cards and a VRAM capacity of 48 GB [26]). The training process for each model lasted several hours at batch-size of 32 and 200 epochs. The quality of the models was tested, and the Average Precision 50 (AP50, Average Precision at Intersection over Union (IoU) = 0.50) and accuracy metrics were used.

3 Results

A total of 4 models were prepared: 2 trained (tuned) with RGB images, and 2 trained (tuned) with thermal images. In Fig. 6, training and validation curves are presented for the adapted EfficientDet D0 model. Similar results for the adapted EfficientDet D1 model are presented in Fig. 7.

The models trained on RGB images achieved AP50 results exceeding 97%, while the models trained on thermal images about 94%. Different methods to prevent overfitting were used (e.g., data augmentation, regularization, etc.). However, during the training, the fixed number of epochs was used without the overfitting test during the

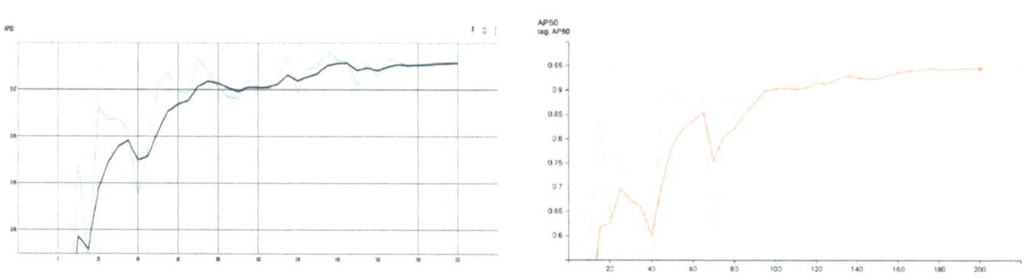

(a) Model trained using RGB images (b) Model trained using thermal images

Fig. 6 Training (pale lines) and validation (solid lines) curves showing relative AP50 values for the EfficientDet D0 model. Maximum epoch (X-axis) is 200

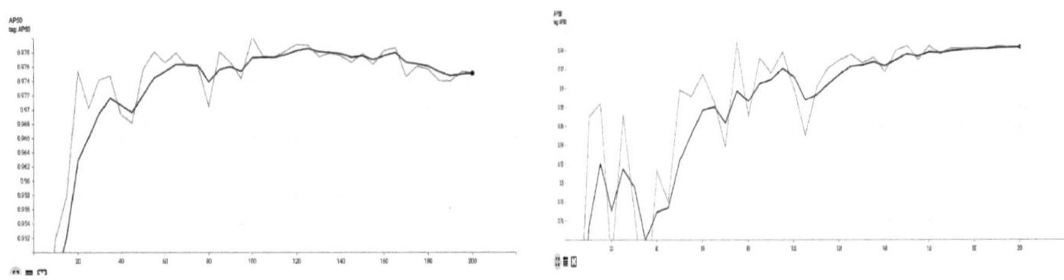

(a) Model trained using RGB images (b) Model trained using thermal images

Fig. 7 Training (pale lines) and validation (solid lines) curves showing relative AP50 values for the EfficientDet D1 model. Maximum epoch (X-axis) is 200

Table 1 Models prediction time and accuracy comparison

	EfficientDet CPU model			
	RGB		Thermal	
	Version 0	Version 1	Version 0	Version 1
Prediction time [ms]	489.95	944.97	481.88	952.81
Accuracy (%)	97.11	97.47	94.23	94.85

Table 2 Coral Dev Board power consumption comparison for the designed application

	Preview model	Full system load
Power consumption [Wh]	10	20

validation process. In Fig. 7a the characteristic small decrease of AP50 is observed. The Early Stopping could be additionally used in the future to improve the generalization properties of the model. In this study, only a limited dataset is available; therefore, the same subset was used for validation and tests. However, tests were performed using data the models had not seen before. The same test was performed for each model. The test consisted of loading a frame and making a prediction, during which processing time was measured. The test has been performed a total of 100 times for each model. These results were recorded in a list from which average values were calculated. Table 1 presents the average performance parameters obtained during tests.

Additionally, the Coral Dev Board's power consumption during the designed application's operation was tested in two cases. The first one is the state when the application runs only with the image preview turned on. The second one is when the application has the preview turned on, and additionally, it collects data and labels them on both cameras. The power consumption values are presented in Table 2.

4 Discussion

The cost-effective data collection system was designed and used to create an original dataset of people swimming in water. Visible light and thermal images were acquired, pre-processed, and used for training deep learning models. The results showed higher detection accuracy for models trained on visible images, for which more high-frequency features can be automatically extracted using CNN models. However, the quality metrics obtained for models trained on low-resolution thermal images were only 3 percent lower. This is a very promising result of this preliminary study. Results could be improved by using high-resolution thermal cameras or dedicated super-resolution algorithms. Models trained with RGB images have better accuracy and could be used in good visibility conditions. However, it could be an important privacy issue since many characteristic features of bathing people are visible. Thermal images could be potentially more acceptable, especially in case of emergency or during the night, or in poor visibility conditions. The obtained detection accuracy for models trained on thermal images is comparable to other (but limited) earlier studies. The authors of [15] obtained 95.58% accuracy for the region outside the swimming pool and 92.44 for the region inside the swimming pool. The results from this study are higher for low-resolution thermal images. Also, compared to other studies, like [16] or [17], results are similar. However, higher-resolution cameras were used in both related works, and the obtained metrics can not be directly compared due to different types of tasks.

In this study, EfficientDet models were adapted. The quality metrics were very similar for both versions of architectures (i.e., version D0 and D1). The number of parameters for the D1 version is about 6.6M, which is much higher than for D0 (about 3.9M). This is probably the reason for the worse performance of the D1 model versus the D0 model (945 ms vs. 490 ms). The obtained average performance values with Coral's CPU are slightly worse than the processing time for the powerful Xeon CPU used in the original EfficientDet paper [24]. For both models, the processing times could be further improved using the Tensor Processing Unit of the Coral board instead of the CPU of this board. It is also important to underline that the power consumption was very low compared to other CPUs or GPUs. This property could be very important for emergency missions when using drones or other battery-operated platforms.

This study has several limitations, including the small size and diversity of the dataset (size, environment, etc.), the limited number of analyzed models, etc. Future works will also focus on data fusion that could combine four data input layers for RGB and thermal data in one deep-learning model. However, this interesting preliminary study shows that promising detection results could be obtained using a mobile edge device with a low-resolution thermal camera. Further studies could potentially improve the performance of the models. The improved models could be used on drones, lifeguard towers, swimming pools, and many other places. Such solutions could potentially save human lives. Additionally, fast rescue time could help decrease the health-related complications in rescued people after a longer lack of oxygen.

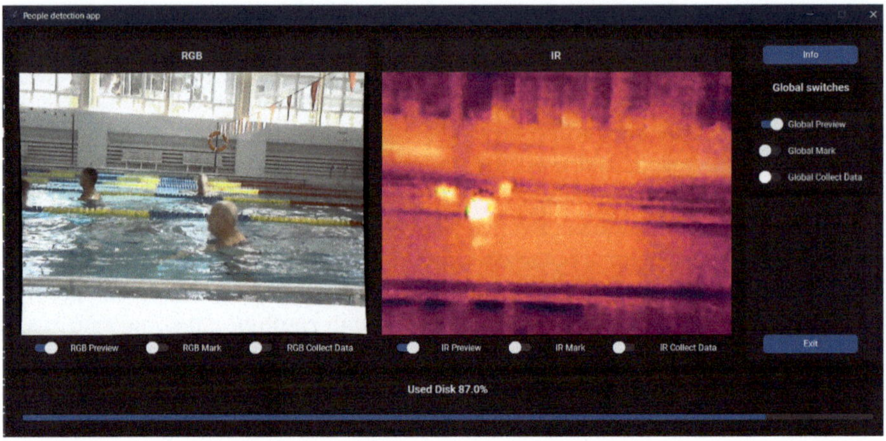

Fig. 8 User interface of the designed application: acquisition of raw video streams

5 Conclusion

In this study, a new dataset containing multimodal images was created. The adapted EfficientDet models achieved relatively high detection results. The models trained on low-resolution thermal images show only 3% lower accuracy results compared to models trained on RGB images. High detection efficiency on a device with low power consumption is a promising result for further studies. The experimental results were obtained for images collected with the designed and implemented platform. It allows data acquisition, image pre-processing, and human detection (Fig. 8). The implemented platform could also be used for other potential applications (e.g., animal studies). It is also relatively easy to replace the thermal camera with a higher resolution version that could allow potentially better detection results in the future.

References

1. Andriluka, M., Roth, S., Schiele, B.: People-tracking-by-detection and people-detection-by-tracking. In: 2008 IEEE Conference on Computer Vision and Pattern Recognition, pp. 1–8 (2008)
2. Hwang, S., Park, J., Kim, N., Choi, Y., Kweon, I.S.: Multispectral pedestrian detection: benchmark dataset and baseline. In: 2015 IEEE Conference on Computer Vision and Pattern Recognition (CVPR), pp. 1037–1045 (2015)
3. Police: public police base on drownings in 2020 in Poland (2020). https://statystyka.policja.pl/st/wybrane-statystyki/utoniecia/48247,Utoniecia-2020.html
4. CDC: drowning facts. In: Centers for Disease Control and Prevention (2022). https://www.cdc.gov/drowning/facts/index.html

5. Paszke, A., Chaurasia, A., Kim, S., Culurciello, E.: Enet: a deep neural network architecture for real-time semantic segmentation (2016). https://arxiv.org/pdf/1606.02147.pdf
6. Google: Coral Dev Board datasheet (2022). https://coral.ai/static/files/Coral-Dev-Board-datasheet.pdf
7. Liu, R., Zhang, H., Scherer, S.: Multiple methods of geometric calibration of thermal camera and a method of extracting thermal calibration feature points, pp. 1–8 (2018). https://henryzh47.github.io/Thermal-Camera-Calibration/
8. Krishnan, A., Saripalli, S.: Cross-calibration of RGB and thermal cameras with a lidar for RGB-depth-thermal mapping. Unmanned Syst. **5**(2), 59–78 (2017)
9. Sun, Y., Zuo, W., Liu, M.: RTFNet: RGB-thermal fusion network for semantic segmentation of urban scenes. IEEE Robot. Autom. Lett. **4**(3), 2576–2583 (2019)
10. Kwasniewska, A., Ruminski, J.: Face detection in image sequences using a portable thermal camera. In: 13th Quantitative Infrared Thermography Conference (QIRT), pp. 493–499 (2016)
11. Luhmann, T., Piechel, J., Roelfs, T.: Geometric calibration of thermographic cameras. In: Kuenzer, C., Dech, S. (eds.) Thermal Infrared Remote Sensing: Sensors, Methods, Applications, pp. 27–42. Springer Netherlands, Dordrecht (2013). https://doi.org/10.1007/978-94-007-6639-6_2
12. Shivakumar, S.S., Rodrigues, N., Zhou, A., Miller, I.D., Kumar, V., Taylor, C.J.: Pst900: RGB-thermal calibration, dataset and segmentation network. In: 2020 IEEE International Conference on Robotics and Automation (ICRA), pp. 9441–9447 (2020)
13. Zhang, Y., Gao, Y., Gu, S., Guo, Y., Liu, M., Sun, Z., Hou, Z., Yang, H., Wang, Y., Yang, J., Ponce, J., Kong, H.: Build your own hybrid thermal/EO camera for autonomous vehicle. In: ICRA 2019-IEEE International Conference on Robotics and Automation. Montreal, Canada (May 2019). https://hal.inria.fr/hal-02051880. To appear in the Proceedings of the IEEE International Conference on Robotics and Automation (2019)
14. Kim, N., Choi, Y., Hwang, S., Park, K., Yoon, J., Kweon, I.: Geometrical calibration of multispectral calibration. In: 2015 12th International Conference on Ubiquitous Robots and Ambient Intelligence (URAI), pp. 384–385 (2015)
15. Wong, W., Hui, J., Chu Kiong, L., Lim, W.S.: Off-time swimming pool surveillance using thermal imaging system. In: 2011 IEEE International Conference on Signal and Image Processing Applications, ICSIPA 2011, pp. 366–371 (Nov. 2011)
16. Rodin, C.D., de Lima, L.N., de Alcantara Andrade, F.A., Haddad, D.B., Johansen, T.A., Storvold, R.: Object classification in thermal images using convolutional neural networks for search and rescue missions with unmanned aerial systems. In: 2018 International Joint Conference on Neural Networks (IJCNN), pp. 1–8 (2018)
17. Kowalski, M.L., Pałka, N., Myńczak, J., Karol, M., Czerwińska, E., Życzkowski, M., Ciurapiński, W., Zawadzki, Z., Brawata, S.: Detection of inflatable boats and people in thermal infrared with deep learning methods. Sensors **21**(16) (2021). https://www.mdpi.com/1424-8220/21/16/5330
18. Redmon, J., Divvala, S., Girshick, R., Farhadi, A.: You only look once: unified, real-time object detection. In: 2016 IEEE Conference on Computer Vision and Pattern Recognition (CVPR), pp. 779–788 (2016)
19. Domingues, A., Barbosa, F., Seixas, A., Borgonovo-Santos, M., Marques, E., Vardasca, R., Mendes, J., Fernandes, R., Vilas-Boas, J.P.: Infrared Thermography in Swimming: Thermal Characterization of Swimming Technique, pp. 795–815 (Jan. 2021)
20. Flir: Lepton 3.5 ir camera (2022). https://www.flir.com/products/lepton/?model=500-0771-01&vertical=microcam&segment=oem
21. Logitech: Logitech c270 hd camera (2022). https://www.logitech.com/en-gb/products/webcams/c270-hd-webcam.960-001063.html
22. Bujnowski, A., Benoit, M., Kaczmarek, M., Przystup, P., Rumiński, J., Pecci, I.: Active and dynamic graphical code for object identification in healthcare. J. Med. Imaging Health Inform. **5**(8), 1631–1639 (2015)
23. Google: get started with the dev board (2020). https://coral.ai/docs/dev-board/get-started

24. Tan, M., Pang, R., Le, Q.V.: Efficientdet: scalable and efficient object detection. In: 2020 IEEE/CVF Conference on Computer Vision and Pattern Recognition (CVPR), pp. 10778–10787 (2020)
25. Tensorflow 2 detection model zoo (2021)
26. Nvidia: Nvidia a40 GPU for visual computing (Oct. 2020). https://www.nvidia.com/en-us/data-center/a40/

Haptic Display of Depth Images in an Electronic Travel Aid for the Blind: Technical Indoor Trials

Piotr Skulimowski, Paweł Strumiłło, Szymon Trygar, and Wacław Trygar

Abstract This paper reports on an assistive system for the visually impaired that presents the environment through haptic stimulation. This wearable system consists of a processing unit attached to the user's chest and a haptic belt placed to the user's abdomen. The custom-made belt consists of 20 vibration actuators arranged in a 4 $\times$ 5 matrix. The depth images captured by the camera are processed to identify the orientation of the ground plane and the location of obstacles. The system's detection range is limited to 2.5 m within a user-defined corridor 0.8–1.5 m wide. The segmented obstacles in the depth images are projected onto the so-called occupancy grid, which is a top view of the scene. Each grid cell corresponds to a haptic actuator that is activated when the cell is occupied by an obstacle. The system is capable of real-time haptic representation of the environment. We present successful indoor technical trials of the system, demonstrating its validity as a potentially useful travel aid for the visually impaired.

Keywords Assistive technologies · Electronic travel aids · Haptic display · Depth maps

1 Introduction

It is estimated that in 2030, one in four Europeans will be of retirement age and, among the elderly, the number of visually impaired people will triple with each decade of life. The results of a survey of European Union residents show that 76% of respondents consider visually impaired people to be the social group with the greatest difficulties in accessing urban infrastructure and services [1].

P. Skulimowski (✉) · P. Strumiłło
Institute of Electronics, Lodz University of Technology, Al. Politechniki 10, 93-590 Lodz, Poland
e-mail: piotr.skulimowski@p.lodz.pl

S. Trygar · W. Trygar
Tyflos Sp. Z O.O, Drzewieckiego 9/4 Str, 39-300 Mielec, Poland

P. Strumiłło et al. (eds.), *The Latest Developments and Challenges in Biomedical Engineering*, Lecture Notes in Networks and Systems 746,
https://doi.org/10.1007/978-3-031-38430-1_34

Developing a device that would effectively assist a blind person in independent mobility is a very difficult challenge. It requires an interdisciplinary research approach—the use of special measurement methods, computational techniques, and the fulfilment of specific ergonomic requirements of the blind user. Despite the rapid development of information and telecommunications technology, electronic travel aids (ETAs) have not replaced or supported the primary travel aids i.e. the white cane or guide dog. Many solutions have been proposed and reported, but none have been fully accepted by the bind community and implemented for wider production and distribution [2]. The two main problems are the discomfort of using the devices that have been developed and the associated discomfort of 'being different'. This situation is also due to limited trust in assistive devices due to their inability to anticipate all the dangerous situations a blind person may encounter. Misinterpretation of the surroundings by the assistive device can put the user at risk. It does not seem possible at this stage to develop an ETA that would completely replace the primary travel aid white cane [3]. With the technologies currently available, we strongly believe that an ETA for the blind should present as concise information about the environment as possible, leaving the interpretation of these signals to the user [4].

In this paper, we demonstrate an ETA which transforms depth images into haptic stimuli generated by the haptic belt prototype we have built. We report on the first technical trials of the system in real-world environments.

2 Related Works

Wearable devices that assist blind people in independent mobility can be classified into two groups: devices using distance sensors (emitting laser or ultrasound beams) and those using imaging devices, e.g. video cameras [5]. The distance sensors work as obstacle detectors and feature narrow receptive field. Within the latter group, devices using depth cameras or stereovision systems can be distinguished, allowing for building a depth map of the surrounding space. Most devices are worn on the head by a blind person, often in the form of a helmet or glasses [5–8]. One example of a commercial device for assisting the blind is the "Orcam MyEye 2" [9]. It is a small camera attached to glasses or a helmet that uses image recognition technology to verbally describe the environment to the wearer, enabling them to read text, recognize faces and identify products in a store.

One of the main benefits of using depth cameras is the ability to scan the surrounding environment and inform the user of the spatial location of scene objects. The person can detect obstacles on the sides without having to move their head or change their direction of movement. However, there are also some drawbacks to this solution. The device adds additional weight to the head, requiring a connection to a controlling computer and a power supply system. Connecting all these elements can also be challenging, particularly if the person is required to wear multiple layers of clothing.

Through our extensive experience in developing devices to aid blind individuals in independent movement, we have found that placing cameras on the head, while acceptable for short testing sessions, is not a practical solution for everyday use [10]. The main concerns are the discomfort of wearing the device and the fear of excessive attention from others. Additionally, blind individuals, particularly those who were born blind, do not have the reflex to move their heads to "look around". Based on our research and understanding of user expectations, we have found that placing the sensors on the body, specifically in the chest area, which does not significantly limit the system's capabilities while greatly increasing user comfort. Our solution involves attaching the camera to a support structure located in the chest area (as seen in Fig. 1) and attaching a lanyard for easy wearing around the neck.

Another important aspect of ETAs is the way in which the environment is presented non-visually to a blind person. Most ETAs use the technique of sensory substitution [11], i.e. replacing one modality of perception (e.g. vision) with another available modality, such as acoustic or haptic stimuli. Due to size and cost, auditory displays are preferred over haptic interfaces in most ETAs. Generation of non-speech sounds is the method mostly used to 'display' the environment to visually impaired people. A comprehensive overview of sonification methods developed to assist blind people can be found in [12]. Haptic interfaces are still at the threshold of fuller development. The sensory properties of the human touch are very well explored. However, user-friendly haptic interface techniques (large devices with a large number of electromechanical components) are still a limitation [13]. Nevertheless, researchers continue to explore this effective way of presenting the environment. See recent research efforts presented in [14].

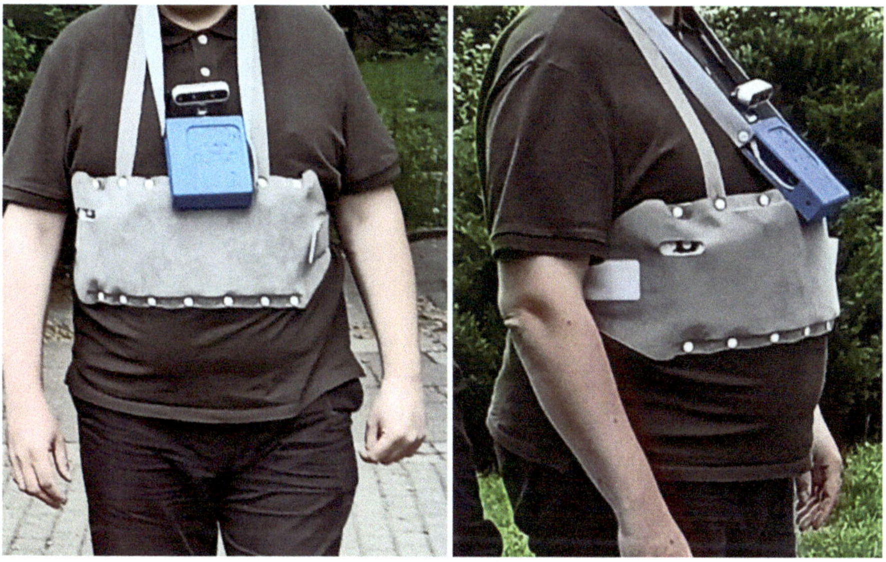

Fig. 1 The method of placing the device on the body of a blind person

3 System Design and Functionality

3.1 Assumptions and System Components

Wearable devices for the blind that aid in independent movement are often designed as an addition to the white cane [15]. This is primarily due to the limited ability of these devices to detect small obstacles and uneven terrain, which can easily be identified with a white cane. Therefore, the following requirements for the developed system were established:

- detection of obstacles at ground level within a distance range of approximately 1 to 2.5–3 m. The minimum distance is determined by the reach of a white cane, while the maximum distance allows for safe movement and reduces false alarms from detecting obstacles at a distance, as well as improves the ability to detect bends in corridors and forks in the path,
- detection of objects up to the height of a person navigating, i.e. those that could potentially collide with a blind person, for example, at head level,
- in basic operating mode, alerts are only given for objects that are obstacles in a blind person's walking path, reducing the amount of information presented and avoiding overloading the senses used to present this information.

The system consists of three parts: Intel® RealSense™ D435i stereovision depth camera, the control computer (Odroid N2 single board ARM based computer), and the haptic belt (Figs. 1 and 2). The camera constructs a depth map by using a built-in laser illuminator, regardless of lighting conditions, whether it is indoors or outdoors. The camera operates at a speed of up to 90 frames per second, which allows for the detection of fast-moving obstacles. It is powered directly from the control computer's USB port and is mounted on a support structure in such a way that the camera's angle of orientation can be adjusted with respect to the ground plane. Stereo vision-based solutions are less expensive than lidar and also provide a visual image that can be used to recognize selected objects (e.g. signs, pedestrian crossings, traffic lights). The control computer's role is to analyze the video images, depth images, and data from the inertial sensors built into the camera, detect obstacles, and preprocess the data on obstacles for presentation to the user. The system also provides an interface for the user to interact with the device. A haptic belt is equipped with actuators in the form of a 5×4 matrix. The belt is worn in the abdominal region and its purpose is to present information about obstacles through vibrating elements with a customizable vibration frequency. The actuators consist of two windings that are wound on a polyamide frame in a push–pull configuration. A cylindrical neodymium samarium magnet is set in motion by a magnetic field. The moving magnet transfers motion to a blunt-ended needle, which transmits a haptic stimulus to the skin. By varying the waveform and frequency of the coil current, different intensities of the haptic stimulus can be obtained at different frequencies. The frequency of the vibratory stimulus can be set over a wide range, typically from 5 to 100 Hz.

Fig. 2 **a** A haptic belt with vibrating elements utilizes a 5 × 4 array of vibrators to present obstacles, **b** The control unit with depth sensor attached (note the keys embossed in the housing plastic of the unit for easier operation by a blind user)

The main unit is equipped with a 14-button keyboard that allows for turning the device on and off, selecting the mode of operation. Each key is clearly marked with a shape that facilitates detection by a blind person. The computer connects wirelessly with a haptic belt.

3.2 *Depth Image Processing*

It is important to note that the ground plane is not considered an obstacle, but its detection is a crucial element as the parameters of objects that can be interpreted as obstacles in the path of a visually impaired person are determined in relation to it. Additionally, the ground plane in the image of the environment is typically located at the bottom of the image and individual rows of stereovision images are nearly parallel to the ground plane. This assumption allows for small deviations of a few degrees of the orientation of the vector normal to the ground plane in the coordinate system associated with the camera.

The algorithm for determining the plane equation and tracking the plane is shown in Fig. 3. The program implementing this algorithm was written in C++ and runs on Odroid N2 single board computer.

The navigation system is initialized by pressing a dedicated key. Then the system calculates the distance of the camera (located at the origin of the coordinate system) from the ground dg, and a vector that defines the camera's tilt relative to the ground ng (ground plane normal vector in camera coordinate system). The above-described initialization should be done once. It is important that the plane's region in the image is significant, that is, that there are as few obstacles on the ground as possible. The ground plane is determined based on the most distant points in the scene whose images are lying on consecutive horizontal lines of the depth map located at the bottom of the depth map image. At this system initialization step, it is assumed that the camera is not rotated with respect to the ground, i.e. the roll angle is close to or

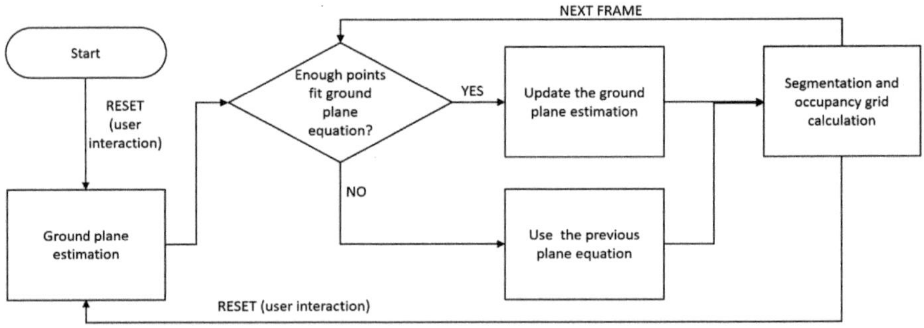

Fig. 3 Algorithm for determining the ground plane equation

equal to zero, while its tilt is such that at least part of the ground plane is visible in the bottom of the depth image. This procedure assumes that the farthest points belong to the ground and identifies non-colinear points. The parameters during this computation step will be used as start parameters in the second optimization phase.

The transition to the ground plane determination phase is possible when the orientation of the camera relative to the ground $n1$ and its distance from the ground $d1$ have been determined. During this phase, the depth image is divided into rectangular triangles with sides of 16 pixels [4, 16] using a regular grid (depth map resolution is 640 × 480 pixels). Knowing the coordinates of these vertices in the scene, the equation of the plane to which these points belong is determined. If the distance of this plane from the origin of the coordinate system differs by no more than 30 cm of the previously determined distance $d1$, and the angle between its normal vector and $n1$ is within 25 degrees, then these points are added to a list. The ground plane is determined based on the points collected in this list. From now on, it is not possible to determine the plane if it is largely obscured by obstacles or large objects like walls or the distance of the newly determined plane from the origin of the coordinate system differs more than 20 cm or its orientation is different by more than 20 degrees. In such a case the earlier saved plane will be used for calculations.

3.3 Obstacle Presentation Using the Tactile Feedback

Knowing the equation of the ground plane, one can convert the coordinates of points to a coordinate system related to the ground [4, 16]. If only the points marked as obstacles in the previous stage are subjected to conversion, it is possible to build the so-called occupancy grid, showing where, the obstacles projected onto the ground plane are located. The user is informed of obstacles located in a selected predefined width range (0.8, 1.2 or 1.5 m). The selected width space is divided into 5 parts (number of columns in the haptic actuator strip). Each part contains eight regions (twice the number of rows in the actuator strip), for each of them the number of pixels belonging to each region is counted. Then, after filtering out any incorrectly found depth map values, the region containing obstacles and closest to the user is determined. The corresponding actuator is activated at a frequency depending on the number of the region in which the presence of obstacles closest to the user has been detected (see Fig. 4d). In each column, if obstacles occur, only one actuator is activated due to the reduction of energy consumption and for easier detection by the user. The closer the obstacle is in each part, the stronger (with higher frequency) is the activation of the actuator, for example, standing in front of a wall at about 2.5 m, the lowest located stimulators are activated, approaching it, successive rows of stimulators are activated in turn, until the highest is located, when the user is about 50 cm away from the wall. Because two distance regions are assigned to one stimulator row, the vibration time depends on the currently represented interval: for the closer one, the vibration time is 100 ms, for the further one, 50 ms. The vibration frequency changes from 30 to 65 Hz in steps of 5 Hz.

See Fig. 4, which illustrates the successive processing steps that convert the depth image into activated haptic belt actuators. The segmented depth map (Fig. 4c) contains regions corresponding to the detected ground plane (coded green), obstacles (coded red), and scene space interpreted as background (coded grey) containing objects at distances greater than 2.5 m. Figure 4d shows the top view of the scene with the front part of the obstacle highlighted (the curved black line). Note the grids

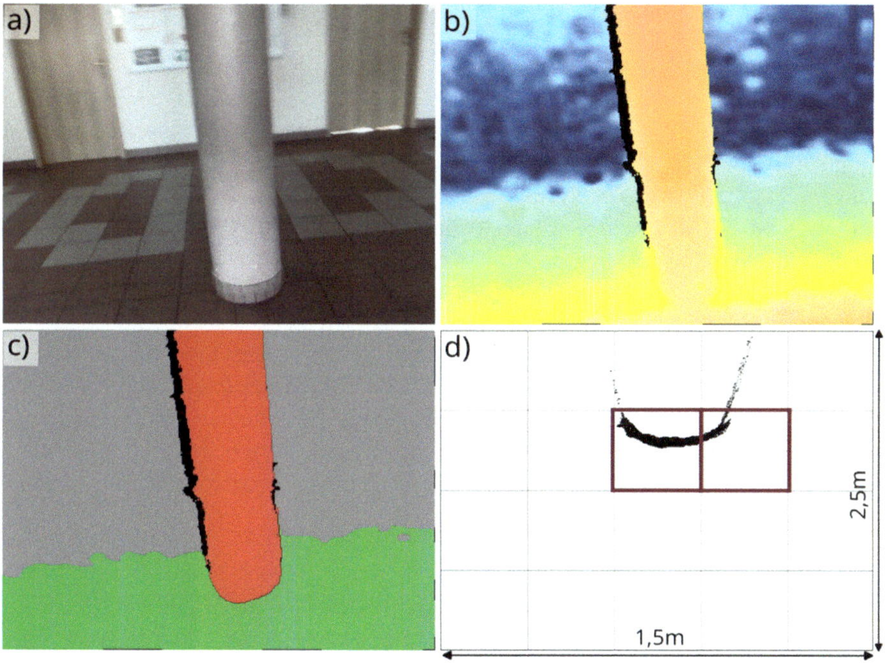

Fig. 4 Obstacle presentation using tactile feedback: **a** scene image, **b** depth map shown in pseudocolors, **c** segmentation result: green–ground plane, red–obstacles, grey–obstacles outside the presentation range, **d** the occupancy grid with two active actuators (note the distance scales shown)

Table 1 The mapping scheme of obstacle distance to the number of the activated row of the haptic belt

Distance [meters]	Row number
> 2.51	0
2.51	1
1.89	2
1.41	3
0.71	4

occupied by the obstacles corresponding to the activated haptic actuators indicated by red borders.

4 Tests and Results

To verify the effectiveness of obstacle detection and their distinguishability using the proposed haptic interface, verification tests were conducted.

4.1 Verifying Distance Detection From a Large, Flat Obstacle (E.g. Wall)

In the first trial, the person equipped with the system slowly walks towards the wall, using a rangefinder. At each activation of consecutive row of actuators the distance to the wall is measured. The average values for 3 trials are shown in Table 1.

In the next trial, the person walked towards the wall at a fast pace, and the distance at which the obstacle was first detected was recorded. The average of 3 trials for the modes: narrow, normal, and wide was: 1.9 m; 1.8 m; 2.1 m. The obtained values are similar, so it can be concluded that they do not depend on the selected mode, but they are significantly lower than in the case of very slow movement towards the obstacle. This is mainly due to the frame rate of the analyzed image (to a lesser extent) and the delays introduced by the activation time of the vibrating elements and wireless transmission. The maximum processing time required to process the depth image and activate the appropriate vibration actuators is approximately 150 ms.

4.2 Verifying The Effectiveness Of Obstacle Discrimination

To verify the effectiveness of obstacle discrimination, the following test was conducted. The person equipped with the system stopped at distances 2.5, 2, 1.5, 1 m in front of a column with a diameter of 40 cm in the first case and 5 cm in the second. For different modes of operation, it was checked how many vibrating elements in a row were activated (see Table 2).

A similar test was carried out to detect open doors 90 cm wide. This time it was noted how many vibrating elements in a row were not activated. The test scene is shown in Fig. 5 and the results are shown in Table 3.

Obstacle detection runs smoothly, especially indoors, where the ground is even and often lighting conditions allows to use of an IR projector which greatly improves the quality of the depth map. For open spaces, a few false detections were eliminated by increasing the obstacle detection threshold to around 20–30 cm. However, transitions between shaded and sunny areas proved to be problematic, resulting in false obstacle detections as it took a moment to automatically adjust the camera parameters.

We encourage you to watch a short film that demonstrates the operation of the system in an indoor environment [17].

Table 2 The number of activated actuators for different operating modes, distances and diameters of the approached columns. *—single, short pulses of one of the vibrating actuators (at the boundary of obstacle detection)

	Column diameter 5 cm			Column diameter 40 cm		
Distance [m]	Narrow	Normal	Wide	Narrow	Normal	Wide
2.5	1	0	0	3	2	2
2.0	1	0*	0*	3	3	2
1.5	1	1	1	3	2	2
1.0	2	2	2	3	3	3

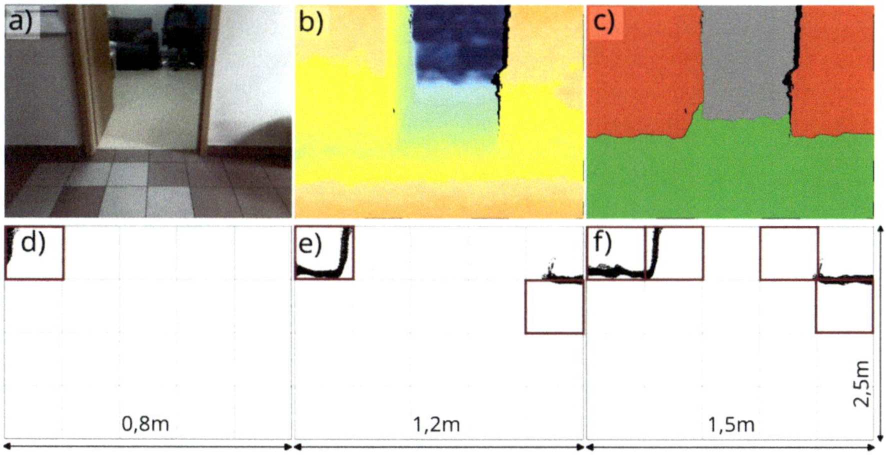

Fig. 5 Obstacle presentation using tactile feedback during door test: **a** scene image, **b** depth map shown in pseudocolors **c** segmentation result: green—ground plane, red—obstacles, grey—obstacles outside the presentation range, active actuators for different operating modes: **d**—narrow, **e**—normal, and **f**—wide mode (note different horizontal scaling for each mode)

Table 3 Number of inactive actuators for different operating modes when standing in front of the open door at different distances

	Door 90 cm wide		
Distance [m]	Narrow	Normal	Wide
2.5	4	3	1
2.0	5 (all)	3	1
1.5	5 (all)	3	1
1.0	5 (all)	3	1

5 Summary

We presented technical trials of a system designed for the visually impaired that uses haptic feedback to present obstacles in real-time. The system includes a stereo vision camera supported by IR Projector (Intel®RealSense™ D435i) for depth estimation.

The system maps the environment onto the occupancy grid, which divides the scene depth into cells and determines whether they are occupied by an object or not. The haptic feedback is generated by a custom-designed tactile belt consisting of 20 vibration actuators arranged in a 4×5 matrix. The bottom row of actuators corresponds to the smallest presented depth, i.e. 1 m. The system can process (i.e., detect the ground plane and segment out obstacles) at a rate several depth images per second, ensuring that the user receives timely information about their surroundings. The system was tested in an indoor real-world scenario, and the results showed that it detects obstacles accurately and provides reliable haptic feedback to the user.

It is important to note that during the system design phase we consulted with the users on design decisions such as the number and spacing of the haptic belt actuators. Following these successful technical tests, we have also carried out trials with blind and partially sighted people. The report of these usability trials is in preparation. In further studies we plan to test whether the integration of verbal commands about the scene geometry together with the haptic display mode would effective in enriching the presentation of the environment and would be accepted by the blind users.

Acknowledgements This work has been partly supported by the European Union and the European Regional Development Fund, as part of the Smart Growth Operational Programme. This paper presents the results of industrial research and development work carried out under the project POIR.01.01.01-00-0091/18-00, which was carried out as part of the 'Fast Track' competition organized by the National Centre for Research and Development (Poland). The aim of the project was to develop an electronic system of non-visual presentation of the environment to support navigation in unfamiliar spaces (ETA—Electronic Travel Aid) for blind and visually impaired people, and its implementation will be the result.

References

1. Helal, A.A., Mokhtari, M., Abdulrazak, B. (eds.): The engineering handbook of smart technology for aging. Disability and Independence. John Wiley & Sons Inc., Hoboken, New Jersey (2008)
2. Cardillo, E., Caddemi, A.: Insight on electronic travel aids for visually impaired people: a review on the electromagnetic technology. Electronics **8**(11), 1281 (2019). https://doi.org/10.3390/electronics8111281
3. Maidenbaum, S., Abboud, S., Amedi, A.: Sensory substitution: closing the gap between basic research and widespread practical visual rehabilitation. Neurosci. Biobehav. Rev. **41**, 3–15 (2014)
4. Skulimowski, P., Owczarek, M., Radecki, A., Bujacz, M., Rzeszotarski, D., Strumillo, P.: Interactive sonification of U-depth images in a navigation aid for the visually impaired. J. Multimodal User Interfaces (2018). https://doi.org/10.1007/s12193-018-0281-3
5. Tapu, R., Mocanu, B., Zaharia, T.: Wearable assistive devices for visually impaired: A state of the art survey. Pattern Recognit. Lett., **137**, 37–52, 2020. ISSN 0167–8655. https://doi.org/10.1016/j.patrec.2018.10.031
6. Tyagi N., Sharma D., Singh J., Sharma B., Narang, S.: Assistive navigation system for visually impaired and blind people: a Review. In: 2021 International conference on artificial intelligence and machine vision (AIMV), Gandhinagar, India, pp. 1–5 (2021). https://doi.org/10.1109/AIMV53313.2021.9670951
7. Leung T. -S., Medioni, G.: Visual navigation aid for the blind in dynamic environments. In: 2014 IEEE conference on computer vision and pattern recognition workshops, Columbus, OH, USA, 2014, pp. 579–586 (2014). https://doi.org/10.1109/CVPRW.2014.89
8. Neto, L.B., Grijalva, F., Maike, V., et al.: A Kinect-based wearable face recognition system to aid visually impaired users. IEEE Trans. Hum.-Mach. Syst. **47**(1), 52–64 (2017). https://doi.org/10.1109/THMS.2016.2604367
9. Orcam MyEye2 device specification, https://www.orcam.com/en/myeye2/specification/. Last accessed 19 Feb 2023
10. Strumillo, P., et al.: Different approaches to aiding blind persons in mobility and navigation in the "Naviton" and "Sound of Vision" projects. In: Pissaloux, E., Velázquez, R. (eds.) Mobility of visually impaired people, pp. 435–468. Springer, Cham (2018). https://doi.org/10.1007/978-3-319-54446-5_15
11. Bach-y, R.P.: Brain mechanisms in sensory substitution. Academic Press, Cambridge (1972)
12. Bujacz, M., Strumillo, P.: Sonification: review of auditory display solutions in electronic travel aids for the blind. Arch. Acoust. **41**(3), 401–414 (2016)
13. Visell, Y.: Tactile sensory substitution: models for enaction in HCI. Interact. Comput. **21**(1–2), 38–53 (2009)
14. Gay, S.L., Pissaloux, E., Romeo, K., Truong, N.-T.: F2T: A novel force-feedback haptic architecture delivering 2d data to visually impaired people. IEEE Access **9**, 94901–94911 (2021). https://doi.org/10.1109/ACCESS.2021.3091441
15. Hersh, M., Johnson, M. (eds.): Assistive technology for visually impaired and blind people. Springer, London (2008). https://doi.org/10.1007/978-1-84628-867-8
16. Skulimowski, P., Strumillo, P.: Interactive travel aid for the visually impaired: from depth maps to sonic patterns and verbal messages. In: Krzhizhanovskaya, V. et al. (eds.) Computational science – ICCS 2020. ICCS 2020. Lecture notes in computer science, vol. 12143. Springer, Cham (2020) https://doi.org/10.1007/978-3-030-50436-6_22. https://link.springer.com/chapter/10.1007/978-3-030-50436-6_22
17. Indoor demo film: http://www.tyflos.pl/pcbbe2023/

Biomaterials and Implants

The Influence of Aging Conditions on the Properties of Polymer Dental Composites

Dariusz M. Bieliński, Maria Rokicka, Tomasz Gozdek, and Katarzyna Klajn

Abstract Long-term exposure of polymer dental composites to selected liquids of different chemical composition and characteristics (pH level, sugar content, etc.), like: coffee, red wine, soy sauce, alkaline mineral water, coke drinks and mouthwash liquid, was studied. The degree of chemical and physical changes to the surface layer of the materials were evaluated based on the results determined by infrared spectroscopy, colorimetry, microhardness, microroughness and wettability. The experiments confirmed suspicions about the liquids studied adversely affect polymer dental composites. It was expected that the material is generally not resistant to the aging agents present in beverages. However, quite unexpectedly, similar behavior was also exhibited by the liquid for oral hygiene. Diet coke, alkaline water and mouthwash liquid had the biggest impact on wear of the composites, contrary to soy sauce. Colorimetry proved the greatest color changes for the samples treated with red wine and coffee, whereas their incubation in alkaline mineral water and mouthwash liquid, made them clearly brighter. Chemical modification of the their surface layer, influenced their microhardness and, together with changes to the surface geometry, wettability. Intriguing differences were observed between regular coke and diet one. Coke with sugar penetrated the composite samples to a smaller depth in comparison to its sugar free version. The observations are important in the context of the selection of polymer composite materials for a specific dental case and the awareness of patients regarding discoloration of teeth and composite fillings, under the influence of liquids present in the daily diet and hygiene.

Keywords Dental composites · Surface layer · Microhardness · Microroughness · Wettability · Discoloration · Friction · Wear

D. M. Bieliński (✉) · T. Gozdek
Faculty of Chemistry, Institute of Polymer and Dye Technology, Lodz University of Technology, Stefanowskiego 12/16, 90-924 Lodz, Poland
e-mail: dariusz.bielinski@p.lodz.pl

M. Rokicka · K. Klajn
Faculty of Medicine, Dental Department, Medical University of Lodz, Pomorska 251, 92-213 Lodz, Poland

P. Strumiłło et al. (eds.), *The Latest Developments and Challenges in Biomedical Engineering*, Lecture Notes in Networks and Systems 746,
https://doi.org/10.1007/978-3-031-38430-1_35

1 Introduction

Many studies have been conducted on the aging of dental polymer composites. They mainly concerned aspects such as discoloration of composite fillings, which in turn translates into an inadequate color in relation to the surrounding dental tissues, strength and resistance of materials to abrasion and the impact of the polymerization process on the shrinkage of the material and its subsequent possible swelling in the environment of the oral cavity. Polymer composite matrices, type and degree of filling as well as environmental conditions or favorable factors affecting the aesthetic, strength and durability parameters of dental polymer composites were studied.

Research conducted by Ceci and coworkers showed that under the influence of i.a. coca-cola, coffee and red wine, there is discoloration in various types of dental light-cured composites [1]. It has been proven that composites with nanofillers are less susceptible to discoloration. Bagheri and colleagues compared the effects of red wine, coffee, tea, soy sauce and cola on discoloration of resin-based composites (RBCs) and glass ionomer cements (GICs) [2]. It was found that every material studied was discolored by the tested media, especially coffee, red wine and tea. In addition, it was proven that the method of finishing the composite material does not affect its resistance to discoloration.

Nagarajan and co-workers addressed the problem of the influence of the degree of filling of dental polymer composites on their abrasive wear [3]. Three medium-filled commercial composites (75–76%) were compared with a high-filled one. The results of SEM, FT-IR and ICP-MS analyzes of plastic samples after friction tests confirmed the occurrence of a complex process involving tribochemical reactions between the filler particles and particles of the reaction medium (distilled water), the formation of a surface film composed of layers consisting of a mixture of filler fragments and reaction products, and delamination of this film. The degree of filling of composites affects the processes mentioned above. The impact of the working environment in the context of the influence of lubrication and temperature on friction and abrasive wear of various friction polymer composites was studied by Jóźwik et al. [4]. It was confirmed that the tested composites are characterized by lower abrasion when working under fluid friction conditions. It is important that the lubricant is chemically inert to the plastic.

In our earlier work, we dealt with the influence of the surface layer morphology on the abrasive wear of polymeric dental materials [5]. We paid attention to the different mechanical properties of the surface layer, which is exposed to environmental factors and is directly involved in the abrasion process, from the inside of the material. The conducted analyzes proved significant differences in the filler content, size and distribution of its particles between the surface layer and the bulk of the composite. The microhardness measurements revealed a correlation between the hardness of the top layer of the composite and its abrasion resistance.

In this work the effects of long-term exposure of a typical polymer dental composite to selected liquids like: coffee, red wine, soy sauce, alkaline mineral water, coke drinks and mouthwash liquid, on chemical modification of its surface layer were studied. The observed changes to microhardness, surface geometry and wettability were analyzed from the point of view of wear resistance of the composite, whereas its discoloration under the influence of liquids present in the daily diet and hygiene was discussed regarding dental applications.

2 Experimental

2.1 *Materials*

Tested samples were made of a popular dental polymer composite intended for filling cavities in the lateral zone of dental arches MegaFill MH of A3 color (Megadenta, Germany). Specimens were prepared using hard dental silicone molds and cured with a UV lamp under conditions recommended by the composite manufacturer. When applying the composite to the mold, it was observed that the thickness of the applied layer should not exceed 0.5 mm, which ensured uniform hardening of the sample.

2.2 *Treatments*

Samples made of the dental polymer composite were incubated in eight different media (3 samples each) corresponding to the food intake and media to which the composite restorations are exposed:

1. artificial saliva—a commercial product of a trade name Kserostemin (Aflofarm, Poland);
2. diet coke drink, sugar-free, containing sweeteners;
3. regular coke drink, containing sugar;
4. dry red wine;
5. instant coffee (in a concentration of 15 g/100 ml);
6. soy sauce;
7. commercial mouthwash liquid;
8. alkaline water of pH 7.3.

The composite samples were kept in the above-mentioned media for 4 weeks at 37 °C, corresponding to the temperature in the oral cavity. Three samples, placed in a refrigerator at a temperature of approx. 3–4 °C, in order to eliminate the possible effect of spontaneous post-polymerization, were not exposed to the above-mentioned media, constituting a reference material.

2.3 Techniques

After this time, the tests were started using the instruments and methodology described below. The artificial saliva was both a separate medium in which the dental composite samples were incubated, and was also used in tribological studies, imitating the oral cavity environment.

Morphology & Surface microroughness. The morphology and microroughness profile of the composite surface exposed to the media, before and after the friction test, were analyzed using a Leica MZ6 stereo optical microscope (Germany) at 100 × magnification and a Keyence VHX-7000 digital optical microscope (Belgium) at magnification of 300x. Both cameras were equipped with image analysis software, enabling the determination of microroughness parameters of the tested surfaces.

Microhardness (NT). The change in the hardness of the surface layer of the composite as a result of the action of the media was investigated using the nanoindentation method [6]. The tests were run using a NanoTest 600 apparatus (Micromaterials Ltd., UK) equipped with a diamond penetrator of Berkovitch geometry. Experimental parameters: load/unload rate of 0.5 mN/s; initial load of 0.05 mN; max. load determined by 5 μm penetration; temperature of 22 ± 2 °C; relative humidity of 60 ± 5%.

Composition of the surface layer (FT-IR ATR). Fourier transform infrared spectroscopy (FT-IR) absorbance spectra of the composite surface subjected to the media were investigated within the 400–4000 cm^{-1} range (64 scans, absorption mode, resolution of 4 cm^{-1}). The experiments were performed with the use of a Nicolet 6700 FT-IR spectrometer (Thermo Fisher Scientific, USA) equipped with a Smart Orbit ATR sampling accessory, operating with a Ge crystal.

Surface wettability. The water contact angle of the rubber surface before exposure to the media was determined by the sessile drop method. Drops of double deionized water with a volume of approx. 2 μl were applied to the test surface using a Hamilton microsyringe. Using a CCD camera coupled with an optical microscope Leica MZ6 (Switzerland), images of drops were recorded. Using a MultiScan 8.0 program (CSS, Poland), the contact angles of the tangents to the drop envelope on its both sides were determined, and each time their average was calculated. The median of the average angle values for five drops placed in different places on the surface studied was taken as the result.

Colorimetry. Colorimetric measurements of the composite surface before and after its exposure to the media were made with a spectrophotometric apparatus KONICA MINOLTA CM-3600A (Germany). The results of the determinations were described in the CIELab color space, using the following parameters: L*—color brightness, a*—color share in the red-green range and b*—color share in the yellow-blue range.

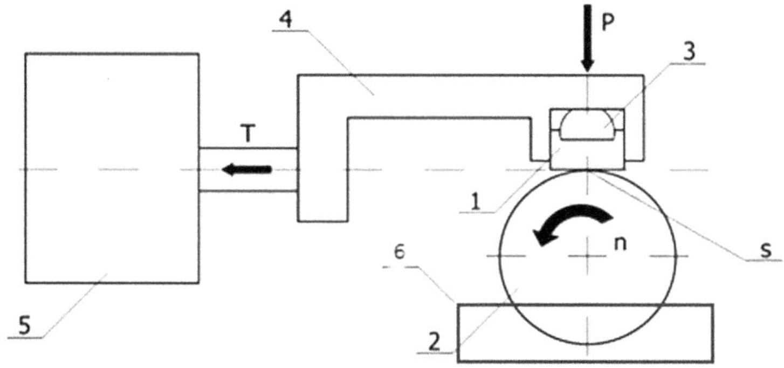

1. composite specimen (block), 2. steel counterface (ring), mounted on a rotating spindle, 3. tensiometer force sensor, 4. load arm, 5. drive motor, 6. container with artificial saliva

Fig. 1 Scheme of the T-05 tribometer with a block (composite)-on-ring (steel) friction pair

Tribological properties. Frictional resistance of the dental composite subjected to the media was tested using a T-05 tribometer with a block-on-ring friction pair (ITE-PIB, Poland)–Fig. 1. Tested specimen (block) was pressed with a force of P = 30 N against a roll made of austenitic stainless steel type 1.4404 (AISI 316), ground to a roughness of Ra 1.25 μm, rotating with a speed of n = 450 rpm. Tests were run for a distance equivalent of 1000 m. The starting temperature of the tests was ca. 21 °C, increasing finally to a max. value of 26 °C as a result of friction. The experiments were carried out in artificial saliva, imitating the real environment of the oral cavity. Each time, the friction force and the temperature of the sample during experiments were measured, and its abrasion wear was determined at the end. Abrasion of the composite was estimated based on the depth of the friction trace, determined on the specimen cross-section using a KeyenceVHX-7000 microscope (Belgium).

3 Results & Discussion

3.1 Composition of the Surface Layer

The FT-IR ATR spectra of the tested composites do not show any spectacular qualitative changes. However, they differ in intensity within individual absorption bands. Changes in the infrared spectra are visible in the wavelength range of 3200–3400 cm^{-1}, indicating on hydroxyl groups derived from water, alcohols or phenols, present in the surface layer of the composite especially in the case of a sample incubated in coffee.

An increase in absorbance, however, to a lesser extent than before, was also noted in the wavelength range of 2800–3000 cm^{-1}, what can be subscribed to C-H vibrations in methylene and methyl groups, the most likely originated from the degradation of the polymer matrix of composite due to its exposure to the liquids studied.

Absorption band disappearance at a wavelength of of ca. 1700 cm^{-1}, originated from carbonyl group vibrations, however, it may be a consequence of the etching of the oxidized surface layer of the composite. This is most evident with the composite incubated in soy sauce or treated with mouthwash liquid.

The occurrence of absorption bands in the range of 1500–1600 cm^{-1} is, in turn, a proof of the presence of C = C double bonds. To sum up, the observed changes in the intensity of the absorption peaks indicate the formation/precipitation of new compounds on the surface of the samples, both from the composite material and from the medium to which it was previously exposed.

3.2 Surface Wettability

Significant differences in the water contact angle values of the composite surface exposed to the media were observed—Fig. 2.

In contrast to the lack of water wettability of the composite sample previously treated with a diet coke or akaline water, complete spreading of the drops over the surface of the coffee-treated sample was noted. The obtained results prove that the surface of the composite samples is energetically differentiated, which coincides with the results of infrared spectroscopy (the presence of various/different concentrations of functional groups), and differences in the microroughness of their surfaces. To

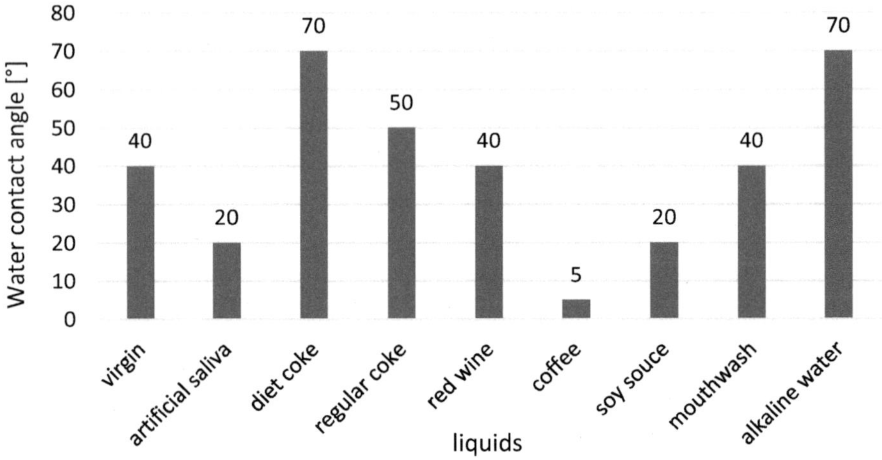

Fig. 2 Influence of various media on the wettability of the dental composite surface

sum up, the incubation of the dental composite samples in selected liquids affects the wettability and surface free energy of the material. In each case, the surface of the material exhibits hydrophilic features, characterized by contact angle value of less than 90°.

3.3 Colorimetry

Microscopic observations clearly indicate the greatest discoloration of the composite samples incubated in coffee and red wine. Parameterized color changes following the incubation in different media are collected in Table 1.

L* parameter clearly indicates the brightening of the composite sample after the action of alkaline water. The similar effect is obtained by incubating the material in mouthwash liquid. The largest decrease in the L* parameter can be observed for the samples treated with red wine, coffee and soy sauce, respectively. These samples are clearly darker than the virgin composite sample, which proves their penetration into the surface layer of the composite material. The color change in the red-green color region—parameter a*, is most visible in the case of the composite sample incubated in red wine, although coffee, diet coke and soy sauce also discolor the material in a similar direction. On the other hand, media such as alkaline water and mouthwash liquid change the color of the composite sample, bringing it closer to green. The color change in the blue-yellow region—parameter b*, is most noticeable in the case of samples incubated in soy sauce and coffee, which become more yellow. Again, the composite samples subjected to the action especially of alkaline water but also of mouthwash liquid showed a change in the opposite direction—towards a blue shade.

Interestingly, the material incubated in red wine exhibited similar behavior. Summing up, the most aggressive character of discoloration was demonstrated by red wine, coffee, alkaline water and mouthwash liquid—changing the color of the material in every color region. The degradation of the surface layer of the polymer

Table 1 Color parameters (D65) of the dental composite surface before (virgin) and after exposure to various media

Composite sample/ Medium treatment	L*	a*	b*
Virgin	69.4	2.7	15.6
Artificial saliva	69.4	2.5	14.4
Diet coke	68.2	3.1	15.2
Regular coke	69.6	3.0	15.6
Red wine	47.8	15.3	5.4
Coffee	55.0	6.9	19.7
Soy sauce	64.9	3.0	21.5
Mouthwash liquid	73.6	0.3	8.9
Alkaline water	84.1	0.6	-0.2

composite additionally promotes the penetration of the material by chemical media as coffee or red wine. As expected, the artificial saliva turned out to be the most neutral for the color of the dental composite, practically not affecting its color. However, the fact that both diet (zero) and regular coke (with sugar) did not significantly change the color of the dental composite should be treated as a surprise. A slightly stronger discoloration effect was obtained for the composite sample incubated in a diet coke, which corresponds to its much higher surface roughness, which resulted in greater penetration of this medium into the material.

3.4 *Microhardness*

The results of hardness and plastic deformation measurements of the surface layer of the dental composite, after incubation in various liquids, are summarized in Table 2. It is worth noting the different behavior of diet coke in comparison to regular one.

The first medium leads to the formation of a hard tarnish on the composite surface, while the second one develops surface microroughness. The micromechanical results confirm previous observations made during FT-IR ATR and microscopic studies of the composite surface.

Based on the results of micromechanical tests, two conclusions can be drawn: firstly—the surface microroughness of the samples may affect the significant dispersion of the microhardness results, and secondly—the dominant plastic character of the deformation of the surface layer is in most cases caused by its degradation (most visible after incubation in red wine), swelling (in artificial saliva or alkaline water) or formation of a tarnish on the surface of the composite samples (significant hardening of the material subjected to incubation in soy sauce or diet coke).

Table 2 Hardness and the plastic deformation yield of the dental composite before (virgin) and after long-term exposure to various media

Composite sample/ Medium treatment	Microhardness (MPa)	Plastic deformation (%)
Virgin	73.1 ± 15.6	91.7 ± 1.7
Artificial saliva	13.5 ± 2.4	96.9 ± 0.8
Diet coke	84.6 ± 22.7	94.9 ± 1.1
Regular coke	57.2 ± 22.9	95.6 ± 0.6
Red wine	18.9 ± 9.3	96.1 ± 1.6
Coffee	57.1 ± 12.7	93.7 ± 0.8
Soy sauce	146.8 ± 34.0	93.7 ± 0.6
Mouthwash liquid	63.8 ± 11.3	94.1 ± 1.8
Alkaline water	10.5 ± 4.1	97.1 ± 1.1

3.5 Tribology

The effect of different media on the friction of the dental composite is compared in Fig. 3. The highest friction showed the samples incubated in alkaline water and diet coke. The smallest friction forces were found in the samples incubated in regular coke and mouthwash liquid. These results do not correspond to the results of the microhardness of the composites. It looks astonishing that the friction force of the composite was relatively high after its incubation in artificial saliva, despite the fact that the observed friction trace was relatively shallow in its case. Most likely, the composite shows swelling in artificial saliva. Friction stability is expressed by the values of the amplitude of the friction force during test. The greatest differences in the values of friction forces occurred in the case of the sample incubated in red wine. This indicates the highest microroughness and/or abrasive wear of the material treated with red wine.

Relatively large differences in friction force amplitudes occurred in the sample incubated in the artificial saliva, which confirms the previously suggested swelling of the surface layer of the material. The sample incubated in soy sauce showed a very high friction stability, which is consistent with the highest microhardness of the composite in this case. The tarnish present on the surface of the sample must play a lubricant role, reducing friction of the composite.

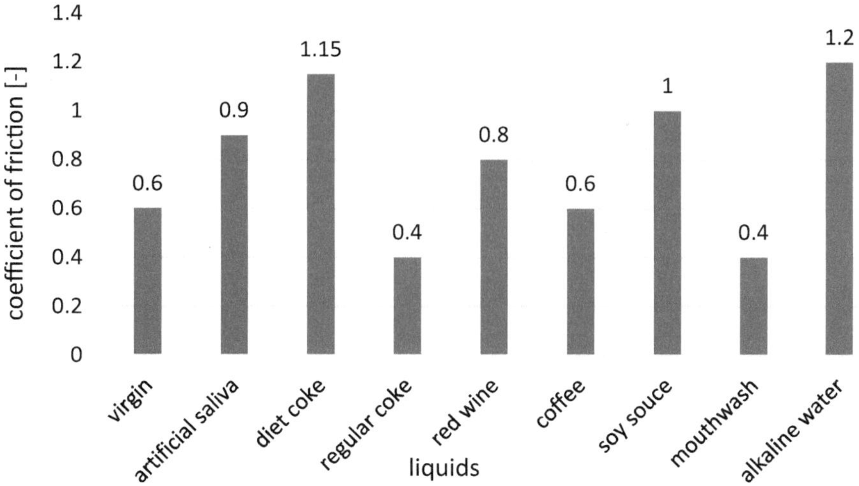

Fig. 3 Influence of various media on the friction of the dental composite

3.6 Surface Morphology and Abrasive Wear

The composite sample after incubation in saliva or alkaline water showed smooth surface of the friction trace. However, the formation of a white ring around the friction trace is clearly visible, which may be the result of permanent deformation of the softest samples, which were practically not subjected to abrasive wear. The microroughness and the depth of abrasive wear traces on the cross-section of the dental composite samples subjected to various media long-time treatment are summarized in Table 3.

The profile of the friction trace of the composite sample incubated in diet coke, unlike for regular one, is not so deep, which can be the result of the difference in hardness, related to their susceptibility to abrasive wear. Regular coke degrades the composite to a much greater extent, contributing to its wear. In the case of the composite surface incubated in sugar-free cola, the chemicals contained in sweeteners, such as aspartame and acesulfame, are much more chemically aggressive than the sugar found in a traditionally sweetened beverage, being responsible for significantly higher surface microroughness.

The greatest discoloration of the composite samples occurs as a result of the action of coffee and red wine. However, after the deeper layers of the composite are exposed by the friction process, the discoloration of the sample persists only for the case of the last medium, which proves its deepest penetration into the material. The high abrasive wear of the composite is observed, but lower than in the case of its exposure to regular coke action. Unlike the latter, however, the microroughness of the composite sample surface is much higher, which may be the result of a lower degree of material degradation in this case. In the case of the sample incubated in coffee, as was the case after regular coke and red wine, we also witness a significant permanent deformation of the composite surface layer after liquid incubation, with a fairly microrough surface of the friction trace surface.

Table 3 Influence of various media on the surface microroughness and the depth of friction trace of the dental composite tested

Composite sample/ Medium treatment	Microroughness R_a (μm)	Max. depth of friction trace (μm)
Virgin	7.95	28.1
Artificial saliva	7.27	7.6
Diet coke	119.70	119.7
Regular coke	15.66	189.9
Red wine	29.98	62.2
Coffee	35.33	36.0
Soy sauce	14.96	21.3
Mouthwash liquid	36.16	36.5
Alkaline water	8.17	8.4

The composite sample incubated in soy sauce was the least affected by the medium. Moderate plastic deformation of the material was observed. The trace of friction is barely noticeable and is characterized by a slight microroughness.

The action of a commercial mouthwash liquid clearly brightened the surface of the composite. The trace of friction is clear, characterized by numerous pits on the surface, although the microroughness of its surface is not as high as in the case of diet coke. The sample incubated in alkaline water has a clear, mineral surface tarnish, which after rubbing revealed the smooth surface of the composite. The alkaline nature of the medium is definitely less aggressive towards the surface of the dental composite tested than the acidic components of both cokes.

4 Summary and Conclusions

The main, albeit rather unexpected, conclusion from the above studies is that dental composites are generally not resistant to the aging factors present in beverages or oral care products. Each of the media used affected the properties of the dental composite after its incubation in their solutions, although the effects of the interaction took slightly different forms.

Samples incubated in some liquids showed signs of etching and discoloration deep into the material. The accompanying changes in absorption in the infrared spectra testify to the aging of the composite, manifested by its degradation. The greatest color darken was observed for samples aged in red wine and coffee. The composite incubated in alkaline water or mouthwash liquid became lighter. Some surprise was the lack of discoloration as a result of the effects of drinks from the coke family, despite significant etching (diet coke) or degradation (regular coke) of the composite surface exposed to them.

Wettability tests of the surfaces of the composite samples after incubation in selected media demonstrated their hydrophilic nature, despite the clear influence of micro-roughness on the obtained results. Wettability correlated mainly with the degree of surface development. Samples after treatment with diet coke and alkaline water showed the least hydrophilic character among the others.

The microhardness increased substantially in the sample treated with soy sauce, resulting in the lowest abrasive wear of the composite. Saliva and alkaline water treatments significantly lowered hardness of the samples in relation to the reference composite, indicating swelling of the material. However, this did not adversely affect abrasion of the material. According to the theory of Bowden and Tabor [7], a softer layer present on a rigid substrate is responsible for reducing the friction resistance. The highest abrasive wear was observed for the composite incubated in regular coke, which was characterized by one of the lowest hardness, caused by the material degradation.

References

1. Ceci, M., Viola, M., Rattalino, D., Beltrami, R., Colombo, M., Poggio, C.: Discoloration of different esthetic restorative materials: A spectrophotometric evaluation. Eur. J. Dent. **11**(2), 149–156 (2017)
2. Bagheri, R., Burrow, M.F., Tyas, M.: Influence of food-simulating solutions and surface finish on susceptibility to staining of aesthetic restorative materials. J. Dent. **33**(5), 389–398 (2005)
3. Nagarajan, V.S., Jahanmir, S., Thompson, V.P.: In vitro contact wear of dental composites. Dent. Mater. **20**(1), 63–71 (2004)
4. Jóźwik, J., Dziedzic, K., Barszcz, M., Pashechko, M.: Analysis and comparative assessment of basic tribological properties of selected polymer composites. Materials. **13**(1), 75 (2019)
5. Kleczewska, J., Bieliński, D., Pacyk A., Wichrowska-Rymarek, K.: Material aspects of abrasion of polymer dental composites. Polish J. Environment. Studies **16**(6A), 280–291
6. Oliver, W.C., Pharr, G.M.: An improved technique for determining hardness and elastic modulus using load and displacement sensing indentation experiments. J. Mater. Res. **7**(6), 1564–1583 (1992)
7. Bowden, F.P., Tabor, D.: The friction and lubrication of solids, 2nd edn. Clarendon Press, Oxford (1954)